Thermodynamics, Statistical Mechanics, and Black Hole Entropy

Book 6 of Physics from Maximal Information Emanation, a seven-book physics series.

ISBN 979-8-89487-009-0

Thermodynamics, Statistical Mechanics, and Black Hole Entropy

by

Stephen Winters-Hilt

ISBN 979-8-89487-009-0

Golden Tao Publishing
Angel Fire, NM
USA

Dedication

This book is dedicated to my family that helped on this lengthy road of discovery: Cindy, Nathaniel, Zachary, Sybil, Eric, Joshua, Teresa, Steffen, Hannah, Anders, Angelo, John and Susan.

Contents

Preface to Physics Series on:

Physics from Maximal Information Emanation

The Road goes ever on and on
Down from the door where it began.
Now far ahead the Road has gone,
And I must follow, if I can,
Pursuing it with eager feet,
Until it joins some larger way
Where many paths and errands meet.
And whither then? I cannot say
– J.R.R. Tolkien, The Fellowship of the Ring

Variation, Propagation, and Emanation

This is a seven book Physics Series that starts with Classical Mechanics (Book 1 [1]), then Classical Field Theory, such as electromagnetism (Book 2 [2]), then Manifold Dynamics, such a General Relativity (Book 3 [3]). The switch to a quantum mechanics description is given in Book 4 [4], and to a quantum field theory, quantum electrodynamics in particular, in Book 5 [5]. A 'quantum manifold theory' would be the obvious next step except it cannot be done (there is not a renormalizable Field theory for Gravitation). Instead a thermal quantum manifold theory is considered, as well as Black Hole thermodynamics in general, in Book 6 [6]. Book 7 [7] describes a new theory, Emanator Theory, that provides a deeper mathematical construct that undergirds quantum theory, much like quantum theory can be shown to provide a deeper (complexified) mathematical construct based on the classical theory.

This is a modern exposition where subtleties of chaos theory are described in Book 1, of Lorentz Invariance in Book 2, of Covariant Derivatives (General Relativity) and Gauge Covariant Derivatives (Yang-Mills Field Theory) in Book 3. Book 4 on Quantum Mechanics provides an extensive review of quantum mechanics, then considers a full self-adjoint analysis on the full general relativistic solution to the spherical shell in-fall system (a result carried over from Book 3). Book 5 considers quantum field theory basics in detail, along with alternate vacua in specific scenarios. Book 6 considers thermodynamics from the basics to the Hamiltonian thermodynamics of some Black Hole systems. Throughout, the odd recurrence of the alpha parameter is noted. In Book 7 we look to a deeper mathematical formulation from which the Quantum Path Integral formulation would result, as well as explaining the odd parameters and structures that have been discovered (such as alpha and Lorentz Invariance).

The physical description starts with the classic formulations of point particle motion. The first approach to doing this is using differential equations (Newton's 1st and 2nd Law); the second is using a variational function formulation to select the differential equation (Lagrangian variation); the third is using a variational functional formulation (Action formulation) to select the variational function formulation. Historically, it wasn't realized until much later that there are two domains for motion in many systems: non-chaotic; and chaotic.

In a description of particle motion, assuming not in a parameter domain with chaotic motion, several important limits are found to exist. Examples include: the universal constants from the aforementioned chaos phenomenon, that are still encountered in non-chaos regimes if driven "to the edge of chaos". Limits are found where scattering is defined in the asymptotic limit and perturbation theory is well-defined in the sense that it is convergent. Overall, if the evolution is described as a 'process' it is often a Martingale process, which has well-defined limits. So, we have descriptions for motion, typically reducible to an ordinary differential equation (ODE), and for which solutions (requiring limit-definitions) are typically found to exist.

The physical description then contends with field dynamics in 2D, 3D, and 4D (in Book 3 [3]). Two-dimensional ("2D") field dynamics can be described as a complex function (that maps complex numbers to complex numbers). A novelty of the 2D complex function is it also shows how to handle many types of singularities (the residue theorem), thus provides important information about fundamental structures in physics as well as fundamental mathematical techniques for solving many integrals. For the 3D field dynamics we do an analysis of the electromagnetic field in 3D. The level of coverage begins at an overview of electrostatics at the level of the graduate text Jackson [155]. Some problems from Jackson Ch's 1-3 are examined closely in developing the theory itself. For some this material (in Book 2 [2]) might provide a useful accompaniment to Jackson's text in a full course on electromagnetism (based from Jackson's text). A quick review of electrodynamics and electromagnetic wave phenomena is then given. In essence, we see many more examples of ODE problems with solutions, such as for the 3D Laplacian, usually involving separation of variables. We then review the famous transform, discovered by Lorentz in 1899 [156], that relates the electromagnetic field as seen by two observers differing by a relative velocity. With the existence of this transform, that brings in the time dimension along with the relative velocity, we effectively have a 4D theory.

From Lorentz Invariance we have, as a point transformation, rotational invariance under SO(3) or SU(2). If Lorentz Invariance is fundamental, then we should see both forms of rotation invariance, one of vector/tensor type from SO(3), and one of spinorial type from SU(2). This is the case, as gauge fields are vectorial and matter fields are spinorial. From Lorenz Invariance as a local invariance we have the Minkowski (flat) spacetime metric, which then generalizes to the Riemannian metric (in General Relativity).

As with the point particle dynamics, for the field dynamics we have three ways to formulate the behavior: (1) differential equation; (2) function variation (on Lagrangian); and (3) functional variation (on the Action). We will see similar limit phenomena as before, but also new phenomena, including (i) inevitable black holes singularity formation (the Penrose singularity theorem); (ii) FRW Universe formation (from homogeneity and isotropy); (iii) the black holes collapse singularity; (iv) the atomic collapse radiative 'singularity'.

Classical dynamics, thus, has two field-like formulations to describe the world: field and manifold. Such formulations can be interrelated mathematically, so what is happening is more a matter of physics emphasis and convenience. The emphasis on this difference, that appears to be no difference (mathematically), is that different physical phenomenologies are at play. Field descriptions appear to work for 'matter', where the fundamental elements are spinorial. Manifold descriptions appear to work best for geometrodynamics (General Relativity), where the fundamental elements are vectorial (or tensorial, such as the metric). Matter fields are renormalizable, thus quantizable in the standard quantum field theory formulation (to be described in Book 5 [5]), while gravitational manifolds are not renormalizable, and have constraints (weak energy condition and positive energy condition given the existence of spinor fields on the manifold).

The presentation in Books 1-3 [1-3], on 'classical' physics, is partly done to make the transition to quantum physics simple, obvious, and in some cases, trivial. Consider the functional variation (Action) formulation of the behavior (whether point-particle or field), this can be captured in integral form, as was done by D'Alembert very early [157] (then by Laplace [158]). Note the use of a large constant to effect a 'highly damped' integral for selection purposes (on variational extremum of the action). To transition to the quantum theory we also have the large constant from 1/h, and so the only difference is the introduction of a factor of 'i', to effect a 'highly oscillatory' integral for selection purposes.

After the transition to a quantum theory, for the point-particle descriptions, the classical collapse problem for atomic nuclei is eliminated. The spectral predictions have excellent agreement with theory, but there is still fine-structure in the spectra not fully explained. The theory is not relativistic and some initial corrections for this are possible (without going to a field-theory) and these indicate closer agreement and explain most of the fine-structure constant discrepancy (and reveal alpha in another place in the theory). It is shown in Book 3 [3] and Book 4 [4], that the General Relativity singularity problem, however, remains partly unresolved (for the test case of spherical dust shell collapse, done in a full General Relativity analysis, then quantized in a full self-adjoint quantization analysis [4]).

In Book 5 [5], the transition to quantum theory is continued to the field theory descriptions. A precise description/agreement of atomic nuclei is now possible with quantum electrodynamics, and within the nuclei themselves (quark confinement) with QCD. The field theories have a small set of bothersome infinities, however, which is eventually solved by renormalization. As mentioned, the quantization of manifold theories, such as General Relativity, does not appear to be possible due to non-renormalizability. Not to be deterred, in Book 6 [6] we consider a Hamiltonian description of a General Relativity system whose quantization would involve an energy spectrum based on that Hamiltonian, if we then use analytic continuation to take us to the thermal ensemble theory based on the partition function that results, we can consider the thermal quantum gravity of such systems.

This last example (from Book 6), showing a consistent thermal quantum gravity theory if we use analyticity, is part of a long sequence of successful maneuvers involving analytic continuations in different settings. What is indicated is the presence of an actual complex structure to the stated theory. There is the trivial complex structure extension mentioned above that brought us from the standard classical physics theory to the standard path integral quantum theory. But we also see actual complex structure at the component level with time complexation (that ties to thermal version of the theory by defining the partition function), and we have complex structure as the dimension-level in the form of the successfully applied dimensional regularization procedure used in the renormalization program.

As well as covering the breadth of core physics topics at both undergraduate and graduate level (for courses taken at Caltech and Oxford), including extensive presentation of problems and their solutions, the Series also examines, in specific cases, the boundaries of the physical world "from the inside" (and then later "from the outside"). To this end exploration of spherical dust collapse to form a singularity is examined in a fully general relativistic formalism, and then carried-over to a quantum minisuperspace (quantum gravity) analysis (in Books 3 and 4 [3,4]). Also examined in-depth are the topics of black hole thermodynamics and quantum field theory with alternate vacua (part of Books 5 and 6 [5,6]). The in-depth material comprises the topics covered in my PhD dissertation [9], portions of which are published [36,91,142,143].

In recent work on machine learning, that includes statistical learning on neuromanifolds [6], we find a possible new source for a foundational element for statistical mechanics (entropy) via seeking a minimal learning process/path on a neuromanifold [6]. By the time the Series reaches thermodynamics in Book 6, therefore, the foundational thermodynamics elements have all been established from the physical descriptions discovered in Books 1-5, they just haven't been put together in a comprehensive analysis that gives us the fundamental constructs of thermodynamics and statistical mechanics. That said, it would seem that thermodynamics is, thus, entirely derivative from other, truly fundamental theories. Not so, in the joining of the parts to make thermodynamics we have something greater than the sum of the parts. In the 'system' descriptions we find that emergent phenomena exist. This, at least, is unique to thermodynamics, so it is fundamental in this "sum greater than the parts' aspect.

In Book 7 (the last) of the Series, we consider the standard physical world, described by modern physics, "from the outside." In doing this we've already eliminated part of the mystery of entropy by the geometric 'neuromanifold' description. If we can understand other oddities of the standard theory, and arrive at them naturally, then we might have an even deeper dive into modern physics, testing the limits of what is possible, and see possible future developments and unifications of the theory. This is what is described in papers [126,132,148,159-162], and organized along with current results into the final Book of the series.

Efforts in the last book of the Series involve choices and concepts identified in the prior six books of the Series, and theoretical maneuvers gleaned from the most advanced courses in physics and mathematical physics taken while at Caltech (as an undergraduate and then as a graduate) and the Oxford Mathematics Institute (as a graduate), and the University of Wisconsin at Milwaukee (as a graduate).

The broad range of topics covered in the Series is, initially, similar to the Landau & Lifshitz graduate textbook series (see [22]), with a similar exposition on classical mechanics at the start of Book 1. Even with well-established classical mechanics, however, there are significant, modern, updates, such as (modern) chaos theory. In the final two books of the Series (Books 6 and 7 [6,7]) we arrive at statistical mechanics and thermodynamics, together with modern topics such as black hole thermodynamics, thermal quantum gravity, and emanator theory.

Key constants and structures of physics, their discovery from the experimental data, and their theoretical placement in the "Grand Scheme," are emphasized throughout the Series. The constant alpha, a.k.a. the fine structure constant, appears in numerous settings so special note of the occurrence of alpha will be made in each chapter. This is the case even at the outset with Book 1, due to fundamental numerical constants appearing from chaos theory. In Book 7 we see the origin of alpha, as a maximal perturbation amount, appears naturally in a formalism for maximal information 'emanation'. But maximal perturbation in what space and in what manner? In Book 7 of the series [7] we will see a possible representation of such an information entity, and its space of existence, in terms of chiral trigintaduonions.

Thus, in the end, this is an effort to tell of a journey to a special place "where many paths and errands meet", giving rise to emanator theory and an answer to the mystery of alpha. Part of this journey is equivalent to 'finding the arkenstone' (alpha) in the most unlikely of places, the trigintaduonion emanation mathematics underpinning the emanator formalism (e.g., Smaug's Lair, described in Book 7 [7]). Why I should have wandered into such an odd place (mathematically speaking), and why I should posit a deeper form of quantum propagation using hypercomplex trigintaduonions, here called emanation, is why there is such extensive background on standard topics. This extensive background even impacts the classical mechanics description via its modern chaos theory material (due to a possible relation between C_∞ and alpha). The critical role of emergent phenomena is only understood at the end, including for manifolds in geometry and neuromanifolds in statistical mechanics, and leads to a Book 6 that goes from very basic (initial thermodynamics) to very advanced (emergent phenomena). Much is made clear with emanator theory, including how reality is both fractal and emergent. At this point in the journey, as with Tolkien, this much I can say: "The Road goes ever on and on … And whither then? I cannot say".

The seven books in the Series are as follows:

Book 1. Classical Mechanics and Chaos
Book 2. Classical Field Theory
Book 3. Classical Manifold Theory
Book 4. Quantum Mechanics and the Path Integral Foundation
Book 5. Quantum Field Theory and the Standard Model
Book 6. Thermodynamics, Statistical Mechanics, and Black Hole Entropy
Book 7. Maximum Information Emanation and Emanator Theory

Overview of Book 1

Book 1 is a modern exposition of classical mechanics, including chaos theory, and including ties to later theoretical developments as well. The exposition consists, throughout, of the presentation of interesting problems with many solved, the others left for the reader. The problems are drawn from classical mechanics and mathematics courses taken at Caltech, Oxford, and the University of Wisconsin. The courses range from undergraduate level to advanced graduate level. The courses had a rich and sophisticated selection of textbook and reference material, as you might expect, and those reference texts are, similarly, drawn on here. Those classical mechanics texts, listed by author, include: Landau and Lifshitz [163]; Goldstein [164]; Fetter & Walecka [165]; Percival & Richards [166]; Arnold (ODE) [167]; Arnold (CM) [168]; Woodhouse [169]; and Bender & Orszag [170]. Notice how the first Arnold reference and the Bender and Orszag reference involve textbooks focused on ordinary differential equations (ODEs). Likewise, an analysis of the excellent, and rapid, exposition by Landau and Lifshitz, reveals that it partly progresses through the material by going through ODEs of increasing complexity (corresponding to more complicated pendulum motion, for example, such as by adding a frictional force). This strong alignment with the underlying mathematics of ODEs is continued in this exposition, so much so that an appendix is provided for a quick review of ODEs from the applied mathematics perspective.

Particle dynamics, with and without forces, are described, with all arriving at descriptions with chaotic motion, with chaos described in the latter half of Book 1 [1]. Universally it is found that systems transitioning to chaotic behavior do so with a remarkable period-doubling process and this will be described both mathematically and with computer results. In the analysis of such dynamical systems we will find that periodic physical systems can be described in terms of repeated "mappings", e.g., classic dynamic mappings [171], and when described in this way the transition to chaos is made much more mathematically evident (as will be shown). The familiar Mandelbrot set is generated by such a repeated mapping, where it's "edge of chaos" is defined by the fractal boundary of the classic Mandelbrot image.

Properties of the classic Mandelbrot set will be relevant to the physics discussed in Book 1 and Book 7, including the property that the fractal boundary has a fractal dimension of 2 (the fractal dimension of the boundary can be between 1 and 2, to get equal to 2 is special). With the Mandelbrot set we also recover the well-studied constants associated with the universal Feigenbaum constants [133]. In the Mandelbrot set we can clearly see the fundamental constant for maximum perturbation that is at maximum antiphase (negative) with magnitude C_{∞}, where the same results hold for a family of basic formulations (for a variety of Lagrangian formulations, for example).

From the Lagrangian variational formulation of 'action' for particle motion we will eventually define the path integral functional variational formulation involving that same Lagrangian to arrive at a quantum description for the non-relativistic quantum particle motion (described in detail in Book 4 [4], and relativistic in Book 5 [5]). From the quantum description we arrive at the propagator formalism for describing dynamics (this exists in the classical formulation too, but typically is not used much in that context). Complex propagators will then be found to have ties to statistical mechanics and thermodynamics properties (Book 6 [6]). The ties to statistical mechanics are further emphasized when at the "edge of chaos" but with the orbit motion still confined. This may be associated with an ergodic regime, thus an equilibrium and martingale regime, the existence of which can then be used at the start of Book 6 [6] statistical mechanics and thermodynamics derivations with the existence of equilibria established at the outset. The existence of the familiar entropy measures are already indicated in the neuromanifold description (Book 3 [3]), thus, together with equilibria, the Book 6 thermodynamics description is able to begin with a well-established foundation that is not claimed by fiat, rather claimed as a direct result of what has already been determined in the theory/experiment described in the previous books of the Series.

Overview of Books 2 & 3

When moving from a theory of point particles to a theory of fields, there's not much discussion in the core physics books on fields in a general sense, it usually just directly jumps to the main field of relevance, electromagnetism. If advanced, it may also cover General Relativity, as with [172]. In what follows we will cover these topics, but we will also cover the more basic fields in 1, 2, and 3D (including fluid dynamics), as well as 4D Lorentzian Field formulations (for Special Relativity), the Gauge Field formulation (thus Yang Mills covered in a classical context), and the General Relativity geometric and gauge formulations. This establishes the foundation for the standard forces, and upon quantization (Books 4 and 5 in the Series), lays the foundation for the standard renormalizable forces (all but gravitation).

The gravitational coupling constant 'G' is a dimensionful coupling (not like with alpha in electromagnetism), and gravitation with manifold construct can be described as a gauge field construct, although not renormalizable. Gravitation, and associated geometry/manifolds, appears to relate to its own emergent structure, as will be discussed in Book 6. From the local Lorentzian geometry and Lorentzian field descriptions we also see the first of many examples where there is system information in the complexification of some parameter, here the time component. If the Lorentzian is shifted to complex time, this shifts it to being a Euclidean field, with formally well-defined convergence properties (as occurs in statistical mechanics). Complex time also shows deep connections between classical motion and associated Brownian motion (where random walk reveals pi). Thus, it should not be surprising that an emergent manifold may have complex structure such that there is also an emergent 'thermal' manifold, possibly the neuromanifold described in Book 3 and the related partition functions examined in Book 6. Just like locally flat space-time is a natural construct in General Relativity, so too are optimization "learning" steps on a neuromanifold such that relative entropy is selected as a preferred measure, and from it Shannon entropy and Boltzmann's statistical entropy. Thus, the manifold construct appearing at Book 3 has far reaching impact into the foundations of the thermodynamic and statistical mechanical theory described in Book 6.

Before we even get to the manifold/geometry complexities of General Relativity, however, we have already established much with the electromagnetic field part of the theory: (i) from 'free' electromagnetism without matter we get the speed of light c, Lorentz invariance, and from that special relativity and locally flat space-time; (ii) from electromagnetism with matter we get the dimensionless coupling constant alpha.

In going over field theories to describe matter, force fields, and radiation we first describe the classical field theories of fluid mechanics, electromagnetism, and General Relativity, with many examples shown. This is then carried over to the quantum field theory description in Book 5. A review of the core mathematical constructs employed in classical field theory and quantum field theory is given in the Appendix. Even as the mathematical physics approach grows in sophistication, we still obtain solutions via variational extrema. Thus, determining the evolution of the system from its variational optimum now becomes the focus of the effort. System 'propagation' from one time to a later time can be described by a propagator. Although a 'propagator' formulation is possible mathematically in classical mechanics and classical field theory, which are shown, this is usually not done, in favor of simpler representations for the experimental application at hand. As we move to descriptions in the quantum realm, however, the use of the propagator formalism becomes typical, and when used in the path integral formulations we arrive at a compact formulation describing both the evolution and stationary-phase solution at once.

In Book 2 the focus is on classical field theory in a fixed geometry, the main physical example is electromagnetism. In this setting alpha appears, for example, in the description of an electron-positron pair: $F = e^2/(4\pi\varepsilon a^2)$ for electron-positron distance 'a' apart, where alpha appears as the coupling constant. Later, in quantum mechanics, both modern and in the early Bohr model, we have that alpha = $[e^2/(4\pi\varepsilon)]/(c\hbar)$. The appearance of alpha in these situations is occurring in bound systems. If we examine electromagnetic interactions that are unbound, on the other hand, such as with the Lorentz Force $F = q(E \times v)$, here there arises no alpha parameter, nor with the early quantum mechanical analysis of such systems such as with Compton scattering. Thus, we see an early role for alpha, but only in bound systems, thus only in systems with (convergent) perturbative expansions in system variables.

In Book 3, classical field theory with *dynamic* geometry, i.e. General Relativity, we don't see alpha at all. Instead we see manifold constructs and the mathematics of differential geometry (and to some extent differential topology and algebraic topology). Manifold constructs are entirely encapsulated in the math background given in Book 3 and the Appendix there. An application in the area of neuromanifolds (see [6]), shows the equivalent of a geodesic path in this setting is evolution involving minimum relative entropy steps. Similar to the description of a locally flat space-time we now have a description of 'entropy' increasing/evolving according to minimum relative entropy.

General Relativity stands apart from the other force fields. All the other force fields are part of an adjoint representation of the standard model vis-à-vis the stability subgroup U(1)xSU(2)$_L$xSU(3). The form of which is derivable from the chiral T one-sided products described in Book 7. The standard model is uniquely obtained in this process, and with no mention of General Relativity. Keep in mind, however, that the adjoint representation has operation on some space (hyperspinorial in case of simple octonion right-products, for example). The 'force' due to gravity is that due to manifold curvature, where the manifold construct is possibly emergent on the space of operation. Thus, the origin of the General Relativity force is entirely different, and it will not allow quantization like the other forces, nor will its singular solutions be resolvable via quantum physics alone, as with electromagnetism in Books 4&5, but will also need thermal physics (as will be described in Book 6).

The existence of singular General Relativity solutions, outside of specially symmetric cases (the classic Black hole solutions), wasn't firmly established until the Penrose singularity theorem [173] (awarded Nobel prize in Physics for this in 2020). Some of this material is covered in Book 3 to show how the mathematical formalism shifts to differential topology methods to describe the singularities, with examples referencing the Hawking and Ellis classic [174] and using Penrose diagrams. This, in turn, will come in handy when describing the classic FRW cosmologies with radiation and matter dominated phases (using notes from Peebles [175], Peebles won the Nobel in Physics in 2019).

The General Relativity development would be remiss if it didn't briefly delve into cosmological models, the classic FRW cosmologies in particular. With the General Relativity tools developed, cosmological results are examined, starting with the entry of the cosmological constant into the formalism (a candidate for Dark energy). Various observational data on galaxy rotations and universe simulations of galaxy cluster formation both indicate the existence of Dark matter. This, then, means we have new matter, non-interacting except gravitationally, and this is actually consistent with the latest observational data on the muon g-2 value [176], where the discrepancy between theory and experiment has grown to 4.2 standard deviations, where an extension in the Standard Model appears to be in the works. This is convenient as Emanator theory (Book 7 [7]), predicts such an extension.

We can thus arrive at field equations for electromagnetism, General Relativity, and Yang-Mills Gauge Fields (Strong and weak). We can obtain wave and vortex phenomena (as hinted in fluid dynamics). We show the classical instability for atomic matter (classical electromagnetic instability) and classical gravitational instability (leading to black hole formation with singularity). From Lagrangian formulations we can then arrive at a quantum field theory formulation (Book 5). The quantum field theory formulation completes the quantum mechanics (Book 4) cure of "non-relativistic atomic instability" with the cure of the fully relativistic atomic description of the radiative-collapse instability. Introduction of quantum field theory also leads to new instability or infinities, but these can be eliminated by renormalization for the electromagnetic and electroweak formulations, and the Yang-Mills strong formulation, but not the General Relativity (gauge) formulation. The current theoretical formulation in modern physics has one glaring gap, therefore: a quantum theory of gravitation. Perhaps this is not a missing element, however, if geometry/General Relativity is a derivative phenomenon, like the field of statistical mechanics and thermodynamics appeared as derivative phenomenon when the complexified quantum propagator gives rise to a real (quantum) partition function. The hint of a deeper emanator theory suggests emergent structures of geometry and thermodynamics are arrived at in the process of emanation, with the information emanated being that of the renormalizable quantum matter fields. In Book 7 [7] a precise mathematical meaning will be found for describing maximal information emanation.

Overview of Book 4

By 1834, with Hamilton's Principle, there was a strong foundation for what is now called classical mechanics. By 1905, with Einstein's publication on the photoelectric effect [177], the rules of classical mechanics were being superseded by the new rules of quantum mechanics. The earliest appearance of quantum mechanics, however, began with the various observations of quantization of light, starting with the strange occurrence of spectral lines for hydrogen. The hydrogen spectrum was made even stranger by a precise fit to a succinct empirical formula by Balmer in 1885 [178]. This is the beginning of an amazing period of discovery. The developments of quantum mechanics from introductory to advanced roughly follows that history.

The early phase of discovery for quantum mechanics moved into the modern quantum mechanics formalism with the discovery of Heisenberg of the successful application of matrix mechanics and the resultant uncertainty principle (1925) [179]. In 1926, Schrodinger showed that the problem of finding a diagonal Hamiltonian matrix in the Heisenberg's mechanics is equivalent to finding wavefunction solutions to his wave equation [180]. An interpretation of the wavefunction was then clarified in 1927 by Born [181]. Dirac developed a manifestly relativistic formalism for the wavefunction and wave-equation for fermionic matter (1928) [182]. An axiomatic reformulation of quantum mechanics was then given by Dirac (1930) [183], laying the foundation for much of modern quantum notation and for critical issues such as self-adjointness. Dirac then described a formulation of a quantum propagation path, with quantum propagator having the familiar phase factor involving the action, in his paper "The Lagrangian in Quantum Mechanics" in 1933 [184]. In essence, Dirac had obtained a single path, in what would eventually be generalized by Feynman to all paths with the invention of the path integral formalism (1942 & 1948) [185,186]. The equivalence of a quantum mechanical formulation in terms of path integrals and the Schrodinger formalism was shown by Feynman in 1948 [186].

In a path integral description, the quantum mixture state, semiclassical physics, and classical trajectories are all given by the stationary phase dominated component. A stationary phase solution that is dominated by a single path is typical for a classical system. Thus, variational methods are fundamental to analysis of physical systems, whether it be in the form of Lagrangian and Hamiltonian analysis, or in various equivalent integral formulations.

Feynman's discovery of the path integral formalism wasn't solely based on the prior work of Dirac (1933) [184], although by appending that paper to his PhD thesis (1946) its importance was clearly emphasized. Feynman also benefited from work going as far back as Laplace [158] for selection process based on highly oscillatory integral constructions that self-select for their stationary phase component. This branch of mathematics eventually became associated with Laplace's method of steepest descents, then to the work of Stokes and Lord Kelvin, then to the work of Erdelyi (1953) [187,188].

Feynman and others then invented quantum field theory for electromagnetism (QED) during 1946-1949 (more on this later). Extension to electroweak occurred in 1959, and to QCD in 1973, and to the "Standard Model" in 1973-1975. Thus, the impact of the path integral revolution in quantum physics was felt well into the 1970's, but this was only the beginning. At their inception path integrals were examined by Norbert Wiener, with the introduction of the Wiener Integral, for solving problems in statistical mechanics in diffusion and Brownian motion. In the 1970's this led to what is now known as "the grand synthesis" which unified quantum field theory and statistical field theory of a fluctuating field near a second-order phase transition, and where use of renormalization group methods enabled significant advances from quantum field theory to be carried over to statistical field theory.

The grand synthesis is one of many instances to come where we see analytic continuation of a constant or a parameter giving rise to familiar physics in the thermodynamic and statistical mechanics domains, showing a deeper connection (still not fully understood, see Book 7). The Schrödinger equation, for example, can be seen to be a diffusion equation with an imaginary diffusion constant. Likewise, the path integral can be seen to be an analytic continuation of the method for summing up all possible random walks.

In Book 4 we also carefully examine the closest gravitational equivalent to the hydrogenic atom (dust shell collapse). What results is an incomplete formulation due to boundary conditions, where to get the time choice you must input that time choice. No specific choice of time is indicated to avoid infall-collapse. The results, however, can show stability and consistency in a "full" thermal quantum gravity description where analyticity is employed. Success in this way, and not others, suggests possible fundamental role of analyticity and thermality (Books 6&7) and also suggests that thermal quantum gravity may 'exist' or be well-formulate-able, while quantum gravity generally might not 'exist'. These results, shown in Book 6, provide the lead-in to the Book 7 discussion on Emanator theory, where core concepts in Books 1-6 that tie to emanator theory are brought together in a new theoretical synthesis.

Overview of Book 5

In Book 5 we show quantum field theory in the gauge field representation, which clearly relates the choice of field theory to a choice of Lie algebra, which, in turn, can be related to a choice of group theory (such as U(1) and SU(3)). From this we can see that non-classical algebraic constructs are ubiquitous in quantum mechanics and quantum field theory, so a review of Group Theory and Lie Algebras is given in the Appendix, as well as a review of Grassman Algebras, and other special algebras needed in quantum mechanics and quantum field theory. Similarly, as regards choice of approach, we find that the Schrodinger and Heisenberg formulations often provide the only tractable way to get a solution for bound systems. In critical theoretical considerations, however, the path integral approach is best (as will be shown). In seeking a deeper theory, the more unified path integral (PI) approach provides important hints as to a deeper theory (see Book 7).

In Book 5 we get the highest precision result for the value of alpha, in its role as perturbation parameter. If a calculation of the electron magnetic moment parameter g-2 is performed, with all of the Feynman diagrams appropriate to expansions up to 5th order, we get a determination of alpha up to 14 digits, where 1/alpha=137.05999..... . This gives us one of the most precise measurements of alpha known. When a similar analysis is done for the muon g-2, given the much larger muon mass, particle production pairs of other particles have a measurable effect, and we are able to probe the lower masses of the standard model that are present. In doing this, in preliminary experiments, there is a discrepancy indicating more particles, e.g. the Standard Model will need to be extended (possibly with a type of 'sterile' neutrino). These missing particles could be the missing "Dark Matter". The prediction of such in Emanator Theory, and why there should be an imbalance between the left and right neutrinos (hint: maximum information transmission) is described in Book 7.

Part of the description of quantum field theory entails use of analyticity and other complex structures to encapsulate more of the physics in a complex extension to the space (or dimension). This often leads to formulations in terms of complex integration, with the choice of complex contour specified, such as with the Feynman propagator. One of the main renormalization methods, for example, is to use dimensional regularization, which entails analytically continuing expressions with dimensionality to dimensionality as a complex parameter. There is also the aforementioned shift to complex and to "Wick rotate" expressions with real time to expressions with pure complex time. In doing this the statistical mechanical partition function for the system is obtained, with well-defined summation. Thus, a connection between 'thermality' and complex structure, in the time dimension at least, is indicated.

The second part of Book 5 describes quantum field theory on curved space-time, where we arrive at an early analysis of Black Hole thermodynamics. Here we find that space-time curvature gives rise to thermality and particle production effects. Black Hole thermality was revealed in Hawking radiation [189], due to the causal boundary at the horizon. Such thermality is even seen in flat space-time (Book 5) if causal boundaries are induced, such as in the case of an accelerated observer [39].

Quantum field theory on curved space-time has one further gift, critical to the statistical mechanics formalism to follow in Book 6, and that's the spin-statistics relation. This relation is usually assumed, along with other critical notions, such as entropy, and the relation between entropy and density of states. These are all shown, with the presentation path chosen in this Physics Series, to be fundamental or derivative to the formalism already established in Books 1-5 (to prepare for Book 6).

The choice of time is related to choice of vacuum, which is related to choice of field geometry or observer motion (such as constant acceleration or expansion). If you have flat spacetime quantum field theory with a boundary, then you have thermodynamic effects (e.g., the Rindler observer). In this setting we can compare the Hawking derivation of Hawking Radiation using the Euclideanization 'trick' vs the Bogoliubov transformations of the field to the Rindler geometry from the

Minkowski geometry (if chosen as the asymptotic vacuum reference). With quantum field theory on curved space-time we also arrive at spin-statistics as mentioned, and get the final extension of the theory by way of Grassman algebras, to arrive at thermodynamically consistent Bose and Fermi statistical descriptions on quantum matter.

Overview of Book 6

Thermodynamics is the oldest of the physics disciplines (fire), with unapologetic use of phenomenological arguments and mysterious thermodynamic potentials (entropy). Obviously, thermodynamics is still prevalent today, including in its more quantified form via statistical mechanics. How is this not a failure of the mechanistic description of the universe indicated by classical mechanics and even quantum mechanics? Concepts that appeared in quantum mechanics, such as probability, are now occurring again. Other new concepts appear as well, including: approximate statistical laws; equations of state; heat as a form of energy; entropy as a variable of state; existence of equilibria; ensembles/distributions; and existence of the partition function. Many of these concepts appear in the path integral descriptions with the analyticity methods/extensions mentioned previously, so there are hints of a deeper theory that arrives at much of thermodynamics/Statistical mechanics foundation from the existing quantum theory.

Book 6 [6] has been placed after the other books in the Series [1-5] to await identification of entropy as fundamental in that it can be identified as an extrinsic system function even before getting to thermodynamics. We also already have experience with many particle systems, via quantum field theory (especially in curved space-time where particle creation is almost unavoidable), without directly tackling that scenario (due to quantum field theory effectively already being many-particle, with analytic determination of many-particle system functions, such as entropy).

With entropy presented at the outset as an important system variable, the derivation of thermodynamic potentials is then a straightforward process, as will be shown. The standard statistical mechanics connections to thermodynamics can then be given. Thus, in covering Thermodynamics and Statistical Mechanics we start with the foundations of the theory mostly established, such as entropy (also with equipartition equivalent to sum on paths with no weightings, etc.), with no assumptions. Everything follows directly from the theoretical discoveries outlined in the preceding books in the Series [1-5]. We don't see new connections to alpha, but we do see new structures/effects, especially manifold constructs (as with General Relativity, where we also saw no role for alpha).

The close ties between quantum mechanics complexified giving rise to the particle ensemble partition function, and quantum field theory complexified and the field ensemble partition function, is a derivative aspect of time-complexation. This complexation will be described in later chapters , where analytic time will allow an exact thermodynamic stability analysis of black hole geometries to be examined.

From atomic physics, described in Book 4, we also obtain the standard rules on electron shell completion (that is encoded in the periodic table). Similarly, we can also understand the origins of the intermolecular quantum chemistry rules. When taken to the statistical mechanics extreme we have thermodynamic equilibrium emergent from (the Law of Large Numbers and reverse Martingale convergence. With completion of application to chemical processes we have clear phase-transition effects, as well as equilibrium and near-equilibrium effects. The familiar chemistry is then observed, with phases of matter.

From chemical equilibrium and near-equilibrium, with 10^{23} elements that interact or not at all, we have two generalizations. The first is to consider chemical near-equilibrium and directly obtain an emergent process at this level, this is the branch that gives us biology/life at its most primitive level. The second is to consider equilibrium and near-equilibrium in general when the elements interact strongly (with10^{10}elements, say), this is the branch that describes biology/life at its most advanced social level and economics. In classic circuit shot noise analysis, the granularity of low-current flow (due to discreteness of electron charge) leads to a noise effect. Thus, as we consider situations with fewer elements, there are more complications, not less, due to granularity noise effects, and we enter the realm of machine learning with sparse data. Noise effects can be significant in complex systems, especially in biology where it is sometimes part of what is selected (such as in hearing, for background noise cancellation).

The second part of Book 6 explores the role of thermodynamics in efforts to extend to thermal quantum field theory and thermal quantum gravity. This is done by exploring Black Hole settings. The recognition of a role for complex structure on system variables becomes apparent in this process. This is done by examining the Hamiltonian thermodynamics of some black hole geometries with stabilizing boundary conditions. In this foray into directly exploring a thermal quantum gravity solution we assume a path integral form for the General Relativity problem and shift directly to a partition function (by 'Wick rotation'). We see that thermal quantum gravity is possible, when positive heat capacity shows stability.

Overview of Book 7

In Books 4,5, and 6 of the Series, we explored examples of quantum mechanics with imaginary time, quantum field theory in curved space-time, Thermal quantum field theory, minisuperspace quantum gravity, and Thermal quantum gravity. In this effort we find the path integral, and PI propagator, to provide the most general representation. In seeking a deeper theory in Book 7 we build on the sum-on-paths with propagator formulation to arrive at a sum-on-emanations with emanator formulation.

Propagation in a complex Hilbert space, in a standard quantum mechanics or quantum field theory formulation, requires the propagator function to be a complex number (not real or quaternionic, etc., [127]). This prohibits what would otherwise be an obvious generalization to hypercomplex algebras. In order to achieve this generalization, we have to introduce a new layer to the theory, one with universal emanation involving hypercomplex algebras (trigintaduonions) that is hypothesized to project to the familiar complex Hilbert space propagation with associated fixed elements (e.g., the emanator formalism projects out the observed constants and group structure of the standard model). The 'projection' is an induced mathematical construct, like having SU(3) on products of octonions, but here it we be the standard model U(1)xSU(2)xSU(3) on products of emanator trigintaduonions. Thus, in Book 7 a unified variational formulation is posed, one that arrives at alpha as a natural structural element, among other things, uniquely specified by the condition of maximal information emanation.

In Book 7 we also make note of the implications of a fundamental mathematical operation on a space that is repeated or added. The non-General Relativity forces are given by the form of the operation (the sequence forming an associative algebra), the General Relativity forces are given indirectly by the form of the space, this leaves the aspect "repeated or added" to be considered with care. If a purely 'repeated' operation, or mapping, occurs we can return to the dynamical mapping discussion of Book 1, where chaos can occur and is ubiquitous. There, the primal 'phase transition', the transition to chaos, is evident. If an operation with addition is involved (in the statistical sense of multiple elements), along with repeated overall steps, we arrive at the general framework of statistical mechanics with effects from the Law of Large Numbers and reverse Martingale convergence, among other things (Book 6). Most notable, however, is the prevalence of a new effect, that of phase transitions and the emergence of new structure (order from disorder), including the remarkable structures of chemistry and biology.

Why the recurring 'Cabbalistic formula'? was a question even in the time of Sommerfeld [190]. Now, the numerological parallel is more exact than realized at that time, so is too much a coincidence to be by chance. The non-coincidence appears to be due to the maximal nature of information transmission in a variety of circumstances (in physics, biology, and even human communication with sufficient optimization) as well as with the fractal-like repetition of key parameter sets that occurs in these different settings {10,22,78,137≅1/alpha}. We see that 10 expresses the dimensionality of propagation (or nodes of connectivity), while 22 corresponds to the number of fixed parameters in the propagation (in Book 7 we explore propagation in a 10 dimensional subspace of the 32 dimensional trigintaduonion space, leaving 22 dimensions at fixed values that appear as parameters in the theory). We will see the number 78 relates to generators of the motion, and that there are 4 chiralities of motion ('doubly chiral'). We will also see that 137 is simply the number of independent tri-octonionic product terms in the general chiral trigintaduonion 'emanation'.

Synopsis – Frodo Lives

Tolkien wrote of eucatastrophes [191], perhaps he anticipated the constructive role of emergent phenomena in maximum information transmission.

Preface to Physics Series, Book #6, on:

Thermodynamics, Statistical Mechanics, and Black Hole Entropy

Thermodynamics is the oldest of the physics disciplines with unapologetic use of phenomenological arguments and mysterious thermodynamic potentials (entropy). Obviously, thermodynamics is still prevalent today, including in its more quantified form via statistical mechanics. This book (sixth in a seven-part physics series) is on the fundamentals of modern thermodynamics, statistical mechanics, and implications about the nature of time and universal thermality. This book's focus on thermodynamics follows the other books in the Series [1-5] to await identification of entropy as fundamental in that it can be identified as an extrinsic system function even before getting to thermodynamics.

This book begins at introductory thermodynamics (undergraduate-level). The first third of the book provides the basics of thermodynamics in a concise description, with many worked-out examples. The second third of the book provides the basics of statistical mechanics (advanced undergraduate and graduate level), again with many examples, and advanced application in emergent phenomena. The last third of the book considers the implication of black hole thermodynamics analysis as to the nature of time (advanced graduate level).

With entropy presented at the outset as an important system variable, the derivation of thermodynamic potentials is then a straightforward process, as will be shown. The standard statistical mechanics connections to thermodynamics can then be given. Thus, in covering Thermodynamics and Statistical Mechanics we start with the foundations of the theory mostly established.

The close ties between quantum mechanics complexified giving rise to the particle ensemble partition function, and quantum field theory complexified and the field ensemble partition function, is a derivative aspect of time-complexation. This complexation will be described in later chapters, where analytic time will allow an exact thermodynamic stability analysis of black hole geometries to be examined. In this a greater understanding of the role of time and the statistical 'forces' due to thermodynamics is then given.

Chapter 1. Introduction

Thermodynamics is the oldest of the physics disciplines starting with practical efforts to produce and control heat (fire). The typical thermodynamics presentation makes unapologetic use of phenomenological arguments and mysterious thermodynamic potentials (entropy). For phenomenological validation, there is the production of steam engines and early industrialization, and much more sophisticated developments since. So it's understandable why the typical thermodynamics text would not feel the need to explore the foundations too much. In this book, however, we have an explanation for why the entropy takes the form it does, and why there should be equilibrium states, and why there should exist a system partitions function that is generative of the entire thermodynamical description of the system. It turns out these are direct consequences of our modern understanding of quantum field theory (for the origin of the partition function from a quantum mechanical context, see Books 4 and 5 of the Series [1-7]); statistical learning theory (for the origin of the preferred measure of entropy chosen, see [8]); and Emanator theory (Book 7 [7]), where a Martingale basis to thermodynamic processes is indicated, which would indicate the existence of equilibria as observed.

This book begins at introductory thermodynamics (undergraduate-level). The first third of the book provides the basics of thermodynamics in a concise description, but with many worked-out examples. The second third of the book provides the basics of statistical mechanics (advanced undergraduate and graduate level), again with many examples, and advanced application in emergent phenomena. The last third of the book considers the implication of Black Hole thermodynamics analysis as to the nature of time (advanced graduate level, based on PhD dissertation [9], with full general relativity derivations for the Black Hole systems given in the appendix).

Part of the power and uniqueness of thermodynamics over the other disciplines is the amazing utility in phenomenological methods. This will be understood in later chapters as partly an aspect of critical-point power law scaling, and other universal structures that occur, making phenomenological arguments well-founded in the overall variational description of the theory. Even so, if the foundational elements of thermodynamics come from quantum mechanics, statistical learning, and emanator theory, etc., as indicated above, then what is thermodynamics exactly, is there a fundamental aspect to thermodynamics at all? Put a different way, is there a 'force' to attribute to thermodynamics?

Consider that the other fundamental areas of physics describe forces, such as the renormalizable field theory forces (electromagnetic, weak, strong) and the dynamical manifold forces (gravitation). For thermodynamics and statistical mechanics we will discover fundamental forces of constraint. These will manifest in the probabilistic interpretation in statistical mechanics in a variety of forms depending on the system description. Such statistical "forces of constraint" include the presumed existence, and evolution towards, equilibrium states. At equilibrium the existence of a generative partition function description occurs. Thus, we get the statistical effects and the probabilistic mathematical foundation itself from these forces of constraint. The asymptotic equipartition property, the law of large numbers, and the variational description with the variety of thermodynamic potentials that can be obtained from the partition function, all derive from the added mathematical underpinnings in the thermodynamic and statistical mechanics frameworks.

General relativity gives a theory for a dynamical manifold, with the appearance of forces on objects according to geodesic motion , and deviation from geodesic motion, in the manifold geometry. Similarly, in many contexts the thermodynamics and statistical mechanics of the system can be captured in an information geometry manifold construct [10-12]. In the information geometry context there is an equivalent analysis, describing motion/optimization, as in modern statistical learning. In the statistical learning on a manifold (neuromanifold) context we find there is locally flat information geometry [10,12], where the shortest distance, selected analogously to Euclidean distance by criteria of local flatness (Riemannian), is relative entropy (technically a shortest 'information divergence'). The selection of relative entropy as measure is further manifest as a form of expectation/maximization algorithm for learning/motion (see [8]), and becomes the standard dimensionless Shannon entropy or dimensionful Boltzmann entropy according to application.

So far we see that thermodynamics becomes fundamental merely by taking all the foundational elements, including existence of equilibrium, as deriving from other fundamental disciplines and simply "adding statistics" and perhaps manifold mathematics (along with the new interpretations necessary when working with statistics). It will be shown in Chapter 5-8, however, that thermodynamics reveals fundamentally new behavior:
(1) local phases of matter and transition between phases
(2) global structure emergence (including existence of equilibrium)
(3) critical exponent universality at critical values of thermodynamic potentials
(4) existence of maximally-extended analytic time and resulting universal thermality

Thus, thermodynamics/statistical mechanics/statical field theory does describe something fundamentally new, and not just a well-defined statistical field theory upon Wick rotation from any physical quantum field theory description. It provides 'force' in an unfamiliar form, that of a statistical mathematical constraint, and exhibits universality and emergence in a variety of forms. Many of the odd effects predicted in statistical field theory have been observed in condensed matter experiments and other areas of physics. The phenomenological analysis of scaling law behavior is an area where computational exploration is now possible, given modern computing power. In the area of scaling for large language model training (Chat GTP) we see similar scaling law behavior. This was considered sufficiently reliable scaling behavior that hundreds of millions were invested based on it to train Chat GTP 3.0 with scaled-up model size and training data, and it worked. The capabilities of the large language model increased to a new level of problem solving capability. This increased scaling was done again, and worked again for Chat GTP 4.5. The process continues and the fit in the learning process to a scaling limit appears to be bounded, but not for a few more iterations of development, probably more than sufficient to see the development of a GAI. Thus, the ideas and applications of thermodynamics continue developing to this day. Even though the oldest of the physics disciplines, it is also the one with the most growth now and in the future, especially given the power of modern computing for simulations and universal learning development.

Chapter 2. Overview of Foundations

2.1 Introduction

Thermodynamics involves the properties of aggregated matter and originally was formulated without use of atomic concepts (not being determined yet). Once atomic concepts are brought into the mix we have (1) kinetic theory; and (2) statistical mechanics.

Thermodynamics introduces new concepts and old concepts in an axiomatic way:
(1) Temperature is a single number (scalar) assigned to each subsystem. This parameter alone can suffice to determine if two subsystems brought into "contact" are in "thermal equilibrium". If you take a system in thermal equilibrium at temperature T and cut it in half, each half will be at temperature T, this makes T an intrinsic variable (versus energy, an extensive variable).
(1a) Two systems are in "contact" if they can exchange energy, entropy, or volume (e.g., they can exchange in the extrinsic variables).
(1b) Two systems are in thermal equilibrium if they are in contact, yet have no exchange of energy, entropy, or volume. Most notably, energy in the form of heat is not exchanged at equilibrium.
(1c) The Zeroth Law of thermodynamics is that the concept of temperature makes sense and has utility as indicated. In Emanator theory [7] and Appendix D, time is analytic with periodic boundary conditions (as in thermal quantum field theory) that, in turn, defines the inverse temperature of the system, thereby eliminating the need for a separate Zeroth Law.
(1d) Temperature can't be negative.

(2) Energy of a system is conserved, the First Law of Thermodynamics, especially on how it can change among its different forms where $dE = dQ + dW$ at the most basic level, where Q is heat, W is work, and E is energy. Note that conservation of energy is well-established in Classical Mechanics, but here we include a new form of energy comprising heat that is distinctive to thermodynamics.
(2a) Aggregated matter has a thermal capacity to hold heat and exchange heat upon thermal contact.

(2b) Thermal capacity describes the property of a large hot body, that is placed in contact with a small cold body, to bring the small body's temperature to the same temperature as the hot body (and thereby reach equilibrium).
(2c) A thermometer is a subsystem with a much smaller heat capacity that that of the body whose temperature is to be measured, where the equilibrium temperature to the combined thermometer-body system is approximately that of the body originally.
(2d) $dQ = TdS$ and $dW = -PdV + \cdots$ where all of the other extrinsic/intrinsic variable pairs of the system (depending on the system) are included in the work definition as well.
(2d) Aggregated matter, a subsystem, has he capacity to perform work on another subsystem.

(3) Entropy for an aggregation of matter is at a maximum when a system is in equilibrium. When a system is out of equilibrium, its entropy will increase as it equilibrates (Second Law of Thermodynamics). Fluctuations do occur, but for aggregations above microscopic level, the fluctuations are greatly suppressed by averaging (effective mean field).
(3a) At the level of thermodynamics entropy is a mysterious concept, a thermodynamic potential that pairs with temperature, also a purely phenomenological parameter (unlike energy in terms of pressure, etc.).
(3b) At the level of statistical mechanics we will have a definition of entropy in terms of the number of possible system states ($S = k \ln \Gamma$), where Γ is the number of possible system states, so it will become more concrete. Even so, there is the mystery of why the relation $S = k \ln \Gamma$ precisely. This can now be understood in a optimal statistical learning context [8], where relative entropy is singled-out as the optimal measure for comparing two distributions. Derivative of that property is the above Boltzmann relation (also known as Shannon entropy in Information Theory). Thus, if we bring in the same information theory that we will use later to explain Black Hole Thermodynamics at this early juncture, we can use it to explain the origin of the entropy relation, and thereby give a better understanding of entropy from the outset – it is a measure of system 'state capacity' which can be seen as 'information', 'complexity', 'order', or even 'disorder', according to context and what the quoted terms are defined to be in those contexts.
(3c) At zero temperature the entropy is zero – the Third Law of Thermodynamics (Nernst's theorem).
(3d) The existence of an entropy thermodynamic potential $S(E, N, V)$ that depends on E, N, V, that is always increasing in its parameters means that

it can be inverted. The inversion for energy in particular means we have the fundamental relation for energy in terms of the other thermodynamic variables: $E = E(S, N, V)$.

The fundamental system variables in a thermodynamics system are then E, S, T and according to the system/application, also kinetic and mechanical parameters such at V, P, N, μ, etc. If we know E as a function of the other extrinsic variables S, V we can deduce everything we nee to know thermodynamically. If we know E as a function of the variables T, V, we generally do not have complete information. Of course, it's much easier to know T than S since even if we use the Boltzmann relation we still need to know the density of states for the system (dependent on E, N, V generally).

Theoretically we have the relation $E = E(S, N, V)$. Regardless of whether it is easy to determine S experimentally, let's consider the theory based solely on the existence of the indicated structure. Let's start with the total differential:

$$dE = \left(\frac{\partial E}{\partial S}\right)_{N,V} dS + \left(\frac{\partial E}{\partial V}\right)_{N,S} dV + \left(\frac{\partial E}{\partial N}\right)_{V,S} dN$$

and by convention we have the relation:

$$dE = TdS - pdV + \mu dN$$

We thereby have relations for temperature, pressure, and chemical potential

$$T = \left(\frac{\partial E}{\partial S}\right)_{N,V}, \quad p = -\left(\frac{\partial E}{\partial V}\right)_{N,S}, \quad \mu = \left(\frac{\partial E}{\partial N}\right)_{V,S}$$

Further developments along these lines (basic thermodynamics) are given in Ch. 3.

2.2 Obtaining the State Function

Since thermodynamics works with conservation of energy, the classical Hamiltonian formulation (in phase space) is needed for much of classical thermodynamics. In the advanced path integral formulation we will need to return to classical mechanics formulations involving the other main variational formalism in terms of the Lagrangian, but that is not needed now. The formalisms are equivalent (classically), regardless, by way of a Legendre transformation (see Book 1 [1] for further details).

System Energy
In classical phase space for a motion that describes a $2d$ dimensional motion with a Hamiltonian $\mathcal{H}(p_j, q_j), j \in 1..f$, we have a system determined by:

$$\dot{p}_j = -\frac{\partial\mathcal{H}}{\partial q_j}, \quad \dot{p}_j = \frac{\partial\mathcal{H}}{\partial p_j}$$

where $\mathcal{H}(p_j, q_j) = E$ is the energy for a conserved system.

In a quantum system the Hamiltonian becomes an operator and the mathematical description of the system switch from a point moving in phase space to a stationary dynamical state in an eigenstate of the Hamiltonian:

$$\widehat{\mathcal{H}}\varphi_l = E_l\varphi_l.$$

Notice that the description of the quantum *energetics* appears simpler where a more clear elucidation of states is necessary in the quantum system description. From the perspective of the system energy description alone (not how those results are obtained), it appears that the quantum descriptions will provide a simpler and more well-defined state elucidation than the the classical description. This is indeed the case. In some cases, statistical mechanical analysis of some quantum system will be as simple as requiring analytic time and Wick rotating the quantum path integral formulation to arrive at the statistical mechanical partition function for the system (done in Chapters 7-9 where Black Hole thermodynamics is examined). From the partition function the entire thermodynamic properties can be determined (much like the entire properties of the quantum system derive from the propagator in the path integral formulation).

System Observables
Statistical mechanics and quantum mechanics both speak of system observables, and they both speak of the probabilities of those observables occurring (typically in an ensemble-context for statistical mechanics and in a wavefunction-context for quantum mechanics). From classical mechanics we have an observable A taking on values in the phase space: $A(q, p) = A$, while in quantum mechanics we have the observable measured for energy eigenstates with eigenvalue l: $A_l = \langle l|A|l\rangle$. Either way, when these systems are considered in a statistical mechanical context, we typically can only see aggregations experimentally that amount to observations of averages of the microscopic or quantum observables over some domain (that is not probed by the sensitivity of the

detector or detection process). In other words, whatever the theoretical observable is, what we can measure, and must determine in our theoretical calculations to follow, is actually the average of the observable:

$$A_{Observable} = \bar{A}.$$

The probability that a set of macroscopic states, $\Delta\Gamma$, occurs is defined as the probability that any one of the microscopic states is seen in the outcome (with all outcome denoted by the set Γ). Thus

$$Prob(\Delta\Gamma) = \int_{\Delta\Gamma} f(p,q)\mathrm{d}\Gamma\,, \quad \Delta\Gamma \subset \Gamma,$$

where $f(p,q)$ is a probability density described by a choice of distribution function. In quantum mechanics it's even simpler:

$$Prob(l) = f(l), \quad l \in \Gamma$$

where $f(l)$ is a probability, again represented by a distribution function when passed into the realm of statistical mechanical analysis.

Regardless of classical or quantum origin, at the level of the statistical mechanical analysis we typically start with an indicated distribution function. For some quantum systems, for example, the choice of distribution involves either the Fermi-Dirac distribution or the Bose-Einstein distribution, it is only in this initial choice of distribution that the quantum effects are passed into the statistical mechanical analysis. As the system becomes large or classical in some manner the FD- and BE-distributions transition to the familiar Boltzmann distribution of classical statistical mechanics. If this clarification on the starting point for the statistical mechanical analysis, focusing on the distributions, seems to suggest that all we need to know at the outset, for Statistical mechanics, are the fundamental distributions, that would be correct. Maxwell famously promote the distributions-as-fundamental hypothesis. Unlike, Maxwell, we will see that the distributions aren't as 'fundamental' as other aspects of the theory. Nonetheless, if there is the impression that all that matters to begin a practical calculation is the system distribution… then that would be correct.

When distribution functions are used we can write, classically:

$$A_{Observable} = \bar{A} = \int_{\Gamma} A(q,p)f(p,q)\mathrm{d}\Gamma$$

or, quantum mechanically we have:

$$A_{Observable} = \bar{A} = \sum_{\Gamma} A_l f(l).$$

More convenient than distribution functions in many instances are use of statistical ensembles. A statistical ensemble is defined by the distribution function it is meant to characterize. In this context we then speak of the observable being the ensemble average $\langle A \rangle$:

$$A_{Observable} = \bar{A} = \langle A \rangle.$$

Ideal Gases, Kinetic Theory, and Maxwell Distributions

Now that we've arrived at the concept of ensemble, let's consider a physical ensemble (not merely a theoretical ensemble) – that of an ideal gas. An ideal gas is where we regard each molecule of the gas as a statistical unit and arrive at an actual physical ensemble in the description. This was the viewpoint of Maxwell in the kinetic theory of gasses (which became the prototype for statistical mechanics historically). The idealized condition for the gas (e.g., non-interacting and no excitable internal structure) is precisely what occurs for dilute gases. At high temperature there is no position dependence, so the distribution function on position and momentum becomes only momentum dependent

$$f(\vec{x}, \vec{p})d\vec{x}d\vec{p} \to f(\vec{p})d\vec{p},$$

where

$$f(\vec{p}) = \frac{1}{(2\pi mkT)^{\frac{3}{2}}} \exp\left\{-\frac{1}{2mkT}\left({p_x}^2 + {p_y}^2 + {p_z}^2\right)\right\},$$

the Maxwell distribution.

Existence of Thermal Equilibrium and Time-Asymmetry

Not only do special states known as equilibrium states exist, but non-equilibrium states evolve to equilibrium states and not the reverse – so, there is already a time-asymmetry from the ubiquitous existence of equilibrium states. In fact, it is a well-known additional aspect of 'time' that if you wait long enough, and enough time passes, a system in isolation will be observed to reach equilibrium. So, we have an arrow of time from the existence of equilibria as a fundamental aspect of our theory [13,14]. This is separate from any arrow of time that might be inferred from the weak interactions or from a choice of time in a general relativistic system and is fundamental. There isn't a 'problem of time' in the thermodynamic context (unlike in quantum theory, see Books 4, 5 [4,5]), instead, there is a problem of equilibration. This was noted as far back as 1897 by Planck [15], that non-equilibrium evolves to equilibrium and not the reverse, as separate from the second law of entropy increase. In 1909 Constantin Cartheodory [16] argued that an asymmetry in time, due to equilibration steps, was present in practical applications such as those involving Carnot Engines.

In Book 7 [7] we will see that all isolated physical systems are predicted to be Martingale, thus, the last independent foundational element of Thermodynamics/Statistical Mechanics is hereby identified, as a Martingale Hypothesis, and that it is already matched with the hypothesis that Emanator Theory requires Martingale projections.

Problem of Time in the Quantum Sense and Quantum Gravity
Having dispensed with the problem of time in the thermodynamic sense (via the Martingale Hypothesis) perhaps thermodynamic concepts can help solve the quantum problem of time? This is, indirectly, what is examined in Chapters 6 and 7 where Black Hole Thermodynamics is examined. We find that a viable theory for Thermal Quantum Gravity can be posited and explored, with systems obtained that are stable (positive heat capacity), even though no theory of Quantum Gravity alone seems to make sense. Fundamentally, quantum gravity may not exist as a theory applicable to non-boundary 4-D spacetime in any direct sense due to a lack of renormalizability for the gravitational field. Thus, geometry is apparatus in the classic quantum measurement device dichotomy of classical device and quantum observable. At the causal boundaries of the spacetime (such as a Black Hole Horizon), however, we may be forced to have a quantum gravity solution in order to have consistency with the quantum fields in the matter (source) description. Here we may be in luck as the AdS/CFT property [17,18] would allow renormalizable gravity on the 3d boundary.

Emergence
If all the foundational elements of Thermodynamics/Statistical Mechanics are derivative of elements identified in other physics disciplines [1-5], information theory [8], and in the hypothesized Emanator Theory [7], then what does Thermodynamics/Statistical Mechanics offer that is theoretically new? Unique to the Thermodynamics/Statistical Mechanics is the phenomenon of emergence, although we see early signs with the symmetry breaking descriptions used to arrive at the Standard Model. In Chapter 5 a brief overview of emergent phenomena will be given, the Ising model in particular. From the existence of emergence we see that thermodynamics will always offer new systems and discoveries.

Application of Thermal Equilibrium – The Principle of Equal Weight
When a macroscopic system is isolated for a long time it will reach a thermal equilibrium state, where the system energy is $E \pm \delta E$, where δE is a small variation that is allowed (required by quantum mechanics but usually have higher fluctuation level than that due to the quantum limit). In the equilibrium state we will always use equal weights on the microscopic states. In essence, if equilibrium describes a system with no discernible state-structure, then this would require a uniform probability of realizing a particular state, thus equal weights as indicated. Feynman avoids this entire topic in [19] where he simply states equal weights (and that he's not going to talk about it).

Microcanonical ensemble
The microcanonical ensemble is used to describe a system at a particular energy which has reached thermal equilibrium -- it is, thus, defined by the notion of equal weight. Let's delve deeper into the mathematical representation of the 'weight'. Consider a macroscopic state described by parameters $E, \delta E, N, V, x$. Let's describe the total number of states with those macroscopic parameters to be: $W(E, \delta E, N, V, x)$.

Quantum mechanically we have:

$$W(E, \delta E, N, V, x) = \sum_{E<E_l(N,V,x)<E+\delta E} 1.$$

For the classical weight function we must assign counts on classical phase space, so how to do this? In the end it is required to match with quantum states by the correspondence principle, so in terms of such a correspondence we could write:

$$W(E, \delta E, N, V, x) = \int_{E<\mathcal{H}(N,V,x)<E+\delta E} \frac{d\Gamma}{h^{3(N_A+N_B+\cdots)} N_A!\, N_B! \ldots}$$

Once we have $W(E, \delta E, N, V, x)$ we have the (Boltzmann) statistical definition of entropy:

$$S = k \log W(E, \delta E, N, V, x).$$

Suppose we consider not just systems at energy E, but all systems up to energy E (and drop the x variable). We then have

$$W(E,N,V) = \sum_{0<E_l(N,V)<E} 1,$$

for quantum and

$$W(E,N,V) = \int_{0<\mathcal{H}(N,V)<E} \frac{d\Gamma}{h^{3(N_A+N_B+\cdots)} N_A!\, N_B! \ldots},$$

for classical. For 'normal' thermodynamics systems we will have asymptotic behavior for such systems such that has number of states has asymptotic behavior according to one of the following forms:

$$W(E,N,V) \sim \exp\left[N\varphi(\frac{E}{N})\right], \sim \exp\left[V\varphi(\frac{E}{V})\right], \sim \exp\left[N\varphi(\frac{E}{N},\frac{V}{N})\right], \sim \exp\left[V\varphi(\right.$$

where $\varphi > 0$, $\varphi' > 0$, $\varphi'' < 0$. Consider the case where

$$W(E,N,V) \sim \exp[N\varphi]$$

then $S = k \log W(E,N,V) = kN\varphi$, and using the relation defining temperature:

$$\frac{\partial S}{\partial E} = \frac{1}{T} \quad \rightarrow \quad T = \frac{1}{k\varphi'} > 0,$$

thus we see that temperature is generally positive in 'normal' thermodynamic systems. Thus, we have come full circle to the beginning of the review where we saw that we could derive thermodynamic properties once we had the Entropy. Further developments along these lines (basic statistical mechanics) are given in Ch. 4.

Chapter 3. Thermodynamics

In this chapter the theoretical and mathematical underpinnings of thermodynamics are explored. As mentioned in Chapter 2, once we have the relation $E = E(S, N, V)$, we have

$$dE = \left(\frac{\partial E}{\partial S}\right)_{N,V} dS + \left(\frac{\partial E}{\partial V}\right)_{N,S} dV + \left(\frac{\partial E}{\partial N}\right)_{V,S} dN$$

and by convention $dE = TdS - pdV + \mu dN$, we thereby have relations for temperature, pressure, and chemical potential:

$$T = \left(\frac{\partial E}{\partial S}\right)_{N,V}, \quad p = -\left(\frac{\partial E}{\partial V}\right)_{N,S}, \quad \mu = \left(\frac{\partial E}{\partial N}\right)_{V,S}.$$

In terms of the thermodynamic variables and potentials for the system there is much more mathematical structure that will follow solely from the above and the axiomatic foundation given in Chapter 2.

3.1 The mathematical structure of thermodynamics

Helmholtz Free Energy

Often it is not convenient to work with $E = E(S, N, V)$ or $E = E(S, V)$, and some other energy-like function is better. Let's consider such functions beginning with Helmholtz Free Energy:

$$F(T, V) = E(S, V) - TS \quad \rightarrow \quad dF = -SdT - pdV,$$

from which we get the relations:

$$S = -\left(\frac{\partial F}{\partial T}\right)_V, \quad p = -\left(\frac{\partial F}{\partial V}\right)_T.$$

Thus, if you have information in the form T, V, then it's best to use Helmholtz Free Energy. Consider, for example, changes made at constant temperature:

$$(\delta F)_T = -p(\delta V)_T,$$

which is the work done on the system. While variation in energy with its variable entropy held constant gives:

$$(\delta E)_S = -p(\delta V)_S,$$

which is the work done when entropy is held constant, applicable to when work is done slowly with the system isolated.

Gibbs Potential
Sometimes we have data in the variables $\{T, P\}$, in which case it is convenient to work with the Gibbs Potential:

$$\Phi(T,p) = F(T,V) + pV = E - TS + pV \;\rightarrow\; d\Phi = -SdT + Vdp$$

with the relations:

$$S = -\left(\frac{\partial \Phi}{\partial T}\right)_p, \quad V = \left(\frac{\partial \Phi}{\partial p}\right)_T.$$

The typical application of the Gibbs Potential is when the size of the system doesn't matter. Consider, for example, examination of conditions under which liquid and gas phases can coexist in equilibrium. This will not be dependent on the amount of material present, but only on the intensive variables T and p.

Enthalpy of heat function

$$W(S,p) = \Phi + TS = E + PV \rightarrow \; dW = TdS + Vdp$$

$$T = \left(\frac{\partial W}{\partial S}\right)_p, \quad V = \left(\frac{\partial W}{\partial P}\right)_S$$

Principle utility: $(dW)_p = T(dS)_p = dQ$ which is why it is called the heat function.

Maxwell Relations
For the Maxwell relations we start with the original energy function: $E = E(S, V)$, and consider any system with a fixed number of particles. For such situations everything is determined in principle if we know any two variables that are not conjugate to each other, and that the system is in equilibrium. This is clear by the definition of the new energy functions but can be expressed independently of them (and is true not only for perfect gas but anything). Four relations become clear from the definitions induced in the four energy-like potentials (defined without adding number and chemical potential variables, so very general):

$$\left(\frac{\partial p}{\partial T}\right)_V = \frac{\partial}{\partial T}\left(-\left(\frac{\partial F}{\partial V}\right)_T\right)_V = \frac{\partial}{\partial V}\left(-\left(\frac{\partial F}{\partial T}\right)_V\right)_T = \left(\frac{\partial S}{\partial V}\right)_T$$

$$\left(\frac{\partial T}{\partial V}\right)_S = -\left(\frac{\partial p}{\partial S}\right)_V$$

$$\left(\frac{\partial V}{\partial S}\right)_p = \left(\frac{\partial T}{\partial P}\right)_S$$

$$\left(\frac{\partial S}{\partial p}\right)_T = -\left(\frac{\partial V}{\partial T}\right)_p$$

Allow Particle number to vary – introduce {N, μ}

The machinery developed so far is for systems with a fixed number of particles, let's now generalise to particle number being free to vary:

$$E = E(S, V, N) \rightarrow \quad dE = TdS - pdV + \mu dN \rightarrow \mu = \left(\frac{\partial E}{\partial N}\right)_{S,V}$$

where μ is called the chemical potential and N is the particle number with that potential. We now have the generalized definitions of the aforementioned energy-like functions:

$$dF = -SdT - pdV + \mu dN$$
$$d\phi = -SdT + Vdp + \mu dN$$
$$dW = Tds + Vdp + \mu dN$$
$$\mu = \left(\frac{\partial F}{\partial N}\right)_{T,V} = \left(\frac{\partial \phi}{\partial N}\right)_{p,T} = \left(\frac{\partial W}{\partial N}\right)_{p,S}$$

Notice that if any new variable (the particle number) is introduced a variation of it for the various energy relations, given their def., satisfies a trivial relationship:

$$(\delta E)_{S,V} = (\delta F)_{T,V} = (\delta\phi)_{T,p} = (\delta W)_{S,p}$$

Like P and T, μ is an intensive variable. Notice that $\phi = \phi(T, P)$ has intensive dependent variables. Notice, if we consider $dE = Tds - pdV + \mu dN$ and scale it by λ (the system is λ times bigger), then

$$d(\lambda E) = Td(\lambda s) - pd(\lambda V) + \mu d(\lambda N)$$

consequently

$$\lambda dE + Ed\lambda = \lambda(TdS - pdV + \mu dN) + (TS - pV + \mu N)d\lambda$$

which is consistent with

$$E = TS - pV + \mu N, \qquad dE = (TdS - pdV + \mu dN)$$

but from the first relation we also have:

$$dE = (TdS - pdV + \mu dN) + SdT - Vdp + Nd\mu$$

Thus, if μ is an intensive variable we must have:

$$SdT - Vdp + Nd\mu = 0 \rightarrow$$
$$d\mu = -\frac{S}{N}dT + \frac{V}{N}dp$$

Thus, μ must be a proper function of T and P. Thus $\phi(P, T) = E - TS + PV = \mu N$.

Notice that Energy E depends on extensive variables only, while the Gibbs potential ϕ depends on intensive variables only.

Landan potential

Of the set of new possible energy function possible with the added number variable, the one most useful is the Landan potential Ω:

$$\Omega = F - \mu N = -pV \rightarrow d\Omega = -SdT - pdV - Nd\mu$$
$$(\delta\Omega)_{T,V,\mu} = (\delta E)_{S,V,N} = (\delta F)_{T,V,N} = (\delta W)_{S,p,N}$$

Thermodynamic derivatives

Entropy cannot, in general, be measured directly. However, changes in the entropy may be measured by putting heat in under specified conditions. In doing so, we are said to be measuring the head capacity:

$$C_V = T\left(\frac{\partial S}{\partial T}\right)_v \quad , \quad C_p = T\left(\frac{\partial S}{\partial T}\right)_p = \left(\frac{\partial E}{\partial T}\right)_V$$

Also,

$$C_V = -T\left(\frac{\partial^2 F}{\partial T^2}\right)_V \quad , \quad C_p = -T\left(\frac{\partial^2 \phi}{\partial T^2}\right)_p$$

Aside from the heat capacities there are two other second derivative relations:

Isothermal compressibility:

$$K_T = -\frac{1}{V}\left(\frac{\partial V}{\partial p}\right)_T = -\frac{1}{V}\left(\frac{\partial^2 \phi}{\partial p^2}\right)_T$$

Adiabatic compressibility:

$$K_S = -\frac{1}{V}\left(\frac{\partial V}{\partial p}\right)_S = \left[V\left(\frac{\partial^2 E}{\partial V^2}\right)_T\right]^{-1}$$

The Perfect Gas (Ideal Gas)

There are three reasons to consider a perfect gas of point atoms.

(1) They take up no volume, obviously, so there is no need to account for the effective volume, $V_{eff} = V_{container} - \sum_1^{number} V_{atom}$, as is done in the Van der Waals approximation.

(2) They collide very rarely

(3) They don't have the added effects of atomic rotation.

Synopsis

The Differentials

$$dE = TdS - pdV$$
$$dF = -pdV - SdT$$
$$d\phi = Vdp - SdT$$
$$dW = Vdp + TdS$$

The Legendre Transformations

$$E = W - pV$$
$$W = \phi + TS$$
$$F = E - ST$$
$$\phi = F + Vp$$

The Maxwell Equilibrium Relations

$$\left(\frac{\partial V}{\partial T}\right)_p = -\left(\frac{\partial S}{\partial p}\right)_T$$
$$\left(\frac{\partial T}{\partial p}\right)_S = \left(\frac{\partial V}{\partial S}\right)_p$$
$$\left(\frac{\partial p}{\partial S}\right)_p = -\left(\frac{\partial T}{\partial V}\right)_S$$
$$\left(\frac{\partial S}{\partial P}\right)_T = \left(\frac{\partial p}{\partial T}\right)_V$$

The Conjugate Equilibrium Relations

For a pair of mechanical variables, X being extensive (such as V) and W being intensive (such as $-p$), we have:

$$\left(\frac{\partial E}{\partial X}\right)_T = Y - T\left(\frac{\partial Y}{\partial T}\right)_X$$
$$\left(\frac{\partial S}{\partial X}\right)_T = -T\left(\frac{\partial Y}{\partial T}\right)_X$$
$$\left(\frac{\partial S}{\partial T}\right)_X = \frac{C_X}{T}$$
$$\left(\frac{\partial C_X}{\partial X}\right)_T = -T\left(\frac{\partial^2 Y}{\partial T^2}\right)_X$$

3.2 Examples

In this section a wide range of thermodynamics examples will be explored. The problems are selected from problems assigned in courses taken while at Caltech, Oxford, and U. Wisconsin. In turn, these courses made use of the best textbook sources available [20-27], thus the problems that follow indirectly derive from these excellent sources as well, notably Morse [20] and Goodstein [21] in thermodynamics and Kubo [26] and McQuarrie [27] in statistical physics and emergent structure and Ising models described by Plischke and Bergersen [23], Chandler [24] and Landau and Lifschitz [22]. Thus, to find more such problems or further discussion along the lines of the physics raised in these problems, refer to those sources for further details.

Example 3.1

Consider an ideal gas with mean kinetic energy per particle given by $f(T)$.

(a) What is an equation relating the heat capacity C_m and N_m and f'?

(b) When there is a temperature differential there cannot be thermal equilibrium by definition, but there can be mechanical equilibrium. What gradient in gas density is needed to achieve mechanical equilibrium given a gradient in temperature (express in terms of density, heat capacity, and internal energy of the gas).

Answer

(a) We have to start:

$$C_m = \frac{C(\text{heat capacity unit mass of body})}{m(\text{unit mass})}$$

$$N_m = \text{number of molecules per unit mass}$$

$$f(T) = \langle K.E. \rangle$$

Thus

$$U = N\langle K.E. \rangle = Nf(T)$$

and for a unit mass sample:

$$U_1 = N_m f(T) \rightarrow \frac{dU_1}{dT} = N_m f'(T)$$

By definition we have

$$\Delta Q = C\Delta T \Rightarrow dQ = CdT$$

thus,

$$\frac{dU_1}{dT} = \frac{dU_1}{dQ} C_m = N_m f'(T).$$

Let's now consider $\frac{dU_1}{dQ}$, starting with $dU = dQ - dW$. Since we are simply adding heat $dQ \neq 0$ and $dW = 0$, thus $dU = dQ$, or:

$$\frac{dU_1}{dQ} = 1$$

and we get:

$$C_m = N_m f'(T)$$

(b) We have a T gradient, with pressure uniform:

$$pV = \frac{2}{3}N\langle K.E.\rangle = \frac{2}{3}Nf(T)$$

Dividing by the mass and introducing density $\rho = m/V$:

$$p\left(\frac{1}{\rho}\right) = \frac{\frac{2}{3}N_m f(T)}{m}$$

So,

$$\nabla P\left(\frac{1}{\rho}\right) = \nabla\frac{2}{3}\frac{N_m f(T)}{m} \rightarrow \quad \nabla\rho = -\frac{\rho C_m}{U}\nabla T$$

At a specific density $\nabla\rho\sim-\nabla T$ does this make sense? Consider $PV = nRT$:

$$\frac{1}{\rho}\sim T \rightarrow \quad -\frac{1}{\rho^2}\nabla P\sim\nabla T \rightarrow \nabla P\sim-\nabla T,$$

thus consistent. So,

$$\nabla\rho = \left(-\frac{\rho C_m}{U}\right)\nabla T$$

Example 3.2

We find for a gas experimentally that:

$$\beta = \frac{1}{V}\left(\frac{\partial V}{\partial T}\right)_P = \frac{RV^2(V-nb)}{RTV^3 - 2an(V-nb)^2}$$

$$K = -\left(\frac{1}{V}\right)\left(\frac{\partial V}{\partial P}\right)_T = \frac{V^2(V-nb)^2}{nRTV^3 - 2an^2(V-nb)^2}$$

Are these results consistent with the equation of state being the Van Der Waals equation?

Answer

From the Van Der Waal's equation:

$$(V-nb)\left(P+\frac{an^3}{V^2}\right) = nRT \rightarrow \quad T = \frac{1}{nR}\left(VP + \frac{an^2}{V} - \frac{abn^3}{V^2} - nbp\right)$$

Using $\left(\frac{\partial V}{\partial P}\right)_T = -\frac{(\partial T/\partial P)_V}{(\partial T/\partial V)_P}$ we have

$$(\partial T/\partial P)_V = \frac{1}{nR}(V-nb)$$

$$(\partial T/\partial V)_P = \frac{1}{nR}\left(P - \frac{an^2}{V^2} - \frac{2abn^3}{V^3}\right)$$

Thus

$$\left(\frac{\partial V}{\partial P}\right)_T = -\frac{(V-nb)}{\left(P-\frac{an^2}{V^2}-\frac{2abn^3}{V^3}\right)} = -\frac{V^3(V-nb)}{(V^3P-an^2V+2abn^3)}$$

which is consistent with the experimental results for the standard constants a and b indicated.

Example 3.3

A gas obeys Van Der Waals equation. Show that there is one critical state with values T_c, V_c, where both $(\partial V/\partial P)_T$ and $\left(\frac{\partial^2 p}{\partial V^2}\right)_T$ are zero. Write the equation of state for $\frac{P}{P_c}$ in terms of $\frac{T}{T_c}$ and $\frac{V}{V_c}$. Consider the cases $\frac{T}{T_c}=\frac{1}{2},1,2$. What does it mean physically when three values of volume are predicted for a single value of P and T?

Answer

From Example above we already have a convenient form for

$$\left(\frac{\partial V}{\partial P}\right)_T = -\frac{V^3(V-nb)}{(V^3nRT-2an^2(V-bn)^2)}$$

Thus,

$$\left(\frac{\partial P}{\partial V}\right)_T = -\frac{(V^3nRT-2an^2(V-bn)^2)}{V^3(V-nb)}$$

where

$$V_c{}^3nRT_c-2an^2(V_c-bn)^2=0$$

Consider the $\left(\frac{\partial^2 p}{\partial V^2}\right)_T$ before solving for the critical values:

$$\left(\frac{\partial^2 P}{\partial V^2}\right)_T = [nRTV^3-2an^2(v-bn)^2] \cdot [+V^{-3}2(v-nb)^{-1}+3V^{-2}(v-nb)^{-2}] + [-V^{-3}(v-nb)^{-2}][3nRTV^2-4an^2(v-bn)]$$

Thus

$$[nRT_cV_c^3-2an^2(v_c-bn)^2][2(v_c-nb)+3V_c] - [3nRT_cV_c^2+4an^2(v_c-bn)]=0$$

$$3nRT_cV_c^2-4an^2(v_c-bn)=0$$

The second relation at the critical point gives:

$$nRT_cV_c^2-\frac{4}{3}an^2(v_c-bn)$$

Putting the two equations together to solve:

$$V_c = 2abn^3/\left(\frac{2}{3}an^2\right) = 3bn$$

and

$$T_c = \frac{8}{27}\frac{a}{Rb}$$

In terms of the critical values, let's rewrite as requested:

$$(3bn - bn)\left(P_c + \frac{an^2}{(3bn)^2}\right) = nR\left(\frac{8}{27}\frac{a}{Rb}\right)$$

$$P_c = \frac{1}{2b}\left(\frac{8}{27}\right)\frac{a}{b} - \frac{an^2}{9b^2n^2} = \left(\frac{a}{b^2}\right)\left[\frac{8}{54} - \frac{1}{9}\right] = \left(\frac{a}{b^2}\right)\left(\frac{1}{27}\right)$$

Using $P = \frac{nRT}{(v-nb)} - \frac{an^2}{V^2}$ we then get:

$$\frac{P}{P_c} = \frac{27b^2}{a}\left[\frac{nRT}{(v-nb)} - \frac{an^2}{V^2}\right] = \frac{27b^2nRT}{a(v-nb)} - \frac{27b^2n^2}{V^2}$$

and using $V_c = 3bn \Rightarrow bn = \frac{1}{3}V_c$:

$$\frac{P}{P_c} = \frac{9bV_cRT}{a\left(V - \frac{1}{3}V_c\right)} - \frac{3V_c^2}{V^2} = \frac{9bRT}{a}\frac{1}{\left({}^{V}/_{V_c} - {}^{1}/_{3}\right)} - 3\left(\frac{V_c}{V}\right)^2$$

Thus

$$\frac{P}{P_c} = 8\left(\frac{T}{T_c}\right)\frac{1}{\left(3\left(\frac{V}{V_c}\right) - 1\right)} - 3\left(\frac{V_c}{V}\right)^2$$

and for the specific cases:

$$T = \frac{1}{2}T_c{:}\ \left({}^{P}/_{P_c}\right) = \frac{4}{\left(3\frac{V}{V_c} - 1\right)} - 3\left(\frac{V_c}{V}\right)^2$$

$$T = T_c{:}\ \left({}^{P}/_{P_c}\right) = \frac{8}{\left(3\frac{V}{V_c} - 1\right)} - 3\left(\frac{V_c}{V}\right)^2$$

$$T = 2T_c{:}\ \left({}^{P}/_{P_c}\right) = \frac{4}{\left(3\frac{V}{V_c} - 1\right)} - 3\left(\frac{V_c}{V}\right)^2$$

When the van den Waals equation indicates three allowed values of V for a single P and T this represents the transition from gas to liquid.

Example 3.4

Consider a thermally isolated cylinder that is divided into two parts that are separated by a moveable piston. Each side has a perfect gas with the same $\{n, P, T, V\}$. Heat is applied slowly and reversibly to the left side (meaning we can use equalities in entropy relations) until left-side

pressure has increased to $P_1 = \frac{243}{32} P_0$ (where P_1 is pressure on the left side, P_2 is pressure on the right side, and P_0 is the initial pressure, etc.).
(a) What is the work done on the gas on the right?
(b) What is the final temperature on the right side?
(c) What is the final temperature on the left side?
(d) What amount of heat was transferred to the gas on the left?

Answer
Note that for a perfect gas of point atoms the system energy is $U = \frac{3}{2} nRT$.

Thermally isolated means transformation described must have occurred with no heat transfer with its environment, thus adiabatic. We know for adiabatic transitions that relation of current volume to prior volume goes as:

$$T_2 V_2^{\gamma-1} = T_0 V_0^{\gamma-1}$$

and for a perfect gas: $\gamma = 5/3$, thus:

$$T_2 V_2^{2/3} = T_0 V_0^{2/3} \quad \rightarrow \quad \frac{P_2 V_2^{5/3}}{nR} = \frac{P_0 V_0^{5/3}}{nR}$$

Thus

$$V_2^{5/3} = V_0^{5/3} \left(\frac{P_0}{P_2}\right) = V_0^{5/3} \left(\frac{32}{243}\right)$$

and

$$V_2 = V_0 \left(\frac{32}{243}\right)^{3/5}, \text{ and note that } \left(\frac{32}{243}\right)^{3/5} = .296 \text{ and } \left(\frac{243}{32}\right)^{2/5} = 2.25.$$

So, we have:

$$P_2 V_2 = nRT_2 \Longrightarrow T_2 = \frac{P_2 V_2}{nR} = \left(\frac{243}{32}\right) \frac{P_0}{nR} V_0 \left(\frac{32}{243}\right)^{3/5} = \frac{P_0 V_0}{nR} \left(\frac{243}{32}\right)^{2/5}$$

$$T_2 = 2.25 T_0$$

For side 2: $\Delta U_2 = \Delta Q - \Delta W_2$ $\qquad$ $\underline{\Delta Q = 0}$

$$\Delta U_2 = \frac{3}{2} nR(T_2 - T_0) = -\Delta W_2$$

$$\Delta U_2 = \frac{3}{2} nR(2.25 - 1)T_0 = -\Delta W_2$$

$$\Delta U_2 = \frac{15}{8} nRT_0 = -\Delta W_2 \qquad \qquad \Delta W_1 = -\Delta W_2$$

(a) The work done on the gas on the right (side 2) is $\Delta W_1 = -\Delta W_2 = \frac{15}{8} nRT_0$.

(b) The final temperature of the gas on the right is T_2, computed above: $T_2 = 2.25T_0$.

(c) Final temperature on left?
We have:
$V_1 + V_2 = 2V_0$

$$V_1 = 2V_0 - V_2 \qquad V_2 = V_0 \left(\frac{32}{243}\right)^{3/5}$$

$$V_1 = 2V_0 - \left(\frac{32}{243}\right)^{3/5} V_0 = V_0 \left[2 - \left(\frac{32}{243}\right)^{3/5}\right] = V_0[2 - .296]$$

$V_1 = 1.70V_0$
$P_1V_1 = nRT_1$

$$T_1 = \frac{P_1V_1}{nR} = \left(\frac{243P_0}{32}\right)\frac{(1.70V_0)}{nR} = 13.9\frac{P_0V_0}{nR} = 13.9T_0$$

So, the final temperature on the left is: $T_1 = 13.9T_0$.

(d) Heat supplied on left?

$\Delta U = \Delta Q - \Delta W$

$\Delta U = \frac{3}{2} nR(T_1 - T_0) = \frac{3}{2} nR \cdot 12.9T_0 = 19.4nRT_0$

$\Delta W = \frac{18}{8} nRT_0 = 1.9nRT_0$

$\Delta Q = \Delta U + \Delta\omega = (19.4 + 1.9)nRT_0 = 21.3nRT_0$

Heat supplied is $\Delta Q = 21.3nRT_0$.

Example 3.5

The standard Carnot Cycle can be described as a 'cycle' of transitions in the P-V place consisting of :
(1) Starting at the high pressure, low volume, and high temperature state (state 1), transitioning isothermally to higher volume and lower pressure (transition 12 to state 2).
(2) From state 2 transition adiabatically to state 3 at even higher volume and lower pressure.

(3) From state 3 transition isothermally to state 4 at lower volume and higher pressure.
(4) From state 4 return to state 1 by transitioning adiabatically.

Compute ΔQ_{12} and ΔQ_{34} for a Carnot cycle based on a perfect gas of point particles, in terms of nR, T_c, and T_h. Show $\Delta W_{23} = -\Delta W_{41}$. Show the efficiency of the cycle is $\eta = \frac{T_h - T_c}{T_h}$.

Answer
We have

$$\Delta Q_{12} = \int P\,dV = T_h nR \int_1^2 \frac{dV}{V} = nRT_h \ln\left(\frac{V_2}{V_1}\right)$$

$$\Delta Q_{34} = \int P\,dV = T_c nR \int_3^4 \frac{dV}{V} = nRT_c \ln\left(\frac{V_4}{V_3}\right)$$

$$\Delta W_{23} = U_2 - U_3 = \frac{3}{2} nR(T_2 - T_3)$$

$$\Delta W_{41} = U_4 - U_1 = \frac{3}{2} nR(T_4 - T_1)$$

$T_3 = T_4 = T_c$ and $T_1 = T_2 = T_h$
So

$$\Delta W_{23} = \tfrac{3}{2} nR(T_h - T_c) = -\tfrac{3}{2} nR(T_c - T_h) = -\tfrac{3}{2} nR(T_4 - T_1) = -\Delta W_{41}$$

and

$$\eta = \frac{\Delta Q_{12} - \Delta Q_{43}}{\Delta Q_{12}} = 1 - \frac{nRT_c \ln\left(\frac{V_4}{V_3}\right)}{nRT_h \ln\left(\frac{V_2}{V_1}\right)} = 1 - \frac{T_c}{T_h}\left[\frac{\ln\left(\frac{V_4}{V_3}\right)}{\ln\left(\frac{V_2}{V_1}\right)}\right]$$

Since transitions1 $\rightarrow$ 4 and 2 $\rightarrow$ 3 are adiabatic, and for a perfect gas of point atoms $\gamma = {}^5/_3$:

$$T_c V_4^{2/3} = T_h V_1^{2/3}, \qquad T_c V_3^{2/3} = T_h V_2^{2/3} \quad \rightarrow \quad \frac{T_c V_4^{2/3}}{T_c V_3^{2/3}} = \frac{T_h V_1^{2/3}}{T_h V_2^{2/3}}$$

Thus

$$\frac{V_4}{V_3} = \frac{V_1}{V_2}$$

and

$$\eta = 1 - {}^{T_c}/_{T_h}$$

Example 3.6

Express the adiabatic thermal expansitivity $\beta_s \equiv \frac{1}{V}\left(\frac{\partial V}{\partial T}\right)_S$ in terms of V, T, β, K, Cv ;that is, in terms of the volume, the absolute temperature, the pressure, the isobaric thermal expansitivity, the isothermal compressibility, and the heat capacity at constant volume. The context is that of the typical U-V-S system, in thermodynamic equilibrium. Then Show that β_s and β always have opposite sign.

Answer

$\beta_s = \frac{1}{v}\left(\frac{\partial V}{\partial T}\right)_s = -\frac{1}{v}\left(\frac{\partial s}{\partial p}\right)_v$

Now, $\left(\frac{\partial S}{\partial X}\right)_T = -\left(\frac{\partial Y}{\partial T}\right)_T = -\left(\frac{\partial Y}{\partial T}\right)_X \quad \left(\frac{\partial S}{\partial T}\right)_X = \frac{C_X}{T}$

$\left(\frac{\partial S}{\partial P}\right)_V = \left(\frac{\partial S}{\partial P}\right)_V \left(\frac{\partial T}{\partial P}\right)_V \leftarrow \left(\frac{\partial T}{\partial P}\right)_V \left(\frac{\partial P}{\partial V}\right)_T = -1$

$\beta_s = -\frac{1}{v}\left(\frac{C_V}{T}\right)\left(\frac{kV}{\beta V}\right) = -\frac{kC_v}{vT\beta}$

Thus

$$\beta_s = -\frac{kC_v}{vT\beta}$$

Given the relation for β_s, and that all the constants on the RHS are positive except for β, then we must have that β_s and β always have opposite sign.

Example 3.7

Consider a gas undergoing the Joule-Thomson throttling process, as shown. (Assume that a steady-state condition has been reached as the gas through the apparatus).

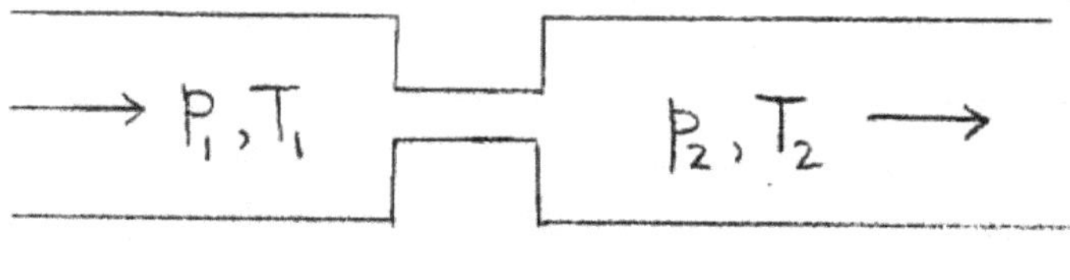

a) Show that the enthalpy, H=E +pV of a given number of moles of the gas is unchanged by this process.
b) Express the Joule-Thomson coefficient $\mu = (\partial T/\partial p)_H$ in terms of V, T, C_p, and the quantity $(\partial V/\partial T)_p$.
c) If the gas is ideal, find the entropy change, ΔS, of ν moles in the above process. Is this positive or negative and why?

Answer

(a) Joule-Thomson throttling:

$$u_2 - u_1 = P_1V_1 - P_2V_2$$
$$u_2 + P_2V_2 = u_1 + P_1V_1 = H$$

which is a constant.

(b) Joule-Thompson coefficient $\mu = (\partial\tau/\partial p)_H$

In terms of

$$C_p = \left(\frac{\partial u}{\partial\tau}\right)_p, \left(\frac{\partial V}{\partial\tau}\right), V, T$$

$$dH = vdp + Tds$$

$$\mu = \left(\frac{\partial T}{\partial p}\right)_H = \frac{-1}{\left(\frac{\partial p}{\partial H}\right)_T\left(\frac{\partial H}{\partial T}\right)_p} = \frac{-\left(\frac{\partial H}{\partial p}\right)_T}{\left(\frac{\partial H}{\partial T}\right)_p}$$

$$\left(\frac{\partial H}{\partial T}\right)_p = V + T\left(\frac{\partial S}{\partial p}\right)_T \qquad \left(\frac{\partial H}{\partial T}\right)_p = T\left(\frac{\partial S}{\partial T}\right)_p = C_p$$

$$\left(\frac{\partial S}{\partial p}\right)_T = -\left(\frac{\partial V}{\partial T}\right)_p \qquad 8.314\,J/k^0$$

$$\mu = \frac{-V - T\left(-\left(\frac{\partial V}{\partial T}\right)_p\right)}{C_p} = \frac{T\left(\frac{\partial V}{\partial T}\right)_p - V}{C_p}$$

Thus

$$\mu = \frac{T\left(\frac{\partial V}{\partial T}\right)_p - V}{C_p}$$

(c) $du = Tds - pdV$

$$Tds = dU + pdV$$

$$u = \frac{3}{0}\mu RT$$

$$pV = VRT$$

$$dV = -\frac{1}{p}dp(VRT)$$

$$ds = \frac{1}{T}\left(\frac{3}{2}VRdT\right) + p\left(-\frac{dp}{p^2}\right)\frac{VRT}{T}$$

$$dS = \frac{3}{2}VRd(\ln T) - VRd(\ln p)$$

$$S = S_0 + VR\ln\left(\left(\frac{T}{T_0}\right)^{3/2}\left(\frac{P_l}{P}\right)\right) \rightarrow \quad \Delta S = uR\ln\left(\left[\frac{T}{T_0}\right]^{3/2}\left(\frac{P_o}{P}\right)\right)$$

$$\Delta S = VR \ln\left[\left(\frac{T_2}{T_1}\right)^{3/2}\left(\frac{P_1}{P_2}\right)\right]$$

$$PV = NkT$$

Perfect gas $V = T\left(\frac{\partial V}{\partial T}\right)_p$

$$\left(\frac{\partial V}{\partial T}\right)_p = \frac{Nk}{P}$$

So, $\left(\frac{\partial T}{\partial P}\right)_H = 0$

Now, $\left(\frac{\partial S}{\partial P}\right)_H = -\frac{V}{T}$

$$\Delta S = \int -\frac{V}{T} = \int -{}^{nR}/_{\beta} = n.\,2\ln\left({}^{R}/_{P_2}\right)$$

Since $P_1 > P_2 \Rightarrow \Delta S > 0$

3.3 Exercises

Exercise 3.1

(a) A liter of water at T = 0C is contained in an insulating vessel in a room at T = 20C. What is the minimum amount of work which must be done by a cooling unit to freeze the water to make ice? (The latent heat of freezing of water at ambient pressure is 3.3×10^5 joules/Kg.)

(b) A turbine engine operates at a temperature of 1500C in and ambient environment at 20C. Kerosene is burned in the engine to release chemical energy. What is the maximum fraction of the chemical energy which can be converted to useful work? How would you expect the actual efficiency to depend on the speed of the turbine? Why?

Exercise 3.2

In zero magnetic field, a magnetic material undergoes a paramagnetic to ferromagnetic transition as the temperature is lowered. The transition is described by a Ginzburg-Landau expansion. For a unit volume of the material, we have

$$G^F = G^P + \frac{1}{2}\alpha_O(T - T_O)M^2 - \frac{1}{4}g_4M^4 + \frac{1}{6}g_6M^6 + \cdots$$

With $\alpha_0 > 0, g_4 > 0, g_6 > 0,$ and $G^P(G^F)$ the free energy of the paramagnetic (ferromagnetic) phase. For the paramagnetic phase, M = 0, while for the ferromagnetic phase, $M \neq 0$. Show that the phase transition is first order (there is a discontinuity in M) and occurs at a Curie temperature $T_C \neq T_O$. Make a sketch of the free energy as a function of M for $T > T_C, T = T_C$, and for $T < T_C$. Derive an expression which relates T_C to T_O in terms of the other given parameters.

Exercise 3.3

A low density gas has the equation of state

$$P = \frac{nRT}{V}\left[1 + \frac{n}{V}B(T)\right]$$

The heat capacity will have the form

$$C_V = \frac{3}{2}nR - \frac{n^2R}{V}f(T)$$

Find the form which $f(T)$ must have in order that these equations be thermodynamically consistent.

Exercise 3.4

A system consists of N identical particles each of which occur in one of two energy states

$E = \Delta$ and $E = -\Delta (\Delta > 0)$. The total energy of the system is determined by the number M of particles with energy Δ and is given by

$$E(M) = M\Delta - (N - M)\Delta = (2M - N)\Delta$$

Using the microcanonical ensemble, find the entropy per particle $\sigma = S/N$ as a function of the energy per particle $\epsilon = E/N$ in the thermodynamic limit $(N \to \infty)$. Make a sketch of σ vs. ϵ. Can you express T in terms of ϵ?

Exercise 3.5

An electric current of 10A is maintained for 1s in a resistor of 25 Ω while the temperature of the resistor is kept constant at 27°C.

(a) What is the entropy change of the resistor?

(b) What is the entropy change of the universe? (Neglect any entropy change in the current source).

The same current is maintained for the same resistor, but now thermally insulated, whose initial temperature is 27^0C. If the resistor has a mass of 0.01 kg and C_p= 0.84 kJ/kg K:

(c) What is the entropy change of the resistor?

(d) What is the entropy change of the universe? (Neglect any entropy change in the current source)

Exercise 3.6

a) Starting from the second law of thermodynamics, derive the four Maxwell relations connecting certain partial derivatives involving S, T, P and V.

b) The coefficient of expansion is defined as

$$\alpha = V^{-1}(\partial V/\partial T)_p$$

And the isothermal compressibility as

$$k = -V^{-1}(\partial V/\partial T)_T$$

Derive an expression for C_p-C_V in terms of $\alpha, k, T, V,$ where C_p and C_V are the heat capacities of a substance at constant pressure and constant volume, respectively.

Chapter 4. Statistical Mechanics

In this chapter the fundamental mathematical structure of statistical mechanics will be described and a wide range of statistical mechanics examples will be explored. The problems are selected from problems assigned in undergraduate and graduate courses taken while at Caltech, and graduate courses taken while at Oxford, and U. Wisconsin. In turn, these courses made use of the best textbook sources available [20-27], thus the problems that follow indirectly derive from these excellent sources as well. The problems cover a wide range of complexity from undergraduate level to advanced graduate level (preliminary exam questions). To find more such problems or further discussion along the lines of the physics raised in these problems, refer to [20-27] for further details.

4.1 The Mathematical Structure of Statistical Mechanics

In additional to the foundational thermodynamics concepts/definitions outlined in Chapter 3, we now add concepts distinct to statistical mechanics, such as ensembles, distributions, particle statistics. But first, let's re-iterate some thermodynamics concepts like adiabaticity, energy partitioning, and equilibrium in this statistical mechanics context. Detailed discussion on systems in general, density of states derivations, the different types of thermodynamic distributions, and more, will then follow, to develop the concepts further.

Quasi-static adiabatic process in statistical mechanics

This is a process involving a very slow change of a parameter x which determines a purely mechanical interaction of a system with an external work source, namely:

$$\mathcal{H}(q,p,x) \rightarrow \mathcal{H}(q,p,x+\Delta x) \ , \quad dx/dt \rightarrow 0.$$

Adiabatic theorem in dynamics: dynamical quantities which are kept invariant in an adiabatic process are called adiabatic invariants. The number of states, $\Omega_0(E)$, is an adiabatic invariant.

$$\Omega_0(E,x) = \Omega_0(E+dE, x+dx)\,, dx/dt \rightarrow 0.$$

Now,

$$dE = \left(\langle X\rangle_{time\ average}\right) dx$$

$$dE = \bar{X} dx$$

Where the last relation is obtained by using the ergodic theorem to equate time averages with ensemble averages.

Adiabatic theorem in Statistical Mechanics

If the number of states is invariant under adiabatic process, then so to is entropy. Thus we have

$$dS/dx = 0 \quad , dx/dt \to 0,$$

for an adiabatic process. Thus

$$dS = \frac{\partial S}{\partial E} dE + \frac{\partial S}{\partial x} dx.$$

Using $\frac{\partial S}{\partial E} = \frac{1}{T}$, $dE = \bar{X} dx$:

$$dS = \frac{1}{T}(dE - \bar{X} dx)$$

Partition of energy between two systems in thermal contact

$$E = E_I + E_{II}$$

$$\Omega(E)\delta E = \iint_{E<E_I+E_{II}<E+\delta E} \Omega_I(E_I)\, \Omega_{II}(E_{II}) dE_I dE_{II}$$

$$= \delta E \int \Omega_I\, (E_I) \Omega_{II}(E - E_I) dE_I$$

The probability that the system 'I' has energy in the range between E_I and $E_I + dE_I$ is given by:

$$f(E_I) dE_I = \frac{\Omega_I(E_I) \Omega_{II}(E - E_I) dE_I \delta E}{\Omega(E)\delta E}$$

The most probable partition of energy

$f(E_1) dE_I$ gives the probability of energy partition when systems I and II are in statistical equilibrium. The probability has a very steep maximum at a certain partition of energy:

$$\Omega_I(E_I) \Omega_{II}(E - E_I) dE_I \delta E = max$$

$$S_I(E_I) + S_{II}(E - E_I) = max$$

$$\frac{\partial S_I}{\partial E_I} = \frac{\partial S_{II}}{\partial E_{II}} \,, \qquad E_I^* + E_{II}^* = E$$

thus

$$T_I(E_I^*) = T_{II}(E_{II}^*).$$

Equilibrium of two systems with a material-transferring contact

$$f(E_I,N_I)dE_I = \frac{\Omega_I(E_I,N_I)\Omega_{II}(E-E_I,N-N_I)dE_I\delta E}{\Omega(E)\delta E}$$

$$\Omega(E,N)\delta E = \delta E\sum_{N_I=0}^{N}\int \Omega_I(E_I,N_I)\Omega_{II}(E-E_I,N-N_I)dE_I$$

Thus,

$$S_I(E_I,N_I)+S_{II}(E_{II},N_{II}) = max$$

and

$$\frac{\mu_I}{T_I}=\frac{\mu_{II}}{T_{II}}\ , \qquad \frac{\partial S}{\partial N}=-\frac{\mu}{T}\ .$$

Equilibrium of two systems with a Pressure contact

$$\frac{P_I}{T_I}=\frac{P_{II}}{T_{II}}\ , \quad \frac{\partial S}{\partial V}=\frac{P}{T}$$

Canonical distributions

Consider a system with volume V and multiple species of particles $\{N_A, N_B\ ....\}$. The system is in equilibrium with a heat bath at a temperature T. The system might be small system with few degrees of freedom, or it might be a large macroscopic system. Either way, let $\Omega(E)$ be the state density of the heat bath, E_t the total energy of the composite system, and E_ℓ the energy of the ℓ -th quantum state, according to the principle of equal weight the probability that the quantum state "ℓ" is realised is proportional to the number of microscopic states allowed, which is equal to $\Omega(E_t-E_\ell)\delta E$. Thus,

$$f(\ell) \propto \Omega(E_t-E_\ell)\delta E \propto \frac{\Omega(E_t-E_\ell)\delta E}{\Omega(E_t)E} = \exp\left\{\frac{S(E_t-E_\ell)-S(E_t)}{K}\right\}.$$

Since the heat bath is very large compared to the system, it may be assumed that $E_t \gg E_l$, then:

$$S(E_t-E_\ell)-S(E_t) = \left\{-E_\ell\frac{\partial S}{\partial E}+\frac{1}{2}E_\ell^2\frac{\partial^2 S}{\partial E^2}+\cdots\right\}|_{E=E_t}$$

$$= -\frac{E_l}{T}\left\{1+\frac{E_\ell}{2CT}+\cdots\right\}|_{E=E_t}$$

where $T=(\partial S/\partial E)^{-1}$ and $C=\partial E/\partial t$. Since $E_\ell \ll CT$:

$$f(\ell) \propto e^{-E_\ell/kT}$$

Thus, the probability of microscopic states is given by the classical state probability:

$$Pr(d\Gamma) = \frac{1}{\pi N_A!\, h^{3N_A}} \frac{e^{-\beta\mathcal{H}_N} d\Gamma}{Z_N}, \quad \beta = 1/kT,$$

$$Z_N = \frac{1}{\pi N_A!\, h^{3N_A}} \int e^{-\beta\mathcal{H}_N} d\Gamma$$

Or by the quantum state probability:

$$Pr(\ell) \equiv f(\ell) = \frac{e^{-\beta E_{N,\ell}}}{Z_N}, \quad Z_N = \Sigma e^{-\beta E_{N,\ell}}$$

More generally, for classical $Z_N = \int_0^\infty e^{-\beta E} \Omega(E) dE$ and Z_N is called the partition function.

$T - \mu$ distribution (Grand canonical distribution)

When a system enclosed in a volume V is in contact with a heat bath at temperature T and with a particle source characterised by the chemical potentials $\mu_A, \mu_\beta, \ldots$ for particle A,B,…., then the number of particles it contains is also indeterminant. The probability that the system has the microscopic state which contains $N_A N_B, \ldots$ particles is given by:

(classical mechanics) $Pr(N, d\Gamma) = \frac{1}{\pi N_A! h^{3N_A}} \frac{e^{-\beta(\mathcal{H}_N - \Sigma_A N_A \mu_A)} d\Gamma}{\phi}$

(quantum mechanics) $Pr(N, \ell) = \frac{e^{-\beta\left(E_{N,\ell} - N_A \mu_A\right)}}{\phi}$

where ϕ is the grand canonical partition function:

(classical mechanics) $\phi = \sum_{N_A=0}^{\infty} \sum_{N_B=0}^{\infty} \cdots \frac{1}{\pi N_A! h^{3N_A}} \int e^{-\beta(\mathcal{H}_N - \Sigma_A N_A \mu_A)} d\Gamma$

(quantum mechanics) $\phi = \sum_{N_A=0}^{\infty} \sum_{N_B=0}^{\infty} \ldots e^{-\beta\left(E_{N,\ell} - \Sigma N_A \mu_A\right)} =$
$\left(\sum_{N_A} \sum_{N_B} \ldots\right) e^{BN_A \mu_B} e^{\beta N_B \mu_B} \ldots Z_N$

The grand canonical partition function for quantum system is rewritten:

$$\phi = \left(\sum_{N_A} \sum_{N_B} \cdots\right) \lambda_A^{N_A} \lambda_B^{N_B} \ldots Z_N$$

where $\lambda_A = e^{\beta \mu_A}$ is known as absolute activity or fugacity.

If we only want the probability of particle numbers:

$$Pr(N_A, N_B, \ldots) = \frac{e^{(N_A \mu_B + N_B \mu_B + \cdots)} Z_N}{\phi} = \frac{\lambda_A^{N_A} \lambda_B^{N_B} \ldots Z_N}{\phi}$$

The T-p distribution function follows similarly.

The distributions can be applied regardless of the size of the system under consideration.

Partition Functions and Thermodynamic Functions
Microcanonical Ensemble ← Distribution for given energy
Canonical Ensemble ← Distribution for given temperature
Grand canonical $(T.\mu)$ Ensemble ← Distribution for given temperature and chemical potential
(T-p) Ensemble ← Distribution for given temperature and pressure

If the system is microscopic, then the thermodynamic function (potential) for each of the prescribed conditions above (i.e., given energy or temperature, etc.) is derived from each partition function, thus:

Distribution function	Partition function	Thermodynamic
Microcanonical	$\Omega(E,V,N)\delta E$	$S(E,V,N) = K\log\Omega\,(E,V,N)\delta E$
Canonical	$Z(T,V,N) = \sum_l e^{-\frac{E_\ell(V.N)}{kT}}$	$F(T,V,N) = -kT\log Z\,(T,V,N)$
Grand canonical	$\phi = \sum_{N=0}^{\infty} e^{N_\mu/kT}\, Z(T,V,N)$	$J = -pV = -kT\log\phi\,(T,V,\mu)$
(T-p)	$Y(T,p,N) = \int_0^\infty e^{-pv/kT}\, Z(T,V,N)dV$	$G(T,p,N) = -kT\log Y(T,p,N)$

In statistical mechanics, the thermodynamic relations between thermodynamic functions are derived as the relations between certain average values obtained from probability laws suitable for the description of the given conditions.

Consider the thermodynamic relations that follow from the thermodynamic function corresponding to the canonical distribution. For the Quantum Mechanics case, the generalised force conjugate to a generalised coordinate, x, follows from correspondence:

$$X = \frac{\partial\mathcal{H}(q,p,x)}{\partial x} \rightarrow X_\ell = \frac{\partial E_\ell(x)}{\partial x}$$

The averages of energy and force in the canonical distribution are:

$$\bar{E} = \frac{\sum_\ell E_\ell \, e^{-\beta E_\ell}}{\sum_\ell e^{-\beta E_\ell}} = -\frac{\partial}{\partial \beta} \log Z(\beta, x) \ , \qquad \beta = \frac{1}{kT}$$

$$\bar{X} = \frac{\sum_\ell X_\ell \, e^{-\beta E_\ell}}{Z} = -\frac{\partial}{\partial x}\frac{1}{\beta} \log Z(\beta, x)$$

In particular,

$$p = \frac{\partial}{\partial V}\frac{1}{\beta} \log Z(\beta, V)$$

Let $-kT \log Z = -\frac{1}{\beta} \log Z = F(T, V, x \ldots)$, then $\bar{E} = \frac{\partial}{\partial \beta}(\beta F)$, $p = -\frac{\partial}{\partial V} F$. So far we haven't named the thermodynamic function $F(T, V, x \ldots)$, but it will soon be proven to be none other than Free energy. Since

$$\frac{\partial}{\partial (1/T)}\left(\frac{E}{T}\right) = E \ , \qquad \frac{\partial F}{\partial V} = -P,$$

from $E = F + ST$ we have $dF = -SdT - PdV \rightarrow \left(\frac{\partial F}{\partial V}\right) = -P$, which is in agreement. Now consider:

$$F + \frac{1}{T}\frac{\partial F}{\partial \left(\frac{1}{T}\right)} = E \rightarrow \ F + \frac{1}{T}\left(\frac{\partial F}{\partial T}\right)\left(\frac{\partial T}{\partial \left(\frac{1}{T}\right)}\right) = E \rightarrow \ \ F - T\left(\frac{\partial F}{\partial T}\right) = E.$$

Since $E = F + ST$, we then have $\left(\frac{\partial F}{\partial T}\right) = -S$ which is also in agreement Thus "F" is the F that usually denotes Helmholtz Free Energy.

Use of reversibility, choice of energy function and of partition function

Reversible

$$dU = \tau d\sigma + (-pdV) + \mu dN$$

Conservation of energy

Fix $N, V \ \rightarrow dU = \tau d\sigma \rightarrow \frac{1}{\tau^2}\left(\frac{d\sigma}{\partial}\right)_{N,v}$, at equilibrium: $\tau_1 = \tau_2 \, , d\sigma = 0, \sigma = max.$

Fix $\sigma, N \rightarrow dU = -pdv \rightarrow p = -\left(\frac{\partial U}{\partial V}\right)_{\sigma,N}$, at equilibrium: $p_1 = p_2$.

Fix $\tau, V \rightarrow d\underbrace{U - \tau\sigma}_{F} = \mu dN \rightarrow \mu = \left(\frac{\partial F}{\partial N}\right)_{\tau,v}$, at equilibrium: $\mu_1 = \mu_2 \, , dF = 0 \, , \ F = min.$

Fix $\tau, p \rightarrow \alpha \underbrace{(U - \sigma\tau + pv)}_{G} = \mu dN \ \rightarrow \mu = \left(\frac{\partial G}{\partial N}\right)_{\tau,p}$, at eq.: $\mu_1 = \mu_2 \, , dG = 0, \ \ G = min.$

Fix $N, P \rightarrow d\underbrace{U + \tau\sigma}_{H} = \tau d\sigma \rightarrow \frac{1}{\tau} = \left(\frac{\partial\sigma}{\partial H}\right)_{p,N}$

General

$dU = dQ + (-pdV) + \mu dN$ is the first law of thermodynamics, which defines dQ to have conservation of energy.

$d\sigma \geq \frac{dQ}{\tau}$ is the second law, where σ never decreases (macroscopically, see later sections for discussion of microscopic fluctuations). In a detailed picture we can use: $\sigma = \ln g$.

Partition functions

Let k specify complete state system with N_k particles;

Fixed N_k → $Prob\ (k) = \frac{e^{-\frac{E_k}{\tau}}}{Z}$; $Z = \sum_k e^{-\frac{E_k}{\tau}} = e^{-F/\tau}$.

Variable N_k → $Prob\ (k) = \frac{e^{\frac{(N_k\mu - E_k)}{\tau}}}{\mathfrak{z}}$; $\mathfrak{z} = \sum_k e^{(N_k\mu - E_k)/\tau} = e^{-G/\tau}$

4.1.1 Fermi-, Bose-, and Boltzmann statistics

Since identical particles are indistinguishable in quantum-mechanics, each quantum state of the particle system is completely specified when the number of one-particle states is given precisely. That is, the set of occupation numbers

$$\ell \equiv \{n_\tau\} = (n_1, n_2 \ldots)$$

gives the quantum numbers of the whole system, and the energy is simply:

$$E_\ell \equiv E_{(n)} = \sum_\tau \varepsilon_\tau n_\tau \,.$$

The occupation numbers are strongly restricted in quantum mechanics such that there can only be the following two cases (see the Spin-Statistic relations in Book5 [5], that establishes this strict relationship both kinematically in flat space-time and dynamically in curved-spacetime). The two cases are (1) Fermi-Dirac statistics: $n = 0\ or\ 1$ (applicable to spinorial matter fields) and (2) Bose-Einstein statistics: $n = 0,1,2 \ldots.$ (applicable to scalar and vector gauge fields).

Using the method of the grand canonical ensemble, and the distribution functions corresponding to the two cases above, we can then determine the average particle numbers present (according to the type of statistics/distribution that underlies the particles involved). For a quantum state $\{n_\tau\} = (n_1, n_2, \ldots)$:

$$E = \sum_{\tau} \varepsilon_\tau n_\tau \quad , \qquad N = \sum_{\tau} n_\tau$$

So, the partition function for a given N is:

$$Z_N = \sum_{\{N_\tau\}} e^{-\beta \sum_\tau \varepsilon_\tau n_\tau}$$

Hence, the grand partition function is:

$$\phi = \sum_{N=0}^{\infty} e^{\beta N_\mu} Z_N = \sum_{N=0}^{\infty} e^{\beta N_\mu} \sum_{\{N_\tau\}} e^{-\beta \sum_\tau \varepsilon_\tau n_\tau}$$

$$\phi = \sum_{N=0}^{\infty} \sum_{\{N_\tau\}} e^{\beta \sum_\tau (\mu - \varepsilon_\tau) n_\tau} = \sum_{n_1} \sum_{n_2} \dots e^{\beta \sum_\tau (\mu - \varepsilon_\tau) n_\tau} = \prod_{\tau} \sum_{n_\tau} e^{\beta (\mu - \varepsilon_\tau) n_\tau}$$

We now account for the statistics in carrying out the summation over n_τ:

Bose-Einstein: $n_\tau = 0,1,2, \dots$

$$\sum_{n=0}^{\infty} e^{\beta(\mu - \varepsilon_\tau) n} = \left(1 - e^{\beta(\mu - \varepsilon_\tau)}\right)^{-1}$$

where $\mu < \varepsilon_\tau$ for all τ.

Fermi-Dirac: $n_\tau = 0,1$

$$\sum_{n_\tau} e^{\beta(\mu - \varepsilon_\tau) n} = 1 + e^{\beta(\mu - \varepsilon_\tau)}$$

Thus, adopting notation where we denote – for B.E. and + for F.D, we can then compactly write:

$$\phi = \prod_{\tau} \left\{1 \mp e^{\beta(\mu - \varepsilon_\tau)}\right\}^{\mp 1}.$$

For the probability that n_τ particles occupy state τ we then have:

$$P(n_\tau) = \frac{e^{\beta(\mu - \varepsilon_\tau) n_\tau}}{\phi} \prod_{\sigma} e^{\beta(\mu - \varepsilon_\sigma) n_\sigma} = \frac{e^{\beta(\mu - \varepsilon_\tau) n_\tau}}{\sum_{n_\tau} e^{\beta(\mu - \varepsilon_\tau) n_\tau}}$$

where $\prod_\sigma$ is the product over all but τ. Hence, the mean number, $\bar{n}_\tau$, of particles in state τ is:

$$\bar{n}_\tau = \sum_{n_\tau = 0}^{\infty} n_\tau P(n_\tau) = \frac{\sum_{n_\tau=0}^{\infty} n_\tau e^{\beta(\mu - \varepsilon_\tau) n_\tau}}{\sum_{n_\tau=0}^{\infty} e^{\beta(\mu - \varepsilon_\tau) n_\tau}} = \frac{1}{\beta} \frac{\partial}{\partial \mu} \log \sum_{n_\tau} e^{\beta(\mu - \varepsilon_\tau) n_\tau}$$

$$\bar{n}_\tau = \mp \frac{1}{\beta} \frac{\partial}{\partial \mu} \log\left\{1 \mp e^{\beta(\mu - \varepsilon_\tau)}\right\} = \frac{1}{e^{\beta(\mu - \varepsilon_\tau)} \mp 1}$$

Thus,

$$\bar{n}_\tau = \frac{1}{e^{(\varepsilon_\tau-\mu)/kT}+1} \qquad (Fermi - Dirac)$$

$$\bar{n}_\tau = \frac{1}{e^{(\varepsilon_\tau-\mu)/kT}-1} \qquad (Bose - Einstein)$$

For the Landan potential:

$$J = -pV = -kT\log\phi = \pm kT\sum_\tau \log\{1 \mp e^{\beta(\mu-\varepsilon_\tau)}\}$$

$$= \mp kT\sum_\tau \log(1 \pm \bar{n}_\tau)$$

Since $dJ = -SdT - pdV - Nd\mu$:

$$S = -\frac{\partial J}{\partial T} = \mp k\Sigma\log\{1 \mp e^{\beta(\mu-\varepsilon_\tau)}\} - \frac{1}{T}\Sigma\frac{\mu-\varepsilon_\tau}{e^{\beta(\varepsilon_\tau-\mu)} \mp 1} = -\frac{J - N_\mu + E}{T}$$

On, using

$$\bar{n}_\tau/(1 \pm \bar{n}_\tau) = e^{\beta(\mu-\varepsilon_\tau)} \to \log\{\bar{n}_\tau/(1 \pm \bar{n}_\tau)\} = \beta(\mu-\varepsilon_\tau)$$

We have

$$S = \pm k\sum[-n_\tau\log\bar{n}_\tau \pm (1 \pm \bar{n}_\tau)\log(1 \pm \bar{n}_\tau)]$$

Also

$$F = J + N_\mu = N_\mu \pm kT\Sigma\log\left(1 \mp e^{\beta(\mu-\varepsilon_\tau)}\right)$$

So, altogether (upper signs are F.D. case, lower signs are B.E. case):

$$\bar{n}_\tau = \frac{1}{e^{(\varepsilon_\tau - N/kT)+1}} \ (F.D.), \quad \bar{n}_\tau = \frac{1}{e^{(\varepsilon_\tau - N/kT)-1}} \quad (B.E.)$$

$$S = k\sum_\tau[-\bar{n}_\tau\log\bar{n}_\tau \mp (1 \mp \bar{n}_\tau)\log(1 \mp \bar{n}_\tau)]$$

$$F = E - TS, \qquad G = N_\mu$$

$$J = -pV = F - G = \pm kT\sum_\tau\log(1 \mp \bar{n}_\tau) = \pm kT\sum_\tau\log\left(1 \pm e^{\frac{(\mu-\varepsilon_\tau)}{kT}}\right)$$

Classical limit (Boltzmann-statistics):

$$\bar{n}_\tau = e^{(\mu-\varepsilon_\tau)/kT} \ll 1$$

For the classical limit we must have $\varepsilon_\tau - \mu \gg kT$:

$$N = \sum_\tau \bar{n}_\tau \cong \sum_\tau e^{(\mu-\varepsilon_\tau)/kT} \cong \int \exp\left(-\frac{\varepsilon-\mu}{kT}\right)\rho(\varepsilon)\,d\varepsilon$$

Note that for a single particle, the number of states in d^3pd^3r is $(d^3pd^3r)/(2\pi\hbar)$, where $d^3p = 4\pi p^2 dp$ and the number of states between p and pdp is $\frac{4\pi V}{(2\pi\hbar)^3}p^2 dp$. Thus

$$p(\varepsilon)d\varepsilon = \frac{4\pi V}{(2\pi\hbar)^3}p^2\frac{dp}{d\varepsilon}d\varepsilon$$

since $\varepsilon = \frac{p^2}{2m}$ for classical kinetics and no interaction we have

$$p(\varepsilon)d\varepsilon = \frac{4\pi\sqrt{2}\,Vm^{3/2}}{(2\pi\hbar)^3}\quad \varepsilon^{1/2}d\varepsilon$$

So,

$$N = \frac{4\pi\sqrt{2}Vm^{3/2}}{(2\pi\hbar)^3}e^{\mu/kT}\int e^{-\varepsilon/kT}\varepsilon^{1/2}d\varepsilon = \frac{V}{\Lambda^3}e^{\mu/kT},\quad \Lambda = \frac{2\pi\hbar}{\sqrt{2\pi mkT}}.$$

Thus,

$$\mu = -kT\log\frac{kT}{P\Lambda^3}$$

And using $PV = NkT$ (from the Landau relation earlier, in the classical approximation) we have

$$\mu = -kT\log\left(\frac{V/N}{\Lambda^3}\right)$$

$$e^{\mu/kT} \ll 1 \rightarrow \frac{\Lambda^3}{(V/N)} \ll 1$$

So,

$$\frac{N}{V} \ll \left(\frac{2\pi mkT}{h^2}\right)^{3/2}$$

is the condition for Boltzmann statistics (and the associated classical approximation with respect to both quantum and relativistic kinematics).

Recall

$$Z_N = \sum_{\{N_\tau\}} e^{-\beta\sum_\tau \varepsilon_\tau n_\tau}$$

thus, for Boltzmann statistics, the one-particle partition function is defined by:

$$Z_1 = \sum_\tau e^{-\beta\varepsilon_\tau} = f$$

Now, $N = \sum_\tau \bar{n}_\tau \cong \sum_\tau e^{(\mu-\varepsilon_\tau)/kT} = \lambda f$ where $\lambda = e^{\mu/kT}$, thus:

$$E = \sum_\tau \varepsilon_\tau \bar{n}_\tau \cong \sum_\tau \varepsilon_\tau \lambda e^{-\beta\varepsilon_\tau} = \lambda\left(-\frac{\partial f}{\partial \beta}\right) = -N\frac{\partial(\log f)}{\partial \beta}$$

$$Z_N = \frac{f^N}{N!}, \qquad F = -NkT - NkT\log\frac{f}{N}$$

$$\phi = \prod_\tau \{1 \mp e^{\beta(\mu-\varepsilon_\tau)}\}^{\mp 1} \cong \prod_\tau e^{(\mu-\varepsilon_\tau)/kT} = e^{\lambda f}$$

the approximate equality, $\cong$, denoting the classical approximation.

4.1.2 Ideal Gas Law from Density of States

Instead of obtaining $PV = NkT$ used in the above by way of the Landau potential approximation, and whatever additional assumptions are needed in a particular application, let's consider a specific case, the ideal gas, and show that we obtain $PV = NkT$ in that context. Let's start with obtaining the microcanonical ensemble by way of determining the density of states $\Omega_0(E)$:

$$\Omega_0(E) = \int_{\mathcal{H}\le E} \frac{d\Gamma}{h^{3N}N!} = \frac{1}{h^{3N}N!}\int_{\mathcal{H}\le E} d^3r d^3p = \frac{1}{h^{3N}N!}V^N \int_{\mathcal{H}\le E} d^3p$$

The last integral involves a 3N-dimensional hyperspherical volume, so let's side-track and compute that:

Consider the n-dimensional volume

$$V_n(p) = C_n p^n, \quad S_n(p) = nC_n p^{n-1}$$

Consider

$$I_n = \int_{-\infty}^{\infty} \dots \int_{-\infty}^{\infty} \exp\{-a(x_1^2 + \cdots x_n^2)\}\, dx_1 \dots dx_n$$

and using the Gaussian integral solution:

$$\int_{-\infty}^{\infty} \exp(-ax^2)\, dx = (\pi/a)^{1/2} \rightarrow I_n = (\pi/a)^{n/2}$$

But we also have

$$I_n = \int_0^\infty e^{-ar^2} nC_n r^{n-1} dr = \frac{1}{2}nC_n \int_0^\infty e^{ay} y^{\frac{1}{2}n-1} dy$$

$$= C_n \Gamma\left(\frac{1}{2}n + 1\right) a^{-\frac{1}{2}n}$$

So,

$$C_n = \pi^{\frac{1}{2}n} \Big/ \Gamma\left(\frac{1}{2}n + 1\right)$$

And recall that $\sum_i \frac{p_i^2}{2m} = \frac{p^2}{2m} = E \to p = \sqrt{2mE}$, thus

$$\Omega_0(E) = \frac{1}{h^{3N}N!}V^N \frac{\pi^{3N/2}}{\Gamma\left(\frac{3N}{2}+1\right)}(2mE)^{3N/2} = \frac{V^N}{h^{3N}N!}\frac{(2\pi mE)^{3N/2}}{\Gamma\left(\frac{3}{2}N+1\right)}$$

We now have an explicit form for $\Omega_0(E)$ from which we can get entropy S and the rest of he thermodynamics relations by way of he Maxwell relations, etc. So we start with

$$S = k\log\Omega_0.$$

Using the large-N approximation:

$$\log(N!) = N\log N - N$$

and

$$\Gamma\left(\frac{3}{2}N+1\right) \approx \left(\frac{3}{2}N\right)!\log\left(\frac{3}{2}N\right) - \frac{3}{2}N$$

and we get:

$$S = k\left\{N\log\left(\frac{(2\pi m^{3/2})}{h^3}\right) + N\log V - [N\log N - N] - \left[\left(\frac{3}{2}N\right)\log\left(\frac{3}{2}N\right) - \left(\frac{3}{2}N\right)\right]\right\}$$

$$S = kN\left\{\log\left(\frac{V}{N}\right) + \frac{3}{2}\log\left(\frac{2E}{3N}\right) + \log\left(\frac{(2\pi m)^{3/2}e^{5/2}}{h^3}\right)\right\} + \frac{3}{2}\log E^N$$

Now

$$dE = Tds - pdV \to \quad T = \left(\frac{\partial E}{\partial S}\right)_V \to \frac{1}{T} = \left(\frac{\partial S}{\partial E}\right)_V = \frac{3}{2}\frac{1}{E}(kN)$$

$$\to \boxed{E = \frac{3}{2}NkT}$$

Similarly,

$$dE = Tds - pdV \to P = -\left(\frac{\partial E}{\partial V}\right)_S$$

and since $\left(\frac{\partial E}{\partial V}\right)_S\left(\frac{\partial V}{\partial S}\right)_E\left(\frac{\partial S}{\partial E}\right)_V = -1$, we can then write:

$$p = \left(\frac{\partial S}{\partial V}\right)_E\left(\frac{\partial E}{\partial S}\right)_V = T\left(\frac{\partial S}{\partial V}\right)_E$$

Thus

$$\frac{P}{T} = \frac{kN}{V} \to \boxed{PV = NkT}$$

4.1.3 Behaviour of N almost independent identical quantum harmonic oscillators

Let's now consider the example of N almost independent identical quantum harmonic oscillators:

Quantum harmonic oscillator: $\varepsilon = \left(n + \frac{1}{2}\right) hv.$

N almost independent oscillators: $E = \frac{1}{2}Nhv + Mhv$, where $n_1 + n_2 \dots n_N = M.$

For density of states (the thermodynamic weight) we ask how many ways we can arrive at meeting the constraint $n_1 + n_2 \dots n_N = M$. Describe as equivalent to having N bins and M balls, which, in turn, is equivalent to having M white balls and (N-1) red balls (the dividers between bins) with the following combinatorics:

$$W_M = \frac{(M + N - 1)!}{M!\,(N - 1)!}$$

Thus, we compute

$$S = k \log W_M = K\{\log(M + N - 1)! - \log M! - \log(N - 1)!\}$$
$$\cong k\{(M + N - 1)\log(M + 1 - 1) - M \log M - (N - 1)\log(N - 1)\}$$

$$\frac{1}{T} = \frac{\partial S}{\partial E} = \frac{\partial S}{\partial M}\frac{\partial M}{\partial E} = k\left\{\log(M + N - 1) - \log M + \frac{M + N - 1}{M + N - 1} - 1\right\}\frac{1}{hV}$$
$$= \frac{k}{hV}\log\left(\frac{M + N}{M}\right) = \frac{K}{hV}\log\left(\frac{E/N + \frac{1}{2}hV}{E/N - \frac{1}{2}hV}\right)$$

Let

$$e^{hv/kT} = \frac{E/N + \frac{1}{2}hv}{E/N - \frac{1}{2}hv}$$

then

$$1 + e^{hv/kT} = \frac{2E/N}{E/N - \frac{1}{2}hv},$$

$$1 - e^{hv/kT} = \frac{-hv}{E/N - \frac{1}{2}hv} \rightarrow \frac{1 + e^{hv/kT}}{1 - e^{hv/kT}} - \frac{2E}{Nhv}$$

Thus

$$E = N\left\{\frac{1}{2}hv + \frac{hv}{e^{hv/kT} - 1}\right\}$$

4.1.4 System with negative temperature regime requiring negative Energy solutions

Notice in the prior derivation we had a positive temperature and continued the calculation as there was a sensible thermodynamic correspondence. Next consider a situation where we consider N independent particles and consider Bose-Einstein combinatorics in their 'binning' while at the same time also constrain to two energy-levels (as if internal state is fermionic). Thus, we have:

(1) N independent particles,
(2) each has only two energy levels, $-\varepsilon_0,\ \varepsilon_0$,
(3) with total energy $E = M\varepsilon_0 = N_+\varepsilon_0 - N_-\varepsilon_0$,
(4) and with density of states simply:

$$W_M = \frac{N!}{N_+!\,N_-!},\quad N_+ = \frac{1}{2}(M+N),\ \ N_- = \frac{1}{2}(N-M)$$

(5) We, thus, compute the entropy:

$$S = k\log W_m = k\left\{\log N! - \log\left[\frac{1}{2}(M+N)\right]! - \log\left[\frac{1}{2}(N-M)!\right]\right\}$$

$$\cong k\left\{N\log N - \left[\frac{1}{2}(M+N)\right]\log[M+N] - \left[\frac{1}{2}(N-M)\right]\log(N-M)\right\}$$

At this juncture computation of system temperature and heat capacity (stability) can be calculated, where

$$\frac{1}{T} = \left(\frac{\partial S}{\partial E}\right)_V,\quad C_v = \left(\frac{\partial E}{\partial T}\right)_v = T\left(\frac{\partial S}{\partial T}\right)_V$$

And if either temperature or heat capacity have negative ranges, the standard correspondence with thermodynamics and statistical mechanics breaks down (specialized interpretations in theses mathematical domains are possible, such as in laser mode inversion, but these exception will not be discussed here). For this situation, let's compute the temperature:

$$\frac{1}{T} = \frac{\partial S}{\partial M}\frac{\partial M}{\partial E} = \frac{K}{\varepsilon_0}\left\{-\frac{1}{2}\log\left[\frac{1}{2}(M+N)\right] + \frac{1}{2}\log\left[\frac{1}{2}(M-N)\right]\right\}$$

$$= \frac{k}{2\varepsilon_0}\log\left(\frac{N-M}{N+M}\right)$$

Thus

$$\frac{1}{T} = \frac{k}{2\varepsilon_0}\log\left(\frac{N-M}{N+M}\right).$$

We see that the temperature is negative for $M > 0$, thus this system requires $E < 0$ initial condition to be thermodynamically well-defined.

4.1.5 The thermodynamics of almost independent systems

Consider a collection of thermodynamic systems A, B, C,... almost independent of each other. What is their combined system partition function in terms of their already known separate system partition functions? And, what of the thermodynamic functions that result in such a (general) circumstance? Starting with the notion of a common heat reservoir (at temperature $T = 1/\beta$), for systems A, B, …, we have partition functions:

$$Z_A = \sum_{\ell} e^{-\beta E_{A,\ell}}, \qquad Z_B = \sum_{m} e^{-\beta E_{B,m}}, \ldots.$$

And the combined partition function is:

$$Z = \sum_{\ell,m} e^{-\beta(E_{A,\ell}+E_{B,m}+\cdots)} = Z_A Z_B \ldots$$

Thus,

$$F = -kT \log Z = -kT \log Z_A - kT \log Z_B - \cdots = F_A + F_B + \cdots$$

Applying the canonical T-p distribution in classical statistical mechanics to an ideal gas of N molecules → let's now derive the thermodynamic functions:

The classical (free) Hamiltonian:

$$\mathcal{H} = \sum_{i=1}^{3N} \frac{{p_i}^2}{2m}$$

Canonical distribution:

$$Z = \frac{1}{h^{3N}N!}\int e^{-\beta\mathcal{H}}\, d\Gamma = \frac{V^N}{h^{3N}N!}\left[\int e^{-(\beta/2m)p^2}\, dp\right]^{3N}$$
$$= \frac{V^N}{h^{3N}N!}\left(\frac{2m\pi}{\beta}\right)^{3N/2}$$

Thus

$$F = -kT \log Z$$
$$= -kT\left\{N\log V + \frac{3N}{2}\log\left(\frac{k2m\pi}{h^2}\right) - N\log N + N + \frac{3N}{2}\log T\right\}$$
$$= -NkT\left\{\log\frac{V}{N} + \frac{3}{2}\log\left(\frac{k2m\pi e^{2/3}}{h^2}\right) + \frac{3}{2}\log T\right\}$$

Since

$$dF = -SdT - pdV \quad \rightarrow \quad p = -\left(\frac{\partial F}{\partial V}\right) \quad \rightarrow \quad pV = NkT$$

Let's now apply the partition function for the T-p distribution:

$$Y(T, p, N) = \int_0^\infty e^{-pV/kT}\, Z_N dV$$

$$= \left(\frac{2m\mu\pi kT}{h^2}\right)^{3N/2} \frac{1}{N!} \int_0^\infty e^{-pV/kT}\, V^N dV$$

Let $V' = \frac{pv}{kT}$, then

$$Y(T, p, N) = \left(\frac{2m\mu\pi kT}{h^2}\right)^{3N/2} \frac{1}{N!} \left(\frac{kT}{p}\right)^{N+1} \int_0^\infty e^{-\alpha v'} V^N dV'$$

with $\alpha = 1$. Let's use the parameterization to do a Feynman-style integral solution:

$$\left(-\frac{\partial}{\partial \alpha}\right)^N \int_0^\infty e^{-\alpha V'}\, dV' = \left(-\frac{\partial}{\partial \alpha}\right)^N \alpha^{-1} = \frac{N!}{\alpha^{N+1}}$$

And if we then set $\alpha = 1$ the expression simplifies to $N!$ And we get:

$$Y(T, P, N) = \frac{1}{N!} \left(\frac{2m\pi kT}{h^2}\right)^{3N/2} \left(\frac{kT}{p}\right)^{N+1} N! \cong \frac{1}{h^{3N}} (2m\pi kT)^{\frac{3}{2}N} \left(\frac{kT}{p}\right)^N$$

where he approximation $N + 1 \approx N$ is used for when $N \gg 1$. Thus,

$$G(T, p, N) = -kT \log Y$$

$$= -kT \left\{N \log\left(\frac{kT}{p}\right) + \frac{3}{2} N \log T + \frac{3}{2} N \log\left(\frac{2\pi mk}{h^2}\right)\right\}$$

$$= -NkT \left\{\frac{5}{2} \log T - \log p + \log\left[\frac{2\pi m l^{3/2} k^{5/2}}{h^3}\right]\right\}$$

Now, using the relation

$$dG = Vdp - sdT \quad \rightarrow \quad V = \frac{\partial G}{\partial p} = NkT\, \frac{1}{p} \rightarrow pV = NkT$$

Also,

$$G = N\mu \quad \rightarrow \quad \mu = kT \log\left[\frac{p}{kT}\left(\frac{h^2}{2\pi mkT}\right)^{3/2}\right]$$

Thus

$$U = -T^2 \frac{\partial [F/T]}{\partial T} = -T^2\left(-\frac{3}{2} Nk \frac{1}{\tau}\right) = \frac{3}{2} NkT$$

and

$$C_v = \left(\frac{\partial U}{\partial T}\right)_V = \frac{3}{2}Nk = T\left(\frac{\partial S}{\partial T}\right)_V$$
$$C_p = T\left(\frac{\partial S}{\partial T}\right)_p = \frac{5}{2}Nk$$

4.1.6 N spin 1/2 particles in a magnetic field

Let's consider a system of N spin $\frac{1}{2}$ particles in a magnetic field, with two energy levels $-\mu\mathcal{H}$ and
$+\mu\mathcal{H}$, at temperature T (thus using canonical distribution to start). The canonical distribution for the collection of independent systems is that of a single-particle system N times:

$$Z_N = Z_1^N\ , \quad Z_1 = e^{\mu\mathcal{H}\beta} + e^{-\mu\mathcal{H}\beta} = 2\cosh(\mu\mathcal{H}\beta)$$

Thus

$$Z_N = [2\cosh(\beta\mu\mathcal{H})]^N \quad \rightarrow \quad F = -NkT\{\log[2\cosh(\beta\mu\mathcal{H})]\}$$

And we, thus, arrive at the Free Energy F for the system. Since

$$dF = -SdT - pdV \rightarrow S = -\frac{\partial F}{\partial T}$$

we have:

$$S = Nk\{\log[2\cosh(\beta\mu\mathcal{H})]\} + NkT\left\{\frac{2\sinh(\beta\mu\mathcal{H})\frac{\mu\mathcal{H}}{k}\left(-\frac{1}{T^2}\right)}{2\cosh(\beta\mu\mathcal{H})}\right\}$$
$$= Nk\left[\log\left\{2\cosh\left(\frac{\mu\mathcal{H}}{kT}\right)\right\} - \left(\frac{\mu\mathcal{H}}{kT}\right)\tanh\left(\frac{\mu\mathcal{H}}{kT}\right)\right]$$

And since

$$U = F + TS = -N\mu\mathcal{H}\tanh\left(\frac{\mu\mathcal{H}}{kT}\right),$$

The magnetization $M = N\bar{\mu}$ is given by:

$$M = -\frac{\partial F}{\partial \mathcal{H}} = N\mu\tanh\left(\frac{\mu\mathcal{H}}{kT}\right),$$

and the system has heat capacity:

$$C = \left(\frac{\partial U}{\partial T}\right)_H = Nk\left(\frac{\mu H}{kT}\right)^2 \Big/ \cosh^2\left(\frac{\mu H}{kT}\right),$$

which is always positive.

4.1.7 The Statistical Ensembles and which to use when

Often we want to know the probability of finding a given state in the system. The approach is indicated by the information stated about the system:

(1) Is there a given energy? then microcanonical, and need a density of states, which gives the entropy:

$$\Omega_0 = \frac{1}{\pi N_A!\, h^{3N_A}} \int_{\mathcal{H} \leq E} d\Gamma\, \Omega_0 \rightarrow S = k \log \Omega_0\, (E, V, N)$$

And the density of states, $\Omega_0 = W(E, \delta E, N, V, x)$ gives:

$$P(\alpha) = \frac{W(\alpha)}{\sum_\alpha W(\alpha)}$$

(2) Is there a given temperature? then canonical, and we have:

$$P(d\Gamma) = \frac{1}{\pi N_A!\, h^{3N_A}} \frac{e^{-\beta \mathcal{H}_N} d\Gamma}{Z_N}, \quad P(\ell) = \frac{e^{-\beta E_{N,\ell}}}{Z_N}, \qquad \beta = 1/kT$$

$$Z_N = \frac{1}{\pi N_A!\, h^{3N_A}} \int e^{-\beta \mathcal{H}_N}\, d\Gamma \qquad Z_N = \sum_\ell e^{-\beta E_{N,\ell}}$$

To relate to density of states we have

$$Z_N = \int_0^\infty e^{-\beta E}\, \Omega(E) dE.$$

Helmholtz Free Energy is useful in this context:

$$F(T, V, N) = -kT \log Z\, (T, V, N).$$

(3) Is there a given temperature and chemical potential? Then Grand canonical:

$$P(d\Gamma, N) = \frac{1}{\pi N_A!\, h^{3N_A}} \frac{e^{-\beta(E_N - \Sigma N_A \mu_A)} d\Gamma}{\phi},$$

$$\phi = \sum_{N_A}^\infty \sum_{N_B}^\infty \cdots \frac{1}{\pi N_A!\, h^{3N_A}} \int e^{-\beta(H_N - \Sigma N_A \mu_A)}\, d\Gamma$$

$$P(N, \ell) = \frac{e^{-\beta(E_{N,\ell} - \Sigma N_A \mu_A)}}{\phi} \qquad \phi = \sum_{N_A}^\infty \sum_{N_B}^\infty \cdots \sum_\ell e^{-\beta(E_{N,\ell} - \Sigma N_A \mu_A)}$$

Consider a state with n adsorptions that has energy $E_n = -n\varepsilon_0$, thus at temperature T we have the canonical weighting factor $e^{n\varepsilon_0/kT}$, also, there is a multiplicity weighing factor due to available adsorption sites:

$$W_n = \frac{N!}{n!\,(N-n)!}$$

So, the partition function (canonical since have specified "N") is:

$$Z_n = e^{n\varepsilon_0/kT} \frac{N!}{n!\,(N-n)!}$$

(useful for describing the 2-D adsorbed part of a gas). Let's now use this to compute the partition function for the $T-\mu$ distribution:

$$\phi = \sum_{n=0}^{N} e^{n\varepsilon_0/kT}\, Z_n = \sum_{n=0}^{N} \frac{N!}{n!\,(N-n)!} e^{\beta(\varepsilon_0+\mu)n} = \left[1+e^{\beta(\varepsilon_0+\mu)}\right]^N$$

So,

$$\phi = \left[1+e^{\beta(\varepsilon_0+\mu)}\right]^N$$

and from this we get:

$$P(n) = \frac{\varepsilon^{n\mu/kT} Z_n}{\phi}\ , \qquad \bar{n} = \Sigma n P(n) = \frac{1}{\beta}\frac{\partial}{\partial\mu}\log\phi$$

Thus

$$\bar{n} = \frac{N}{\beta}\frac{\partial}{\partial\mu}\log\left[1+e^{\beta(\varepsilon_0+\mu)}\right] = \frac{N}{\beta}\left\{\frac{\beta e^{\beta(\varepsilon_0+\mu)}}{1+e^{\beta(\varepsilon_0+\mu)}}\right\} = N\frac{1}{1+e^{-\beta(\varepsilon_0+\mu)}}$$

$$\theta = \frac{\bar{n}}{N} = \frac{1}{1+e^{-\beta(\varepsilon_0+\mu)}}$$

For external non-adsorbed gas:

$$Z = \frac{1}{h^{3N}N!}V^N(2\pi mkT)^{\frac{3}{2}N} \rightarrow F = -kT\log Z$$

$$F = kT\left\{N\log\frac{V}{N} + N + \frac{3}{2}N\log\left(\frac{2\pi mkT}{h^2}\right)\right\}$$

$$F = -kTN\left\{\log\frac{V}{N} + 1 + \frac{3}{2}\log\left(\frac{2\pi mkT}{h^2}\right)\right\}$$

$$\mu = (\partial F/\partial N)_{T,V}$$

$$= -kT\left\{\log\frac{v}{N} + 1 + \frac{3}{2}\log\left(\frac{2\pi mkT}{h^2}\right)\right\} - kTN\left\{\frac{-V/N^2}{V/N}\right\}$$

$$\boxed{\mu = kT\log\frac{N}{V}\left(\frac{h^2}{2\pi mkT}\right)^{3/2}}$$

Assuming Boltzmann statistics holds, we write

$$e^{\mu/kT} = \frac{p}{kT}\left(\frac{h^2}{2\pi mkT}\right)^{3/2}$$

And substitute $\theta = \frac{e^{\beta\mu}}{e^{\mu\beta}+e^{-\beta\varepsilon_0}}$ to get:

$$\theta = \frac{P}{P + kT\left(\frac{2\pi mkT}{h^2}\right)^{3/2} e^{-\beta\varepsilon_0}} = \frac{P}{P+P_0(T)}$$

where θ is known as Langmuir's isotherm, and where

$$P_0(\tau) = \left(\frac{h^2}{2\pi mkT}\right)^{3/2} e^{-\varepsilon_0/kT} kT$$

The Equipartition of Energy Law

Consider a classical system in contact with a heat bath (canonical distribution). Show the equipartition of energy law. Consider:

$$\overline{P_i \frac{\partial \mathcal{H}}{\partial P_j}} = \int P_i \frac{\partial \mathcal{H}}{\partial P_j} e^{-\beta \mathcal{H}} d\Gamma = \int P_i \left(-kT \frac{\partial e^{-\beta \mathcal{H}}}{\partial P_j}\right) d\Gamma$$

$$= kT \int \dots \int \left(\left[-P_i e^{-\beta \mathcal{H}}\right]_{P_j=-\infty}^{P_j=\infty} + \int_{-\infty}^{\infty} dP_j \frac{\partial P_i}{\partial P_j} e^{-\beta \mathcal{H}} \right) d\Gamma_{(j)} = kT\delta_{ij}$$

$$\overline{P_i \frac{\partial \mathcal{H}}{\partial P_j}} = kT\delta_{ij}$$

When the momenta only appear in a K.E. term in the Hamiltonian, K is a homogenous quadratic function of momenta, and, accord to Euler's theorem:

$$\sum_{i=1}^{f} P_i \frac{\partial K}{\partial P_i} = 2k \ and \ \frac{\partial K}{\partial P_i} = \frac{\partial \mathcal{H}}{\partial P_i}$$

Therefore

$$\bar{K} = \frac{1}{2}\sum_{i=1}^{f} \overline{P_i \frac{\partial K}{\partial P_i}} = \frac{1}{2}\sum_{i=1}^{f} \overline{P_i \frac{\partial \mathcal{H}}{\partial P_i}} = \frac{1}{2} fkT.$$

Thus, if

$$\mathcal{H} = \sum_i \frac{1}{2} a_i(q) P_i^2 + U(q)$$

we have

$$\frac{1}{2} a_i(q) P_i^2 = \frac{1}{2} \overline{P_i \frac{\partial \mathcal{H}}{\partial p_i}} = \frac{1}{2} kT$$

which shows equipartition.

Statistical thermodynamics of Gases

The partition function of an Ideal Gas has the form:

$$Z_N(T,V) = \left(\frac{2\pi mkT}{h^2}\right)^{3N/2} \frac{V^N}{N!} j(T)^N$$

where $j(T)$ is the internal partition function of a single particle.

Internal degrees of freedom and internal partition functions

Let's write the internal partition function as

$$j(T) = \sum_{\ell} g_{\ell}\, e^{-\varepsilon_{\ell}/kT}$$

where:
ε_{ℓ}: energy of a single molecule
g_{ℓ}: the degeneracy
Where the internal motion of a molecule consist of:

(1) Electronic state - g_e
(2) State of nucleus - g_n
(3) Rotational state of molecule as a whole - $j_{rot}(T)$
(4) Vibrational states within the molecule - $v(T)$

The approximate formula for $j(T)$ for various gases
Monatomic molecules: $He, Ne. A$ ... electrons usually from closed shells, thus energy difference is very large between ground state and first excited state:

$$j_{rot}(\tau) = 1, \qquad v(T) = 1, \qquad g_e = 1$$

and

$$g_n = 2S_n + 1$$

for nuclear spin of magnitude S_n. So, $j = 1 \times 1 \times 1 \times g_n = (2S_n + 1)$.

Diatomic molecules: In many cases the lowest electronic state is non-degenerate and separated by fairly large energy from the next excited state, so suppose $g_e = 1$. Provided that the temperature is not high, vibration and rotation may be separated as a first approximation. Thus, the nuclear state and the molecular rotation are independent in heteronuclear molecules. Due to spin-statistics, nuclear state and the molecular rotation are closely coupled in homonuclear molecules through the restriction of F.D. or B.E. statistics. Thus,

$$j = g_e g_{nuc} r(T) v(T)$$

if a heteronuclear molecule AB. Alternatively,

$$j = g_e j_{nuc-rot}(T) v(T)$$

if a homonuclear molecule AA, and

$$j \cong \frac{1}{2} g_e g_{nuc} r_c(\tau) v(\tau)$$

if a homonuclear molecule AA at high temperature, where:
$g_e = 1$ ordinarily
g_{nuc} is the degeneracy due to nuclear spin:
For (AB) case:
$g_{nuc} = (2S_A + 1)(2S_B + 1)$
For (AA) case:
$g_{nuc} = (2S_A + 1)^2$

We have

$$V(\tau) = \frac{e^{-\theta_v/2T}}{1 - e^{-\theta_v/2T}} \equiv \left[2 \sinh \frac{\theta_v}{2T}\right]^{-1} \quad , \quad \theta_v = hv/k$$

which is the vibrational part of the partition function with frequency v. For the rotational part of the partition function with moment of Inertia I.

$$r(\tau) = \sum_{\ell=0}^{\infty} (2\ell + 1) \exp\left\{-\ell(\ell + 1) \frac{h^2}{8\pi^2 IkT}\right\}$$

$$= \sum_{\ell=0}^{\infty} (2\ell + 1)\, e^{-\ell(\ell+1)\theta_r/\tau} \,, \theta_r = h^2/8\pi^2 IK$$

The semiclassical rotation partition function is:

$$r_c(\tau) = \frac{8\pi^2 IkT}{h^2}$$

For a homonuclear molecule AA the nuclear-rotational partition function is given by:

(1) When the nucleus A is a Fermi-particle (obeying F.D. statistics):

$$J_{nuc-rot} = S_n(2S_n + 1) r_{even} + (S_n + 1)(2S_n + 1) r_{odd}$$

Example: $H_2 \,, S_n = {}^1/_2$

(2) When A is a Bose-particle (obeying B.E. statistics):

$$J_{nuc-rot} = (S_n + 1)(2S_n + 1) r_{even} + S_n(2S_n + 1) r_{odd}$$

Example: $D_2, S_n = 1$.

Where for both cases:

$$r_{even} = \sum_{\substack{\ell=0,2,4,\dots \\ (even)}}^{\infty} (2\ell + 1) \exp\left\{\frac{-\ell(\ell + 1)h^2}{8\pi^2 IkT}\right\}$$

$$r_{odd} = \sum_{\substack{\ell=1,3,\dots \\ (odd)}}^{\infty} (2\ell + 1) \exp\left\{\frac{-\ell(\ell + 1)h^2}{8\pi^2 IkT}\right\}$$

Although $J_{nuc-rot} \cong g_{nuc} \frac{1}{2} r_c$ usually, the hydrogen molecule has rotational characteristic temperature that is high. This allows formation of orthohydrogen molecules ($j = odd$), para hydrogen molecules ($j = even$), ortho-deuterium molecules ($j = even$) and para-deuterium molecules ($j = odd$).

Fermi-Dirac and Bose-Einstein statistics are exhibited when

$$\frac{N}{V} \geq \frac{(2\pi mkT)^{3/2}}{h^3}$$

(i.e. low temperature and high densities). Average occupation number of a one-particle state τ is given by:

$$\bar{n}_\tau = \frac{1}{e^{(\varepsilon_\tau - \mu)/kT} + 1},$$

where ε_τ is the energy of the quantum state τ. So,

$$E = \sum_\tau \varepsilon_\tau \bar{n}_\tau = \sum_\tau \left(\frac{\varepsilon_\tau}{e^{(\varepsilon_\tau - \mu)/kT} + 1}\right)$$

$$N = \sum_\tau \bar{n}_\tau = \sum_\tau \left(\frac{1}{e^{(\varepsilon_\tau - \mu)/kT} + 1}\right),$$

$$F = N\mu - kT \sum_\tau \log\left(1 + e^{-(\varepsilon_\tau - \mu)/kT}\right)$$

Interpretations:

(1) When T and μ are given: $\bar{n}_\tau$ is the average occupation number of each 1-particle state, and E is the average total energy, and N is the average total number of particles.
(2) When T and N are given: μ is determined as a function of T and N. $\bar{n}_\tau$ and E as before.
(3) When E and N are given, T and μ determined and $\bar{n}_\tau$ as before.

Thermodynamics from one-particle state density

Let V be the volume of the space in which the particles are confined. When V is made larger and larger, the one-particle levels become more and more densely distributed. For a volume sufficiently large the number of states which have energies between ε *and* $\varepsilon + \Delta\varepsilon$ may be written as

$$D(\varepsilon)\Delta\varepsilon$$

which defines the one-particle state density $D(\varepsilon)$. Using the state density we have:

$$\bar{n}_\tau \to f(\varepsilon) = \frac{1}{e^{\beta(\varepsilon - \mu)} + 1}$$

(the Fermi distribution function). We have:

$$E = \int \varepsilon f(\varepsilon) D(\varepsilon) d\varepsilon$$

$$N = \int f(\varepsilon)\, D(\varepsilon) d\varepsilon$$

$$F = N\mu - kT \int D(\varepsilon)\, d\varepsilon \log\left(1 + e^{-(\varepsilon - \mu)/kT}\right)$$

Let's study Ideal Gas thermodynamics laws to obtain an expression for entropy:

$$dU = TdS - pdV \rightarrow \quad TdS = \left(\frac{dU}{dT}\right)_V dT + \left(\frac{dU}{dV}\right)_T dV + pdV$$

$$dS = \frac{1}{T}\left(\frac{dU}{dT}\right)_V dT + \frac{1}{T}\left[\left(\frac{\partial U}{\partial v}\right)_T + p\right] dV$$

Since dS is an exact differential, have Riemann-type relation (from complex analysis):

$$\left(\frac{\partial}{\partial V}\right)_T \left(\frac{1}{T}\left[\frac{\partial U}{\partial T}\right]_V\right) = \left(\frac{\partial}{dT}\right)_V \left(\left(\frac{1}{T}\left[\frac{\partial U}{\partial V}\right]_T\right) + \frac{p}{T}\right)$$

$$\frac{1}{T}\frac{\partial^2 U}{\partial V \partial T} = -\frac{1}{T^2}\left(\frac{\partial U}{\partial V}\right)_T + \frac{1}{T}\frac{\partial^2}{\partial T \partial V} - \frac{p}{T^2} + \frac{1}{T}\left(\frac{\partial p}{\partial T}\right)_V$$

From which we get the obscure relation:

$$\left(\frac{\partial U}{\partial V}\right)_T = T\left(\frac{\partial p}{\partial T}\right)_V - p$$

Using the above relation we can then write:

$$Tds = C_v dT + T\left(\frac{\partial p}{\partial T}\right)_V dV$$

For a classical ideal gas we have already determined:

$$pV = nRT\,, \quad C_v = \alpha nR$$

where $\alpha = \frac{3}{2}$ for monoatomic, and $\alpha = \frac{5}{2}$ for diatomic. Thus,

$$\left(\frac{\partial p}{\partial T}\right)_V = \frac{nR}{V} \rightarrow dS = \frac{\alpha nR}{T} dT + \frac{nR}{V} dV$$

$$S = S_0 + \alpha nR \ln\left({}^{T}/_{T_0}\right) + nR\ln(V/V_0)$$

$$S = S_0 + nR\ln\left[\left(\frac{T}{T_o}\right)^{\alpha} \frac{V}{V_o}\right]$$

Using $S = S_0 + nR\ln\left[\left(\frac{T}{T_o}\right)^{\alpha} \frac{V}{V_o}\right]$, we can rewrite $\exp\left(\frac{S-S_o}{\alpha nR}\right) = \left(\frac{T}{T_o}\right)\left(\frac{V}{V_o}\right)^{1/\alpha}$, thus:

$$T = \left(\frac{V_o}{V}\right)^{1/\alpha} T_o \exp\left(\frac{S - S_o}{\alpha nR}\right).$$

We also have

$$C_v = \alpha nR = \left(\frac{\partial U}{\partial T}\right)_V \quad \rightarrow \quad U = \alpha nR(T - T_o) + U_o.$$

$$U = U_o + \alpha nRT_o\left[\left(\frac{V_o}{V}\right)^{1/\alpha} \exp\left(\frac{S - S_o}{\alpha nR}\right) - 1\right]$$

4.1.8 First and second Order Phase Transitions

First Order Phase Transition

In the vicinity of a first order phase transition show

$$\frac{dP}{dT} = \frac{\ell}{T\Delta V}$$

Consider a P-T phase diagram

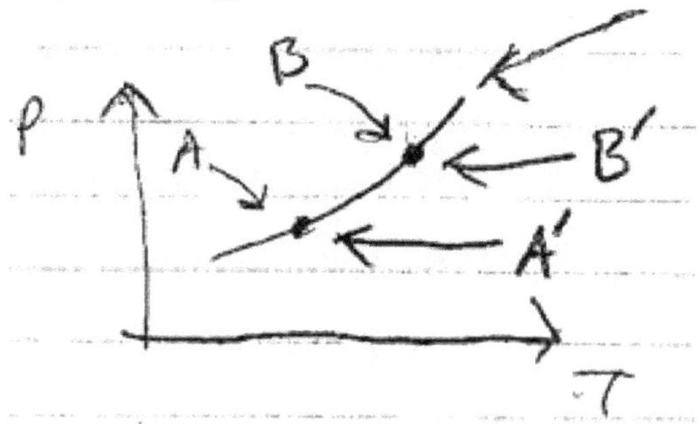

Where the line separates one phase from another, and $P_B - P_A = dP$, $T_B - T_A = dT$, with the slope of curve $\frac{dP}{dT}$. On the curve: $\mu_A = \mu_{A'}$ and $\mu_B = \mu_{B'}$, thus $\mu_B - \mu_A = \mu_{B'} - \mu_{A'}$. We also have:

$$d\mu = \mu_B - \mu_A = -sdT + vdp$$

$$d\mu' = \mu_{B'} - \mu_{A'} = -S'dT + V'dp$$

So,

$$(S - S')dT = (V - V')dp$$

$$\frac{dp}{dT} = \frac{(S' - S)}{(V' - V)} = \frac{\Delta S}{\Delta V}$$

And if we define $\ell = T\Delta S$, we have:

$$\frac{dp}{dT} = \frac{\ell}{T\Delta V}$$

Second Order Phase Transition

For a second phase transition:

$$\frac{dp}{dT} = \frac{\Delta\alpha}{\Delta kT} = \frac{1}{vT}\frac{\Delta C_p}{\Delta\alpha}$$

The equation representing constant Gibbs-Duhem function during the phase transition is:

$$(V' - V)dP = (S' - S)dT$$

for discontinuous change in ΔS we have a first order phase transition, with $C_p \to \infty$ at transition.

For a second order phase transition there is no discontinuity in either volume or entropy, however, any such situation we must then have continuity of the Gibbs function and its rate of change along the transition curve. Thus, we can take particles of the Gibbs function either to P or T to get the 2nd order phase transition relations:

$$\left[\left(\frac{\partial V}{\partial p}\right)_T dp - \left(\frac{\partial S}{\partial p}\right)_T dT\right]_2 = \left[\left(\frac{\partial V}{\partial p}\right)_T dp - \left(\frac{\partial S}{\partial p}\right)_T dT\right]_1$$

$$\beta = -\frac{1}{V}\left(\frac{\partial V}{\partial T}\right)_p = -\frac{1}{V}\left(\frac{\partial S}{\partial p}\right)_T$$

$$K = -\frac{1}{V}\left(\frac{\partial V}{\partial P}\right)_T \quad , \qquad C_p = T\left(\frac{\partial S}{\partial T}\right)_p$$

$$-k_2 dp - \beta_2 dT = -k_1 dp + \beta_1 dT \rightarrow \frac{dP}{dT} = \frac{\beta_2 - \beta_1}{k_2 - k_1} = \frac{\Delta\beta}{\Delta k}$$

Similarly, $\frac{\partial}{\partial T}$ gives other relation.

4.2 Examples

Example 4.1 Ideal Gas in Cylindrical Container

Consider an Ideal gas of N particles, mass m, enclosed in a cylindrical container (infinitely tall) in a uniform gravitational field. Let's find the thermodynamic potentials of interest ($Z = ?, F = ?\ , \bar{\varepsilon} = ?\ , C_v = ?$):

$$\mathcal{H} = \sum_{i=1}^{3N} P_i^2/2m + \sum_{j=1}^{N} mgz_j$$

$$Z_1 = \int e^{-\mathcal{H}/kT}\, d\Gamma = \left(\frac{2\pi mkT}{h^2}\right)^{3/2} \sigma \int_0^\infty e^{-mgz/kT}\, dz$$

$$= \frac{(2\pi mkT)^{3/2}}{h^3}\sigma\left(\frac{kT}{mg}\right)$$

$$Z_N = \frac{Z_1^N}{N!} = \left(\frac{2\pi mkT}{h^2}\right)^{3N/2}\left[\sigma\left(\frac{kT}{mg}\right)\right]^N \frac{1}{N!}$$

$$F = -kT\log Z_N$$

$$= -kT\left\{N\log\left(\frac{(2\pi)^{3/2} mk^{5/2} e\sigma}{h^3 g}\right) + \frac{5}{2}N\log T - N\log N\right\}$$

$$U = -T^2\frac{\partial\left(\frac{F}{T}\right)}{\partial T} = \frac{5}{2}NkT$$

$$C_v = \left(\frac{\partial u}{\partial T}\right)_V = \frac{5}{2}Nk$$

Note that $C_v = \frac{5}{2}Nk$ is greater that $C_v = \frac{3}{2}Nk$ for gas in a constant volume. This is because with increasing temperature molecules can rise and increase their potential energy.

Example 4.2 Ideal Gas in Rotating Cylindrical Container

Next, consider a rotating cylinder radius R, length L, angular velocity ω, density distribution =? In the rotating frame we have the centrifugal force $F = mr\omega^2$, thus a potential

$$F = -\nabla U \rightarrow P.E. = -\frac{1}{2}mr^2\omega^2$$

So, a canonical ensemble would thus give

$$\rho(r) = C\exp\left\{+\frac{\frac{1}{2}mr^2\omega^2}{kT}\right\}$$

And normalizing to solve for C:

$$\int \rho(r) rdr\, d\theta dz = M = Nm = 2\pi L\int_0^R \rho(r)\, rdr$$

Thus

$$\frac{Nm}{2\pi L} = C\int_0^R \exp\left(+\frac{\frac{1}{2}mr^2\omega^2}{kT}\right) rdr = C\left(\frac{2kT}{m\omega^2}\right)\int_0^{R\sqrt{\frac{m\omega^2}{2kT}}} \exp(+r'^2)\, r'dr'$$

$$C = \left(\frac{2kT}{m\omega^2}\right)\left\{\exp\left(+\frac{m\omega^2}{2kT}R\right)+1\right\}2$$

$$C = \frac{Nm}{2\pi L}\frac{1}{\left(\frac{2kT}{m\omega^2}\right)\left\{\exp\left(\frac{m\omega^2}{2kT}R\right)-1\right\}}2$$

$$p(r) = \left(\frac{Nm}{\pi k^2 L}\right)\frac{\exp\left(\frac{mr^2\omega^2}{2kT}\right)}{\left(\frac{2kT}{m\omega^2 R^2}\right)\left\{\exp\left(\frac{m\omega^2}{2kT}R^2\right)-1\right\}}$$

For a canonical distribution show $\overline{(E-\bar{E})^2} = kT^2C_v$:

$$\bar{E} = \frac{1}{Z}\int e^{-\beta E}\, E\Omega(E)dE = -\frac{\frac{\partial}{\partial\beta}Z}{Z} = -\frac{\partial}{\partial\beta}\log Z$$

$$\frac{\partial E}{\partial \beta} = -\frac{\partial^2}{\partial \beta^2}\log Z = -\frac{Z''}{Z} + \left(\frac{Z'}{Z}\right) = -\overline{E^2} + \bar{E}^2 = -\overline{(E-\bar{E})^2}$$

Also,

$$\frac{\partial \bar{E}}{\partial \beta} = -kT^2\frac{\partial \bar{E}}{\partial \beta} = -kT^2 C_v$$

So,

$$\overline{(E-\bar{E})^2} = kT^2 C_v$$

Example 4.3 Box with Partition with small hole

A box of 2 liters volume contains 10^{20} molecules of gas, which will be treated as distinguishable particles. The box is divided in equal parts by a partition pierced by a small hole, with the parts identified by A and B. assume the particles are noninteracting and move back and forth between sides A and B through the hole in a statistically independent fashion.

a) If each distinguishable particle is considered to have two states, A and B, depending on its position in sides A or B, what is the total number of different states of all the particles considered together?
b) Approximately, how many different states have exactly equal numbers of particles on the two sides?
c) What is the probability there will be exactly equal numbers of particles in the two sides of the box? What is the probability there will be more in side A than B but fewer than 100 more? 10^{10} more?

Answer

(a) Each particle has two possible states, the total number of states corresponds to

$2x2x2\ldots = 2^N = 2^{10^{20}}$.

(b) The multiplicity function is $g(N) = \frac{N!}{N_A!N_B!}$ For exactly equal numbers in A and B we have

$N_A = N_B = \frac{1}{2}N$ so

$$g(N) = \frac{N!}{\left(\frac{N}{2}\right)^2}$$

Approximating this with the Stirling expansion we get a more useful expression:

$$N! \cong (2\pi N)^{1/2} N^N \exp\left(-N + \frac{1}{12N}\cdots\right)$$

$$\frac{N!}{\left(\frac{N}{2}\right)^2} = \sqrt{\frac{2}{\pi N}} \cdot 2^N \cdot \exp\left(\frac{1}{12N} - \frac{1}{3N}\right)$$

Thus, there are approximately $\sqrt{\frac{2}{\pi N}} \cdot 2^N$ different states having exactly equal numbers of particles on the two sides. Since $N = 10^{20}$ we know this value: Number of states $\approx 10^{-10} \times 2^{10^{20}}$.

(c) P(exactly equal numbers) = Number of equal states/Number of states

$$\cong \frac{10^{-10} 2^{10^{20}}}{2^{10^{20}}} \cong 10^{-10}$$

Fewer than 100 imbalance?

$P(A > B + \alpha)\ for\ 0 < \alpha < 100$

$$P(A > B + \alpha) = \frac{\int_1^{49} \left(\frac{2}{\pi N}\right)^{1/2} 2^N \exp(-2S^2/N)\, ds}{2^N}$$

$$= \int_1^{49} \left(\frac{2}{\pi N}\right)^{1/2} 2^N \exp(-2S^2/N)\, ds$$

$$= \frac{1}{\sqrt{\pi}} \int_{\sqrt{\frac{2}{N}}}^{49\sqrt{\frac{2}{N}}} e^{-x^2}\, dx$$

Since $\operatorname{erf}(x) = \frac{2}{\sqrt{\pi}} \int_o^x e^{-u^2}\, du$

$\operatorname{erf}(x) = \frac{2}{\sqrt{\pi}}\left(x - \frac{x^3}{3.1!} + \frac{x^5}{5.2!} - \frac{x}{7.3!} + \cdots\right)$

For every small x:

$\operatorname{erf}(x) \sim \frac{2}{\sqrt{\pi}} x$

So $\sqrt{\pi} \int_{\sqrt{\frac{2}{N}}}^{49\sqrt{\frac{2}{N}}} e^{-x^2}\, dx \cong \sqrt{\pi}\left[49\sqrt{\frac{2}{N}} - \sqrt{\frac{2}{N}} = 48\sqrt{\frac{2\pi}{N}}\right]$

$P(A > B\ but\ by\ less\ than\ 100) = 48\sqrt{\frac{2\pi}{10^{20}}} = 48\sqrt{\frac{2}{\pi}}.10^{-10} 3.83x10^{-9}$

$P(A > B\ but\ by\ less\ than\ 10^{10}) = \sqrt{\pi} \int_{1\sqrt{\frac{2}{N}}}^{\frac{1}{2}10^{10}\sqrt{\frac{2}{N}}} e^{-x^2}\, dx$

$$= \sqrt{\pi}\left[\int_o^{\sqrt{2}/2} e^{-x^2}dx - \int_o^{\sqrt{2}10^{-10}} e^{-x^2}dx\right] = \sqrt{\pi}\left(\frac{\sqrt{2}}{2} - \frac{2\sqrt{2}}{8}.\frac{1}{3} + \frac{4\sqrt{2}}{32}.\frac{1}{10} - \frac{8\sqrt{2}}{128}\frac{1}{42} + \cdots\right)$$

$$\approx \frac{1}{\sqrt{\pi}}(.605) \approx .341$$

Example 4.4 Box with Partition with small hole and external pressure

Consider the same box and particles in Example 4.3 as a perfect gas at room temperature, exerting pressure according to the usual kinetic model.

a) Estimate the probability that the pressure will increase in side A by as much as one part in 10^{10} because more of the particles are in side A. (Use estimate from normal error curve or tables.)
b) Estimate the probability that the pressure will increase in side A by as much as one part in 1000000 (measurable) because more of the particles are in side A than in side B.
   Note: $\int_x^{\infty} \exp\ (-y^2)dy = exp(-x^2\ )/(2x)$ for $x \gg 1$.

Answer

T=room temperature = $25^oC = 298.15K$

$N = 10^{20}$

$V = 2liters$

$pV = Nk_BT$

$$p = \frac{10^{20}(1.38066x10^{-23}JK^{-1})(298.15k)}{(2\ liters)} = .206Nm^{-2} = .206Pa$$

$\Delta p = 10^{-10}$ holding V , T constant.

$$\Delta N = \frac{\Delta pV}{k_BT} = \frac{10^{-10}(2)}{(1.38x10^{-23})(298.15k)} = 5\ x\ 10^{10}$$

$$P(A > B, by\ at\ least\ 5\ x\ 10^{10} or\ greater) = \frac{1}{\sqrt{\pi}}\int_{\frac{5}{2}x10}^{\frac{1}{2}x10^{20}\sim\infty} e^{-x^2}dx$$

$$P(A > B, by\ at\ least\ 5x10^6\ or\ greater) = \frac{1}{\sqrt{\pi}}\int_{\frac{5}{2}x10}^{\frac{1}{2}x10^{20}\sim\infty} e^{-x^2}dx$$

Example 4.5 Particles in a box with a partition

A total of N distinguishable particles are contained in a box divided by a partition. Particles on side A are in state A and on side B in state B. All particles have zero energy so that the total energy remains zero in all states.

a) In situation 1, all particles are on side A, and there is only one possible combined state. In situation 2, the particles can pass through a hole to side B. Find the number of possible states in situation 2, and the entropy σ_1 and σ_2 in the two situations.

b) The states A and B are subdivided so that a particle in side A can be in state A_1 or state A_2, depending on the location of the particle as shown, and correspondingly for side B. Find the number of possible states in situation 2, and the entropy σ_1 and σ_2 in the two situations, as before.

c) Generalize (b) by increasing the subdivision to $A_1, A_2, \ldots A_n$, and $B_1, \ldots B_n$, and find σ_1 and σ_2. Then find the change in entropy caused by making the hole for (a), (b) and (c).

Answer

(a) Number of possible states in situation 2: each particle can either be in state A or state B allowing 2 states, there are N particles: There are 2^N states for situation 2.

$\sigma_I = \log g_I = \log(1) = 0$

$\sigma_{II} = \log g_{II} = \log 2^N = N \log 2$

(b) Now, $g_I = 2^N$, $g_{II} = 4^N$

$\sigma_I = N \log 2$

$\sigma_{II} = N \log 4$

(c) Generalise for n subdivisions for situations 1 and 2

$\sigma_I = N \log n$

$\sigma_{II} = N \log 2n$

Change of entropy for (a) $\sigma_{II} - \sigma_I = N \log 2$; for (b) $\sigma_{II} - \sigma_I = N \log 2$; and for (c) $\sigma_{II} - \sigma_I = \log 2$.

Example 4.6 Particle in a box with a heat reservoir

One particle of mass M is in a one dimensional box of length L in contact with a heat reservoir at temperature τ. Typical wavefunctions corresponding to the energy eigenstates are as shown below:

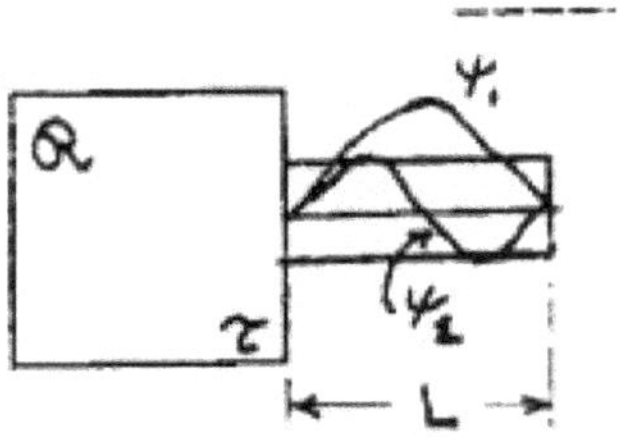

a) Write an expression for the energy E_n for the quantum state Ψ_n.
b) Find the partition function Z, as a sum, and then approximate the value by means of an integral.
c) Find the average energy <KE>=U of the particle.
d) If N distinguishable particles are placed in the box, what is the partition function Z and the average energy per particle U/N?

Answer

(a) $\hat{H} = \frac{\hat{P}^2}{2m}$, $\Psi_n = \sin\left(\frac{n\pi x}{L}\right)$, $\hat{H}\Psi_n = E_n\Psi_n$, $\hat{P} = -i\hbar\frac{\partial}{\partial x}$

$\hat{H}\Psi_n = \frac{\hat{P}^2}{2m}\Psi_n = +\left(\frac{n\pi}{L}\right)^2\frac{\hbar^2}{2m}\sin\left(\frac{n\pi x}{L}\right) \quad \rightarrow \quad E_n = +\frac{\hbar^2}{2m}\left(\frac{n\pi}{L}\right)^2$

(b)

$$Z = \sum_{n=0}^{\infty}\exp(-E_n/\tau) = \sum_{n=0}^{\infty}\exp\left[-\frac{n^2\pi^2\hbar^2}{2mL^2\tau}\right]$$
$$= \int_o^{\infty}\exp\left[-x^2\left(\frac{\pi^2\hbar^2}{2mL^2\tau}\right)\right]dx = \frac{1}{2}\sqrt{\frac{\pi}{a}}$$

where $a = \frac{\pi^2\hbar^2}{2mL^2\tau}$. Thus

$$Z = \sqrt{\frac{mL^2\tau}{2\pi\hbar^2}}$$

(c) $u = \tau^2\left(\frac{\partial \log Z}{\partial\tau}\right) = \tau^2\frac{\left(\frac{\partial Z}{\partial\tau}\right)}{Z} = \tau^2\cdot\frac{\frac{1}{2}\left(\frac{mL^2\tau}{2\pi\hbar^2}\right)^{-1/2}\frac{ML^2}{2\pi\hbar^2}}{\left(\frac{mL^2\tau}{2\pi\hbar^2}\right)^{1/2}} = \frac{1}{2}\tau^2\frac{\left(\frac{mL^2\tau}{2\pi\hbar^2}\right)}{\left(\frac{mL^2\tau}{2\pi\hbar^2}\right)} = \frac{1}{2}\tau$

(d) $Z_N = N\sqrt{\left(\frac{mL^2\tau}{2\pi\hbar^2}\right)}$; $u = \frac{1}{2}N\tau$; $u/N = \frac{1}{2}\tau$

Example 4.7 Photons in a box with a heat reservoir

Photons that can travel only in the $\pm x$ direction are in a box of length L in contact with a heat reservoir at temperature τ. The mode frequencies are determined by the classical electric field required to vanish at the ends.

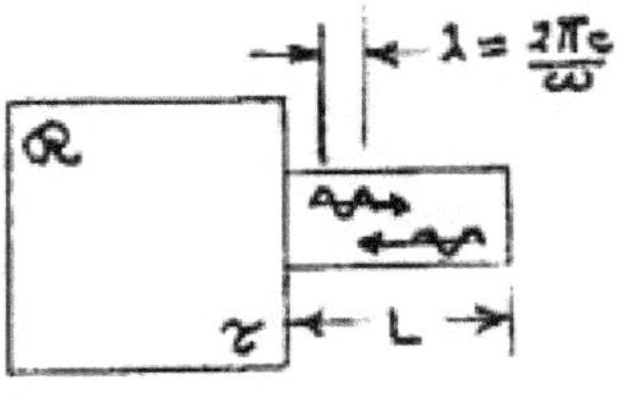

a) What are the energies E_n of the photon states?
b) What is the average number of photons in the state of frequency ω_n? (There are 2 polarizations and 2 directions for each frequency.)
c) Find the total energy U in the box.
d) Find the entropy σ of the photons in the box assuming $\sigma = 0$ at $\tau = 0$. Note $\int_0^\infty \frac{xdx}{e^x-1} = \frac{\pi^2}{6}$.

Answer

(a) $\Psi_n = \sin\left(\frac{n\pi x}{L}\right)$

$\hat{H}\Psi_n = E_n\Psi_n \qquad E_n = Ln\hbar w_o \qquad w_o = \pi/L$

(b) $Z = \frac{1}{1-\exp(-\hbar w/\tau)}$

Thermal average value of the number of photons in a mode is

$< S >= Z^{-1}\sum S\exp(-S\,\hbar w/\tau)$

$< S >= \frac{\exp(-\hbar w/\tau)}{1-\exp(-\hbar w/\tau)}$

The average number of photons in the state of frequency w_n is

$$4 < S >= \frac{4\exp(-\hbar w/\tau)}{1-\exp(-\hbar w/\tau)}$$

(c) $U = C\int_o^\infty 4 < \varepsilon > dw = \int_o^\infty \frac{4\hbar w}{\exp(-\hbar w/\tau)-1} dw = \int_0^\infty \frac{4\tau x\frac{\tau}{\hbar}dx}{\exp(x)-1} = \frac{4\tau^2}{\hbar}\int_0^\infty \frac{xdx}{e^x-1}$

$$U = \frac{4\tau^2}{\hbar}\frac{\pi^2}{6} = \frac{2}{3}\frac{(\tau\pi)^2}{\hbar}$$

(d) $F = -\tau\log Z$, $\sigma = -\left(\frac{\partial F}{\partial\tau}\right) = \frac{\partial(\tau\log Z)}{\partial\tau}$

Using $dU = \tau\, d\sigma$ easier:

$$\sigma = \frac{\tau\hbar w}{\tau^2}\frac{\exp(-\hbar w/\tau)}{[1-\exp(-\hbar w/\tau)]} = \frac{(\hbar w/\tau)}{\exp(-\hbar w/\tau)-1}$$

Example 4.8 Circuit elements in contact with a heat reservoir

1. A resistor R in thermal contact with a reservoir at temperature τ. The mean square noise voltage when the resistor is open-circuited is given by $4\tau R\Delta f$ for the sinusoidal components in any frequency interval Δf.

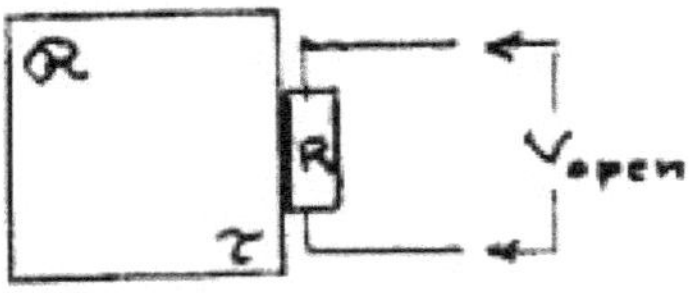

a) A capacitor C is connected across the resistor as shown. Find the mean square noise voltage $<V_C^2>$ across the capacitor in the interval w to $w + \Delta w$.

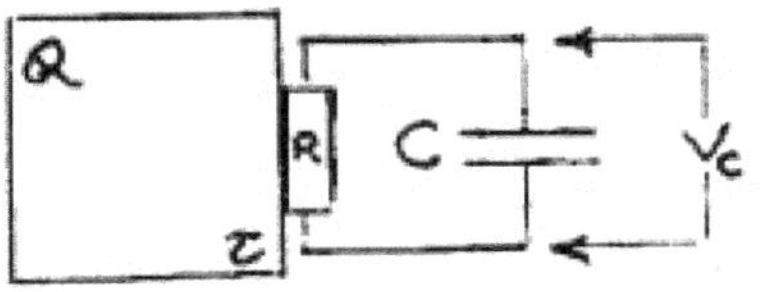

b) Find the average U_C stored in the capacitor including all frequency components. (The energy stored in a capacitor C at the voltage V is $\frac{1}{2}CV^2$.)

c) Compare U_c with the average energy U of the single particle in a box calculated in Prob. 1.
Note $\int_o^{\infty}\frac{dx}{1+x^2} = \frac{\pi}{2}$.

Answer

(a)

$$V_{c(rms)} = V_{rms}\left|\frac{\frac{1}{SC}}{R+\frac{1}{SC}}\right| = \frac{V_{rms}}{|s(R+1)|}$$

$$\sqrt{< V_C^2 >} = V_{c(rms)} = \frac{V_{rms}}{\sqrt{(wcR)^2+1^2}}$$

$$< V_C^2 > = \frac{< V^2 >}{1+(wcR)^2} = \frac{\left(\frac{2\tau R\Delta w}{\pi}\right)}{1+(wcR)^2}$$

(b) $U_c = \int_o^{\infty}\frac{c}{2} < V^2 > = \int_o^{\infty}\frac{c}{2}\frac{2\tau R\Delta w}{\pi(1+(wcR)^2)} = \frac{\tau}{\pi}\int_o^{\infty}\frac{cRdw}{(1+(wcR)^2)} = \frac{\tau}{\pi}\cdot\frac{\pi}{2} = \frac{\tau}{2}$

$$U_c = {}^{\tau}/_{2}$$

(c) They are equal.

Example 4.9 Box with particle exchange and heat reservoir

1. A cubical box of side L contains an ideal gas of noninteracting particles of mass M that can flow freely in or out of a 1-dimensional system of length L connected to one wall of the box. Both the box and

the 1-d system are in thermal equilibrium with a reservoir at the temperature τ.

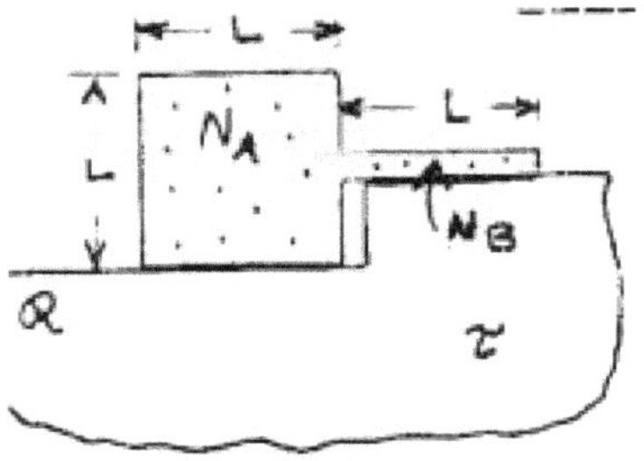

a) Consider first only the cubical box. Write an expression for the chemical potential μ_A in terms of the number of particles N_A and the single particle partition function Z_{1A}.
b) Now consider only the 1-dimensional system with N_B particles, and write an expression for the potential μ_B in a similar form to that in (a)
c) With the 1-d system in diffusive contact, find the ratio of the number of particles in the 1-d system to the particles in the box $<N_B>/<N_A>$ when in thermal equilibrium as a function of L,M,τ and $\hbar$. Note: in a real system, particles in a 3-d solid may be in diffusive contact with the 2-d system represented by the particles on the surfaces.

Answer

(a) $\mu_A = \tau \log(N_A/Z_A)$

$Z_{1A} = (M\tau/2\pi\hbar^2)^{3/2}L^3$

$$\mu_A = \tau \log\left(\frac{N_A}{(M\tau/2\pi\hbar)^{3/2}L^3}\right)$$

(b) $Z_{1B} = (M\tau/2\pi\hbar)^{1/2}L$

$$\mu_B = \tau \log\left(\frac{N_B}{(M\tau/2\pi\hbar)^{1/2}L}\right)$$

(c) $$\frac{<N_B>}{<N_A>} = \frac{(M\tau/2\pi\hbar)^{1/2}L\,\exp(\mu B/\tau)}{(M\tau/2\pi\hbar)^{3/2}L^3\,\exp(\mu B/\tau)}$$

$$< N_B >/< N_A > = \frac{\exp\left(\frac{\mu_B-\mu_A}{\tau}\right)}{\left(\frac{M\tau L^2}{2\pi\hbar^2}\right)}$$

but there is one other condition in thermal equilibrium namely $\mu_A = \mu_B$:

$$< N_B >/< N_A > = \frac{2\pi\hbar^2}{M\tau L^2}$$

Example 4.10 Two-level energy system in thermal and diffusive equilibrium

A two-energy level system with levels –E/2 and +E/2 is in thermal and diffusive equilibrium with a reservoir of identical particles at the temperature τ. A difference of $\Delta\mu_{int}$ exists between the internal chemical potentials of the system and the reservoir. A typical state, N=2 and U=0, is shown with $\Delta\mu_{int} = \mu_{system} - \mu_R$:

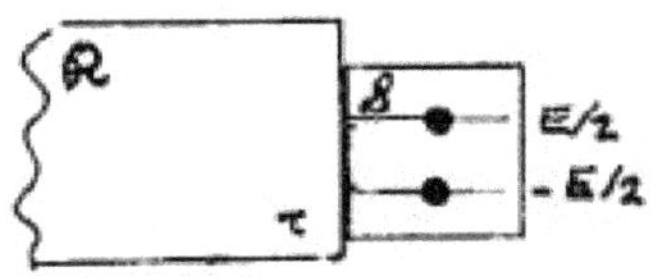

a) By means of diagrams similar to that above, show all of the states N<=2, writing, U,N, and a Gibbs factor for each state.
b) What is the ratio of the probability of finding the state N=2 and U=0 to the probability of the state N=1 and U=E/2?
c) Write the grand partition function both as an indicated sum and explicitly for all terms N<=2.
d) What is the probability of finding a state with N=2?

Answer

(a) States for N<=2, and associated Gibbs Factor:

$N = 2\,, U = E\,;\, \exp[(2\mu - E)/\tau]$

$N = 2\,, U = 0\,;\, \exp[2\mu/\tau]$

$N = 2\,, U = -E\,;\, \exp[(2\mu + E)/\tau]$

$N = 1\,, U = {}^{E}/_{2}\,;\, \exp[(\mu - {}^{E}/_{2})/\tau]$

$N = 1\,, U = -{}^{E}/_{2}\,;\, \exp[(\mu + {}^{E}/_{2})/\tau]$

$N = 0\,, U = 0\,;\, \exp(0) = 1$

(b) probability of state $N = 2$ and $u = 0$ is $\frac{1}{2}$.

Probability of star $N = 1$ and $u = {}^{E}/_{2}$ is $\frac{1}{2}$

.

(c) $\mathfrak{z}(\mu, \tau) = \sum_{N=0}^{2} \sum_{S(N)=1}^{2} \exp[(N\mu - \varepsilon_{S(N)})/\tau]$

$= 1 + \exp[(\mu + {}^{6}/_{2})/\tau] + \exp[(\mu + {}^{6}/_{2})/\tau] + \exp[(2\mu + E)/\tau]$

$+ \exp[2\mu/\tau] + \exp[(2\mu - E)/\tau]$

(d)

$$P(2,\varepsilon) = \frac{\exp[(2\mu + E)/\tau] + \exp[2\mu/\tau] + \exp[(2\mu - E)/\tau]}{\mathfrak{z}}$$

$$= \frac{1 + 2\cosh(E/\tau)}{\exp[-2\mu/\tau] + \exp[-\mu/\tau][2\cosh E/2\tau] + 2\cosh(E/\tau) + 1}$$

Example 4.11 System in Example 4.10 with change

2. The particles in Prob. 2 are given a charge q, and the system is held at the voltage V with respect to the reservoir by means of a battery as shown. Assume that no more than one particle per state is permitted. The reservoir temperature is τ.

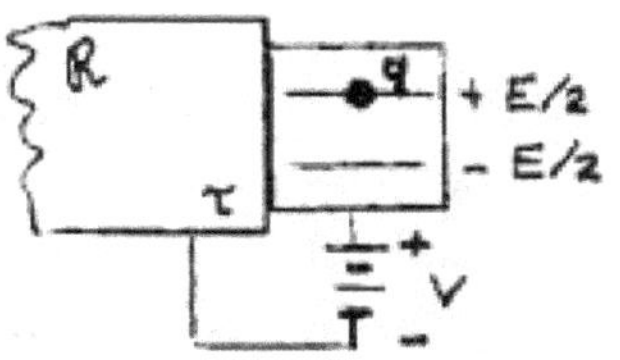

a) Find the probability there will be no particles in the system.
b) Find the average energy U of the system.
c) Find the average number of particles <N> in the system.

Answer

(a) $P(0,0) = \frac{1}{\mathfrak{z}}$

$\mathfrak{z} = \sum_{ASN} \exp\left((N\mu - \varepsilon_{s(N)})/\tau\right) = 1 + \exp(\mu/\tau)\left[2\cosh\left(\frac{E}{2\tau}\right)\right] + \exp\left(\frac{2\mu}{\tau}\right)$

$$= \exp(\mu/\tau)\left[2\cosh\left(\frac{\mu}{\tau}\right) + 2\cosh\left(\frac{E}{2\tau}\right)\right]$$

$Z_1 = 2\cosh\left(\frac{E}{2\tau}\right)$

$\mathfrak{z} = 1 + \exp[-\log Z_1 + qV]\left[2\cosh\left(\frac{E}{2\tau}\right) + \exp\left(\frac{2}{\tau}\right)(2\log Z_1 - \log 2 - \log Z_1)\right]$

$= 1 + e^{qv/\tau} + \frac{Z_1^2 e^{2qv/\tau}}{4} = 1 + e^{qv/\tau} + \cosh^2\left(\frac{E}{2\tau}\right)e^{2qv/\tau}$

$$P(0,0) = \frac{1}{1 + \exp\left({}^{qV}/_{\tau}\right) + \left[\cosh\left(\frac{E}{2\tau}\right)\exp\left({}^{qV}/_{\tau}\right)\right]^2}$$

(b) $U = \tau^2(\partial \log Z_1/\partial\tau) = \tau^2 \frac{\frac{\partial Z_1}{\partial \tau}}{Z_1} = \frac{\tau^2 .2\sinh\left(\frac{E}{2\tau}\right)\cdot\frac{E}{2}\left(\frac{-1}{\tau^2}\right)}{2\cosh\left(\frac{E}{2\tau}\right)} =$
$-(E/_2)\tanh(E/2\tau)$

(c) $\mathfrak{z} = 1 + \exp(\mu/\tau)\left[2\cosh\left(\frac{E}{2\tau}\right)\right] + \exp(2\mu/\tau) = 1 + \lambda 2\cosh\left(\frac{E}{2\tau}\right) + \lambda^2$

$$< N >= \lambda \frac{\partial}{\partial\lambda}\log\mathfrak{z} = \lambda\frac{\frac{\partial\mathfrak{z}}{\partial\lambda}}{\mathfrak{z}} = \frac{\lambda\left(2\cosh\left(\frac{E}{2\tau}\right) + 2\lambda\right)}{1 + \lambda 2\cosh\left(\frac{E}{2\tau}\right) + \lambda^2}$$

Example 4.12 Monatomic ideal gas is in thermal and diffusive equilibrium

1. A monatomic ideal gas is in thermal and diffusive equilibrium, with the same type of atom in the solid phase. The pressure is p in the gas system, and the temperature is τin both the gas and solid systems.

a) Find the chemical potential μ_{gas} in the gas system as a function of the temperature τ and the pressure p.

b) Assume the temperature τ is low enough that the entropy contribution to the free energy can be neglected and also that the only internal energy is the binding energy. The free energy per atom in the solid is then the binding energy ε_B of an atom in the crystal lattice of the solid. Find the chemical potential μ_{solid} of the solid system in terms of ε_B.

c) Find an expression for the vapor pressure of the gas in equilibrium with the solid in terms of the mass M of the atoms and τ, ε_B and h

Answer

(a) $\mu_{gas} = \tau\log(n/n_Q)$; $PV = N\tau$; $p = \frac{N}{V}\tau$

$n/n_Q = \frac{N}{V}\frac{1}{n_Q} = \frac{p}{\tau}\frac{1}{(M\tau/2\pi\hbar^2)^{3/2}}$

$$\mu_{gas} = \tau\log\left(\frac{P}{\tau^{5/2}\left(\frac{M}{2\pi\hbar^2}\right)^{3/2}}\right) = \tau\log\left(\frac{P}{\tau^{5/2}}\right) - \tau\frac{3}{2}\log\left(\frac{M}{2\pi\hbar^2}\right)$$

(b) $\mu = \mu_{ext} - \mu_{ini} \cong \mu_{ini} = -\varepsilon_B$

(c) $\mu_{solid} = \mu_{gas}$

$$-\varepsilon_B = \tau\log\left(\frac{P}{\tau^{5/2}\left(\frac{M}{2\pi\hbar^2}\right)^{3/2}}\right)$$

$P = \tau^{5/2}\left(\frac{M}{2\pi\hbar^2}\right)^{3/2} \exp(-\varepsilon_B/\tau)$

Note, we can arrive at this answer much more quickly by considering the zero of the energy scale shifted by ε_B and then no longer considering the solid we simply have a gas with

$\mu = \varepsilon_B + \tau\log(n/n_Q)$, we want this to be in equilibrium with a ground state solid so

$\varepsilon_B + \tau\log(n/n_Q) = 0$. Thus $-\varepsilon_B + \tau\log\left(\frac{P}{\tau^{5/2}\left(\frac{M}{2\pi\hbar^2}\right)^{3/2}}\right)$, thus

$$P = \tau^{5/2}\left(\frac{M}{2\pi\hbar^2}\right)^{3/2} \exp(-\varepsilon_B/\tau)$$

Example 4.13 Gas in container subjected to three stage cycle

N atoms of an ideal monatomic gas are in a container that is subjected to a cycle that carries it through three states: A, B and C . In state A, the volume is V_A and the temperature is τ_A. The gas is compressed isentropically (slowly and without any heat flow into the system) to state B where the volume is $V_A/2$. The piston compressing the gas then shatters and the gas suddenly returns to the original volume. In state C the pressure has equalized throughout the container but no heat has flown into the system. Heat is the added or subtracted at a constant volume to bring the gas back to the original temperature and complete the cycle.

a) Find ΔU, $\Delta\sigma$ and ΔQ, in the three steps of the cycle and make a table of these values.
b) Find $\Sigma\Delta U$, $\Sigma\Delta\sigma$ and $\Sigma\Delta Q$, the sum of the individual changes around the cycle for the gas system. What is $\Delta\sigma_R$ the change in the entropy of the reservoir that supplied heat to the gas? Which of the variable U,σ,and Q are functions of N, τ, and V only?

Answer

(a) $A \rightarrow B$ reversible Isotropic compression

$\Delta U = +\frac{3}{2}N\tau_1\left[1 - \left(\frac{V_1}{V_2}\right)^{2/3}\right]$

$\Delta\sigma = 0$

$\Delta Q = 0$

$B \rightarrow C$ irreversible compression

$\Delta U = 0$

$\Delta\sigma = N\log\left(\frac{Va}{Va/2}\right) = N\log 2$

$\Delta Q = 0$

$C \to A$ heat added

Note that $\sigma(\tau, V) = N\left(\log\tau^{3/2} + \log V + C\right)$ and $\Delta U = \Delta Q + \Delta W$

$\Delta W = 0.\ \ \Delta U = \Delta Q = -C_V(\tau_2 - \tau_1) = -\frac{3}{2}N(\tau_2 - \tau_1)$

$\tau_1^{3/2}V_1 = \tau_2^{3/2}V_2 \to \tau_1/\tau_2 = (V_1/V_2)^{3/2}$

$\Delta U = -\frac{3}{2}N\tau_1\left[1 - \left(\frac{V_1}{V_2}\right)^{3/2}\right]$

$\Delta Q = -\frac{3}{2}N\tau_1\left[1 - \left(\frac{V_1}{V_2}\right)^{3/2}\right]$

$\Delta\sigma = N\log\tau_2^{3/2} - N\log\tau_1^{3/2} = N\log\left(\frac{\tau_2}{\tau_1}\right)^{3/2} = N\log\left(\frac{V_2}{V_1}\right) =$
$N\log\left(\frac{Va/2}{Va}\right)$

$\Delta\sigma = -N\log 2$

(b) $\sum\Delta u = 0$

$\sum\Delta\sigma = 0$

$\sum\Delta Q = -\frac{3}{2}N\tau_a\left[1 - (2)^{2/3}\right]$

$\Delta\sigma_R = \frac{\Delta u}{\tau} + \frac{pdV}{\tau_1} = -\frac{\Delta Q}{\tau_a} = -\frac{3}{2}N\left[1 - (2)^{2/3}\right]$

U and Q are functions of N, V, and temperature only.

Example 4.14 Fermion Deposition

2. Identical fermions (He^3 atoms) are deposited on flat surface at a low temperature to form a 2-dimensional system with N atoms in an L by L square. The chemical potential of the system is μ.

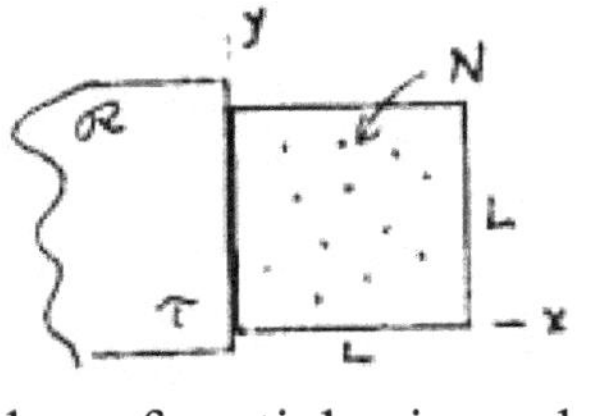

What is the average number of particles in each energy state defined by the quantum numbers n_x and n_y?

Answer

(a) Guess an eigenfunction $\sin\left(\frac{n_x\pi x}{L}\right) = \varphi_x$

$\frac{-\hbar^2}{2m}\frac{\partial^2}{\partial x^2}\varphi_x = E_{nx}\varphi_x$

$E_{nx} = \frac{-\hbar^2}{2m} - \left(\frac{n_x\pi}{L}\right)^2 = \left(\frac{\hbar n_x}{L}\right)^2 \frac{1}{2m}$

$\varepsilon_s = \left(\frac{\hbar\pi}{L}\right)\frac{1}{2m}\left[n_x^2 + n_y^2\right]$

$$f(\varepsilon_s) = \frac{2}{\exp[(\varepsilon_s - \mu)/\tau] + 1}$$

Example 4.15 N distinguishable particles in a harmonic oscillator potential

A total of N distinguishable particles are in a harmonic oscillator potential. The energy of a single-particle state is $sh\omega$ for state s, measured from E>0 at the ground state s=0. The system $\mathcal{S}$ (no connection to heat reservoir $\mathcal{R}$ yet) is in thermodynamic equilibrium at a temperature τ. Assume $N \gg 1$.

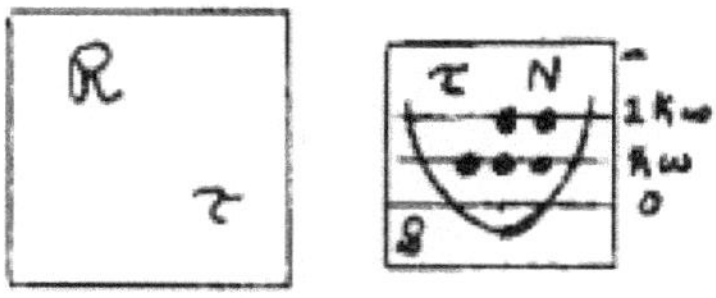

(a) Write exact expressions for the multiplicity g and the entropy σ. Let n be defined in terms of the unknown total energy U_A by $U_A = nh\omega$, and the approximation $d(\ln N!)/dN = \ln N$, to find $d(\sigma)/dN$. Find U_A in terms of $h\omega$, N, and τ.

(b) The system is placed inContact with a large reservoir at he same temperature. Find the partition function Z as a series, and then sum the series. Then find the total energy of the system U_B for the system of oscillators directly from Z for the given temperature. Then compare U_A and U_B.

Answer

(a) $g(N,n) = \frac{(N+n-1)!}{n!(N-1)}$

$\sigma(N,n) = \ln g(N,n) = \ln[(N+n-1)!] - \ln(n!) - \ln[(N-1)!]$

$U_A = n\hbar w$

$\frac{\partial\sigma}{\partial n} = \frac{\partial\{\ln[(N+1-1)]\}}{\partial n} - \frac{\partial\{\ln(n!)\}}{\partial n} - \frac{\{\ln[(N-1)!]\}}{\partial n} = \ln(N+n-1) - \ln n = \ln\left[\frac{N+n-1}{n}\right]$

$$\frac{1}{\tau} = \frac{\partial\sigma}{\partial n}\cdot\frac{\partial n}{\partial u_A} = \ln\left[\frac{N+n-1}{n}\right]\frac{1}{\hbar w}$$
$$\tau = \hbar w/ln\left[\frac{N+n-1}{n}\right]$$
$$ln\left[\frac{N+n-1}{n}\right] = \frac{\hbar w}{\tau}$$
$$\frac{N+n-1}{n} = \exp(\hbar w/\tau)$$
$$n = \frac{N-1}{\exp(\hbar w/\tau)-1}$$
$$U_A = n\hbar w = \frac{(N-1)\hbar w}{\exp(\hbar w/\tau)-1}$$
Since N>>1 we have
$$U_A = \frac{N\hbar w}{\exp(\hbar w/\tau) - 1}$$
(b)
$$Z(\tau) = \textstyle\sum_s \exp(-\varepsilon_s/\tau) = 1 + \exp(-\hbar w/\tau) + \exp(-2\hbar w/\tau) + \cdots$$
$$Z(\tau) = \frac{1}{1-\exp(-\hbar w/\tau)}$$
$$U = \tau^2\left[\frac{\partial \log Z}{\sigma\tau}\right] = \tau^2\frac{\partial Z/\partial\tau}{Z} = \tau^2\left(1-\exp(-\hbar w/\tau)\frac{\partial(1-\exp(-\hbar w/\tau))}{\partial\tau}\right)$$
$$= \tau^2(1-\exp(-\hbar w/\tau))(-1)(1-\exp(-\hbar w/\tau))^{-2}\left(-\exp(-\hbar w/\tau)\left(\frac{+\hbar w}{\tau^2}\right)\right)$$
$$= \frac{\hbar w(1-\exp(-\hbar w/\tau))\exp(-\hbar w/\tau)}{(1-\exp(-\hbar w/\tau))^2} = \frac{\hbar w(\exp(-\hbar w/\tau))}{(1-\exp(-\hbar w/\tau))} = \frac{\hbar w}{\exp(\hbar w/\tau)-1}$$
$U_B = NU$ since there are N particles, thus
$$U_B = \frac{\hbar w}{\exp(\hbar w/\tau) - 1}$$
and we see that
$$U_A = U_B.$$

Example 4.16 N distinguishable particles in thermal contact with a heat reservoir

N distinguishable particles have single-particle wave functions corresponding to energies +E and -E only. This system of particles is placed in thermal contact with a large reservoir $\mathcal{R}$ at a temperature τ.

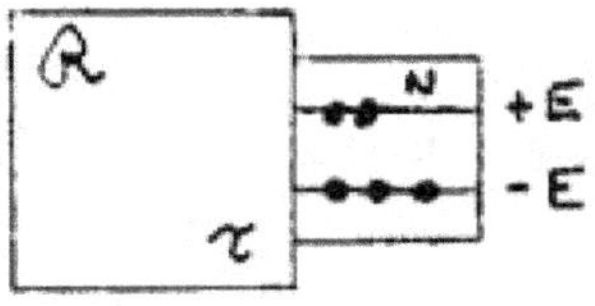

(a) Write the partition function Z, and from this calculate the average total energy U of the system of N particles after thermal equilibrium is established at the temperature τ.
(b) Find the change dU in the energy if the temperature of the reservoir is changed by $d\tau$, in terms of τ, N, E.
(c) Find the free energy F and the entropy σ as a function of τ, N, E.

Answer

$Z = \exp(-E/\tau) + \exp(E/\tau) = 2\cosh(E/\tau)$

$$U = N\left(\frac{\sum \varepsilon_s \exp\left(-\frac{\varepsilon_s}{\tau}\right)}{Z}\right) = N\left(\frac{-E\exp\left(\frac{E}{\tau}\right) + E\exp\left(-\frac{E}{\tau}\right)}{Z}\right)$$

$$= N\frac{-E2\sinh(E/\tau)}{2\cosh(E/\tau)} = -NE\tanh(E/\tau)$$

(b) $\frac{dU}{d\tau} = -E\frac{\left(\tanh\left(\frac{E}{\tau}\right)\right)N}{d\tau} = -E\,\mathrm{sech}^2(E/\tau)\cdot\frac{d(E/\tau)N}{d\tau} = -E^2\,\mathrm{sech}^2(E/\tau)\cdot\left(\frac{-1}{\tau^2}\right)N$

$$du = N\left[\left((E/\tau)\,\mathrm{sech}(E/\tau)\right)\right]^2 d\tau$$

(c) $F \equiv U - \tau\sigma$ *also* $F = -\tau\log Z\cdot N$

$$F = -\tau\log[2\cosh(E/\tau)\cdot N]$$

$\tau\sigma = U - F$

$$\sigma = \frac{U-F}{\tau} = \frac{-NE\tanh\left(\frac{E}{\tau}\right) + \tau\log\left[2\cosh\left(\frac{E}{\tau}\right)\right]\cdot N}{\tau}$$

$$= N\left(\log[2\cosh(E/\tau)] - \left(\frac{E}{\tau}\right)\tanh(E/\tau)\right)$$

Example 4.17 Work done by a rubber band

A rubber band is idealized by N links of length d, each link extending the band either to the right or to the left. The succession of choices gives the combined state of the system Assume the energy is independent of the configuration and that N>>1.

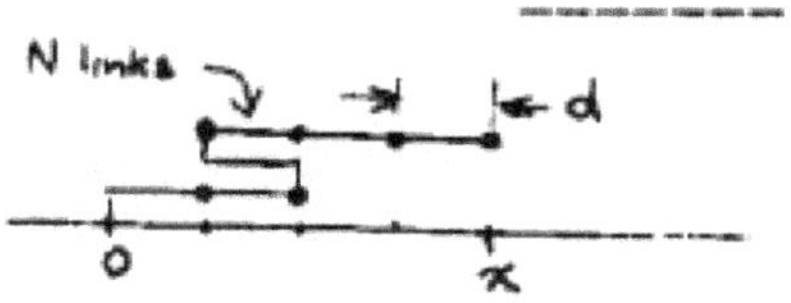

(a) Find the number of states with the right end at position x.

(b) Find he corresponding entropy.
(c) The rubber band is place in contact with a heat reservoir at temperature τ, and then stretched from x=0 to x=L. Find the mechanical work W done by the rubber band (should be negative).

Answer
(a) We have:

$$g(N,S) = \frac{N!}{(N/2+S)!\,(N/2-S)}, S = \frac{1}{2}\left(N_{right} - N_{left}\right) = \frac{x}{2d}$$

Thus

$$g(N,x) = \frac{N!}{\left({}^{N}\!/_{2} + \frac{x}{d2}\right)!\left({}^{N}\!/_{2} - \frac{x}{d2}\right)!}$$

(b) $\sigma(N,x) = \log g(N,x)$ (convention k=1 adopted throughout):
$\sigma(N,x) = \log[N!] - \log\left[\frac{1}{2}\left(N + \frac{x}{d}\right)\right]! - \log\left[\frac{1}{2}\left(N + \frac{x}{d}\right)\right]!$
Since $N \gg 1,\ g(N,S) = 2\left(\frac{2}{\pi N}\right)^{1/2} 2^N \exp(-2S^2/N)\ ,\ \ S = \frac{x}{2d}$
So

$$\sigma(N,x) = \log\left[2\left(\frac{2}{\pi N}\right)^{1/2} 2^N\right] - \frac{2\left(\frac{x}{2d}\right)^2}{N} = \log[2g(N,0)] - \frac{x^2}{2Nd^2}$$

(c) Let f = the external force, since energy is independent of configurations f must also be the force applied to the rubber band to conserve energy.

$$\frac{-f}{\tau} = \left(\frac{\partial\sigma}{\partial x}\right) = -\frac{2x}{2Nd^2} = -\frac{x}{Nd^2} \rightarrow\ f = \frac{x\tau}{Nd^2}$$

$$W = -\int_{o}^{L} f dx = -\frac{L^2\tau}{Nd^2}$$

Example 4.18 System of Fermi Particles
Consider a system of non-interacting spin-½ Fermi particles, characterised by a temperature T and chemical potential μ. The volume of the system is V, and the energy of the particle of momentum $\vec{p}$ is given by $E = c\sqrt{p^2 + m^2c^2}$.

a) Show that the average value of the total number of particles in the system is given by

$$< N >= \frac{8\pi m^3 c^3}{h^3} V \int_0^\infty \frac{\sinh\theta \cosh\theta\, d\theta}{e^{-\beta\mu + \beta mc^2 \cosh\theta} + 1}$$

Where θ is defined through the relation $|\vec{p}| = \mathrm{mc}\sinh\theta$ and $\beta = 1/\mathrm{kT}$.

b) Find a similar expression for the average total energy <E>.
c) Find a similar expression for the average pressure <p>.

Answer

(a) $E_p = c\sqrt{p^2 + m^2c^2}$

Occupational number for states P: $(fermions)$

$$\bar{n}_p = \frac{1}{1 + e^{-(\mu - E_p)/kT}}$$

$$< N >= \int \bar{n}_p \frac{d\Gamma}{h^3} = \frac{V}{h^3} 2(4\pi) \int_0^{\infty} \bar{n}_p\, p^2 dp$$

Let $p = mc\sinh\theta$, $\beta = {}^1/_{kT}$, then

$E_p = mc^2(\sinh^2\theta + 1)^{1/2} = mc^2\cosh\theta$

$dp = mc\cosh\theta d\theta$

and

$$< N >= \frac{8\pi V}{h^3}\int_0^{\infty} \frac{(mc)^2\sinh^2\theta (mc)\cosh\theta d\theta}{1+\exp\left\{-\frac{mc^2}{kT}\cosh^2\theta\right\}} =$$

$$\frac{8\pi(mc)^3}{h^3} V \int_0^{\infty} \frac{\sinh^2\theta\cosh^2\theta d\theta}{\exp\{-\beta\mu+\beta mc^2\cosh\theta\}+1}$$

(b)

$$< E >= \int E_p \bar{n}_p \frac{d\Gamma}{h^3} = \frac{8\pi(mc)^4 cV}{h^3}\int_0^{\infty} \frac{\sinh^2\theta\cosh^2\theta d\theta}{\exp\{-\beta\mu + \beta mc^2\cosh\theta\} + 1}$$

(c)

$$pV = N_\mu - F = kT\sum \log\left(1 + e^{-\beta(\varepsilon-\mu)}\right)$$

$$= 8\pi\left(\frac{mc}{h}\right)^3 \int_0^{\infty} \frac{1}{\beta}\sinh^2\theta\cosh\theta d\theta \log[1 + \exp\{-\beta(mc^2\cosh\theta - \mu)\}]$$

Example 4.19 Dilute gas composed of heavy particles

Consider a dilute gas composed of heavy particles (of mass m_1 and light particles of mass m_0) in thermal equilibrium, with $m_0 \ll m_1$. The temperature T is chosen such that $m_0c^2 \ll kT \ll m_1c^2$. that is, the heavy particles are nonrelativistic, whereas the light particles are ultra-relativistic. Show that each ultra-relativistic particle has twice as much kinetic energy, on the average, as each non-relativistic particle. (Hint: $\int_0^\infty e^{-x}x^n dx = n!$)

Answer

Heavy particles: $m_1c^2 \gg kT$

$$Z_1 = \frac{1}{h^3}\int e^{-E/kT}\, d\Gamma = \frac{V}{h^3}\left[\int_{-\infty}^{\infty} e^{-P_x^2/2mkT}\, dp_x\right]^3 = \frac{V}{h^3}x\left(\sqrt{2\pi mkT}\right)^3$$

$$Z = \frac{Z_1^N}{N!} = \frac{V^N}{N!}\left(\frac{2\pi mkT}{h^2}\right)^{3/2N}$$

Light particles: $m_0c^2 \ll kT$ → ultrarealistic $E \approx pc$:

$$Z_1 = \frac{1}{h^3}V4\pi\int_0^\infty e^{-pc/kT}\, p^2dp = \frac{4\pi V}{h^3}\left\{\frac{-2}{-\left({}^{c}/_{kT}\right)}\right\}\int_0^\infty e^{-pc/kT}\, pdp$$

$$= \frac{4\pi V}{h^3}2\left(\frac{kT}{c}\right)\left(\frac{+kT}{c}\right)\int_0^\infty e^{-PC/kT}\, dp = \frac{8\pi V}{h^3}\left(\frac{kT}{c}\right)^3$$

So

$$Z = \frac{V^N}{N!}\pi^N\left(\frac{2kT}{hc}\right)^{3N}$$

Non-relativistic:

$$< K.E. > = < \frac{p^2}{2m} > = -\frac{\partial \log Z}{\partial \beta}$$

Relativistic:

$$< K.E. > = -\frac{\partial \log Z}{\partial \beta}$$

Non-relativistic:

$$\log Z = \log\left(\frac{V^N}{N!}\right) + \frac{3N}{2}\log\left(\frac{1}{\beta}\right) + \frac{3N}{2}\log\left(\frac{2\pi m}{h^2}\right)$$

$$-\frac{\partial \log Z}{\partial \beta} = -\frac{3N}{2}\frac{-{}^{1}/_{\beta^2}}{{}^{1}/_{p}} = \frac{3}{2}NkT$$

$$< K.E. > = \frac{3}{2}NkT$$

Relativistic:

$$\log Z = \log\left(\frac{V^N}{N!}\right) + \log \pi^N + \log\left(\frac{2}{hc}\right)^{3N} + 3N\log\left(\frac{1}{\beta}\right)$$

$$-\frac{\partial \log Z}{\partial \beta} = -3N\left(\frac{-1/\beta^2}{1/\beta}\right) = 3NkT$$
$$\langle K.E.\rangle_{rel.} = 3NkT = 2\langle K.E.\rangle_{non.rel.}$$

Example 4.20 Equilibrium with Particle Reservoir

Consider a molecule m in equilibrium with a reservoir which contains both energy ("heat") and particles of a certain kind (to be called "reservoir particles" or R-particles); these R-particles move freely in and out of the reservoir and may form bound states or complexes with M. Let M-nR denote a bound state of M with n R-particles, and consider four non-degenerate quantum levels as follows:

$$\begin{cases} level\ 1: M - R;\ 0\ eV, \\ level\ 2:\ \ M;\ 0.02\ eV, \\ level\ 3: M - 3R;\ 0.03\ eV, \\ level\ 4: M - 4R;\ 0.05\ eV \end{cases}$$

a) In a system of many such molecules, in equilibrium with the reservoir, find the expected population at level 4 if we are given the following information on the expected populations of the other three levels:

$$\begin{cases} 2^{20}\ \ at\ level\ 1, \\ 2^{17}\ \ at\ level\ 2, \\ 2^{19}\ \ at\ level\ 3. \end{cases}$$

b) Also find the absolute temperature T and chemical potential μ of the R-particles. (Data: 1eV=1.602 x 10^{19} Joule, k=1.381 x 10^{-23} Joule/ Kelvin).

Answer

$$Z_1 = e^{-E_1/kT}e^{-\mu/kT} + e^{-E_2/kT} + e^{-3\mu/kT} + e^{-S_u/kT}e^{-4\mu/kT}$$
$$= e^{-\mu/kT} + \alpha^2 + \left(\alpha e^{-\mu/kT}\right)^3 + \alpha^5 e^{-4\mu/kT}$$
$$Z_N = Z_1^N$$
$$P(1) = e^{-\mu/kT}/Z_1 \qquad NP(1) = 2^{20}$$
$$P(2) = \alpha^2/Z_1 \qquad NP(2) = 2^{17}$$
$$P(3) = \left(\alpha e^{-\mu/kT}\right)/Z_1 \qquad NP(3) = 2^{19}$$
$$P(4) = \left(\alpha^5 e^{-4\mu/kT}\right)/Z_1$$
$$\frac{NP(1)}{NP(2)} = \frac{2^{20}}{2^{17}} = 2^3 = \boxed{8 = \frac{e^{-\mu/kT}}{\alpha^2}} \rightarrow \begin{array}{c} 8\alpha^2 = e^{-\mu/kT} \\ \boxed{8^3\alpha^7 = 4} \end{array}$$

$$\frac{NP(1)}{NP(3)} = \boxed{2 = \frac{e^{-\mu/kT}}{\alpha^3 e^{-3\mu/kT}}}$$

$$\boxed{\alpha = 4e^{3\mu/kT}}$$

$$\frac{NP(3)}{NP(2)} = 4 = \frac{\alpha^3 e^{-3\mu/kT}}{\alpha^2} = \alpha e^{-3\mu/kT} \rightarrow \alpha^2 = 16e^{6\mu/kT}$$

$$8 = \frac{1}{16}e^{-7\mu/kT}$$

$$\frac{NP(4)}{NP(3)} = \frac{\alpha^5 e^{-4\mu/kT}}{\alpha^3 e^{-3\mu/kT}} = \alpha^2 e^{-\mu/kT}$$

$$\alpha^7 = \frac{4}{8^3} = \frac{4}{8}\frac{1}{64} = \frac{1}{128} = \frac{1}{2^7} \quad 2^7 = 4.4.4.2$$

$\boxed{\alpha = {}^1/_2}$!

$e^{-\mu/kT} = 2 \rightarrow \boxed{\mu = -kT\ln 2}$

(6)

$$\frac{NP(4)}{NP(3)} = \left(\frac{1}{2}\right)^2 2 = \frac{1}{2}$$

$$NP(4) = 2^{19}\left(\frac{1}{2}\right) = 2^{18}$$

(a) Expected population at level 4 is 2^{18}

$e = 2.7$

$\ln 2 = y$

$2 = e^y$

(b) $absolute\ temp.: \frac{1}{2} = \alpha = e^{-(.01)eV/kT}$

$\ln 2 = .673 = .7$

$-\ln 2 = -(.01)\, eV/kT$

$$T = \frac{1}{\ln 2}(.01)\frac{1.602x10^{-19}J}{1.381x10^{-23}\, J/k}$$

$\boxed{T = .167\ {}^0k}$

Chemical potential:

$\mu = -(1.381x10^{-23}\, J/k^0)(167^0k)(.693)$

$\boxed{\mu = -1.6x10^{-21}J}$

Example 4.21 Para versus Ortho Hydrogen

The two protons in an H_2 molecule may be in a singlet or a triplet spin state giving rise to a "para" or "ortho" type H_2 molecule, respectively.

a) Derive an expression for N_{ortho}/N_{para}, the equilibrium population ratio between the two types of molecules in an H_2 gas, at moderately low temperature T (so it is only necessary to consider the rotational energy levels. Denote by I the moment of inertia of H_2).
b) Given the following in SI units: Proton rest mass = 1.7×10^{-27}, separation between protons = 0.74×0^{-10} h=6.63×10^{-34} , k=1.38×10^{-23}. Calculate $\frac{N_{ortho}}{N_{para}}$ at $T = 30^0 K$, to 1% accuracy.

Answer

For the internal partition function we have

$j(\tau) = g_e x \Gamma_{nc-rot} v(T)$

$r_{nuc-rot}(\tau) = (S_n + 1)(2S_n + 1)r_{odd} + S_n(2S_n + 1)r_{even}(\tau)$

$$r_{even} = \sum_{leven} (2l + 1)\exp\left\{\frac{-l(l+1)\hbar^2}{2IkT}\right\} r_{odd} \sum_{lodd} (2l + 1)\exp\left\{\frac{-l(l+1)\hbar^2}{2IkT}\right\}$$

(a) $$\frac{N_{ortho}}{N_{pus}} = \frac{3\sum_l (2l+1)\exp\left\{\frac{-l(l+1)\hbar^2}{2IkT}\right\}}{\sum_{l=12}(2l+1)\exp\left\{\frac{-l(l+1)\hbar^2}{2IkT}\right\}}$$

(b) $m_p = 1.7x10^{-27} kg$

$r_e = 0.74x10^{-16} m$

$h = 6.63x10^{-34} m^2 kg/s$

$K = 1.38x10^{-23} J/k^o$

$T = 30^o k$

SI units not egs

$KT = E \;\; k = J/k^o$

$J = kg\, m^2/s^2$

$T = 3$

$I = 2mp\left(\frac{r_e}{2}\right)^2 = \frac{1}{2} mpr_e^2$ etc.

$(10^{-27}(10^{-10})^2(10^{-23})10^2)/10^3(10^{-34})^2$

$10^{-68}/10^2 10^{-68} \approx 10^{-2}$

$\exp\{-l(l+1)10^2\} \approx 1 - l(l+1)10^2$

$\frac{N_{ortho}}{N_{par}} \cong 9\exp\left\{\frac{-\hbar^2}{IkT}\right\} \approx 0$!

Example 4.22 Gas of Rigid Diatomic Particles

A box of volume V contains a gas consisting of N rigid diatomic molecules. Each molecule is free to rotate about its center of mass, and the center of mass is free to undergo translational motion. Assume that the kinetic energy of rotation about the center of mass, ϵ_{rot}, and the kinetic energy of translation of the centre of mass, ϵ_{trans}, are addictive and independent. At typical temperatures, the translation can be treated classically, but the rotation of a molecule must be treated quantum mechanically. The rotational energy levels of a molecule are given by

$$\epsilon_{rot}(\ell) = \frac{\ell(\ell+1)\hbar^2}{2I} \quad (\ell = 0,1,2,\dots)$$

where I is the rotational inertia. For any particular value of the quantum number ℓ there are $2\ell + 1$ microstates of the molecule having the same energy $\epsilon_{rot}(\ell)$.

a) Write the rotational part of the partition function, ϵ_{rot}, of a molecule. (Do not perform the summation).
b) At very high temperatures, one can regard ℓ as a continuous variable running from 0 to ∞ and replace the sum by an integral. Perform the integration to obtain ϵ_{rot} at very high temperature.
c) At very high temperature, find the mean energy of the gas, including the contribution of the translational kinetic energy.
d) Obtain the heat capacity (at very high temperatures) of the gas at constant volume.

Answer

$V, N\ diatomic \rightarrow rot, indep\ of\ trans$

(a) $Z_{rot} = \sum_{l=0}^{\infty}(2l+1)\, e^{-\{l(l+1)\hbar^2/2IkT\}}$ $\qquad \alpha = \frac{\hbar^2}{2IkT}$

(b) $Z_{rot,high} = \int_0^{\infty}(2l+1)\, e^{-\alpha l(l+1)} dl$

$(2l+1)^2 = 4l^2 + 4l + 1$

$Z = \int_0^{\infty}(2l+1)\, e^{\left\{-\frac{\alpha}{4}(2l+1)^2 + \frac{\alpha}{4}\right\}} dl$

$L = 2l + 1$

$dL = 2dl$

$Z = \int_0^{\infty} \frac{1}{2} L e^{-\frac{\alpha}{4}L^2} e^{\frac{\alpha}{4}} dL$

$= \frac{1}{2} e^{\alpha/4} \int_1^{\infty} e^{-(\alpha/4)L^2}\, L dL$ $\qquad L' = \sqrt{\left(\frac{\alpha}{4}\right)} L$

$= \frac{1}{2} e^{\alpha/4} \left(\frac{4}{\alpha}\right) \frac{1}{2} \int_{r_4}^{\infty} e^{-(L\prime)^2}\, d(L')^2$

$= \frac{e^{\alpha/4}}{\alpha} e^{-\alpha/4} = \left(\frac{1}{\alpha}\right)$

$$\boxed{Z_{high} = \frac{2IkT}{\hbar^2}}$$

(c) Very high temp: $\bar{\varepsilon}_{tans} = \frac{3}{2} NkT$

$$\mathcal{Z}_{rot} = \sum_{l=0}^{\infty}(2l+1)\, e^{-\varepsilon\beta} - \frac{\partial \mathcal{Z}_{rot}}{\partial \beta} = \sum_{l=0}^{\infty}(2l+1)\, \varepsilon e^{-\varepsilon\beta}$$

$$\bar{\varepsilon}_{rot} = -\frac{\partial \log \mathcal{Z}_{rot}}{\partial \beta}$$

(3)

(c) $\bar{E}_{rot} = -\frac{\partial}{\partial \beta}\log\left(\frac{2I}{\hbar^2 \beta}\right) = -\frac{\left(\frac{2I}{\hbar^2}\right)\left(-\frac{1}{\beta^2}\right)}{\left(\frac{2I}{\hbar^2 \beta}\right)} = \frac{1}{\beta} = kT$

N molecules $E_{rot,total} = NkT$, so

$\bar{E}_{Total} = \frac{5}{2}NkT$

(d) $C_v = \frac{\partial \bar{E}}{\partial T} = \frac{5}{2}Nk$

Example 4.23 N atoms of mass m is contained in a volume V at temperature T

An ideal gas of N atoms of mass m is contained in a volume V at temperature T. Some of those atoms are absorbed on the surface of the container on which they are free to move and can form a two dimensional ideal gas. Call ε_0 the binding energy which holds a molecule on the surface.

Find at equilibrium the mean number n' of molecules adsorbed per unit area of the surface when the mean pressure of the gas in the container is p. You may treat the three dimensional and two dimensional ideal gas classically but do not forget to take into account the indistinguishability of the particles. (Hint: Make use of chemical potentials)

Answer

We have

$\mathcal{Z}_{voi} = \int e^{-\mathcal{H}/kT} \frac{d\Gamma}{h^{N'}N'!}$ $\qquad\qquad N' = N - n'$

$= \frac{1}{h^{3N'}}\frac{1}{N'!}V^{N'}(2\pi mkT)^{3N'/2}$

$\mathcal{Z}_{surface} = \frac{1}{h^{2n'}n'!}S^{n'}(2\pi mkT)^{n'}e^{\varepsilon_0 n'/kT}$

$\mathcal{Z} = \mathcal{Z}_{0t}\mathcal{Z}_{sur} = \frac{1}{h^{3N'}h^{2n'}}\frac{1}{N'!}\frac{1}{n'!}V^{N'}S^{n'}(2\pi mkT)^{\frac{3N'}{2}+n'}e^{\varepsilon_0 n'/kT}$

At equil. $F = min, so$

$\frac{\partial \log \mathcal{Z}}{\partial n'} = 0$

$(3N' + 2n')\log h = -(3N - 3n' + 2n)\log h = (-3N + n')\log h$

$\log Z = -\log(h^{3N'} h^{2n'}) - \log N'! - \log n'! + N' \log V + n' \log S +$
$\left(\frac{3}{2}N' + n'\right) \log(2\pi mkT) - [N' \log N' - N'] \qquad - [n' \log n' -$
$n'] \qquad + \varepsilon_0 n'/kT$

$\frac{\partial \log Z}{\partial n'} = -\{-\log N' + (-1)(-1)\} - \{\log n' + 1 - 1\} = \log V + \log S +$
$\left(-\frac{3}{2} + 1\right) \log(2\pi mkT) + \varepsilon_0/kT$

$= \log N'/n' + \log s/v - \frac{1}{2}\log\left(\frac{2\pi mkT}{h^2}\right) + \frac{\varepsilon_0}{kT} = 0$

$\log\left(\frac{\bar{n}'}{N-\bar{n}'}\right) = \log\left(\frac{Sh}{v(2\pi mkT)^{1/2}}\right)^{1/2} e^{\varepsilon_0/kT}$

$\frac{mrV}{m^2V^2} = [L]$

$\frac{\bar{n}'}{N-\bar{n}'} = \frac{S}{V}\left(\frac{h^2}{2\pi mkT}\right)^{\frac{1}{2}} e^{\varepsilon_0/kT}$

$\bar{P}_v = NkT$

$\frac{N-\bar{n}'}{\bar{n}'} = \frac{V}{S}\left(\frac{2\pi mkT}{h^2}\right) e^{-\varepsilon_0/kT} = \frac{N}{\bar{n}} - 1 \qquad \frac{N}{\bar{n}} = 1 + \frac{V}{S}(\quad)^{1/2} e^{-\varepsilon_0/kT}$

$$\boxed{\bar{n} = N\left\{1 + \frac{V}{S}\left(\frac{2\pi mkT}{h^2}\right)^{1/2} e^{-\varepsilon_0/kT}\right\}^{-1}}$$

$\bar{p}V = NkT \quad \rightarrow N = \frac{\bar{p}V}{kT}$

$\bar{n} = \frac{\bar{p}V}{kT}\left\{1 + \frac{V}{S}\left(\frac{2\pi mkT}{h^2}\right)^{1/2} e^{-\varepsilon_0/kT}\right\}^{-1}$

Or

$\bar{n} = N\left\{1 + \frac{NkT}{\bar{p}S}\left(\frac{2\pi mkT}{h^2}\right)^{1/2} e^{-\varepsilon_0/kT}\right\}^{-1}$

Example 4.24 A, B atoms on a line

A, B atoms occupy a+1 sites along a line, providing n nearest neighbor contacts. (Each site is occupied by either an A atom or a B atom.) For each AB or BA adjacency, there is an energy of ε (two cases: binding if $\varepsilon < 0$, repulsion if $\varepsilon > 0$), independently of what happens elsewhere; AA and BB adjacencies count for no energy.

(a) Give an argument that demonstrates that in thermodynamic equilibrium, the mean number of sites that are occupied by A atoms is equal to the mean number of sites that are occupied by B atoms (each number = ½ (n+1)).

(b) Find the specific heat C as a function of $\beta = \frac{1}{T}$.
(c) Show that C satisfies the third law of thermodynamics, both if $\varepsilon > 0$ and also if $\varepsilon < 0$.

HINT: By not requiring either an excess of A's or B's, examiner has made it unnecessary for you to introduce a chemical potential, thereby making the problem easier, thus part (a) is there to help you see how easy it is. Indeed, if you introduce ε bonds at random, you will be well on your way.

Answer

$w = \frac{n}{n_\uparrow! n_\downarrow!}$ $\quad possibilities \quad n = n_\uparrow + n_\downarrow$

$E = n_\uparrow = E/\varepsilon \;\; \rightarrow n_\uparrow = E/\varepsilon$

$n_\downarrow = n - E/\varepsilon$

$Z = \sum_{E=0}^{n\varepsilon} \frac{n!}{(E/\varepsilon)!(n-E/\varepsilon)!} e^{-E/kT}$

$= \frac{n!}{n!0!}\left(e^{\varepsilon/kT}\right)^n + \frac{n!}{(n-1)!1!}\left(e^{-\varepsilon/kT}\right)^{n-1}$

$Z = \left(e^{-\varepsilon/kT} + 1\right)^n$

$F = -kT \log Z = -nkT \log\left(e^{-\varepsilon/kT} + 1\right)$

$dF = -SdT - pdV$

$dF = -SdT - PdV$

$S = -\left(\frac{\partial F}{\partial \tau}\right)$

(a) Consider an external bath of A and B atoms outside the line, now for either $\varepsilon < 0 \;\; or\; \varepsilon > 0$, the result is simple to explain:
The number of states $w = \frac{n!}{n_A! n_B!}$ is maximized when $\bar{n}_A = \bar{n}_B$, one can only presume that for a macroscopic n in thermodynamic equilibrium the density of states weighing factor that follows for outweighs the weighing factor due to the energy from the commercial term $\left(e^{-\varepsilon/kT}\right)$.

Thus, $\bar{n}_A = \bar{n}_B = \frac{1}{2}(n+1)$

(b)

$c_V = T\left(\frac{\partial S}{\partial \tau}\right)_V$

$S = -\left(\frac{\partial F}{\partial T}\right) = nk \log\left(e^{-\varepsilon/kT} + 1\right) + nkT\left(\frac{-\frac{\varepsilon}{k}\left(-\frac{1}{\tau^2} e^{-\varepsilon/kT}\right)}{e^{-\varepsilon/kT}+1}\right)$

$\frac{\partial S}{\partial T} = nk\left(-\frac{\varepsilon}{k}\right)\left(-\frac{1}{T^2}\right)\frac{e^{-\varepsilon/kT}}{e^{-\varepsilon/kT}} - nk\left(\frac{\varepsilon}{k}\right)\left(\frac{1}{T^2}\right)(\ldots) + 1[\,]$

$$= \frac{n\varepsilon}{T}\left(\frac{\partial}{\partial T}\right)\left(\frac{1}{1+e^{\varepsilon/kT}}\right) = \frac{n\varepsilon}{T}\left[-\left(\frac{1}{k}\frac{-1}{T^2}\right)\left(1+e^{\varepsilon/kT}\right)e^{\varepsilon/kT}\right]$$

$$u = -T^2\frac{\partial(F/\tau)}{\partial\tau} = -T^2 nk\frac{\partial\log(e^{-\varepsilon/kT}+1)}{\partial\tau}$$

$$= nkT^2\left[\frac{-\frac{\varepsilon}{k}\left(-\frac{1}{T^2}\right)e^{-\varepsilon/kT}}{e^{-\varepsilon/kT}+1}\right]$$

$$= n\varepsilon\left(1+e^{\varepsilon/kT}\right)^{-1}$$

$$C_v = \left(\frac{\partial u}{\partial\tau}\right)_v = n\varepsilon\left[-\left(1+e^{-\varepsilon/kT}\right)^{-2}e^{\varepsilon/kT}\left(\frac{\varepsilon}{k}\right)\left(-\frac{1}{\tau^2}\right)\right]$$

$$C_V = \frac{n\varepsilon^2}{kT^2}\frac{e^{\varepsilon/kT}}{\left(1+e^{\varepsilon/kT}\right)^2} = nk\varepsilon^2\beta^2\left(\frac{e^{\varepsilon\beta}}{\left[1+e^{\varepsilon\beta}\right]^2}\right)$$

(c) $\lim_{T\to 0} C_V = 0$ is an expression of the 3rd law $T \to 0$

So, consider $\varepsilon > 0\,, then\, T \to 0 \Rightarrow \beta \to \infty$ $\quad C_v \cong nk\varepsilon^2\beta^2\left(\frac{e^{\varepsilon\beta}}{e^{2\varepsilon\beta}}\right)$

$\cong nk(\varepsilon\beta)^2 e^{-\varepsilon\beta} \to 0$

For $\varepsilon < 0\ (T \to 0\,, \beta \to \infty)$ $\quad C_v = nk(\varepsilon\beta)^2 e^{\varepsilon\beta} \to 0$

Example 4.25 System with three non-interacting particles

Consider a system consisting of three non-interacting particles each of which can assume the energies $\epsilon_i = \epsilon_0 n_i$, $n_i = 0,1,2,3,\ldots$

If $\beta\epsilon_0 = \frac{1}{2}$ determine the most probable value and the expectation value of energy of the system. Here $\beta = \frac{1}{kT}$ where k is Boltzmann constant and T is the temperature.

Answer

$$Z_1 = \sum_{n_1} e^{-\varepsilon_1/kT} = \sum_{n_1=0}^{\infty} e^{-n_1\varepsilon_0/kT} = \frac{1}{1-e^{-\varepsilon_0/kT}}$$

$$Z^{(3)} = \left(\frac{1}{1-e^{-\varepsilon_0/kT}}\right)^3$$

$$F = -kT\log Z$$

$$dF = -PdV = sdT$$

$$u = -T^2\frac{\partial(F/T)}{\partial T} \qquad \left(\frac{\partial F}{\partial T}\right) = -S$$

$$u = +\tau^2 k\frac{\partial\log Z}{\partial T} = -3kT^2\frac{\partial\log[1-e^{-\varepsilon_0/kT}]}{\partial T}$$

$$= -3kT^2\left\{\frac{\frac{\varepsilon_0}{k}\left(-\frac{1}{T^2}e^{-\varepsilon_0/kT}\right)}{1-e^{-\varepsilon_0/kT}}\right\}$$

$$u = \left(\frac{3\varepsilon_0}{e^{\varepsilon_0/kT}-1}\right) \leftarrow expectation\ of\ the\ energy \qquad \boxed{u = \frac{3\varepsilon_0}{\sqrt{e}-1}}$$

$P(\varepsilon) = \frac{g_\varepsilon e^{-\varepsilon/kT}}{Z^{(3)}}$

$P(\varepsilon) = g_\varepsilon e^{-\varepsilon_0/kT}\,\frac{1}{Z}$

$g_\varepsilon = ?$
$n_1 = 0\ to\ N$
$n_2 + n_3 = N - n_1 \qquad n_2 = \{0 \ldots N - n_1\}$

$\varepsilon = \varepsilon_1 + \varepsilon_2 + \varepsilon_3$
$= n_1\varepsilon_0 + n_1\varepsilon_0 + n_3\varepsilon_0$
$= (n_1 + n_2 + n_3)\varepsilon_0$
$\varepsilon = N\varepsilon_0$
$n_1 = 0 \ldots j \ldots . N$
$n_2 + n_3 = N - j$

$$g_N = \sum_{j=0}^{N} (N - j + 1)$$

$$\sum_{j=0}^{N} j(N - j) + N = N^2$$

$$= N(N+1)^2 - \sum_{j=0}^{N} j$$

$= (N+1)^2 - \frac{1}{2}N(N+1) = \frac{1}{2}N^2 + \frac{3}{2}N + 1$
$= \frac{1}{2}(N+1)(N+2)$

$g_{\varepsilon_{\square}} = \frac{1}{2}(N+1)(N+2)$
Most prob. Value of energy for
$\frac{\partial P(\varepsilon)}{\partial \varepsilon} = 0$ (for large N, treating the differentive eqn. etc., as a differential eqn.)
$\varepsilon = N\varepsilon_0$
$\frac{\partial}{\partial N}\{(N+1)(N+2)\exp[N\,\varepsilon_0/kT]\} = 0$
$(2N+3)\exp\{\ldots\} - \frac{\varepsilon_0}{kT}(N+1)(N+2)\{\ldots\} = 0$
$(2N+3) - \frac{1}{2}(N^2 + 3N + 2) = 0$
$\varepsilon_0\beta \equiv \frac{1}{2}$
$-\frac{1}{2}N^2 + 2N - \frac{3}{2}N + 3 - 1$
$-\frac{1}{2}N^2 + \frac{1}{2}N + 2 = 0$
$N^2 - N - 4 = 0 \quad \rightarrow \quad N = \frac{1 \pm \sqrt{1-4(-4)}}{2} = \frac{1+\sqrt{17}}{2}$
$N \cong \frac{5}{2} \qquad \left[\frac{5}{2}\right] = 3\ \ since\ \sqrt{17}\ \ not\ \sqrt{16}$
So, the most probable energy is
$\boxed{E = 3\varepsilon_0}$

Example 4.26 Plasma Ionization

By heating a gas of N cesium atoms to sufficiently high temperature one-may generate plasma consisting of an appreciable number of dissociated positive ions and electrons. The ionization energy of the cesium atoms is 3.89eV and its atomic weight is 132.9. Assume that the atoms of cesium form an ideal gas and that the interaction between the ions and electrons after dissociation can be neglected.

a) Using the canonical distribution and the ratio of the probability P_D that an atom is found dissociated into an electron and an ion, to the probability that it is found P_U undissociated.

b) Express the degree of association $\frac{n}{N}$, where n is the number of atoms of dissociated and assumed to be much smaller than N in terms of T and the mean presence of the gas.

(Hint: Show that $\frac{P_0}{P_0} = \frac{n^2}{N-n}$)

c) Suppose that one heats the cesium to 1200K and pressure of $10^2\ \frac{N}{m^2}$. Calculate the percentage of gas ionized under these conditions.

Answer

$$Z_{undissociated} = \int e^{-\mathcal{H}/kT} \frac{d\Gamma}{h^N N!} \qquad \mathcal{H} = \frac{P^2}{2m}$$

$$= \frac{V^N}{h^N N!}\left[\int_0^\infty e^{-\frac{P_x^2}{2\pi mkT}}\, dp_x\right]^{3N}$$

$$= \frac{V^N}{h^N N!}(2\pi mkT)^{3N/2}$$

$$Z_{dissociated} = Z_{C_s} Z_e = \left[\frac{V^2}{h^2}(2\pi kT)^3 (m_e n_s)^{3/2} e^{-\emptyset/kT}\right]^n \frac{1}{(n!)^2}$$

$$\mathcal{H} = \frac{\vec{P}_{C_S}^0}{2m_{C_S}} + \frac{P_\emptyset^2}{2m_C} - \emptyset$$

$$P_0 = \frac{Zd}{Z_u \cdot Z_d}\ ;\ P_u = \frac{Z_u}{Z_u \cdot Zd} \qquad for\ N = 1\ , n = 1$$

$$\frac{P_o}{P_u} = \frac{Zd}{Z_u} = \left[\frac{V^2}{h^2}(2\pi kT)^3 \left(m_e - m_{C_S^+}\right)^{3/2} e^{-\emptyset/kT}\right]$$

$$\left[\frac{V}{h}(2\pi kT)^{3/2} 2m_{C_s}^{3/2}\right] \qquad m_{CS} \cong m_{CS}^T$$

(a) $\frac{P_o}{P_u} = \frac{V}{h}(2\pi kTm_c)^{3/2} \exp\{-\emptyset/kT\}$

(b) *degree of dissociation*

$$\frac{\partial}{\partial n}[\log(Z_u Zd)] = 0$$

$$Zu = \left[\frac{V}{h}(2\pi mkT)^{3/2}\right]^{(N-n)}_{\frac{1}{(N-n)!}}$$

$$\frac{\partial}{\partial n}\left\{(N+n)\log\left[\frac{V}{n}(2\pi mkT)^{3/2}\right] - (N-n)\log(N-n) + (N-n)\right\}$$

$$+N\log\left[\frac{V^2}{h^2}(2\pi kT)^3(m_e m_{Cs})^{3/2}\right] - \frac{n\emptyset}{kT} - n\log n + 2n = 0$$

$n \to \vec{n}$ that satisfies relation

$$-\log\left[\frac{V}{h}(2\pi mkT)^{3/2}\right] + \log\left(\frac{N-n}{n^2}\right) + \log\left[\frac{V^3}{h^2}(2\pi kT)^3(m_e m_{cs})^{3/2}\right] - \frac{\emptyset}{kT}$$

$$\frac{\vec{n}^2}{N-\vec{n}} = \frac{V}{h}(2\pi m_e kT)^{3/2}\exp(-\emptyset/kT)$$

So, $\frac{P_0}{P_u} = \frac{\vec{n}^2}{N-\vec{n}^2}$

$\vec{n} \ll N$ so,

$$\frac{\vec{n}^2}{N} = \frac{V}{h}$$

$$\left(\frac{\vec{n}}{N}\right) \cong \sqrt{\frac{V}{Nh}}(2\pi m_e kT)^{3/4}\exp(-\emptyset/2kT)$$

$$PV \cong NKT \qquad (\vec{n} \ll N)$$

$$V = \frac{NkT}{P}$$

$$\left(\frac{\vec{n}}{N}\right) = \sqrt{\frac{kT}{hP}}(2\pi m_e kT)^{3/4}\exp(-\emptyset/2kT)$$

Example 4.27 A string of elastomer

A string of elastomer can be considered as a single macromolecule consisting of N sections of length a oriented in any way whatever in space. The number of possible orientations within a solid angle dΩ. What is the length of a strip as a function of the applied tension τ? (Hint: it is a good approximation along the direction of the strip and are equal in magnitude to the applied tension τ)

Answer

$$Y = \int_{l=0}^{Na} e^{+BlF}\,dl \underbrace{\int \dots \int d\omega_1 \dots d\omega_n}_{l=\sum_{j=1}^{N} n<l+dl}$$

$$= \int \dots \int e^{+\beta(\sum_i a\cos\theta_i)F}\,dw_1 dw_2 \dots dw_N = \left[\int e^{+\beta\, a\cos\theta F}\sin\theta d\theta d\emptyset\right]^N$$

$$= \left[2\pi - \frac{1}{a\beta F}2\sinh(a\beta F)\right]^N \qquad \begin{aligned} u &= +a\beta F\cos\theta \\ du &= a\beta F\sin\theta d\theta \end{aligned}$$

$$= \left[\frac{2\pi}{a\beta F}2\sinh(a\beta F)\right]^N$$

$$Y = \left[4\pi \frac{\sinh(\alpha\beta F)}{\alpha\beta F}\right]^N$$

$$F = \tau$$

$$\bar{L} = \frac{1}{(-\beta)} \frac{1}{Y} \frac{\partial Y}{\partial F} = -\frac{1}{\beta} \frac{\partial}{\partial F} \log Y = NkT \frac{\partial}{\partial F}\left[\log\left\{\frac{\sinh(\alpha\beta F)}{\alpha\beta F}\right\}\right]$$

Example 4.28 Partition function of simple harmonic oscillator

(a) Calculate the partition function Z of a one-dimensional simple harmonic of frequency ω. Treat the oscillator as a quantum mechanical system.

(b) Use (Einstein's) single frequency model of a solid containing N atoms, and the result of part (a), to find an expression for the heat capacity of the solid at temperature T.

(c) Suppose that the frequency ω of the oscillations in the above solid depends on the volume V of the solid. Find the equation of state of the solid if $\omega = \omega_0 + b(V - V_0)^6$, where $\omega_0, b,\ V_0$ are constant.

Answer

a) $Z = \sum_l e^{-E_l/kT}$ $\qquad E_l = \left(l + \frac{1}{2}\right) hv$

$$Z = e^{-\frac{1}{2}hv/kT} \quad \sum_n e^{nhv/kT} = e^{-\frac{1}{2}hv/kT}\left\{\frac{1}{1-e^{hv/kT}}\right\}$$

$$= \left[2\sinh\left(\frac{hv}{2kT}\right)\right]^{-1}$$

$$hv = \hbar\omega$$

$$Z = \left[2\sinh\left(\frac{hv}{2kT}\right)\right]^{-1}$$

b) Solid with N atoms: $Z = \left[2\sinh\left(\frac{h\omega}{2kT}\right)\right]^{-N}$

$$F = -kT\log Z = +NkT\log\left[2\sinh\left(\frac{hv}{2kT}\right)\right]$$

$$u = -T^2 \frac{\partial\left(\frac{F}{\tau}\right)}{\partial(T)} = -NkT^2\left\{\frac{2\cosh\left(\frac{\hbar\omega}{2kT}\right)\left(-\frac{\hbar\omega}{2kT^2}\frac{1}{}\right)}{2\sinh\left(\frac{h\omega}{2kT}\right)}\right\}$$

$$u = \frac{1}{2}N\hbar w\coth\left(\frac{h\omega}{2kT}\right)$$

$$C_v = \left(\frac{\partial u}{\partial T}\right)_v = \frac{1}{2}N\hbar w\left\{1 + \coth^2\left(\frac{\hbar\omega}{2kT}\right)\right\}$$

$$U = -\frac{\partial}{\partial\beta}\log Z = N\frac{\partial}{\partial\beta}\log\left[\sinh\left(\frac{\hbar\omega}{2kT}\right)\right] = \frac{N\hbar\omega}{2}\coth\left(\frac{\hbar\omega}{2kT}\right)$$

c) $du = \tau ds - pdV$ $\qquad dF = -sdT - pdV$

$\left(\frac{\partial u}{\partial v}\right) = -P = N\hbar$ $\qquad \left(\frac{\partial F}{\partial v}\right) = -P$

$$w = \omega_0 + b(V - V_0)^6$$

$-P = NkT = \cosh\left(\frac{\hbar\omega}{2kT}\right)\left(\frac{\hbar}{2kT}\right)\frac{\partial\omega}{\partial v} = \frac{1}{2}N\hbar\coth\left(\frac{\hbar\omega}{2kT}\right)b6\,(V - V_0)^6$

$\boxed{P = -3N\hbar b(V - V_0)^5}\coth\left(\frac{\hbar\omega}{2kT}\right)$

$\frac{w}{v}\hbar = \frac{[s]^{-1}}{[L]^3}[K_g][L]^2[s]^{-1}$

$= \frac{[K_g]}{[s][L]^2}$

$P = \frac{F}{A} = [K][-]_{[s]^2}$

Example 4.29 Gas in spherical container with potential

A gas of N classical particles of mass m is at temperature T and is inside a spherical container of radius R. The gas particles within the sphere are subjected to a potential $V(r) = Ar^3$ where A is a constant and r is the distance from the center of the sphere.

(a) Find the density of particles at r.

(b) Find the total energy of the gas in terms of R, T, N and A.

(c) Suppose the sphere is made to rotate at frequency ω. Let n_{max} and n_{min} be the maximum and minimum density of particles, respectively, within the sphere. Find n_{max} and n_{min} for the case where A is negative.

(d) Suppose the gas consisted of indistinguishable fermions rather than classical particles. Would your analysis in parts (a) and (b) be applicable to the fermion problem? (Explain briefly) if your analysis is not applicable, briefly describe a correct method of solving parts (a) and (b) for fermions.

Answer

$v(r) = Ar^3$

$p(r)\alpha\ e^{-v(r)/kT} = e^{-Ar^3/kT}$

$\int p(r)\,r^2 dr d\Omega = 4\pi\int_0^R Q\,e^{-Ar^3/kT}T^2 dr$

$= 4\pi\left(\frac{-kT}{3A}\right)Q\int_0^{u_o} e^u du$ $\qquad u = -Ar^3/kT,\quad du = -\frac{3A}{kT}r^2 dr$

$= 4\pi\left(\frac{-kT}{3A}\right)\{-e^{-AR^3/kT} - 1\}Q$

$u_o = -AR^3/kT$

$= \frac{4}{3}\pi\left(\frac{kT}{A}\right)\{1 - e^{-AR^3/kT}\}Q = Nm$

(a) $p(r) = \frac{Nme^{-AR^3/kT}}{\frac{4}{3}\pi\left(\frac{kT}{A}\right)\{1-e^{-AR^3/kT}\}}$

(b) $E_\tau = \int (Ar^3) p(r) r^2 dr d\Omega + \frac{3}{2} kT(N)$ $\qquad \frac{1}{2}\dot{q}\frac{\partial}{\partial q} = \frac{1}{2}kT$

$= \frac{AQ}{Nm}\int_0^R r^3\, e^{-AR^3/kT} r^2 dr + \frac{3}{2}NkT$

$= \frac{AQ}{Nm}\left\{R^3\left(\frac{-kT}{3A}\right)e^{-AR^3/kT} - 3\left(\frac{-kT}{3A}\right)\frac{1}{Q}\right\}$

$= \frac{kT}{Nm}\left[1 - \frac{1}{3}QR^3 e^{-AR^3/kT}\right] + \frac{3}{2}kTN$

(c) IF= $mr\sin\theta\,\omega^2 \rightarrow RE = f(r,\theta) = ?$

$-\nabla f = F$

$\vec{F} = (+mr\sin\theta\,\omega^2)\{\sin\theta\,\hat{r} + \cos\theta\,\hat{\theta}\}$

$-\frac{\partial}{\partial r} f = +mr\, sin^2\theta\,\omega^2$ $\qquad -\frac{1}{r}\frac{\partial}{\partial r} f =$

$-mr\sin\theta\cos\theta\,\omega^2$

$f = -\frac{1}{2}mr^2\sin^2\theta\,\omega^2$ $\qquad f = -\frac{1}{2}mr^2\sin^2\theta\,\omega^2$

So, P.E. $= -\frac{1}{2}mr^2\sin^2\theta\,\omega^2$

$p(r)\alpha \exp\left\{|A|r^3 + \frac{1}{2}mr^2w^2\sin^2\theta\right\}/kT$

Min. max of $f(r)$ give aperture min. max of $p(r)$:

Maximum for $\theta = \frac{\pi}{2}$ $\qquad r = R$

Maximum for $\theta = 0$ $\quad , \quad r = 0$

$p\left(\theta = \frac{R}{2}, r = R\right)$ $\quad \alpha \exp\left\{\frac{|A|R^3 + \frac{1}{2}mR^2\omega^2}{kT}\right\}$

$p(\theta = 0, r = 0)$

(d) Depends on T

4.3 Exercises

Exercise 4.1

Three distinguishable noninteracting particles A,B and C are in a simple harmonic oscillator. potential with single–particle energies $\left(s + \frac{1}{2}\right)h\omega$. Measure energy E from the zero point energy hw/2 so that the three-particle state in the example has a total energy $E = 3h\omega$:

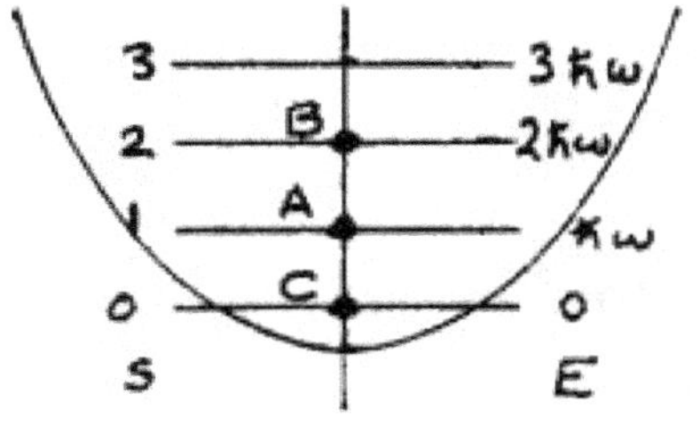

a) Show by means of diagrams similar to the one above all of the different possible such states with energy $E = 3\hbar\omega$.
b) Make sure you have all of the states in (a) by calculating the number possible of such states.

Exercise 4.2

A box contains 50 noninteracting distinguishable particles. Each particle is in one of two energy levels, E=0 or E=1, define the ``spin excess`` $S=(N_1-N_0)/2$, Where N_1 particles have the energy E=1 and N_0 have E=0.

a) Make a table with headings S, U, g, where U is the total energy, g is the multiplicity of states,
and σ is the entropy. Fill in the table for s=0 to s=5, using a calculator with a factorial button to make the calculations.
b) Sketch the shape of the function g versus s over the allowed range of s, keeping the scale accurate where you can.
c) Compare the entries in the table for σ with those obtained using the Stirling approximation.

Exercise 4.3

Consider two boxes A and B each identical to the box in Exercise above with 50 particles and two energy states. The two boxes are in thermal contact so that energy can transfer from one box to the other with the total energy U remaining constant during the transfer.

a) find the relation between S_A and S_B, the spin excess in box A and in box B , to maximize g(N,U), the combined multiplicity of the system.
b) Find $g(N, S_A,S_B)$ as function of U for the combined system, where $g(N,S_A,S_B)$ is for the maximum condition in part (a).
c) Find the temperature τ corresponding to U=25.
d) Why does the temperature become negative for U>50 , whereas in two boxes with the particles in a harmonic-oscillator potential the temperature τ defined in the same way is always positive?

Exercise 4.4

Four distinguishable particles have three energy states E=-1, E=0, and E=1. A typical state with total energy U=-1 is shown at left.

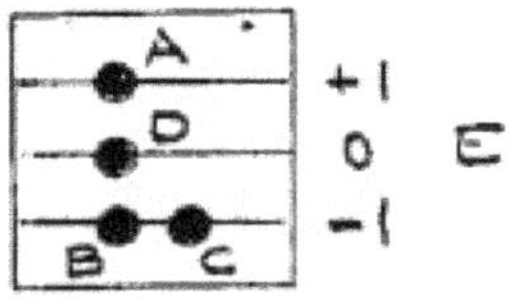

a) Find the multiplicity g and the entropy σ for U=-4,U=-3, and U=-2.
b) Sketch σ as a function of U over the allowable range and estimate a temperature τat the energy U=-3.
c) The 3- state system above is placed contact with a large reservoir at a temperature $\tau = .87$. Find the partition function Z_1 for one particle in the system.
d) From Z for the 4- particle system, find the total energy U.

Exercise 4.5

A spin-j particle in a uniform magnetic field B has energy levels $E_m = m\gamma B$, where γ is a constant and m is an integer in the range $-j<m<j$. Assume j>>1. A single particle in the field B is placed in thermal contact with a reservoir of temperature τ.

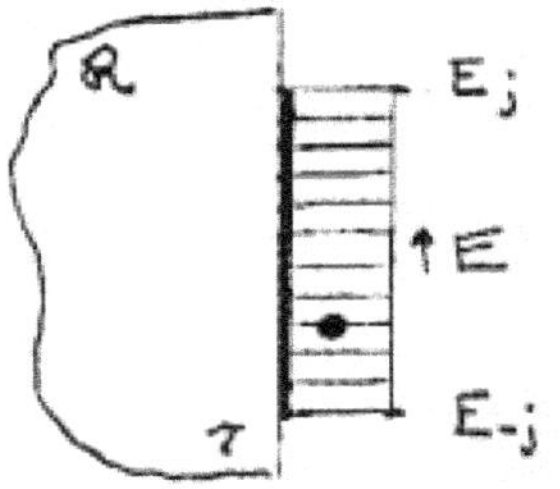

a) Find the partition function as a sum, and approximate this sum by means of an integral.
b) Find the average magnetization $\langle M \rangle$ defined in terms of the average energy by $\langle M \rangle = -\langle U \rangle / B$.
c) Find the value of $\langle M \rangle$ in the limits $\tau \to 0$ and $\tau \to \infty$. Describe the change in the distribution of the probabilities of different quantum states as τ increases and the corresponding change in magnetization.
d) Sketch $\langle M \rangle$ as a function of τ/τ_0 for fixed B, where τ_0 is a scale constant. Specify τ_0 in terms of given constants for your plot.

Exercise 4.6

A box of length L contains a one-dimensional photon gas with photons of both polarizations. The box is in thermal equilibrium with a large reservoir at temperature τ.

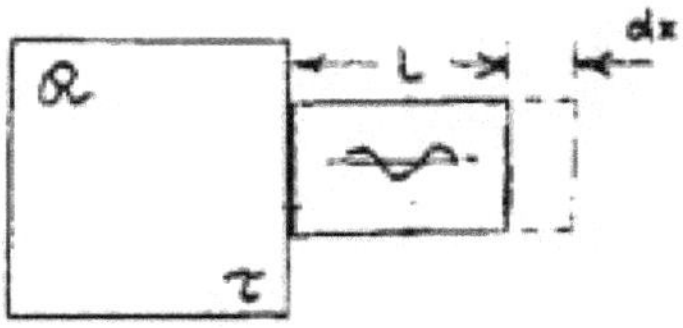

a) Find the free energy F in the form of sum, and approximate this sum by an integral.
b) The box is increased in length by dx with the photons doing the work fdx, where f is the force on the moving end of the box . Find the change in free energy dF/dx.
c) Find the force F on the end of the box in terms of τ, $\hbar$ and σ.

Note 1: $Z = \prod_n \left(1 - e^{-\hbar\omega_n/\tau}\right)^{-2}$, with the exponent -2 corresponding to 2 polarizations.

Note 2: $-\int_0^\infty \ell n\,(i - e^{-x})dx = \frac{\pi^2}{6}$

Exercise 4.7

A system contains N identical noninteracting fermions. The total energy of the system is

$$U = \sum \varepsilon_s N_s,$$

where ε_s is the energy of level s and N_s is the number of particles in level s. Denote $N = \sum N_s$.

(a) What is the partition function $Z(N)$?

(b) Now consider the system with level s=x removed. Denote the new partition function by $Z_x(N)$. Express $Z(N)$ in terms of $Z_x(N)$, $Z_x(N-1), \tau$, and ε_s.

(c) Assume N>>1 so that $\mu = F(N) - F(N-1) = \partial F/\partial N$. Assume also $Z_x(N) \approx Z(N)$ and $F(N) = -\tau \ln Z(N)$. Find the fractional occupancy $\langle N_s \rangle$ in terms of ε_s, μ, and τ.

Exercise 4.8

Consider a gas of Bose particles in a box in thermal contact with a reservoir at temperature τ and chemical potential μ. The average number of particles at equilibrium is N.

(a) What is the relation between μ and N?

(b) Find the average number of particles N_0 in he ground state.

(c) Find the total energy U assuming the ground state energy is zero and an appreciable fraction is in the ground state.

(d) Find the specific heat C_V and compare $C_V/(N - N_0)$ with the value of C_V/N for the ideal gas at high temperature. (Note: $\int_0^\infty \frac{x^{3/2}}{e^x - 1} = 1.783$)

Exercise 4.9

An engine operates using a perfect gas with a three step cycle: Step 1, increase pressure at constant volume; Step 2, decrease pressure isentropically to starting pressure; Step 3, decrease volume to starting volume at constant pressure. Let $\gamma = C_P/C_V$.

(a) Find Q and W for each step of the cycle in terms of p_1, V_1, p_2, V_2, and γ.

(b) The efficiency $\eta = \sum W / \sum Q$ can be written as

$$\eta = 1 - \gamma \left[\frac{(V_1/V_2) - 1}{(V_1/V_2)^{4\gamma} - 1} \right]$$

where $\sum Q$ includes only positive heat flow into the system. Find η for a monatomic gas with a volume ratio $(V_1/V_2) = 8$, which is similar to that of an automobile engine.

Exercise 4.10

Two identical tubes are fitted with freely moving pistons of masses 2M and M and connected by means of a thin tube at the bottom. At the start the 2M piston is at the topmost position with N molecules of ideal gas at a pressure p_1 and volume V_1 on the left side. The gas is slowly forced through the tiny tube at the bottom until all is displaced to the right side (state 2). The motion is very slow and the pressure remains uniform throughout the process. No heat transfers into the system.

(a) Compute $H_2 - H_1$, the change in enthalpy of the gas, $\tau_2 - \tau_1$, the change in temperature, and $\sigma_2 - \sigma_1$, the change in entropy.

(b) The process has proceeded slowly and adiabatically throughout. Is it revesible in the adiabatic sense?

(c) A force of attraction exists between the molecules of he gas, and the equation of state becomes $pV + N^2 \frac{a}{V} = N\tau$, which gives an energy $U = \frac{3}{2} N\tau - N^2 \frac{a}{V}$. Find $\tau_2 - \tau_1$ in terms of N, a, V_1, V_2.

Exercise 4.11

A perfect gas AB is partly dissociated into its components A and B. The equation of the reaction is A+B=AB. The internal free energy of AB is ε_{AB}, while the internal energy of the A and B states is zero.

(a) Write the equilibrium constant $K(\tau)$ in terms of the quantum concentrations n_A, n_B, n_{AB}, ε_{AB}, and τ.

(b) Let AB be a Ca atom in the chromosphere of the sun, B an electron, and A the Ca+ cation. Assume the gas initially neutral so that the concentration $n_A = n_B$, and assume the quantum concentrations $n_{AB} \cong n_A$. Find n_A/n where $n = n_A + n_{AB}$ is the total number of atoms, in terms of n_B, n, τ, and ε_{AB}. The temperature is about 6000K and the pressure $10^{-10} atm$.

Exercise 4.13

The isotherms on he p-V plot below show a region of gas-liquid equilibrium in the flat part of the curves. Consider a reversible cycle in 4 steps: (A) Change all liquid to all gas at temperature T by adding the latent heat L, (B) reduce temperature adiabatically by ΔT, (C) condense gas backlot liquid by subtracting L (approximately). Assume ΔT is small so that the shape of the B and D steps is unimportant.

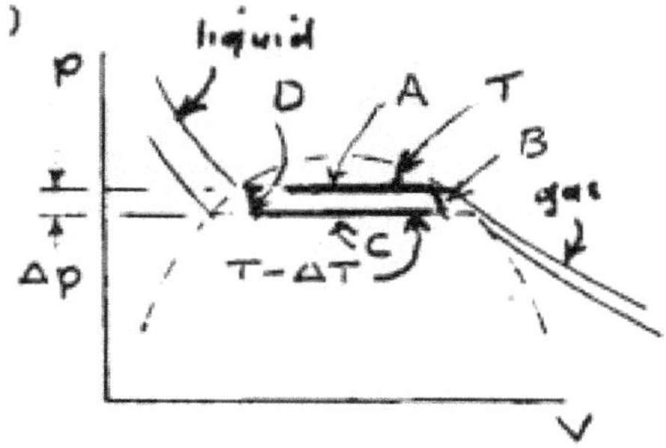

(a) Find the network done in the cycle in terms of Δp, the difference in pressure between the two isotherms in the cycle, and $V_{gas} - V_{liquid}$, the difference of volume between T and ΔT.
(b) Find the efficiency in terms of T and ΔT.
(c) Find $\Delta p/\Delta T$ in terms of $V_{gas} - V_{liquid}, L, T$.

Exercise 4.14

A molecule has two states (or "isomers"), of volumes V_1 and V_2. Both states have the same total angular momentum, and negligibly different internal energies, temperature T, pressure P. If the population in a sample of such molecules at state 1 is x_1, find the population x_2 at state 2, in terms of V_1, V_2, T, P, x_1, and Boltzmann's k. Note: the volumes V_1, V_2 pertain to the molecule itself, not to the experiment's container: think of state 1 as "collapsed" and state 2 as "puffed up".

Exercise 4.15

A finite length string, at an absolute temperature T, is under a tension F (in the x direction, say, as shown). Certain properties of this string can be studied statistical-mechanically by the following model. (1) The string consists of N massless rigid rods each of length λ. (2) The rods are hinged so that each rod can be oriented in any direction in three dimensions. (3) The tension F is constant in magnitude and direction throughout the string. Using this model, find the mean distance L between two ends of the string in terms of T, F, N and λ (neglect all interactions between the rods)

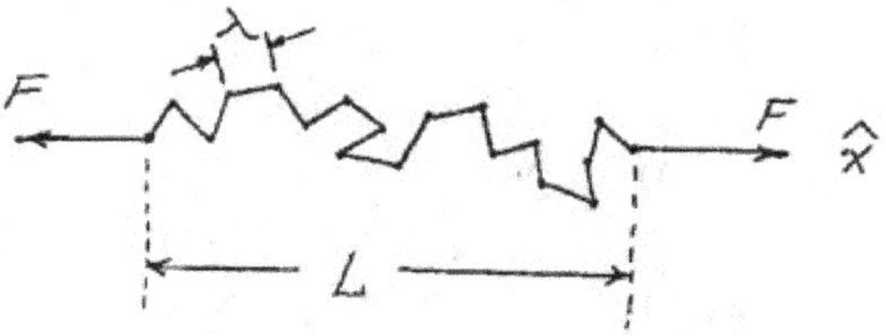

Exercise 4.16

Consider a gas of N weakly interacting identical particles obeying Fermi-Dirac statistics. The gas is in equilibrium at temperature T in a volume V. Let E_r denote the energy of a particle in a single particle state r, and let n_r denote the number of particles in state r.

a) Calculate the partition function Z (to good approximation) for this system. In your calculation, explain how the chemical potential was introduced.
b) Obtain the average number $\bar{n}_s$ of particles in the single particle states, and write the equation relating the chemical potential to the total number N.
c) Show how the dispersion is obtained from lnZ and $\bar{n}_s$.

Exercise 4.17

Show that $pV = Nk'T$ holds for blackbody radiation (a photon gas in a cavity) where, however, k' is not Boltzmann's constant k. Find an expression for k'/k and estimate k'/k at least well enough to show whether it is > or < unity. N is the expected total number of photons in the cavity. (Hint: it is useful to know that the pressure is one-third the energy density for a photon gas.)

Exercise 4.18

The lowest possible energy of a conduction electron in a metal is $-V_0$ below the energy of a free electron at infinity. The conduction electrons have a Fermi energy (or chemical potential) μ. The minimum energy needed to remove an electron from the metal is then $\phi = V_0 - \mu$ and is called the work function of the metal.

Consider an electron gas outside the metal in thermal equilibrium with the electrons in the metal at the temperature T. The density of electrons outside the metal is extremely small at all laboratory temperatures where $kT \ll \phi$.

a) Find the mean number of electrons per unit volume outside the metal.

b) Calculate the number of electrons from the exterior region that strike a unit area of the metal surface in one second.

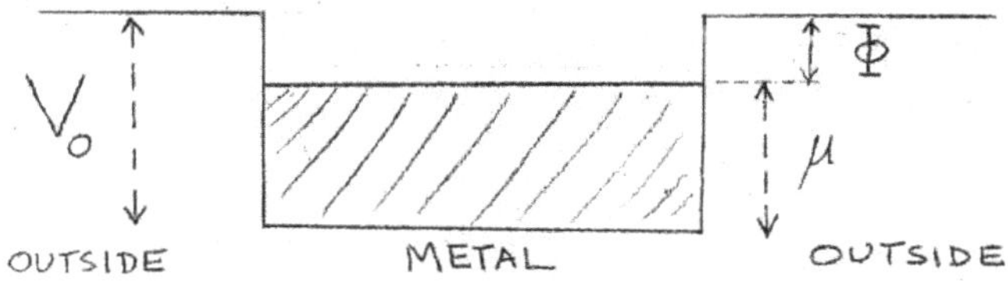

Exercise 4.19

Consider the harmonic longitudinal vibrations of a one-dimensional lattice with two ions per primitive cell (masses M_A and M_B with an equilibrium separation of d). The equilibrium separation of the A type ions in adjacent cells is a. the coupling (spring constant) is G between the ions (A and B) in the same cell. In addition each of the ions in a particular cell only couples to the like ions in the nearest neighbor cells: A-A coupling is k, and B-B coupling is also k.

a) Obtain the dispersion relations for the vibrational modes of this lattice.
b) There are two limiting situations where the vibrational modes of the lattice are easy to predict. These are the situations where G>>k and G<<k. describe the modes of the system in these limiting cases, and check that your dispersion relations behave properly in these limits (it will suffice to demonstrate this for the special case where M_A=M_B=M).

Exercise 4.20

Suppose we have 10^{10} small spherules (little spheres) on a smooth flat table which is one meter square at a temperature of 300°k. The spherules have a diameter of 10^{-6} m and a mass of 2 x 10^{-16} kg. Assume that, because of gravity, their motion is essentially horizontal (i.e. neglect the vertical motion). After the spherules have some into equilibrium with table,

a) What is their root-mean-square horizontal velocity?
b) What is their (2 dimensional) speed distribution?
c) What is the force per unit length they exert on the rail at the edge of the table?
d) What are the mean free path and collision frequency of a spherule?

e) For general temperature, mass, and number of spherules, what "law" do they obey (analogous to the ideal gas law)? Define all symbols and state their dimensions.
f) Show that the assumption that we can neglect their vertical motion is a good approximation.

4.4 Statistical Learning

Amari's Dually Flat Formulation [10-12]

The application of differential geometry methods to the study of statistical models traces back to C.R. Rao in 1945. Rao noted that families of probability distributions could be described by a manifold and that the Fisher information matrix might be taken as a metric on that manifold.

Probability Distributions:

$$p(x) \geq 0 \ \forall x \in \chi \text{ and } \sum_{\{x \in \chi\}} p(x) = 1.$$

Family of Probability Distributions:

$$S = \{p_\theta = p(x;\theta) \mid \theta = [\theta^1, \ldots, \theta^n] \in \Phi\}$$

θ parametrizes a n-dimensional statistical model.

Fisher Information Matrix:

$$G(\theta) = [g_{ij}(\theta)];\ g_{ij}(\theta) = E_\theta[\partial_i l_\theta \partial_j l_\theta],$$

$$where\ \partial_i = \frac{\partial}{\partial \theta^i}\ and\ l_\theta(x) = \log p(x;\theta)$$

Sufficient Statistic:

Consider

$$p(x;\theta) = p(x|y;\theta)q(y;\theta).$$

Suppose $p(x;\theta) = p(x|y;\theta)\, q(y;\theta)$, with associated spaces $S \equiv \{p(\cdot, \theta)\}$ and $S_F \equiv \{p(\cdot\,;\theta)\}$, and y is related to x by the constraint $y = F(x)$. If $\forall x \in \chi$ $p(x|y;\theta)$ does not depend on θ, then F is a "sufficient statistic" w.r.t. S and there results $p(x;\theta) = p(x|y)\, q(y;\theta)$. Now, in order to estimate θ, it suffices to know y, and

$$\frac{\partial}{\partial \theta^i} \log p(x;\theta) = \frac{\partial}{\partial \theta^i} \log p(x;\theta) \Rightarrow g_{ij}\ same\ for\ both\ S\ and\ S_F$$

G(θ) is symmetric and positive semi-definite (positive definite with linear independence in $\partial_1 p, \ldots \partial_n p$). From this we can define inner product on the natural basis $[\theta^i]$ and obtain the "Fisher Metric." First reveled in an obscure paper by Chentsov [28] it was shown that the Fisher metric could be uniquely singled out based on invariance with respect to "sufficient statistics." *Also invariant w.r.t. sufficient statistics was a class of connections that Amari refers to as α-connections.*

"Invariance" w.r.t. sufficient statistics:
Here the invariance result of Chentsov is restated following the presentation of Amari. In essence, the Fisher metric and α-connections are uniquely characterized by invariance w.r.t. sufficient statistics on arbitrarily reduced probability spaces (reminiscent of Khinchin's reducibility axiom that entropy $H(p_1,p_2,\ldots p_n,0) = H(p_1,p_2,\ldots p_n,)$ in establishing the uniqueness of Shannon's entropy).

The formulation is done for a discrete set and then extended to (continuum) probability densities:

Consider events: $X_n \equiv \{0, 1,\ldots,n\}$; probabilities: $P_n \equiv P(X_n)$; and pairs (g_n,∇_n) of metric and connection on $S \equiv \{P_n\}$. Consider $S_F \equiv \{P_m\}$ where $F: \chi_n \rightarrow \chi_m$ ($n \geq m$, and F is surjective). If F is sufficient w.r.t. S, then g_{ij} and Γ_{ijk} on S and S_F are the same. If sufficiency holds for all n, m, s and F, then g_n is the Fisher metric on P_n and ∇_n is an α-connection on P_n.

Recall Dual Connections:

$$X\langle Y,Z\rangle = \langle\nabla_X Y,Z\rangle + \langle Y, \nabla^*_X Z\rangle.$$

$$\partial_k g_{ij} = \Gamma_{kij} + \Gamma^*_{kji}$$

$$\Gamma^{(\alpha)}_{ijk} = E_\theta\left[(\partial_i\partial_j l_\theta + \frac{1-\alpha}{2}\partial_i l_\theta \partial_j l_\theta)(\partial_k l_\theta)\right], \qquad and\ l_\theta = \log p(x;\theta)$$

Note : The α-connections and (-α) connection are dual w.r.t. the Fisher metric.

The α-connections are defined as follows:
Consider the Exponential Family of distributions in this context:

$$p(x;\theta) = exp\left[c(x) + \textstyle\sum \theta^i F_i(x) - \psi(\theta)\right]$$
$$\partial_i l = F_i - \partial_j \psi$$
$$\partial_i \partial_j l = -\partial_i \partial_j \psi \longrightarrow no\ x\ dependence$$

for α=1 we then have $\Gamma_{ijk}^{(1)} = -\partial_i\partial_j\psi E_\theta[\partial_k l_\theta] = 0$. So, $[\theta^i]$ is a $\nabla^{(1)}$–affine coordinate system, S is $\nabla^{(1)}$–flat. Amari refers to $\nabla^{(1)}$ as the exponential connection, or "e-connection".

Consider the Mixture Family of probability distributions in this context:

$$p(x;\theta) = \sum\theta^i p_i + (1 - \Sigma\theta^i)p_0(x)$$

$$\partial_i l = \frac{p_i - p_0}{p};\ \partial_i\partial_j l = -\frac{(p_i - p_0)(p_j - p_0)}{p^2}$$

$$\Rightarrow \partial_i\partial_j l + \partial_i l\partial_j l = 0 \Rightarrow \Gamma_{ijk}^{(-1)} = 0$$

So, $[\theta^i]$ is a $\nabla^{(-1)}$–affine coordinate system, S is $\nabla^{(-1)}$–flat. Amari refers to $\nabla^{(-1)}$ as the "mixture connection" or "m-connection".

Divergence: A triplet (g,∇,∇^*) can be defined locally from a "divergence" [12], where a divergence D is characterized by:

$$D(\bullet \| \bullet) : S \times S \rightarrow \Re, \text{ where } \forall p, \forall q \in S \times S\text{: } D(p\|q)\geq 0, \text{ and } D(p\|q)=0 \text{ iff } p=q.$$

Dually flat spaces: If ∇ and ∇^* are both symmetric then ∇ flat $\leftrightarrow$ ∇^*-flat. Since the α connections are symmetric, S is α-flat $\leftrightarrow$ S is $(-\alpha)$-flat, and, in particular, if $\nabla^{(1)}$ is flat then so is $\nabla^{(-1)}$.

If (S, g, ∇, ∇^*) is dually flat, then there exists ∇-affine coordinates $[\theta i]$ and ∇^*-affine coordinates $[\eta_j]$. The inner product between the elements of the tangent spaces is a constant on S, from which it is possible to get the relations:

$$\frac{\partial\eta_i}{\partial\theta^k} = g_{ik} \quad and \quad \frac{\partial\theta^i}{\partial\eta_j} = g^{ij}$$

Introduce Potentials: Suppose $\partial_i\psi=\eta_i$, then $\partial_i\partial_j\psi=g_{ij}$, and since g_{ij}, is a metric tensor the partial derivative must describe a positive definite matrix, which in turn implies that ψ is strictly convex. Likewise for $\partial^i\varphi=\theta^i$. In terms of the potentials ψ and φ thus introduced, it is easy to show that the two coordinate systems $\{\theta^i,\eta_j\}$ can be related by Legendre transformation:

$$\varphi = \theta^i \eta_i - \psi$$

Convexity in potentials that define a Legendre transformation leads to some interesting constructions. If we take our dually flat space (S, g, ∇,∇^*) with coordinate systems $\{[\theta^i], [\eta_i]\}$ and their potentials $\{\psi, \varphi\}$, then it can be shown that

$$\varphi(q) = \lim_{p \varepsilon S}\{\theta^i(p)\eta_i(q) - \Psi(p)\}$$

and

$$\Psi(q) = \lim_{q \varepsilon S}\{\theta^i(p)\eta_i(q) - \varphi(p)\}$$

Since we have:

$$\Psi(q) - \lim_{q \varepsilon S}\{\theta^i(p)\eta_i(q) - \varphi(p)\} = 0$$

we have:

$$\Psi(q) - \{\theta^i(p)\eta_i(q) - \varphi(p)\} > 0$$

call this D(p||q):

$$D(p//q) = \psi(p) + \varphi(q) - \theta^i(p)\eta_i(q),$$

and it is easily shown that:

$$D(p//q) \geq 0, and\ D(p//p) = 0 \Leftrightarrow p = q$$

Thus, the Exponential and Mixture Families induce a dually-flat space which leads to the fundamental difference measure on distributions being a divergence. The special form of "distance" function is also indicated in the Link formalism (Ch. 9). Although the divergence family is singled out as fundamental at this point, the selection of a specific divergence (like Euclidean distance in the case of metrics) is yet to be determined. In [11] we will see that the "simplest" divergence, the Kullback-Leibler divergence, is selected when maximizing log likelihood during learning, and that this is, fundamentally, because of the shortest path, or projection theorem, described in [11]. ***Implicit in this result is that the proper way to measure the difference between distributions is not the Euclidean distance between them but the Kullback-Leibler Divergence between them.***

Generalization of Pythagorean Theorem

Let p, q and r be three points in S. Let γ_1 be the ∇- geodesic connecting p and q, and let γ_2 be the ∇^*-geodesic connecting q and r. If at the intersection q, the curves γ_1, and γ_2 are orthogonal (w.r.t. g), then

$$D(p||r) = D(p||q) + D(q||r)$$

To show this, first consider the γ_1 geodesic:

$$\theta_t^i = t\theta^i(p) + (1-t)\theta^i(q)$$
$$\frac{d}{dt}\theta_t^i\partial_i = [\theta^i(p) - \theta^i(q)] * \partial_i$$

$$\eta_{ti} = t\eta_i(q) + (1-t)\eta_i(r)$$
$$\frac{d}{dt}\eta_{ti}\partial^i = [\eta_i(q) - \eta_i(r)]\partial^i$$

So, prove relation with:

$$D(p||q) + D(q||r) - D(p||r)$$
$$= \Psi(\rho) + \varphi(q) + \theta^i(p)[\eta_i(r) - \eta_i(q)] - \theta^i(q)\eta_i(r)$$
$$= [\theta^i(p) - \theta^i(q)][\eta_i(r) - \eta_i(q)] = <$$
$$\left(\frac{d\gamma_1(t)}{dt}\right)_i \left(\frac{d\gamma_2(t)}{dt}\right)^j >= 0$$

Projection Theorem and relation between Divergence and Link Formalism

Suppose M a, submanifold of S, is ∇^*-autoparallel, then

> D(p||q) = $\min_{\{r\in M\}}$ D(p||r) when the ∇-geodesic connecting p and q is orthogonal to M at q.

Relation to Link formalism, start with:

$$D(p||q) = \psi(p) + \varphi(q) - \theta^i(p)\eta_i(q)$$

Use Legendre transformation:

$$\varphi(p) = \theta^i(q)\eta_i(q) - \psi(q)$$

to get:

$$D(p||q) = \psi(\rho) - \psi(q) + \left(\theta^i(q) - \theta^i(p)\right)\eta_i(q), \text{ where } \eta_i = \frac{d\psi}{d\theta^i}$$

Thus, with shift in notation $\{f=\frac{\partial F}{\partial\omega},\omega\} \rightarrow \{g, \theta\}$ we get the Link formalism used in Ch. 9:

$$D(\omega||\omega') = F(\omega) - F(\omega') - (\omega-\omega')\frac{\partial F}{\partial\omega}\Big|_{\omega=\omega'}, \; where\ \frac{\partial D}{\partial\omega} = f(\omega) - f(\omega').$$

The $\{f,\omega\}\leftrightarrow\{g, \theta\}$ duality corresponds precisely to the dually flat connection construction with potentials φ and ψ that are related via Legendre Transformation to the "coordinates" θ and ω. The link function $f(\omega) = \omega$, is associated with square loss and the gradient descent (GD) learning rule. The link function $f(\omega) = \ln(\omega)$, is associated with divergence loss and the exponentiated gradient descent (EG) learning rule. These will be explored in Ch. 9 in the link formalism context, along with an interpolating learning rule between GD and EG given by the link function $f(\omega) = \sinh^{-1}(\omega)$.

Neuromanifolds [10-12]
The information geometry methods described for families of probability distributions (parametrized by $\{\theta^i\}$) can just as easily be applied to neural networks; where now the parameters are the connection weights (further detailed description of neural nets is in Ch. 9 and Ch. 13). The statistical arguments also carry over and are applicable to stochastic neural nets, i.e., neural networks with noisy input or non-deterministic behavior. As Amari states, "even when a network is deterministic, it is sometimes effective to train it as if it were a stochastic network." [12]. With a stochastic network we then have probability distribution $p(x;\theta)$ and/or conditional probability distribution $p(y \mid x;\theta)$.

Complications associated with repeated observation:
For a single observation we have the distribution $p(x;\theta)$, an element of the family (space) S. For repeated independent observations there is the joint conditional distribution where θ generally changes from one observation to the next. The joint distribution, thus, is an element of a larger family (space)

$$S_T^* = S_1 \times S_2 \times \ldots \times S_T.$$

Multiple observations thus lead to a joint distribution whose manifold dimension (under direct product) increases with the number of observation, while the underlying parameterization is fixed—being the manifold dimension of the family of distributions considered for the individual distribution. This generally leads to a Curved Exponential family description on the joint distribution manifold (where the individual distribution was in an exponential family) ,i.e., a submanifold of the joint distribution manifold is specified. It is possible to describe repeated observations within the framework of the manifold S without referring to the product space S_T^*, but this holds only in the i.i.d. case. In general

$$\theta_T^* = (\theta_1, \theta_2, \ldots, \theta_T),$$

and

$$\theta_T^* = \theta(x_t, u_t),\ t=1,\ldots,T.,$$

where x_t's are given and the u_t's are the only free parameters (the parameters of the underlying neural network). Suppose all x_t are subject to $p(x;\theta)$, $\theta = \theta, = \ldots = \theta_T$, where $p(x;\theta)$ is the exponential family $p(x;\theta) = \exp(\theta \cdot x - \psi)$.The joint distribution is

$$p(x_1, \ldots x_T;\theta) = \exp\{\Sigma x_t \cdot \theta - T\psi\}$$

$$p(\bar{x};\theta) = \exp\{\bar{x} \cdot \theta - \psi\}$$

The maximum likelihood estimator (m.l.e.) $\hat{\theta}$ from the observed data $x_1, \ldots, x_T$ is given by maximizing $p(\bar{x};\theta)$:

$$\bar{x} = \frac{\partial}{\partial \theta} \psi(\theta)|_{\theta = \hat{\theta}} = \hat{\eta}$$

The observed data are then represented by the m.l.e $\hat{\eta}$ in S in the η-coordinate system.

Suppose that all x_t are subject to $p(x;\theta)$ where $p(x;\theta)$ is a curved exponential family. Again the observed data $x_1, \ldots, x_t$ are represented by $\bar{x} = \hat{\eta}$, now, however, we have that $\hat{\eta}$ does not necessarily belong to M. The m.l.e. $\hat{u}$, or corresponding distribution $\theta(\hat{u}) \in M$, is given by maximizing the log likelihood $x \cdot \theta(u) - \psi(\theta\ (u))$ w.r.t. u. Maximizing the log likelihood is equivalent to simply m-projecting $\hat{\theta}$ to M in the Amari differential geometry formalism, and this is equivalent to minimizing the Kullback-Leibler (KL) divergence $K(\hat{\theta} \,||\, \theta(\hat{u}))$ from $\hat{\theta}$ to $\theta(\hat{u}) \in M$. Thus, the KL Divergence is used in the Amari neuromanifold learning process.

EM Algorithm

The following EM/em discussion closely follows Amari [11].

Consider $M = \{p(r;\theta(u))\}$ a curved exponential family from which data is regenerated. Data r is observed. Consider $r = r\ (S_v, S_h)$, a sufficient statistic that includes hidden part S_h. Need to estimate unknown part of r information. Can do this based on observed S_v , and some candidate distribution u′, via the conditional expectation:

E: $$\hat{r}(u') = E[r|s_v;\theta(u')]$$
Now estimate $\log p(\check{r};\theta(u))$ by conditional expectation also:
E:
$$LLH(s_v;\theta(u')) = E[\log p(\check{r};\theta(u)|\, s_v;\theta(u')]$$
$$= \theta(u) * \check{r}(u') - \psi(\theta(u))$$

Now search for better candidate u by maximizing LLH, or equivalently, by minimizing the KL divergence $D(\hat{r}(u')||\theta(u))$ from the guessed data point $\hat{\eta} = \hat{r}(u')$ to μ w.r.t. u.

The algorithm:

Step 0: Initialization step. Guess u_0, the initial guessed distribution $P_0 \in M$ is given by $\theta(u_0)$. Then repeat the following:
Step 1: E-step. Based on candidate probability distribution $P_i \in M$, calculate the conditional expectation of r. This gives the ith candidate for the observed point $Q_i \in D$, whose η-coordinate are $\eta_{(i)}$.
Step2: M-step. Calculate the mle $u_{(i+1)}$ from $Q_i \in D$.

em Algorithm
Search for the pair of points $P \in M$, $Q \in D$ that minimizes the divergence between D and M, that is (originally proposed in1984 [11]):

$$D(\hat{Q}||\hat{P}) = \lim_{P\in M, Q\in D} D(Q||P)$$

(i) point $\hat{P} \in M$ that minimizes $D(Q||P)$ is given by the m-projection of Q to M (i.e., by the m-geodesic connecting $\hat{P}$ and Q that is orthogonal to M at $\hat{P}$).
(ii) point $\hat{Q} \in D$ that minimizes $D(Q||P)$ is given by the e-projection of P to D (i.e., the e-geodesic connecting P and $\hat{Q}$ that is orthogonal to D at $\hat{Q}$).

The algorithm:

Step 0: Initialization Step. Guess $\hat{u}_0 \Rightarrow \hat{P}_0 \in M$. Then repeat:
Step 1: e-step. e-project $\hat{P}_i$ to D, gives $\hat{Q}_i$.

Step 2: m-step. m-project $\widehat{Q}_i$ to M, gives $\widehat{P}_{i+1}$.

Relation between EM and em:
(1) M-step and m-step are the same
(2) E-step and e-step differ depending on the conditional expectation of r_h given r_v:

Note: Divergence need not be symmetric or satisfy the triangle inequality, but given "orthogonal" learning steps, a Divergence satisfies a generalized Pythagorean theorem, leading to a the same critical Chapman-Kolmogorov-like propagation rule whether learning step assumes a Euclidean notion of distance (GD) or a KL divergence notion of distance (EG).

Which is Better?
EM is more natural from statistical point of view, but the representation via a neural network does not correspond exactly to statistical inference, therefore em serves better as the approximator ideally suited to the neural network representation of the input-output relation. The two algorithms are asymptotically equivalent when T is large (in the framework of exponential families). If framework extended from exponential family to function space they are exactly equivalent. It may be that a hybrid of GD and EG, or of EM and em type learning, is best, and this will be explored further in Ch 9 in an explicit loss bounds analysis on the learning process.

Amari's Dually Flat formulation is a more natural structure for representing exponential families and neural nets than it might appear at first sight. In particular, the Kullback-Leibler measure is naturally represented in terms of the dual coordinates and their relations via Legendre transformation (with the introduction of appropriate potentials). The potentials, in turn, provide a natural formulation that is precisely that exhibited when using the "link formalism." Similarly, the duality between the coordinates (via Legendre transformation) is precisely that exhibited in the neural net learning algorithms.

Maximizing the KL measure, on either argument, is conveniently expressed in the dually flat formulation, it is just the e-projection or m-projection, appropriately. This provides a natural, geometric, interpretation for the EM algorithm. It also provides an algorithm more suited to the neural net implementation of the input-output relation—Amari's "em" algorithm. The expression of the EM or em algorithm, in terms of alternating e-projections and m-projections, is then directly modifiable to a learning algorithm.

While the e-projection and m-projection are natural geometric notions and certainly a strength of Amari's program, they also represent a weakness. This weakness is perhaps most clearly illustrated in the learning algorithm version where the projection is determined and then the actual update is interpolated (between old data and estimator and new, projected data and estimator). In this instance we must fall back on the gradient descent algorithm, or some such arguments, in order to have stable updates. One approach might be to use the SA link algorithm since it results in a formalism that interpolates between GD and EG learning according to the weight magnitude. Further details on this approach are given in Ch. 9.

Chapter 5. Emergent Phenomena

5.1 The Universal properties of phase transitions – critical exponents

Let's now consider the Statistical Mechanics of a system that changes phase as a function of temperature, pressure, or other intensive variable. We will first examine the one-dimensional Ising model with no external field – even though it does not exhibit a phase transition it is foundational to more complex Ising models that do have phase transitions. Even with the one-dimensional Ising model we will encounter the concepts: mean field theory, Landau theory, Weiss molecular field theory, Bragg-Williams approximation, the Bethe approximation, and the critical behavior of mean field theories.

The examination of the two-dimensional Ising model is done using the results from the one-dimensional analysis. The Onsager solution for the two-dimensional Ising model [29] will be briefly described using modern notation and techniques [23,24]. The key to Onsager's solution, or statistical cluster analysis in general, or to renormalization group methods to follow in Section 5.12, is a system equation, such as a partition function or Hamiltonian, that has a recalling self-similarity relation (which defines the renormalization group). The Onsager solution will be given exactly, and then approximately solved using renormalization group methods that will give a reasonably close answer to the Onsager solution with significantly less effort. Qualitatively the system examined will have the same phase-transition behavior at a critical temperature. Quantitatively, the system will have the same critical exponent behavior upon approach in a system parameter to the critical value (for the phase transition), e.g., there is universal behavior in critical exponents revealed regardless of approximation as long as that approximation sufficiently captures long-range coupling effects.

5.2 The Ising Model Overview

The section that follows is described in much more detail in [23,24].

Recall that the Hamiltonian for two particles interacting according to their mutual alignment can be expressed in terms of their inner product:

$$H = -c \sum_{N.N.} s_\alpha \cdot s_\beta \,,$$

Where "$N.N.$" refers to the Nearest Neighbor spin interactions in the sum only. Recall that the local spin operators s_α do not commute with H and that this is often resolved by performing measurements (quantum state preparation) of spin in the z-direction. The focus is then on representation where the $(s_\alpha)_z \rightarrow \sigma_i = \pm 1$ representation is diagonal. Similarly, here, we will add a magnetic field along the z-direction, which results in addition of a Zeeman term to the energy:

$$H = -J \sum_{N.N.} \sigma_i \, \sigma_j - h \sum_i \sigma_i \,,$$

where h is proportional to the magnetic field (this can always be set to zero to get the case of no magnetic field).

Let's now explore the temperature behavior of the system where a simple mean field approximation is used (implied) in the form of assuming an overall system magnetization, m, exists (as observed in many natural settings) and describe that system magnetization as the expectation on σ_i:

$$m = \langle \sigma_i \rangle \,.$$

Thus an "order parameter", magnetization, is hypothesized to exist – this is a distinctive characteristic of mean field theory methods.

Writing the mean field system Hamiltonian as a function of a particular spin, here denoted σ_0, we have:

$$H(\sigma_0) = -\sigma_0 \left[J \sum_{N.N.} \sigma_j + h \right] = -\sigma_0 [JNm + h], \quad N = \sum_{N.N.} 1$$

Using H we can now directly determine $m = \langle \sigma_i \rangle = \langle \sigma_0 \rangle$, an establish a constitutive relation for determining m:

$$m = \langle \sigma_0 \rangle = \frac{Tr[\sigma_0 \exp(-\beta H(\sigma_0))]}{Tr[\exp(-\beta H(\sigma_0))]},$$

which is solved in the Examples to give:

$$m = \tanh[\beta(JNm + h)] \,.$$

This must be solved numerically to get a general solution.

For the one-dimensional Ising model the hypothesis of (spontaneous) magnetization given by writing $m = \langle \sigma_0 \rangle \neq 0$ means that there is a critical temperature above which the magnetization is zero and below which the magnetization grows, with two solutions (upward or downward). Near this critical temperature (from below) the above expression for magnetization can make use of Taylor expansion on the tanh, and the small magnetization would then be approximately:

$$m \propto \left(\frac{T_c}{T} - 1\right)^{\frac{1}{2}}.$$

There is, thus, power law behavior (consistent with Landau theory) and the power law, or 'critical', exponent is ½ in this approximation. Thus, if $0 < T_c < \infty$, then we will have a phase transition, and

$$T_c = JN/k_B$$

is indicated by the Bragg-Williams approximation where the spins are taken to be independent. This turns out to be incorrect, however, as $T_c = 0$, as will be shown in exact derivations, and again later using renormalization group methods. The discrepancy arises because the mean field and Bragg-Williams approximations have effectively arrived at a description with infinite-range interactions, and this is what is permitting the false phase transition to appear. Furthermore, the mean-field derivation would suggest that the number of nearest neighbors is all that matters and not lattice structure or dimensionality, both very wrong in practice. So, knowing the limits of mean field applications will be necessary.

In [30,31], a very clever approach to testing the low temperature regime of the one-dimensional Ising model can be made. Let's suppose that the system is in such a low energy state that there is only on transition between blocks of spin. Thus, for $i \leq l$ we have spin up, and for $i > l$ we have spin down. For length N Ising chain (with free ends) there are $N - 1$ such states. At temperature T the free energy change due to these fluctuations is:

$$\Delta G = 2J - k_B T ln(N - 1)$$

which is always negative for large N, thus, the magnetization should be zero. A similar argument can be done for he two-dimensional Ising problem, where now the lowest order fluctuation is a separating line between two regions of block spin (one up and one down). The boundary length L between the two spin regions (counted according to the number of spins in relevant, anti-aligned, nearest neighbor sums) then allow the change in energy in the domain wall (over zero ground state) to be simply written as:

$$\Delta E = 2LJ.$$

The number of possible boundary walls of length L goes as 2^L (ignoring edge effects). For a $N \times N$ lattice there are N stating points for the boundary on the left, and there are N steps across, but half the time the boundary moves transversely, thus the average length L is $2N$. Putting this together we have for free energy change due to boundary wall fluctuations:

$$\Delta G = 4NJ - k_B T ln(N2^{2N}).$$

This system is stable against forming the domain indicated if less than a critical temperature, where:

$$T_c = \frac{2J}{k_B ln2} = 2.885\ \frac{J}{k_B}.$$

This is remarkably close to the exact answer due to Onsager:

$$T_c = 2.269\ \frac{J}{k_B}.$$

5.3 The Bethe Approximatio

In this section a description of Bethe's approximation will be given, as well as a description of the Betha lattice (Cayley tree) that is also used in approximations.

Recall that the mean field method has $m = \langle \sigma_i \rangle$ and assumes no correlation information even on nearest neighbors: $\langle \sigma_i \sigma_j \rangle = \langle \sigma_i \rangle \langle \sigma_j \rangle$. By elimination of all correlations we've fundamentally altered the problem in the approximation such that a phase transition is then seen even in 1-D Ising models, when better approximations, and the full solution, show no such phase transition. What is surprising is how well even the simplest approximation will work at identifying critical values and properly identifying the existence or non-existence of associated phase transitions. What is happening is that even a little bit of correlation information can reacquire critical behavior for given dimensionality and connectivity of a lattice and, in essence, is reacquiring part of the power-law universality of the system described at its critical points. When applied with renormalization group arguments in Section X, we will find that estimates of critical parameters in complex systems can be obtained via simple approximations in much more simplified process than from direct solution, and it will build on this understanding of approximation (only needing to be sufficient to capture correlation behavior) in obtaining solutions very quickly.

In the Bethe approximation we retain nearest-neighbor correlation information. For a 2-D square lattice of spins, for example, a given σ_i will have four nearest neighbors to interact with (there is no connectivity on the diagonals in the square lattice). The energy of a cluster in this setting (with central reference spin denoted σ_0) is:

$$H = -J\sigma_0 \sum_{i=1}^{q} \sigma_i - h\sigma_0 - h' \sum_{i=1}^{q} \sigma_i\ .$$

The partition function is:

$$Z = \sum_{\sigma_i=\pm1} e^{-\beta H} = e^{-\beta h}\{2\cosh[\beta(J+h')]\}^q + e^{-\beta h}\{2\cosh[\beta(J-h')]\}^q$$

and we can thus determine:

$$\langle\sigma_0\rangle = \frac{1}{Z}(e^{\beta h}\{2\cosh[\beta(J+h')]\}^q - e^{-\beta h}\{2\cosh[\beta(J-h')]\}^q)$$

and

$$\langle\sigma_i\rangle = \frac{1}{Z}\begin{pmatrix} 2e^{\beta h}\sinh[\beta(J+h')]\{2\cosh[\beta(J+h')]\}^{q-1} \\ -2e^{-\beta h}\sinh[\beta(J-h')]\{2\cosh[\beta(J-h')]\}^{q-1}\end{pmatrix}$$

Let's now set $h = 0$ and use the relation $\langle\sigma_0\rangle = \langle\sigma_i\rangle$ that must exist if there is translational invariance to get the relation:

$$e^{2\beta h'} = \frac{\cosh^{q-1}[\beta(J+h')]}{\cosh^{q-1}[\beta(J-h')]}$$

The condition for solutions, other than the trivial $h = 0$, defines critical temperature as:

$$\coth\beta_c J = q - 1.$$

Thus, for the square lattice with $q = 4$ we have:

$$\beta_c J = \coth^{-1} 3 = 0.34657 \ldots$$

or

$$T_c = (2.885\ldots)\frac{J}{k_B}$$

which is very close to the exact answer by Onsager $T_c = 2.269\, J/k_B$, which will be obtained using sophisticated methods to simply over the approach of Onsager, and even so, it is very complicated. Yet the approximation approach above indicates much can be accomplished with appropriate choice of approximation. This will be important in later analysis for the 3-D lattice, and further complications, where exact solution will be very unlikely, but understanding when a good approximation is possible will suffice, and that is part what is explored in the following.

5.4 The Bethe Lattice

In a Bethe lattice you have a tree-like connectivity where each node has z connections to adjacent nodes. If approaching a node from one connection, the ability to 'propagate' indefinitely through the other connections outgoing, $(z-1)$ in number, must be such that the ptrobability of jumping between nodes, p, is greater than $(z-1)$ such that $p(z-1) > 1$. In other words, there is percolation through the tree with a cluster solution if

$$p > \frac{1}{(z-1)}.$$

Bethe Lattice with coordination number Z=3

Find as many of the critical exponents as possible for percolation on the Bethe lattice with Z=3 (coordination number). Show that these "mean field exponents" satisfy the Josephson scaling law $2\beta = \gamma = \nu d$ for $d = 6$. What does this say about the minimum dimension for validity of the mean field theory of percolation?

Let's consider Bethe lattice with coordination $Z = 3$ with the following definitions:

1. P=the probability that a bond is unblocked
2. R(p) = the probability a branch is restricted to a finite size.
3. P(p) = the probability that a randomly chosen bond belongs to an infinite cluster. Thus,
   P(p) = percolation probability.

For a branch to be finite, either the first bond of the branch is blocked (with probability 1-p), or, if it is unblocked (probability p), then all of the Z-1 branches which lead out from the end of the first bond must be dead-ended.

$$R = 1 - p + pR^{Z-1} = 1 - p + pR^2$$

Solving this equation for R: $R = \frac{1-p}{p}$ or $R = 1$, and:

$$p \geq \frac{1}{2}, \qquad p_c = \frac{1}{2}$$

To obtain $P(p)$, we understand that the bond must be unblocked (probably p) and at least one of the branches leading outward from the two ends of the bond must be infinite. For $2(Z-1)$ outward bound branches the probability of them all being blocked is simply $R^{2(Z-1)}$, thus:

$$P(p) = p\left(1 - R^{2(Z-1)}\right) = p(1 - R^4) = p\left(1 - \frac{(1-p)^4}{p^4}\right)$$

$$= p - \frac{(1-p)^4}{p^3}$$

Let $p - p_c = \delta$:

$$P(p) = \frac{1}{2} + \delta - \frac{\left(\frac{1}{2} - \delta\right)^4}{\left(\frac{1}{2} + \delta\right)^3} \approx \frac{1}{2} + \delta - \frac{1}{2} = (p - p_c)^1 \to \beta = 1$$

Next consider:

$$S_{av} = \frac{\sum_{s=1}^{\infty} S^2 n(S)}{\Sigma S n(S)} = \frac{\sum_{s=1}^{\infty} S b_s\, p^s (1-p)^{s+3}}{p}$$

where S=open bonds and S+3=closed bonds.

Let

$$A = \sum_s b_s x^s y^{s+3} \quad , \quad x = p \quad , y = 1-p$$

$$B = \sum_s b_s{}' \mathfrak{z}^s \quad , \quad b_s^1 = \frac{b_s}{s} \quad , \quad \mathfrak{z} = xy$$

Then $S_{av} = \frac{\partial A}{\partial x}|_{x=p\,,y=1-p}$ gives:

$$A(x,y) = xy^4 \frac{\partial B}{\partial \mathfrak{z}} \rightarrow p(1-p)^4 \frac{\partial B}{\partial \mathfrak{z}} = p \rightarrow \frac{\partial B}{\partial \mathfrak{z}} = \frac{1}{(1-p)^4}$$

$$\mathfrak{z} = xy = p(1-p) \rightarrow p = \frac{1 \pm \sqrt{1-4\mathfrak{z}}}{2} \quad , \qquad 1-p = \frac{1 \mp \sqrt{1-4\mathfrak{z}}}{2}$$

Therefore, for $S_{av} < \infty \;\; : p < \frac{1}{2} \;, 1-p > \frac{1}{2}$, thus

$$p = \frac{1-\sqrt{1-4\mathfrak{z}}}{2} \quad \& \, 1-p = \frac{1+\sqrt{1-4\mathfrak{z}}}{2} = \frac{1+\sqrt{1-4xy}}{2}$$

$$A(x,y) = xy^4 \left(\frac{1+\sqrt{1-4xy}}{2}\right)^{-4}$$

$$S_{av} = \frac{\partial A}{\partial x}|_{x=p,y=1-p} = y^4 \left(\frac{1+\sqrt{1-4xy}}{2}\right)^{-4} + xy^4(-4)\left(\frac{1+\sqrt{1-4xy}}{2}\right)^{-5} \frac{1}{4}(1-axy)^{-\frac{1}{2}} \cdot (-4y)$$

$$= 1 + \frac{4P}{\sqrt{1-4p(1-p)}} = 1 + \frac{4p}{1-2p} = 1 + \frac{2\left(\frac{1}{2} - \left(\frac{1}{2} - p\right)\right)}{\frac{1}{2} - p}$$

$$= 1 + \frac{1}{p_c - p} - 2$$

Thus,

$$\gamma = 1$$

$$l_{av} \sim \sqrt{S_{av}} \sim \frac{1}{(P_c - P)^{1/2}} \quad \rightarrow \nu = \frac{1}{2}$$

Josephson's scaling law gives

$$2\beta = \gamma = \nu d \rightarrow \; d = 6.$$

Thus, $d = 6$ is the critical dimensionality of the mean field theory.

River Crossing calculation approximated by Bethe Lattice

A river bed is randomly littered with round boulders. The diameter of these boulders is distributed according to a Gaussian distribution with mean d=1 meter and standard deviation of $\sigma = 0.2\ meters$. On average, there is one boulder for every $3m^2$ of river bed. A man can jump from a given boulder to another 2m away. For what water level will it just be possible for the man to cross the river bed without getting wet. An exact answer is not required. Make the best estimate you can.

Answer

If on average there is one boulder for every $3m^2$ of riverbed and a man can jump from a given boulder to another 2m away then the man can cover an area of $(2)^2\pi \cong 12m^2$ and in that area there are 4 boulders. This is effectively a Bethe lattice with $Z = 3$. Let's use percolation theory on a Bethe lattice with $Z = 3$ to get solutions. The probability that any site is occupied (p) is governed by the Gaussian distribution of boulder size and the river's depth. The probability of crossing the river should be thought of as the probability of having an infinite size cluster.

For percolation in a Bethe lattice the critical probability of a site being occupied that allows the existence of an infinite size cluster is

$$p_c = \frac{1}{Z-1}$$

For this case $p_c = \frac{1}{2}$. This implies using the Gaussian distribution of boulder such that 50% of the boulders are larger than 1m in diameter. Since 1m is the mean then when the river is 1m deep 50% of the boulders are uncovered by water. This is the "critical depth" as the water level goes down the probability of crossing increases in the following fashion.

$$P(p) = p - \left(\frac{1-p}{p}\right)^4 p$$

(this is the probability of crossing).

Let's approximate.

$$P(p) \cong (p - p_c)^p \ , \ \beta = 1$$

It is easy to see that for $p \leq \frac{1}{2}$ the probability of crossing is zero. It is important to point out that this "percolation" approximation to the problem is assuming the distance between the river banks is infinite.

Obviously, for finite, the less jumps that have to be made the better the chances are of crossing.

5.5 1-D Ising Model – Exact Solution

Let's now obtain the exact solution to the 1-D Ising Model:

$$H = -J \sum_{i=1.} \sigma_i \, \sigma_{i+1} \, .$$

We obtain the partition function:

$$Z_N = \sum_{\sigma_1 = \pm 1} \cdots \sum_{\sigma_N = \pm 1} e^{\beta J \sum_{i=1} \sigma_i \, \sigma_{i+1}} .$$

Making use of

$$\sum_{\sigma_N = \pm 1} e^{\beta J \, \sigma_N \, \sigma_{N+1}} = 2 \cosh \beta J$$

we have:

$$Z_N = 2[2 \cosh \beta J]^{N-1} .$$

Now that we have the partition function, we can obtain the free energy as:

$$F = -N k_B T \ln(2 \cosh \beta J) \, .$$

Let's now add a magnetic field as previously but also have our 1-D Ising Chain form a ring to eliminate edge effects (now more complicated with the magnetic field). So,

$$H = -J \sum_{i=1..N} \sigma_i \, \sigma_j - h \sum_{i=1..N} \sigma_i \, ,$$

with $N + i = i$:

$$H = - \sum_{i=1..N} \left\{ J \, \sigma_i \, \sigma_j + \frac{h}{2} (\sigma_i + \sigma_{i+1}) \right\} .$$

This is typically solved by introducing a transfer matrix. To begin, we have

$$Z_N = \sum_{\sigma_1 = \pm 1} \cdots \sum_{\sigma_N = \pm 1} e^{\beta \sum_{i=1..N} \left\{ J \, \sigma_i \, \sigma_j + \frac{h}{2} (\sigma_i + \sigma_{i+1}) \right\}} .$$

Or,

$$Z_N = \sum_{\sigma_1 = \pm 1} \cdots \sum_{\sigma_N = \pm 1} P_{\sigma_1 \, \sigma_2} P_{\sigma_2 \, \sigma_3} \cdots P_{\sigma_N \, \sigma_1} = Tr \, P^N ,$$

where

$$P_{1,1} = e^{\beta(J+h)}; \;\; P_{-1,-1} = e^{\beta(J-h)}; \;\; P_{-1,1} = P_{1,-1} = e^{-\beta J}$$

If we diagonalize P in terms of eigenvalues $\lambda_{1,2}$ and then evaluate $Tr\ P^N$ we get:

$$Z_N = (\lambda_1)^N + (\lambda_2)^N,$$

where

$$\lambda_{1,2} = e^{\beta J} \cosh \beta h \pm \sqrt{e^{2\beta J} \sinh^2 \beta h + e^{-2\beta J}}.$$

The free energy is now determined to be:

$$F = -k_B T \ln((\lambda_1)^N + (\lambda_2)^N) = -k_B T \left\{ N \ln \lambda_1 + \ln \left[1 + \left(\frac{\lambda_2}{\lambda_1} \right)^N \right] \right\}.$$

Since $\lambda_1 > \lambda_2$ we have in the thermodynamics limit as $N \to \infty$ that:

$$F = -N k_B T \ln \left(e^{\beta J} \cosh \beta h + \sqrt{e^{2\beta J} \sinh^2 \beta h + e^{-2\beta J}} \right)$$

We can now evaluate the magnetization:

$$m = \langle \sigma_0 \rangle = -\frac{1}{N} \frac{\partial G}{\partial h} = \frac{(\sinh \beta h)}{\sqrt{\sinh^2 \beta h + e^{-4\beta J}}}.$$

This is the exact solution and it shows that as $h \to 0$ there is no spontaneous magnetization (thus no phase transition).

1D Ising Specific Heat

Using the exact solution to the one dimensional Ising problem with

$$F = -k_B T \ln Z = -N K_B T \ln \lambda_m$$

where λ_m is the larger of the two eigenvalues

$$\lambda_\pm = e^K \cosh B \pm \{e^{2K} \sinh^2 B + e^{-2K}\}.$$

Let's derive an expression for the specific heat in zero field, $C_{H=0}$, in the limit where $T \to 0$. Also, show that the field magnetic susceptibility $X_{H \to 0}$ diverges in the limit where $T \to 0$. What is the nature of this singularity?

Partial Answer

We have

$$F = -K_B T \ln Z = -N k_B T \ln \left\{ e^4 \cosh B + (e^{2k} \sinh^2 B + e^{-2k})^{1/2} \right\}$$

Thus

$$\frac{J}{k_B T} = k \ ; \frac{\beta H}{k_B T} = B.$$

Using the relations

$$C_{H=0} = T \left(\frac{\partial S}{\partial T} \right), \qquad \left(\frac{\partial F}{\partial T} \right) = -S$$

We start with:

$$F_{H=0} = -N k_B T \{e^k + e^{-k}\} = -N k_B T \ln\{2 \cosh k\}$$

$$\left(\frac{\partial F}{\partial T}\right) = -Nk_BT\ln\{2\cosh k\} - Nk_BT\left(\frac{2\sinh k}{2\cosh k}\right)\left(-\frac{J}{k_B}\frac{1}{T^2}\right)$$
$$= -Nk_BT\ln\{2\cosh k\} + NJ(\tanh k)\left(\frac{1}{T}\right)$$

Thus

$$S = Nk_B\ln(2\cosh k) - NJ(\tanh k)\left(\frac{1}{T}\right)$$

and

$$\left(\frac{\partial S}{\partial T}\right) = Nk_B(\tanh K)\left(-\frac{J}{k_BT^2}\right) - NJ\tanh k\left(\frac{-1}{T^2}\right) - \frac{NJ}{T}(\text{sech}^2 k).$$

So

$$C_{H=0} = -NJ\,\text{sech}^2\left(\frac{J}{k_BT^2}\right) \rightarrow \quad \lim_{T\to 0} C_{H=0} = 0$$

5.6 1D Ising Solution by constructing entropy $S(E,N)$

Let's solve the 1 dimensional Ising problem by direct construction of the entropy function $S(E,N)$. Use this entropy function to derive the Helmholtz free energy $F(T,N)$ and show that result agrees with the result obtained by summing the partition function Z.

$$F = E - TS$$

$$E_m = -2NJ + 2mJ \quad ; S(m) = k_B\ln\left[\binom{N}{m}\right]$$
$$F = (2J)(m-N) - Tk_B\ln\left[\frac{N!}{m!\,(N-m)!}\right]$$
$$F = 2J(m-N) - k_BT\{\ln N! - \ln m! - \ln(N-m)!\}$$
$$F = 2J(m-N) - k_BT\{N\ln N - m\ln m - (N-m)\ln(N-m)\}$$

The derivation obtained by summing the partition function Ƶ has the Hamiltonian going from $-NJ$ to NJ as we go from order to disorder. Observe that E_m defined thus far is shifted by NJ in that it goes from$-2NJ$ to 0. Lets make the model consistent by making the energy scales consistent, so redefine: $E_m = -NJ + 2mJ$

From Ƶ derivation:

$$F = -Nk_BT\ln\{e^k + e^{-k}\} = -N\left(\frac{J}{K}\right)\ln\{2\cosh k\}$$
$$= -Nk_BT\ln(2\cosh k)$$

For entropy we then get:

$$S(m) = k_B\{N\ln N - m\ln m - (n-m)\ln(N-m)\}$$

$$E = -NJ + 2mJ \rightarrow m = (E + NJ)/2J$$

So,

$$\left(\frac{\partial S}{\partial E}\right) = \frac{1}{T} = \left(\frac{\partial S}{\partial m}\right)\left(\frac{\partial m}{\partial E}\right)$$

$$= \left(\frac{1}{2J}\right) k_B \left\{\frac{\partial}{\partial m}[-m \ln m - (N-m)\ln(N-m)]\right\}$$

$$= \left(\frac{K_B}{2J}\right)\left\{-\frac{m}{m} - (N-m)\frac{(-1)}{(N-m)} - \ln(N-m)(-1) - \ln m\right\}$$

$$= \frac{K_B}{2J}\left(\ln\left(\frac{N-m}{m}\right)\right) = \frac{1}{T}$$

Thus,

$$\frac{N-m}{m} = e^{2k} \rightarrow m = \frac{N}{e^{2k}+1} \rightarrow \quad m = \frac{2Ne^{-k}}{\cosh k}.$$

Using this we then write:

$$F = \left[-NJ + 2J\left(\frac{2Ne^{-k}}{\cosh k}\right)\right] - Tk_B\left\{N \ln N - \left(\frac{2Ne^{-k}}{\cosh k}\right)\ln\left(\frac{2Ne^{-k}}{\cosh k}\right) - \left(N - \frac{2Ne^{-k}}{\cosh k}\right)\ln\left(N - \frac{2Ne^{-k}}{\cosh k}\right)\right\}$$

$$F = -JN + 2JN - \frac{NJ}{K}\ln(e^{2k}+1) = JN - \frac{JN}{K}\ln\left[\frac{e^k + e^{-k}}{e^{-k}}\right]$$

$$= -\frac{JN}{k}\ln(2\cosh k) = -Nk_B T \ln(2\cosh k)$$

5.7 Evaluation of Magnetic Susceptibility for 1D Ising Model

Let's evaluate the magnetic susceptibility

$$\chi = \left[\frac{\partial M}{\partial H}\right]_{H\rightarrow 0, T\rightarrow 0}$$

For the one-dimensional Ising problem and show that χ is singular in the limit where $T \rightarrow 0$ (and $H \rightarrow 0$)

Solution

By definition:

$$M_I = \langle \sum_{i=}^{N} s_i \rangle = kT \cdot \frac{1}{Z_I} \cdot \frac{\partial Z_I}{\partial H} = -\frac{\partial F_I}{\partial H}$$

where

$$Z_I = \sum_{s_1=\pm 1} \sum_{S_2=\pm 1} \dots \sum_{S_{N=\pm 1}} e^{\beta V \Sigma s_i \cdot s_j + B \Sigma S_1}$$

where

$$B = \frac{aH}{k_B T} \;, \qquad \beta = \frac{1}{k_B T}.$$

We have:

$$F_I = -Nk_B T ln\lambda_+$$

$$\lambda_+ = e^k \cosh B + \sqrt{e^{2k} \sinh^2 B + e^{-2k}}, \qquad k = \beta V$$

$$M_I = -\frac{\partial F_I}{\partial H} = \frac{Nk_B T}{\lambda_B}\left(\frac{\alpha}{k_B T}\right)\frac{\partial \lambda_+}{\partial B}$$

We need to evaluate $\frac{\partial \lambda_+}{\partial B}$ and $\frac{\partial^2 \lambda_+}{\partial B^2}$:

$$\frac{\partial \lambda_+}{\partial B} = e^k \sinh B + \frac{1}{2}\frac{2 \sinh B \cosh B\, e^{2k}}{2\sqrt{e^{2k}\sinh^2 B + e^{-2k}}}$$

$$\frac{\partial^2 \lambda_+}{\partial B^2} = e^k \cosh B - \frac{1}{4}(e^{2k}\sinh^2 B + e^{-2k})^{-3/2}(2e^{2k}\sinh B \cosh B)^2 + \frac{1}{2}(e^{2k}\sinh^2 B + e^{-2k})^{-\frac{1}{2}}[2e^{2k}\cosh^2 B + 2e^{2k}\sinh^2 B]$$

So,

$$\chi_T = \frac{\partial M}{\partial H} = \frac{\alpha}{k_B T}\cdot\frac{\partial M}{\partial B} = \left(\frac{\alpha}{k_B T}\right)Nk_B T\left[\left(\frac{-1}{\lambda_+^3}\right)\cdot\left(\frac{\partial \lambda_+}{\partial B}\right)^2 + \frac{1}{\lambda_+}\frac{\partial^2 \lambda_+}{\partial B^2}\right]$$

As $H \to 0 \to \frac{\partial \lambda_+}{\partial B} \to 0, \frac{\partial^2 \lambda_+}{\partial B^2} \to e^k + \frac{1}{2}e^k \cdot 2e^{2k} = e^k + e^{3k}$, and $\lambda_+ \to e^k + e^{-k}$. So

$$\chi_T|_{H=0} = \left(\frac{\alpha^2}{k_B T}\right)N\cdot\left[\frac{1}{(e^k + e^{-k})^2}\cdot 0 + \frac{1}{(e^k + e^{-k})}\cdot(e^k + e^{3k})\right]$$

Thus

$$\chi_T|_{H=0} = \frac{N\alpha^2}{k_B T}\cdot e^{2k} = \left(\frac{N\alpha^2}{k_B T}\right)e^{\left(\frac{2V}{k_B T}\right)}$$

And as $T = 0$ we see that $\chi_T \to \infty$.

5.8 1D Ising Model derived from microcanonical ensemble

Let's compute the solution of the one dimensional Ising problem using the microcanonical ensemble. Find an expression for $S(E, H = 0)$. Also find $F(T, H = 0)$.

Solution

$$S(E) = k_B \ln w(E)$$

and $w(E)$ is the density of states. So, for this situation

$$S(E) = k_B ln \frac{2N!}{m!\,(N-m)!} = k_B[ln2 + lnN! - lnm! - ln(N-m)!]$$

Using the Stirling approximation:

$$S(E) \cong k_B[ln2 + NlnN - mlnm - (N-m)ln(N-m)]$$

Now we have the relations $\frac{\partial S}{\partial E} = \frac{1}{T}$ and $E = -(N-2m)V \rightarrow \frac{\partial E}{\partial m} = 2V$, so:

$$\frac{\partial S}{\partial m} = \frac{\partial S}{\partial E} \cdot \frac{\partial E}{\partial m} = \frac{2V}{T} = k_B ln\left(\frac{N-m}{m}\right), \quad m = \frac{E+NV}{2V}$$

Thus,

$$E = NV \tanh \frac{V}{k_B T}$$

$$S(E) = k_B \left[ln2 + NlnN - \left(N - \frac{E+NV}{2V}\right) ln\left(N - \frac{E+NV}{2V}\right) - \left(\frac{E+NV}{2V}\right) ln\left(\frac{E+NV}{2V}\right)\right]$$

$$= k_B \left[\begin{array}{l} ln2 + NlnN - \frac{N}{2}\left(1 + \tanh\frac{V}{k_B T}\right) ln\frac{N}{2}\left(1 + \tanh\frac{V}{k_B T}\right) \\ -\frac{N}{2}\left(1 - \tanh\frac{V}{k_B T}\right) ln\frac{N}{2}\left(1 - \tanh\frac{V}{k_B T}\right) \end{array}\right]$$

$F = E - TS = \frac{N}{2}\tanh\frac{V}{k_B T} - k_B T[\ldots]$, with above expression for $[\ldots]$.

5.9 Bragg-Williams Theory

Recall the simple Bragg Williams theory of alloys of two elements A and B on a three dimensional simple cubic lattice. The atoms interact through nearest neighbour interactions $H_{AA} = H_{BB}$, and H_{AB}. Assume that the heat of mixing parameter

$$\delta = H_{AB} - (H_{AA} + H_{BB})/2 > 0$$

(such that the the alloy tends to phase separate). Find the critical temperature, T_c, for phase separation. Assuming that $T \ll T_c$, find an expression for the solubility of B in A. Show that the solubility goes to zero exponentially in the limit where T goes to 0. If Ge and Si have a positive heat of mixing, what do you suppose happens as an alloy of the two is cooled to T=0? Does complete phase separation actually occur? If not, why? Why is it so difficult to obtain elemental materials such as silicon with ultrahigh purity?

Answer

In Bragg-Williams theory, elements A and B are in a simple cubic lattice.

$$H_{AA} = H_{BB}, \qquad \delta = H_{AB} - \frac{(H_{AA} + H_{BB})}{2}, \qquad \delta > 0$$

(such that the ground state will be phase separate). From Section X we have the following:

$$\Delta G_{mix} = \underbrace{ZN\delta C_A(1 - C_A)}_{enthalpy\ part} + \underbrace{Nk_B T[C_A \ln C_A + (1 - C_A)\ln(1 - C_A)]}_{entropy\ part}$$

We have the following graphical pictures of ΔG_{mix} at various temperatures.

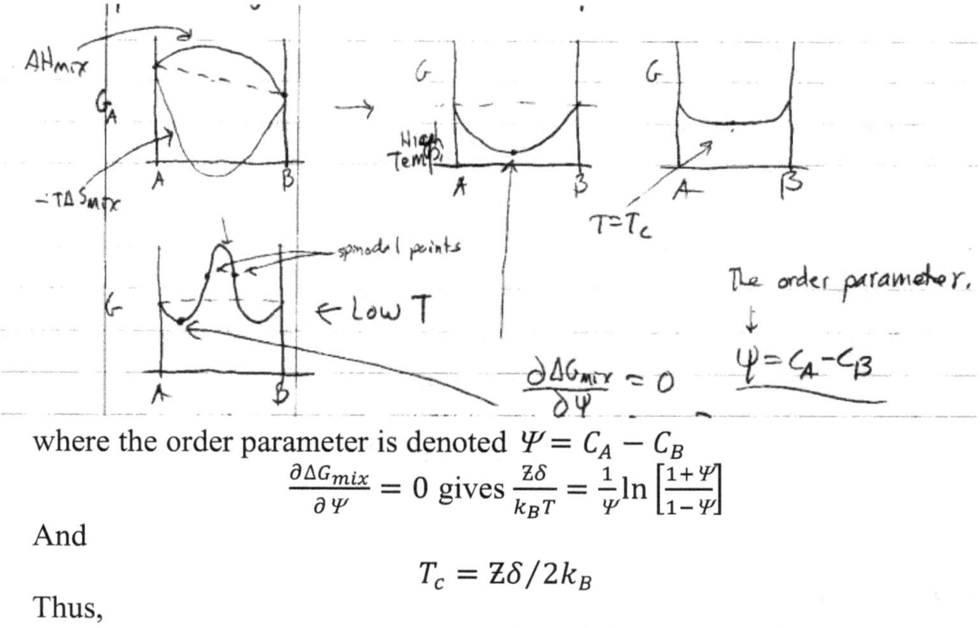

where the order parameter is denoted $\Psi = C_A - C_B$

$$\frac{\partial \Delta G_{mix}}{\partial \Psi} = 0 \text{ gives } \frac{Z\delta}{k_B T} = \frac{1}{\Psi}\ln\left[\frac{1+\Psi}{1-\Psi}\right]$$

And

$$T_c = Z\delta/2k_B$$

Thus,

$$(2T_c/T) = \frac{1}{\Psi}\ln\left[\frac{1+\Psi}{1-\Psi}\right]$$

Recall that $\tanh^{-1} x = \frac{1}{2}\ln\left(\frac{1+\Psi}{1-\Psi}\right)$,$(-1 < x < 1)$. So, $(2T_c/T)\,\Psi = 2\tanh^{-1}\Psi$ and

$$\Psi = \tanh\left(\frac{T_c}{T}\,\Psi\right)$$

Now, as $T \to 0$ we can simplify our expression since $\frac{T_S}{T}\Psi$ becomes large.
Since $(2T_c/T) = \frac{1}{4}\ln\left[\frac{1+\Psi}{1-\Psi}\right]$

$$1 - \Psi = (1 + \Psi)e^{-\frac{2T_C}{T}\Psi}$$

Thus

$$1 - \Psi_o = e^{\left(-\frac{2T_C}{T}\right)}$$

for $T \ll T_C$.

Solubility of B in A is C_B when calculating the positive root for Ψ:

$$\Psi_o = C_A - C_B = 1 - \exp\left(\frac{-2T_c}{T}\right)$$

Thus

$$C_A \cong 1, \qquad C_B = \exp\left(\frac{-2T_c}{T}\right)$$

If Ge and Si have a positive heat of mixing, as $T \to 0$ the alloy phase separates.

Complete phase separation does not occur because some impurity is always "frozen-in the cooling process and, also, δ is often fairly close to zero. There are also complications for when $H_{AA} \neq H_{BB}$ and, say, $H_{AA} < H_{AB} < H_{BB}$ etc. From the G-L theory we know the slope of $\Delta G\ at\ C_A = 1$ is infinite.

The mobility of dopants in silicon is greatly restricted at T=0. They random walk on diminishing length scales as $T \to 0$ such that they are never in a situation where the physics of the heat of mixing parameter is important and besides, δ is often close to zero (but positive) for the case of Si anyway.

5.10 Dense Gases and Liquids – The Virial Expansion

Consider the classical partition function for a simple atomic gas or liquid:

$$Z = \frac{1}{N!\,h^{3N}}\int d^{3N}p\, d^{3N}r\ e^{-\beta H}$$

where

$$H = \sum_i \frac{p_i{}^2}{2m} + \sum_{i<j} U(r_i - r_j),$$

where only central two-body forces are considered. Of course, this approximation is wrong, especially for the dense gases and liquids under consideration. However, when considering critical exponents and phase transition behavior in the Ising lattice-based analysis, we saw that as long as some correlation structure was retained in the approximation, then correct identification of the universal critical parameter behavior could be obtained (aided by the universality behavior itself). That will be seen to occur again here in the non-lattice setting. Thus, the interactions in the Hamiltonian for three-body interactions, etc., will be dropped from the analysis. This is a very convenient approximation to be able to perform because we can then directly integrate out the momentum variables in the above form:

$$Z = \lambda^{-3N} \frac{1}{N!} \int d^{3N}r \; e^{-\beta \sum_{i<j} U(r_i - r_j)}, \qquad \lambda = [h^2/(2m\pi k_B T)]^{1/2}.$$

The dimensionless part dependent on configuration is known as the configuration integral:

$$Q(V,T,N) = \frac{1}{N!} \int d^{3N}r \; e^{-\beta \sum_{i<j} U(r_i - r_j)}, \quad Z = \lambda^{-3N} Q$$

Recall that the ideal gas law gives us a simple relation between pressure and density. Let's now consider pressure described as a power series in density, this is known as the virial relation.

Virial Expansion

The viral expansion modifies the ideal gas law to describe pressure as a power series in density:

$$\frac{P}{k_B T} = \frac{N}{V}\left(1 + B_2(T)\left(\frac{N}{V}\right) + B_3(T)\left(\frac{N}{V}\right)^2 + B_4(T)\left(\frac{N}{V}\right)^3 \ldots\right).$$

The objective is to obtain the virial coefficients, $B_2(T)$, $B_3(T)$, for the particular gas model of interest. In developing the theoretical framework a little further, it will also be advantageous to compare with experiment to demonstrate that the critical behavior approximations are good, and get better as higher virial coefficients are considered.

Since working with N it is easiest to work with the grand canonical ensemble Z_{Grand} (following the method discovered by J.E. Mayer [40] as summarized by Plischke [23]):

$$\frac{P}{k_B T} = \frac{1}{V} \ln Z_{Grand} = \frac{1}{V} \ln \sum_{N=0}^{\infty} e^{\beta \mu N} \lambda^{-3N} Q.$$

Following the Mayer derivation [ref], we don't to work with $e^{\beta U}$ terms in the configuration integral since it doesn't have the desired power law expansion, instead we want to shift to functions $f_{ij} = f(r_{ij}) = e^{-\beta U(r_{ij})} - 1$, known as Mayer functions. Thus:

$$Q(V,T,N) = \frac{1}{N!}\int d^{3N}r \prod_{i<j}(1+f_{ij}).$$

This can then be expanded

$$Q(V,T,N) = \frac{1}{N!}\int d^{3N}r \left(1 + \sum_{i<j} f_{ij} + \sum_{i<j}\sum_{l<m} f_{ij}\, f_{lm} + \cdots\right).$$

What results is a diagrammatic analysis similar to a Feynman Diagram analysis to various orders. This will not be shown here (see Plischke [23]), we skip to the results for the virial coefficients:

$$B_2(T) = -\frac{1}{2}\int d^3r\, f_{12}(r)$$

$$B_3(T) = -\frac{1}{3V}\int d^3r_1 d^3r_2 d^3r_3\, f_{12}f_{23}f_{31}$$

In the case of the Lennard-Jones potential with parameters tuned for Argon:

$$U(r) = 4E\left[\left(\frac{\sigma}{r}\right)^{12} - \left(\frac{\sigma}{r}\right)^{6}\right], \frac{E}{k_B} = 120K, \sigma = 0.34nm,$$

we get for critical temperature:

Order of Approximation	$k_B T/E$
3rd-order (using B_3)	1.445
4th-order (using B_4)	1.300
5th-order (using B_5)	1.291
Argon Experimental	1.260

Thus, we do see that there is a (weak) convergence to the correct critical temperature as the virial expansion is considered to higher order, with a pretty good approximation by 3rd or 4th order.

Virial Expansion coefficients for hard sphere gas

Consider the Virial Expansion:

$$P = \rho kT\left(1 - \sum_{m=1}^{\infty}\frac{m}{m+1}\beta_m \rho^m\right)$$

Where $\beta_m = \frac{1}{m!}\sum^{(m)}\int \ldots \int \{m\}\pi \prod f_{ij}$ is a sum over irreducible clusters of particle 1 plus m other particles, and

$$f_{ij} = \exp\left[-\frac{u(r_{ij})}{kT}\right] - 1$$

Compute the value of β_1 and β_2 for a gas of interacting particles with the pair potential given by:

$$u(r) = \begin{cases} \infty & r < R \\ 0 & r \geq R \end{cases}$$

Starting with

$$P = \rho kT - \rho kT\left\{\frac{1}{2}\beta_1\rho\right\} - \rho kT\left\{\frac{2}{3}\beta_2\rho^2\right\} - \cdots$$

$$P = \rho kT[1 + \rho B(T) + \rho^2 C(T) + \cdots]$$

where

$$B(T) = -\frac{1}{2}\beta_1 \ , \qquad C(T) = -\frac{2}{3}\beta_2$$

$$\beta_1 = \int_0^\infty \left[\exp\left(-\frac{u(r_{12})}{kT}\right) - 1\right] d^3r_{12}$$

$$\beta_2 = \frac{1}{2}\int_0^\infty\int_0^\infty f_{12}f_{13}f_{23}\,d^3r_{12}d^3r_{13}$$

Using $u(r) = \begin{cases} \infty & r < R \\ 0 & r \geq R \end{cases}$:

$$\beta_1 = -\int_0^R [e^{-u/kT} - 1]4\pi r_{12}^2 dr_{12} + \int_R^\infty 4\pi r_{12}^2 dr_{12} = -\frac{4}{3}\pi R^3$$

Thus,

$$B(T) = \frac{2}{3}\pi R^3$$

$$\beta_2 = \frac{1}{2}\int_0^\infty\int_0^\infty \left[\exp\left(\frac{-u(r_{12})}{kT}\right) - 1\right]\left[\exp\left(\frac{-u(r_{13})}{kT}\right) - 1\right]\left[\exp\left(\frac{-u(|\vec{r}_{12}-\vec{r}_{13}|)}{kT}\right) - 1\right] d^3r_{12}d^3r_{12}$$

The simple potential makes the problem tractable:

$$\left[\exp\left(\frac{-u(r_{12})}{kT}\right) - 1\right] \to (-1) \ , \ r_{12} \leq R \text{, zero otherwise.}$$

Similarly,

$$\left[\exp\left(\frac{-u(r_{13})}{kT}\right) - 1\right] \to (-1) \ , \ r_{13} \leq R \text{, zero otherwise.}$$

$$\left[\exp\left(\frac{-u(|\vec{r}_{12}-\vec{r}_{13}|)}{kT}\right) - 1\right] \to (-1) \ , \ |\vec{r}_{12} - \vec{r}_{13}| \leq R \text{, zero otherwise.}$$

Thus, the total product is (-1) when $r_{12} \leq R, r_{13} \leq R$ and $|\vec{r}_{12} - \vec{r}_{13}| \leq R$ and is zero otherwise. This is satisfied while in the intersection region of radius R centered $|r_{12}|$ apart .

So, we have:

$$\beta_2 = -\frac{1}{2}\int_0^R dV_I$$

where V_I is the intersection volume of two radius R spheres a distance r_{12} apart:

Let $Z = \sqrt{R^2 - \left(\frac{r_{12}}{2}\right)^2}$ then:

$$V_I = \int_0^Z 2\left(\sqrt{R^2 - y^2} - \frac{r_{12}}{2}\right) 2\pi y dy$$

Now substitute $y = R\sin\theta$ and use $Z = \sin\theta' = \sqrt{1 - \left(\frac{r_{12}}{2}\right)^2}$ to get:

$$V_I = \int_0^{\theta\prime} 4\pi R^3 \cos^2\theta \sin\theta d\theta - 2\pi r_{12} \cdot R^2 \int_0^{\theta\prime} \cos\theta \sin\theta d\theta$$

$$= \frac{4}{3}\pi\left\{R^3 - \frac{3}{4}R^2 r_{12} + \frac{1}{16}r_{12}^3\right\}$$

Thus,

$$\int_0^R V_I d^3 r_{12} = \frac{4}{3}\pi\int_o^R 4\pi r_{12}^2\left\{R^3 - \frac{3}{4}R^2 r_{12} + \frac{1}{16}r_{12}^3\right\}dr_{12} = \frac{\pi^2}{3}\left[\frac{15}{6}\right]R^6$$

$$= \frac{15}{18}\pi^2 R^6$$

So,

$$\beta_2 = -\frac{1}{2}\left[\frac{15}{18}\pi^2 R^6\right]$$

and

$$C(T) = -\frac{2}{3}\beta_2 = \frac{15}{18}\pi^2 R^6$$

Virial Expansion coefficients for hard sphere gas with repulsive core

Let's now recompute for pair potential slightly trickier. Compute the value of β_1 and β_2 for a gas of interacting particles with the pair potential given by (spherical square well with repulsive core):

$$u(r) = \infty, \quad r < R$$
$$u(r) = -V, \quad R < r < R + a$$
$$u(r) = 0, \quad r > R$$

Again, from

$$P = pk_B T\left\{1 - \sum_{m=1}^{\infty}\frac{m}{m+1}\beta_m p^m\right\}$$

$$\frac{p}{kT} = p + B(T)p^2 + C(T)p^3 + \cdots$$
$$B(T) = -\frac{1}{2}\beta_1 \; ; C(T) = -\frac{2}{3}\beta_2$$
$$\beta_1 = \frac{1}{v}\iint f_{12} dr_1 dr_2 = \int \left[e^{-\beta u(r)} - 1\right] d\vec{r}, \qquad \beta = {}^{1}/_{kT}$$

Let's express the potential in terms R: $R + a = \lambda R$, $\lambda = 1 + \frac{a}{R}$, so that:

$$u(r) = \infty, \qquad r < R$$
$$u(r) = -V, \qquad R < r < \lambda R$$
$$u(r) = 0, \qquad r > \lambda R$$

We now have:

$$\beta_1 = 4\pi \int \left(r^2 e^{-\beta u(r)} - r^2\right) dr = 4\pi \left\{ \int_0^R -r^2 dr + \int_R^{\lambda R} \left(r^2 e^{V\beta} - r^2\right) dr \right\}$$
$$= -\frac{4\pi R^3}{3}\{1 - (e^{V\beta} - 1)(\lambda^3 - 1)\}$$

Thus,

$$\beta_1 = -\frac{4\pi R^3}{3}\left\{1 - \left(e^{V\beta} - 1\right)\left(\left(1 + \frac{a^3}{R}\right) - 1\right)\right\}$$

Let's now compute

$$C(T) = -\frac{2}{3}\frac{1}{2V}\iiint f_{12} f_{13} f_{23} d\vec{r}_{12} d\vec{r}_{13}$$
$$\beta_2 = \frac{1}{2}\iint f_{12} f_{13} f_{23} d\vec{r}_{12} d\vec{r}_{13}$$

Let $x = \exp(V\beta) - 1$:

$$C(T) = 2\left(\frac{\pi}{6}R^3\right)^2 [5 - (\lambda^6 - 18\lambda^3 + 32\lambda^3 - 15)x$$
$$+ (-2\lambda^6 + 36\lambda^4 + 8\lambda^3 + 8\lambda^2 + 16)x$$
$$- (6\lambda^6 - 18\lambda^4 + 18\lambda^2 - 6)x^3]$$

For $\lambda \leq 2$ and

$$C(T) = 2\left(\frac{\pi}{6}R^3\right)[5 - 17x + (32\lambda^3 - 18\lambda^2 - 48)x^2$$
$$- (5\lambda^6 - 32\lambda^3 + 18\lambda^2 + 26)x^3]$$

For $\lambda \geq 2$. Thus, using $\beta_2 = -\frac{3}{2}C(T)$, $\lambda = 1 + \frac{a}{R}$, $x = e^{v\beta} - 1$:

$$\beta_2 = -3\left(\frac{\pi}{6}R^3\right)$$
$$\times \begin{bmatrix} 5 - 17x + (\lambda^6 - 18\lambda^4 - 32\lambda^3 - 15)x \\ -(-2\lambda^6 + 36\lambda^4 + 18\lambda^2 + 16)x^2 - (6\lambda^6 - 18\lambda^4 + 18\lambda^2 - 6) \end{bmatrix}$$

For $a \le R$, and

$$\beta_2 = -3\left(\frac{\pi}{6}R^3\right)[5 - 17x + (32\lambda^3 - 18\lambda^2 - 48)x^2 - (5\lambda^6 - 32\lambda^3 + 18\lambda^2 + 26)x^3]$$

For $a \ge R$.

Determining Gas parameters in the Virial Approximation

Let's find the expression for the entropy function in the dilute gas approximation (where, for example, only the first term in the virial expansion is retained, i.e. B(T)=-1/2 $\int f_{12} d^3 r_{12}$ *and* $u(r)$ is the spherical square well potential. Suppose that you were to design an experiment to determine the parameters V,R and a. Discuss how you would design this experiment and what properties you would measure. Explain how you would extract the values of the parameters. Would the second order terms in the expansion of the free energy be of any use in this endeavour?

Answer#1

Starting with an expression for the Helmholtz free energy in this situation:

$$\frac{F}{N} = \frac{F_{IG}}{N} - pKTB(T)$$

where F_{IG} is the F for a ideal gas:

$$F_{IG} = -KTN\log\frac{V}{N\Lambda^3} - NkT\ ,\ \Lambda = \frac{2\pi\hbar}{\sqrt{2\pi mkT}}.$$

Thus

$$F = -KTN\ln\left(\frac{V}{N\Lambda^3}\right) - NKT - pKNT\,B(T)$$

$$S = -\left(\frac{\partial F}{\partial T}\right)_V = KN\ln\left(\frac{V}{N\Lambda^3}\right) - KN - pKN\left(B(T) + T\frac{\partial B(T)}{\partial T}\right)$$

$$S = KN\left\{\ln\left(\frac{V}{N\Lambda^3}\right) - 1 - p\left(B(T) + T\frac{\partial B(T)}{\partial T}\right)\right\}$$

where $B(T) = \frac{2\pi R^3}{3}\left\{1 - (e^{v\beta} - 1)\left[\left(1 + \frac{a}{R}\right)^3\right]\right\}$ and

$$\frac{\partial B(T)}{\partial T} = \frac{2\pi R^3}{3}\left\{-\left[\left(1 + \frac{a}{R}\right)^3 - 1\right]e^{vB}\frac{v}{K}\left(-\frac{1}{T^2}\right)\right\}$$

Thus

$$S = KN\left\{\ln\left(\frac{V}{N\Lambda^3}\right) - 1 - p\left(\frac{2\pi R^3}{3}\right)\left\{1 - (e^{vB} - 1)\left[\left(1+\frac{a}{R}\right)^3 - 1\right] + \left[\left(1+\frac{a}{R}\right)^3 - 1\right]e^{vB}(v\beta)\right\}\right\}$$

$$= KN\left\{\ln\left(\frac{V}{N\Lambda^3}\right) - 1 - p\left(\frac{2\pi R^3}{3}\right)1 + e^{vB}\left[\left(1+\frac{a}{R}\right)^3 - 1\right][v\beta - 1] + \left[\left(1+\frac{a}{R}\right)^3 - 1\right]\right\}$$

$$= KN\left\{\ln\left(\frac{V}{N\Lambda^3}\right) - 1 - p\left(\frac{2\pi R^3}{3}\right)\left[e^{v\beta}\left[\left(1+\frac{a}{R}\right)^3 - 1\right](v\beta - 1) + \left(1+\frac{a}{R}\right)^3\right]\right\}$$

Answer#2

Dilute gas approximation:

$$P = pkT[1 + pB(T)], \qquad pB(T) \ll 1$$

where the $u(r)$ in the $B(T)$ calculation is due to a spherical square well. Thus, from Example X we have:

$$B(\tau) = \frac{2\pi R^2}{3}\left\{(e^{V\beta} - 1)\left[1+\frac{a}{R}\right]^3 - 1\right\}, \quad \beta = {}^{1}/_{kT}$$

where

$$u(r) = \begin{cases} \infty & r < R \\ -V & R < r < R + a \\ 0 & r > R + a \end{cases}$$

$$\frac{\partial B(T)}{\partial T} = \frac{2\pi R^3}{3}\left\{\left(1+\frac{a}{R}\right)^3 - 1\right\}e^{V\beta}\frac{\partial(V\beta)}{\partial T}$$

$$= \frac{2\pi R^3}{3}\left\{\left(1+\frac{a}{R}\right)^3 - 1\right\}e^{V\beta}\frac{V}{K_B}\left(\frac{-1}{T^2}\right)$$

Note that V is not volume. Let`s change the symbol for the potential energy of the well from –V to –t. Use $V = \frac{NKT}{P}[1 + \rho B(T)]$, in the connection term with $\rho = {}^{P}/_{KT}$, thus:

$$V = \frac{NKT}{P} + NB$$

Now use the Maxwell's relation $\left(\frac{\partial S}{\partial P}\right)_T = -\left(\frac{\partial V}{\partial T}\right)_P$

$$S = NKln({}^{p_0}/_{p}) + N\left(\frac{\partial B}{\partial T}\right)(P_0 - P)$$

$$S = NKln(P_o/P) + N\frac{2\pi R^3}{3}\left[\left(1+\frac{a}{R}\right)^3 - 1\right]e^{t\beta}\left(\frac{t}{k_B}\right)\left(-\frac{1}{T^2}\right)(P_0 - P)$$

A simple way to determine V, R and a is by comparison to a Van der Waals equation of state at various T. Thus, starting with the Van der Waals form:

$$\left(P + \frac{aN^2}{V^2}\right)(V - Nb) - NkT$$

$$P = a\frac{N^2}{V^2} + \frac{NkT}{V}\frac{1}{\left(1-\frac{N}{V}b\right)}$$

$$P = \frac{NkT}{V}\left\{1 + \left(b - \frac{a}{kT}\right)\left(\frac{N}{v}\right) + b^2\left(\frac{N}{V}\right)^2 + \cdots\right\}$$

$$P = pkT\left\{1 + \left(b - \frac{a}{kT}\right)p + b^2p^2 + \cdots\right\}$$

Thus

$$P \cong pkT\left\{1 + \left(b - \frac{a}{kT}\right)p\right\}$$

for the dilute gas. Let's compare forms:

$$\left(b - \frac{a}{kT}\right) = \frac{2\pi R^3}{3}\left\{1 - \left(e^{\nu\beta} - 1\right)\left[\left(1 + \frac{a}{R}\right)^3 - 1\right]\right\}$$

So,

$$b = \frac{2\pi R^3}{3}$$

And

$$a = V\left[\left(1 + \frac{a}{R}\right)^3 - 1\right]\frac{2\pi R^3}{3}$$

5.11 The 2-D Ising Model – Exact Solution

The exact solution to the 2-D Ising model was obtained by Onsager in 1944 [29]. Even using a modern approach to simplify things as with Schultz [33] and then using a clearly written synopsis by Plischke [23], it is still too much to consider in detail here. For a description of this amazing derivation see Plischke [23]. To have a sense of the maneuvers involved in the mathematics employed by Schultz refer to Chapter 10 where Euclideanized time is described. In that section we see that a common maneuver in quantum field theory in 3+1 Lorentzian spacetime is to shift to imaginary time to get a 4D Euclidean spacetime, where the integrals become well-defined. The Wick rotation to go from real time to imaginary time essentially takes a (N-1)+1 dimensional Lorentzian field theory over to a N dimensional (Euclidean) statistical field theory. This can be done in reverse (Wick rotation the other way) as long as the Lorentzian theory is well-defined. If working in a quantum mechanics formulation (not quantum field theory form), the Lorentzian description will automatically be well-defined. Thus, a maneuver used in quantum field theory (Wick rotation) will be used by Schultz in reverse to take a difficult 2D theory (with angular momentum commutation) over to a much less complicated 1+1D quantum system description (with canonical commutation). The simplified description is then solved to get he critical temperature:

$$\sinh\frac{2J}{k_B T_c} = 1 \rightarrow \frac{k_B T_c}{J} \cong 2.269\ldots$$

2D Ising Model has Specific Heat with logarithmic singularity

Using the exact solution to the 2-dimensional Ising problem discussed in Landau [22], let's show that the specific heat exhibits a logarithmic singularity (as a function of temperature) at the critical temperature T_c:

$$\phi = -NT\ln 2 + NT\ln(1-x^2)$$

$$-\frac{NT}{2(\pi)^2}\int_0^{2\pi}\int_0^{2\pi}\ln[(1+x^2)^2 - 2x(1-x^2)(\cos w_1 + \cos w_2)]\,dw_1 dw_2$$

where

$$x = \tanh(J/T)$$

The function $\phi(T)$ has a singularity when the argument of the logarithm can vanish.

The minimum value for a zero in the argument, in terms of $x(T)$, will correspond to the phase transition point. We want the minimum value, k_c, for a zero in the argument of the log. So, we must minimize as a function of w_1 and w_2 of:

$$(1+x^2)^2 - 2x(1-x^2)(\cos w_1 + \cos w_2) = 0$$

Which is found to occur when $\cos w_1 = \cos w_2 = 1$, with positive root of interest $x_c = \sqrt{2} - 1$.

What is vital thus far is that $\cos w_1 = \cos w_2 = 1$ provides x_c. Near the transition point not only is the argument of the log no longer considered zero (for the singularity) but we get contributions from the different $w_1's\ and\ w_2's$. Still, the most important contributions are with those that have $x \sim x_c$, or, in terms of an expansion about the minimum and truncating the rest, we get the approximate integrand:

$$A = (1+x^2) - 2x(1-x^2)(2 + w_1^2 + w_2^2)$$

Now if we expand A about x_c, transform $x \rightarrow T$ and pick off leading order behaviour we get integrand of the form:

$$A = C_1 t^2 + C_2(w_1^2 + w_2^2)$$

thus

$$\phi \cong \int_0^{2\pi}\int_0^{2\pi} A dw_1 dw_2 = a + \frac{1}{2} b(T - T_c)^2 \ln|T - T_c|$$

and

$$C = \frac{\partial^2 \phi}{\partial T^2} = -bT_c \ln|T - T_c|.$$

Thus, the specific heat exhibits a logarithmic singularity.

5.12 Renormalization Group

1D Ising Renormalization Group

Recall the partition function for the theory:

$$Z_N = \sum_{\sigma_1=\pm 1} \cdots \sum_{\sigma_N=\pm 1} e^{\beta J \sum_{i=1} \sigma_i \sigma_{i+1}}.$$

At the level of the partition function we now perform a very different kind of approximation maneuver. For contrast, recall that in mean field theory we average over all but a few degrees of freedom, an infinity of contributions is reduced to a finite number of degrees of freedom. For renormalization the objective is to remove an infinite number of degrees of freedom, but still leave an infinite number of degrees of freedom. Suppose we remove half an infinite 1D lattice by averaging out every other position. We can then describe a relation between the original partition function and that of the renormalized theory (both being 1D infinite lattices). In doing this we arrived at a renormalization group relation, from which very useful results can be obtained easily.

Let's regroup the above expression:

$$Z_N = \sum_{\sigma_1=\pm 1} \cdots \sum_{\sigma_N=\pm 1} e^{\beta J[\sigma_1\sigma_2 + \sigma_2\sigma_3]} e^{\beta J[\sigma_3\sigma_4 + \sigma_4\sigma_5]} \cdots.$$

and then sum out the even numbered indices:

$$Z_N = \sum_{\sigma_1=\pm 1} \cdots \sum_{\sigma_{odd}=\pm 1} \begin{matrix} \left\{e^{\beta J[\sigma_1+\sigma_3]} + e^{-\beta J[\sigma_1+\sigma_3]}\right\} \\ \times \left\{e^{\beta J[\sigma_3+\sigma_5]} + e^{-\beta J[\sigma_3+\sigma_5]}\right\} \cdots \end{matrix}.$$

As indicated above, we want to relate this to the form of he original 1D Ising partition function with half the lattice density. To do so let's rewrite in the form:

$$e^{K[s+s']} + e^{-K[s+s']} = f(K) e^{-K'ss'}, \qquad K = \beta J,$$

for which we have:

$$Z_N = \sum_{\sigma_1=\pm 1} \cdots \sum_{\sigma_{odd}=\pm 1} f(K) e^{K'[\sigma_1\sigma_3]} f(K) e^{K'[\sigma_3\sigma_5]} \cdots$$

Thus,

$$Z_N(K) = [f(K)]^{\frac{N}{2}} Z_{N/2}(K').$$

From the latter form, known as a Kadanoff transformation [34,35,24] we can quickly obtain some useful results. To use this form, however, we must verify it exists, i.e., we must obtain the unknowns $f(K)$ and K'.

If $s = s' = \pm 1$ we have:

$$e^{2K} + e^{-2K} = f(K)e^{K'}.$$

If $s = -s' = \pm 1$ we have:

$$2 = f(K)e^{-K'} \rightarrow f(K) = 2e^{K'}$$

Thus

$$f(K) = 2\cosh^{1/2}(2K)$$
$$K' = 1/2\, ln \cosh(2K)$$

and we see that the transformation exists.

If writing the partition function in relation to free energy we separate a N factor, leaving the remaining factor an intensive variable, independent of system size:

$$\ln Z_N(K) = Ng(K)$$

Thus,

$$g(K') = 2g(K) - \ln\left[2\cosh^{\frac{1}{2}}(2K)\right].$$

The $f, g,$ and K' relations are the renormalization group equations. Successive application of these equations will result in flow to a stable fixed point of the theory (or away from an unstable fixed point). For the 1D Ising result the flow will take the K parameter to zero or infinity, indicating there is no phase transition in the theory.

5.12.1 The 2D Ising Renormalization Group

For the 2D problem we will repeat the strategy outlined in the 1D case. Starting with the partition function for the theory:

$$Z_N = \sum_{\sigma_1=\pm 1} \cdots \sum_{\sigma_N=\pm 1} e^{\beta J \sum_{i=1} \sigma_i \sigma_{i+1}}.$$

except now the pairwise couplings occur according to the nearest neighbor in a square lattice configuration. Suppose the four nearest neighbors of σ_5 are σ_1, σ_2, σ_3, and σ_4,we can write:

$$Z_N = \sum_{\sigma_1=\pm 1} \cdots \sum_{\sigma_N=\pm 1} \ldots e^{\beta J \sigma_5[\sigma_1+\sigma_2+\sigma_3+\sigma_4]} e^{\beta J \sigma_{10}[\sigma_6+\sigma_7+\sigma_8+\sigma_9]} \ldots.$$

where alternating diagonals are separated out (σ_5 and σ_{10} might be on the same diagonal, for example). This is done to then sum out the spins on alternating diagonals in the lattice (upon removal of the alternating diagonals on the lattice you have a new square lattice with half the node density and turned at 45 degrees relative to the original lattice). In this manner we partly sum the partition function to get:

$$Z_N = \sum_{\sigma_1=\pm 1} \cdots \sum_{\sigma_{odd\ diagonal}=\pm 1} \cdots \times \left\{e^{K[\sigma_1+\sigma_2+\sigma_3+\sigma_4]} + e^{-K[\sigma_1+\sigma_2+\sigma_3+\sigma_4]}\right\} \times \ldots$$

in a similar manner to before, we now want to obtain the form:

$$e^{K[\sigma_1+\sigma_2+\sigma_3+\sigma_4]} + e^{-K[\sigma_1+\sigma_2+\sigma_3+\sigma_4]} = f(K)e^{-K'[\sigma_1\sigma_2+\sigma_2\sigma_3+\sigma_3\sigma_4+\sigma_4\sigma_1]}.$$

The inequivalent choice of spin configuration now gives rise to four equations, not two, and guessing that a solution exists in the form:

$$e^{K[\sigma_1+\sigma_2+\sigma_3+\sigma_4]} + e^{-K[\sigma_1+\sigma_2+\sigma_3+\sigma_4]} = f(K)\exp\left[\begin{array}{c}\left(\frac{1}{2}\right)K_1(\sigma_1\sigma_2+\sigma_2\sigma_3+\sigma_3\sigma_4+\sigma_4\sigma_1) \\ +K_2(\sigma_1\sigma_3+\sigma_2\sigma_4) \\ +K_3(\sigma_1\sigma_3\sigma_2\sigma_4)\end{array}\right]$$

we find that it does with:

$$K_1 = \frac{1}{4}\ln\cosh(4K), \qquad K_2 = \frac{1}{8}\ln\cosh(4K)$$

$$K_3 = \frac{1}{8}\ln\cosh(4K) - \frac{1}{2}\ln\cosh(2K)$$

$$f(K) = 2[\cosh(4K)]^{1/8}[\cosh(2K)]^{1/2}.$$

Consider the implications of the form now shown to exist by these solutions, it indicates that our nearest neighbor coupling, upon renormalization as indicated, results in a new coupling that is not only nearest neighbor, but also next nearest neighbor, and also has a 4-coupling involving spin couplings that complete a square in the lattice. Notice that if we try to approximate the renormalization equation on the form above by dropping the K_2 and K_3 terms we get:

$$e^{K[\sigma_1+\sigma_2+\sigma_3+\sigma_4]} + e^{-K[\sigma_1+\sigma_2+\sigma_3+\sigma_4]} = f(K)\exp\left[\left(\frac{1}{2}\right)K_1(\sigma_1\sigma_2+\sigma_2\sigma_3+\sigma_3\sigma_4+\sigma_4\sigma_1)\right]$$

which is precisely the same form as in the 1D Ising problem, with similar renormalization group flow results, i.e., no phase transition. Thus, to capture the local correlations sufficiently to demonstrate that there is a phase transition we must at least show the theory at complexity greater than the 1D Ising form. So, let's now drop only the K_3 term but let's still approximate the K_2 contribution describing next nearest neighbor interactions as a modification to the nearest neighbor interaction such that we have: $K_1 \to K'(K_1, K_2)$. Thus, we have a very similar renormalization group equation form to the 1D Ising case:

$$Z_N(K) = [f(K)]^{\frac{N}{2}} Z_{N/2}\big(K'(K_1, K_2)\big).$$

If we again let

$$\ln Z_N(K) = Ng(K)$$

we get

$$g(K') = 2g(K) - \ln\left[2[\cosh(4K)]^{1/8}[\cosh(2K)]^{1/2}\right].$$

When all spins are aligned we have the relation

$$K' \cong K_1 + K_2 = \frac{3}{8}\ln\cosh(4K).$$

There is a fixed point at

$$K = \frac{3}{8}\ln\cosh(4K) \to \quad K \cong 0.50698\,...$$

The fixed point is associated with a critical temperature:

$$\frac{k_B T_c}{J} \cong 1.972\,...$$

where the exact answer (with a lot more work) is:

$$\frac{k_B T_c}{J} \cong 2.269\,...\,.$$

5.13 Exercises

5.1 The one dimensional lattice gas

Consider a one dimensional lattice gas consisting of sites occupied by atoms or by vacancies.

Nearest neighbour atom pairs have an attractive interaction $-V$ $(V > 0)$ while nearest neighbour vacancy pairs or atom/vacancy pairs do not interact at all. Suppose this system is in contact with a large reservoir of atoms at chemical potential μ and at temperature T (i.e. the number of atoms is variable but the chemical potential is fixed). What will be the fraction of occupied sites (density) of the lattice gas as a function of temperature? Is there a phase transition to a "condensed phase"? (Hint – this is a 1 dimensional Ising problem in disguise – the chemical potential can be thought of as being like the applied magnetic field H in the usual Ising problem.)

The following questions relate to critical phenomena. Give a short answer to each question.

(a) Explain the principle of the renormalization group method.

(b) Give an example of a scaling law for critical exponents.

(c) Define the term "percolation threshold".

(d) Explain what is meant by "fractal dimensionality" of a percolating cluster.

(e) Can you give a simple argument showing why one dimensional systems exhibit no phase transitions?

5.2 Alternating Bethe Lattice

Consider a Bethe lattice which has the property that, as you move along any pathway from one site to the next, the coordination number of the site alternates between Z=3 and Z=4. The lattice is illustrated:

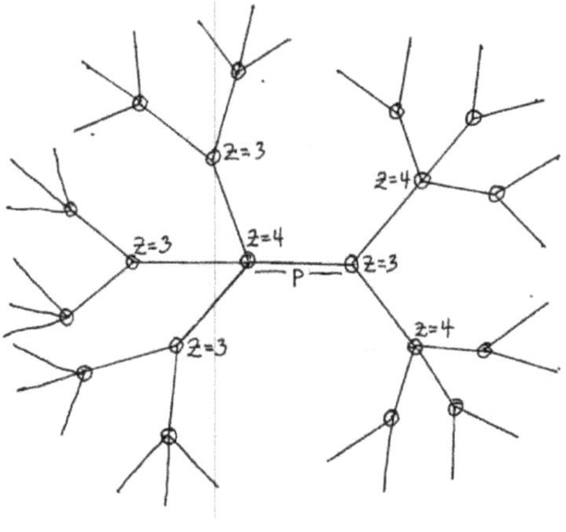

What is the percolation threshold, Pc, for such a lattice?

5.3 Using 1D Renormalization Group

How can you use the renormalization group result for the 1D problem:

$$c(K) = 1/2 \ln f(K) + 1/2\, c(K')$$
$$K' = 1/2 \ln \cosh(2K)$$
$$f(K) = 2 \cosh^{1/2}(2K)$$

To determine the nature of the behaviour of the zero field heat capacity $C_{H=0}$, in the limit where $T \to 0$. Does this agree with the exact solution?

5.4 Show convergence to solution for critical value

We obtained a good approximate renormalization of the 2D problem using the recursive law.

$$K' = K_1 + K_2 = (3/8)\ln\cosh(4K)$$

and

$$c(K') = 2c(K) - \ln\left[2\cosh^{1/2}(2K)\cosh^{1/8}(4K)\right]$$

This can be used to numerically determine Z as in the 1D problem. Prove that there is a critical value of $K = K_c$ for which the renormalization does not change K, i.e. for which $K = K'$. This gives the critical point. Find the transition temperature in units of J/k_B. First use a calculator or computer to numerically determine K_c. Make a plot of $c(k)$ vs K. Make a plot of F and comment on the nature of the singularity at T_c. How might you try to improve upon this approximate renormalization? Using the recursion relations should give $k_c \cong 0.50698$, where

$$k_c = \frac{J}{k_B T_c} \rightarrow T_c = \left(\frac{J}{k_B}\right)\left(\frac{1}{.507}\right).$$

5.5 Show 2D Ising phase transition temperature is finite

Show that the two dimensional Ising problem has a phase transition at a finite temperature T_c. Also determine the nature of the specific heat anomaly at T_c. (You may use the solution for F given by Onsager). 0.46 is exact solution.

Onsager's solution:

$$F(T, H = 0) = -k_B T\left[ln(2\cosh 2k) + \frac{1}{2\pi}\int_0^{\pi} d\varphi\, ln\left\{\frac{1}{2}\left(1 + \sqrt{1 - k^2\sin^2\varphi}\right)\right\}\right]$$

$$k = \frac{2\sinh 2k}{\cosh^2 2k}$$

Chapter 6. Reissner-Nordstrom Anti-deSitter Black Holes

6.1 Introduction

In this chapter a description is given for the Hamiltonian dynamics and thermodynamics of spherically symmetric Einstein-Maxwell spacetimes with a negative cosmological constant [36]. We impose boundary conditions that enforce every classical solution to be an exterior region of a Reissner-Nordstrom-anti-de Sitter black hole with a nondegenerate Killing horizon, with the spacelike hypersurfaces extending from the horizon bifurcation two-sphere to the asymptotically anti-de Sitter infinity. The constraints are simplified by a canonical transformation in the spherically symmetric vacuum Einstein theory, and the theory is reduced to its true dynamical degrees of freedom. After quantization, the grand partition function of a thermodynamical grand canonical ensemble is obtained by analytically continuing the Lorentzian time evolution operator to imaginary time and taking the trace. A similar analysis under slightly modified boundary conditions leads to the partition function of a thermodynamical canonical ensemble. The thermodynamics in each ensemble is analyzed, and the conditions that the (grand) partition function be dominated by a classical Euclidean black hole solution are found. When these conditions are satisfied, we recover in particular the Bekenstein-Hawking entropy. The limit of a vanishing cosmological constant is briefly discussed.].

Background

Hawking's celebrated result of black hole radiation [37] and related developments [38-40] made it possible to consider thermodynamical equilibrium systems involving black holes in the manner first anticipated by Bekenstein [41,42]. At a semiclassical, "phenomenological," level, a black hole thermodynamical equilibrium system can be introduced by simply immersing a radiating black hole in a heat bath such that the outgoing Hawking radiation balances the radiation that falls in from the bath [43-47]. At a deeper level, one aspires to construct a full thermodynamical equilibrium ensemble by starting from a quantum theory of gravity for black hole type geometries [48-53]. For reviews, see for example Refs. [44,52,54-56]. At the semiclassical level, the thermodynamical equilibrium configurations involving black holes tend to be unstable against thermal fluctuations [43,44]. The classic example is a Schwarzschild black hole in equilibrium with an asymptotically flat heat bath, in the approximation where the back-reaction of the radiation on the geometry is neglected: the heat capacity in this instance is $-(8\pi T^2)^{-1}$, where T is the temperature measured at the infinity, and the fact that this heat capacity is negative indicates thermodynamical instability. While such instabilities are not unexpected in self-gravitating systems, they do pose an obstacle to constructing thermodynamical equilibrium ensembles from quantum gravity. This is because the existence of a thermodynamical ensemble implies the positivity of certain response functions associated with that ensemble [57]. For example, in the canonical ensemble the heat capacity is necessarily positive; consequently, a canonical ensemble of the usual kind does not appear to exist for Schwarzschild black holes in asymptotically at space [58] (see Appendix A for details).

To construct a thermodynamical ensemble appropriate for black hole geometries from a quantum theory of gravity, one thus needs to choose the boundary conditions for the ensemble in a judicious manner, motivated by the stability of the corresponding semiclassical equilibrium situations. One possibility is to replace an asymptotic infinity by a finite "box" at which the local temperature is then fixed [50,59-70]. The possibility on which we shall concentrate in this analysis is to include a negative cosmological constant [54,71-75].

A negative cosmological constant makes classical black hole solutions asymptotically anti-de Sitter. We shall consider spherically symmetric spacetimes, and as the only matter field we include the spherically symmetric Maxwell field. All the relevant classical solutions then belong to the Reissner-Nordstrom-anti-de Sitter (RNAdS) family [76-79]. The temperature of the Hawking radiation is redshifted to zero at the asymptotically anti-de Sitter infinity, but from the rate at which the local Hawking temperature approaches zero one can extract a \renormalized" Hawking temperature, and this renormalized Hawking temperature can then be taken as one fixed quantity in the thermodynamical ensembles [54,71,72,74]. We shall consider both the canonical ensemble, in which the electric charge is fixed, and the grand canonical ensemble, in which the electric potential difference between the event horizon and the infinity is fixed.

To quantize the theory and to build the equilibrium ensembles, we shall adapt the method introduced in [68] in the context of spherically symmetric vacuum geometries in the presence of a finite boundary. We shall first set up a classical Lorentzian Hamiltonian theory in which, on the classical solutions, the right end of the spacelike hypersurfaces is at the asymptotically anti-de Sitter infinity in an exterior region of a black hole spacetime, and the left end of the hypersurfaces is at the bifurcation two-sphere of a nondegenerate Killing horizon. We then canonically quantize this theory, and obtain the thermodynamical (grand) partition function by suitably continuing the Schrodinger picture time evolution operator to imaginary time and taking the trace. A crucial input is how to handle the analytic continuation at the bifurcation two-sphere. As in [68], we shall see that a continuation motivated by smoothness of Euclidean black hole geometries yields a (grand) partition function that is in agreement with that obtained via path integral methods.

To implement the method used in [68], one must be able to canonically quantize the Lorentzian theory in some practical fashion. In [68] this was achieved by using canonical variables that were first introduced by Kuchar under asymptotically flat, Kruskal-like boundary conditions [80]. In these variables the constraints of the vacuum theory become simple, and the classical Hamiltonian theory can be explicitly reduced into an unconstrained Hamiltonian theory with just one canonical pair of degrees of freedom. We shall show that an analogous set of canonical variables exists for our system, and the classical Hamiltonian theory can again be explicitly reduced into an unconstrained Hamiltonian theory. Under boundary conditions tailored to the grand canonical ensemble, the reduced Lorentzian Hamiltonian theory has two canonical pairs of degrees of freedom; under boundary conditions tailored to the canonical ensemble, the reduced Lorentzian Hamiltonian theory has just one pair of canonical degrees of freedom. Using these variables, it will be possible to construct a quantum theory and a (grand) partition function in close analogy with [68].

It will turn out that both the canonical ensemble and the grand canonical ensemble for our system are well defined. In particular, the appropriate thermodynamical response functions are positive. We shall also be able to give the conditions under which the (grand) partition function is dominated by a classical Euclidean solution. The grand canonical ensemble exhibits a transition from a region where a classical solution dominates to a region where no classical solution dominates, in close analogy with what happens with the spherically symmetric boxed vacuum canonical ensemble [60,61]. As in [60,61] one may see this as evidence for a phase transition between a black hole sector and a topologically different "hot anti-de Sitter space" sector. In the canonical ensemble we find evidence for this kind of a phase transition only in the special case when the charge vanishes. When the charge is nonvanishing, there occurs a different kind of phase transition in which the dominating contribution to the partition function shifts from one classical solution to another as the boundary data changes.

Some facts about the RNAdS solutions are collected in Appendix B. Such as the classical Hamiltonian analysis and the quantization of the reduced Hamiltonian theory in the case where the cosmological constant vanishes and the asymptotically anti-de Sitter falloff conditions are replaced by asymptotically flat falloff conditions. With asymptotic flatness, neither the partition function nor the grand partition function turns out to be well-defined, and we recover neither a canonical ensemble nor a grand canonical ensemble.

The reduction used in the minisuperspace quantization in Book 4 [4], for spherical dust shell collapse, and was also a full general relativity derivation. Let's compare at a high level to see what lies ahead.

The Minisuperspsace Quantization for the Dust Shell Collapse [4]

(i) Reduced phase space formalism for spherical dust shell in spherically symmetric, asymptotically flat geometry is obtained.

(ii) For prescribed M_-:

$$S = \int dt\,(p\dot{\hat{r}} - h)$$

$$p = \sqrt{2M_-\hat{r}} - \sqrt{2M_+\hat{r}} + \hat{r}\ln\left(\frac{\hat{r} + \hat{p} + \sqrt{\hat{p}^2 + m^2} + \sqrt{2M_+\hat{r}}}{\hat{r} + \sqrt{2M_-\hat{r}}}\right)$$

$$\hat{p}\colon M_+ + M_- = \sqrt{\hat{p}^2 + m^2} + \frac{m^2}{2\hat{r}} - \hat{p}\sqrt{\frac{2M_+}{\hat{r}}}$$

$$h = M_+ + M_-$$

(iii) Time Reparameterization

Proper Time:

$$H(\hat{r}, P) = m\cosh\left(P/m\right) - \frac{m^2}{2\hat{r}} + M_-\left[2 - \left(\frac{m}{\hat{r}}\right)e^{-P/m}\right]$$

"Minkowski" Time:

$$\tilde{h}(\hat{r}, P) = \sqrt{P^2 + m^2} + \frac{m^2}{2\hat{r}} + M_-\left[2 - \frac{1}{\hat{r}}\left(\sqrt{P^2 + m^2} - P\right)\right]$$

(iv) Quantization
  (a) Vacuum case (M_- dynamical)
  (b) $M_- = 0$
  (c) $M_- \neq 0$

Overview of Black Hole Reduced Quantization Thermodynamics for RNAdS

(i) Canonical formulation in metric variables

$$S = \frac{1}{16\pi}\int d^4 x\sqrt{-{}^{(4)}g}\left({}^{(4)}R + 6\ell^{-2} - F^{\mu\nu}F_{\mu\nu}\right) +$$

(boundary terms)

(a) bulk → equations of motion

(b) boundary terms → complete variational formulation → Hamiltonian Thermodynamics

(ii) Canonical transformation: constraints are $M' = 0, Q' = 0, P_R = 0$.

(iii) Reduction:

$$S = \int dt\left(P_m\dot{m} + P_q\dot{q} - h\right)$$

$$h = -\frac{1}{2}R_h^2\widetilde{N}_0^M + \widetilde{N}_t m + \left(\widetilde{\Phi}_t - \widetilde{\Phi}_0\right)q$$

(iv) Quantization → time evolution operator

(v) Analytically continue to imaginary time, trace → grand partition function

$$Z(\beta,\phi) = N\int \tilde{\mu}\, dR_h dq \exp\left[-\frac{1}{2}\beta R_h(R_h^2\ell^{-2} + 1 + q^2R_h^{-2}) - \beta\phi q - \pi R_h^2\right]$$

(ϕ is the electric potential difference between horizon and infinity, in the curvature coordinates and in an electromagnetic gauge that makes $A = \frac{Q}{R}dT$ invariant under Killing time translations; $\phi = QR_0^{-1}$ on a classical solution)

$$S = \left(1 - \beta\frac{\partial}{\partial\beta}\right)\ln(Z) \approx \pi(R_h^t)^2$$

(vi) Neither ensemble exists when $\ell \to \infty$ (asymptotically flat).
  (a) Davies has shown stability of RN-flat when $q^2 > \frac{3}{.4}m^2$.
  (b) Existence of ensemble may be tied to global properties.

6.2 Metric Formulation

In this section we present a Hamiltonian formulation of spherically symmetric electrovacuum spacetimes with a negative cosmological constant, with boundary conditions appropriate for the exterior of a RNAdS black hole with a nondegenerate horizon. Some relevant properties of the RNAdS metric are reviewed in Appendix B.

The general solution of Einstein's elements equations for spherically symmetric spacetimes, that have negative cosmological constant and an electromagnetic field, is locally isometric to the line element.

$$ds^2 = -F(R)dT^2 + F^{-1}(R)dR^2 d\Omega^2,$$

which is written in curvature coordinates (T,R), where the notation of [80] is used throughout the derivation. Here,

$$F(R) = 1 - \frac{2M}{R} + \frac{q^2}{R^2} + \frac{R^2}{l^2},$$

where $G = C = 1$ and the cosmological constant is set to $\Lambda = -{}^3/_{l^2}$.

For R positive, F(r) has one zero, which corresponds to the bifurcation 2-sphere boundary of the region studied, leading to the (eventual) Euclidean analysis.

The geometrodynamical approach generates the above solution by using a Hamiltonian formulation of Einstein's equations to evolve spherically symmetric asymptotically Anti- de Sitter spacetimes in the presence of an electromagnetic field. The ADM form for the line element used in the ADM action is

$$ds^2 = -N^2 dt^2 + \Lambda^2(dr + N^r dt)^2 + R^2 d\Omega^2.$$

In the ADM formulation the canonical variables are $\Lambda, R, P_\Lambda, P_R$. The variables N and N^r enter as Lagrange multipliers.

The electromagnetic potential is taken to be described by the spherically symmetric one-form

$$A = \Gamma dr + \Phi dt$$

where Γ and $\Phi$are functions of t and r only. The fact that this one-form is globally defined makes the electromagnetic bundle trivial, and will preclude the black hole from having a magnetic charge. The coordinate r takes the semi-infinite range $[0, \infty)$. Unless otherwise stated, we shall assume both the spatial metric and the spacetime metric to be nondegenerate. In particular, Λ, $R$and $N$are taken to be positive. We shall work in natural units, $\hbar = c = G = 1$.

Following the method of [80], a canonical formulation is sought that uses canonical variables that are more manifestly geometric in content, permitting spacetime intuition to be a more prominent guide.

The canonical formulation that is sought is indicated by the mass equation (derivable from a grouping of the constraints). For the Schwarzschild line element the equation was

$$\text{Constant} = M = \frac{P_\Lambda^2}{2R} + \frac{R}{2} - \frac{RR'^2}{2\Lambda^2}$$

If no longer viewed as arising from constraints, the variable M may be considered a function of $R, R', \Lambda, P_\Lambda,$ which is constant when the constraints are imposed. The elevation of M(r) to a function on the phase space now allows for the possibility of its being a canonical variable in its own right given the appropriate transformation. Since it is desirable to retain R(r) as variable invariantly defined through $4\pi R^2$ being the area of the 2-spheres, what is sought is a new set of canonical variables $\{\mu, \Re, P_M, P_\Re\}$, where $\Re = R$ but $P_\Re$ is not equal to P_R. Ultimately, the hypersurfaces in the ADM analysis must be imbedded in a spacetime with line element given above. Following [80] the hypersurface is then considered a leaf of a foliation

$$T = T(t,r)\ , R = R(t,r)$$

Upon substitution:

$$ds^2 = -\left(F\dot{T}^2 - F^{-1}\dot{R}^2\right)dt^2 + 2\left(-F\dot{T}T' + F^{-1}\dot{R}R'\right)dtdr + \left(-FT'^2 + F^{-1}R'^2\right)dr^2 + R^2 d\Omega^2$$

Compare with:

$$ds^2 = -\left(N^2 - \Lambda^2 N^{r2}\right)dr^2 + 2\Lambda^2 N^r dtdt + \Lambda^2 dr^2 + R^2 d\Omega^2$$

Thus,

$$\Lambda^2 = -FT'^2 + F^{-1}R'^2$$

$$\Lambda^2 N^r = -F\dot{T}T' + F^{-1}\dot{R}R'$$

The N^r eqn.:

$$N^r = \frac{-F\dot{T}T' + F^{-1}\dot{R}R'}{-FT'^2 + F^{-1}R'^2}$$

The N eqn.:

$$N^2 = \frac{\left(-F\dot{T}T' + F^{-1}\dot{R}R'\right)}{\left(-FT'^2 + F^{-1}R'^2\right)} + \left(F\dot{T}^2 - F^{-1}\dot{R}^2\right)$$

$$= \left\{\begin{matrix} F^2\left(\dot{T}T'\right)^2 + F^{-2}\left(\dot{R}R'\right)^2 - 2\left(\dot{T}T'\right)\left(\dot{R}R'\right) - F^2\left(\dot{T}T'\right)^2 - F^{-2}\left(\dot{R}R'\right)^2 \\ +T'^2\dot{R}^2 + \dot{T}^2R'^2 \end{matrix}\right\}$$

$$\times \left(-FT^2 + F^{-1}R'^2\right)^{-1}$$

$= (\dot{T}^2 R' - T'\dot{R})(-FT'^2 + F^{-1}R'^2)^{-1}$

$$N = \frac{R'\dot{T} - T'\dot{R}}{\sqrt{-FT'^2 + F^{-1}R'^2}}$$

(proper choice of radial label r then ensures that N is positive).

6.3 Canonical Formulation

The action of the Einstein-Maxwell theory with a negative cosmological constant is

$S = \frac{1}{16\pi}\int d^4 x\sqrt{-(4)_g}((4)_R + 6^{-2} - F^{\mu\nu}F_{\mu\nu}) +$ (boundary terms) ,
where $(4)_g$ is the determinant of the four-dimensional metric, $(4)_R$ is the fourdimensional Ricci scalar, and $F_{\mu\nu} = \partial_\mu A_\nu - \partial_\nu A_\mu$ is the electromagnetic field tensor. The cosmological constant has been written as $-3\ell^{-2}$, where$\ell > 0$. Using the spherically symmetric fields and integrating over the two-sphere we obtain, up to boundary terms, the action

$$\begin{aligned} & S[, R,, N, N^r, \Phi] \\ & = \int dt \int_0^\infty dr \Big[-N^{-1}(R(-\dot{\Lambda} + (\Lambda N^r)')(-\dot{R} + R'N^r) \\ & \quad + \tfrac{1}{2}\Lambda\left(-\dot{R} + R'N^r\right)^2\Big) \\ & \quad + N\Big(\Lambda^{-2}RR'\Lambda' - \Lambda^{-1}RR'' - \frac{1}{2}\Lambda^{-1}R'^2 + \frac{1}{2}\Lambda^{-1}R'^2 + \frac{1}{2}\Lambda \\ & \quad + \frac{3}{2}\ell^{-2}\Lambda R^2\Big) + \frac{1}{2}N^{-1}\Lambda^{-1}R^2(\dot{\Gamma} - \Phi')^2\Big] \end{aligned}$$

The equations of motion derived from local variations are the full Einstein-Maxwell equations for the spherically symmetric fields. A generalized Birkhoff's theorem can be proven using the same techniques as in the case of a vanishing cosmological constant [84]: every classical solution is locally either a member of the extended RNAdS family (see Appendix B), or a spacetime that generalizes the Bertotti-Robinson solution to accommodate a negative cosmological constant [77-79,84]. We shall address the boundary conditions and boundary terms that are needed to make the variational principle globally well-defined after passing to the Hamiltonian formulation.

The momenta conjugate to the configuration variables Λ, R,and T are

$$P_\Lambda = -N^{-1}R(\dot{R} - R'N^r)$$

$$P_R = -N^{-1}\left(\Lambda(\dot{R} - R'^{N^r} + R(\dot{\Lambda}(\Lambda N^r)')\right)$$

$$P_\Gamma = N^{-1}\Lambda^{-1}R^2(\dot{\Gamma} - \Phi')$$

A Legendre transformation gives the Hamiltonian action

$$S_\Sigma[\Lambda, R, \Gamma, P_\Lambda, P_R, P_\Gamma; N, N^r, \tilde{\Phi}] = \int dt \int_0^\infty dr(P_\Lambda\dot{\Lambda} + P_R\dot{R} + P_\Gamma\dot{\Gamma} - NH - N^R H_r - \tilde{\Phi}G),$$

where the super-Hamiltonian constraint H, the radial supermomentum constraintH_r, and the Gauss law constraint G are given by

$$H = -R^{-1}P_R P_\Lambda + \frac{1}{2}R^{-2}\Lambda(P_\Lambda^2 + P_\Gamma^2) + \Lambda^{-1}RR'' - \Lambda^{-2}RR'\Lambda' + \frac{1}{2}\Lambda^{-1}R'^2 - \frac{1}{2}\Lambda - \frac{3}{2}\ell^{-2}\Lambda R^2,$$

$$H_r = P_R R' - \Lambda P_\Lambda' - \Gamma P_\Gamma$$

$$G = -P_\Gamma'$$

We have written the electric potential Φ in terms of the quantity

$$\tilde{\Phi} := \Phi - N^r\Gamma, \tag{2.8}$$

which now acts as the Lagrange multiplier associated with the Gauss constraint. It would be possible to proceed retaining Φas the Lagrange multiplier, and the supermomentum constraint would then be the same as without the electromagnetic field (see, for example, Ref. [87]). However, using $\tilde{\Phi}$ has the technical advantage that the supermomentum constraint generates spatial diffeomorphisms in both the gravitational and electromagnetic variables.

The Hamiltonian equations of motion are obtained from local variations. The constraint equations are:

$$H = 0,$$

$$H_r = 0, G = 0,$$

and the dynamical equations of motion read

$$\dot{\Lambda} = N(R^{-2}\Lambda P_\Lambda - R^{-1}P_R) + (N^r\Lambda)',$$

$$\dot{R} = -NR^{-1}P_\Lambda + N^r R',$$

$$\dot{\Gamma} = N\Lambda R^{-2}P_\Gamma + (N^r)' + \tilde{\Phi},$$

$$\dot{P}_\Lambda = \frac{1}{2}N[-R^{-2}(P_\Lambda^2 + P_\Gamma^2) - (\Lambda^{-1}R')^2 + 1 + 3\ell^{-2}R^2] - \Lambda^{-2}N'RR' + N^r P_\Lambda'$$

,

$$\dot{P}_R = N[\Lambda R^{-3}(P_\Lambda^2 + P_\Gamma^2) - R^{-2}P_\Lambda P_R - (\Lambda^{-1}R')' + 3\ell^{-2}\Lambda R] (-\Lambda^{-1}N'R') + N^r P_R{}'$$

$$\dot{P}_\Gamma = N^r P_\Gamma'$$

It is easy to verify that the Poisson bracket algebra of the constraints closes, and we thus have a first class constrained system [86].

We now wish to adopt boundary conditions that enforce every classical solution to be an exterior region of a RNAdS spacetime with a nondegenerate horizon (see Appendix B), such that the constant *t* hypersurfaces begin at the horizon bifurcation two-sphere at $r = 0$ and reach the asymptotically anti-dc Sitter infinity as $r \to \infty$.

Consider first the left end of the hypersurfaces. At $r \to 0$, we adopt the conditions

$$\begin{gathered}
\Lambda(t,r) = \Delta_0(t) + O(r^2)\,, \\
R(t,r) = R_0(t) + R_2(t)r^2 + O(r^4)\,, \\
P_\Lambda(t,r) = O(r^3)\,, \\
P_R(t,r) = O(r)\,, \\
N^r(t,r) = N_1(t)r + O(r^3)\,, \\
N^r(t,r) = N_1^r(t)r + O(r^3)\,, \\
\Gamma(t,r) = O(r)\,, \\
P_\Gamma(t,r) = Q_0(t) + Q_2(t)r^2 + O(r^4)\,, \\
\widetilde{\Phi}(t,r) = \Phi_0(t) + O(r^2)\,,
\end{gathered}$$

where Λ_0 and $R_0$are positive, and$N_0 \geq 0$. Here $O(r^n)$stands for a term whose magnitude at $r \to 0$ is bounded by r^ntimes a constant, and whose $k'th$ derivative at $r \to 0$ is similarly bounded by r^{n-k} times a constant for $1 \leq k \leq n$. It is straightforward to verify that these falloff conditions are consistent with the constraints $H = H_r = G = 0$, and that they are preserved by the time evolution equations. Some of the metric falloff conditions, which are identical to those introduced in [68] in the context of the Schwarzschild black hole, guarantee that the classical solutions have a nondegenerate horizon, and that the constant *t* hypersurfaces begin at $r = 0$ at a horizon bifurcation two-sphere in a manner asymptotic to hypersurfaces of constant Killing time. The coordinates become thus singular at $r \to 0$, but this singularity is quite precisely controlled. In particular, on a classical solution the future unit normal to a constant *t* hypersurface defines at $r \to 0$ a future timelike unit vector $n^a(t)$ at the bifurcation two-sphere, and the evolution of the constant *t* hypersurfaces boosts this vector according to

$$n^a(t_1)n_a(t_2) = -\cosh\left(\int_{t_1}^{t_2} \Lambda_0^{-1}(t)\, N_1(t)dt\right)$$

The falloff conditions for the electromagnetic field variables are motivated by our thermodynamical goal, and they will be discussed further in later sections.

Consider then the right end of the hypersurfaces. At $r \to \infty$, we assume that the variables have asymptotic expansions in integer powers of $(1/r)$, with the leading order behavior

$$\begin{aligned}
\Lambda(t,r) &= \ell r^{-1} - \frac{1}{2}\ell^3 r^{-3} + \lambda(t)\ell^3 r^{-4} + O^\infty(r^{-5}), \\
R(t,r) &= r + \ell^2{}_{p^{(t)r^{-2}}} + O^\infty(r^{-3}), \\
P_\Lambda(t,r) &= O^\infty(r^{-2}), \\
P_R(t,r) &= O^\infty(r^{-4}), \\
N(t,r) &= \Lambda^{-1}R'^{\left(\widetilde{N}_+(t)+O^\infty(r^{-5})\right)}, \\
N^r(t,r) &= O^\infty(r^{-2}), \\
\Gamma(t,r) &= O^\infty(r^{-2}), \\
P_\Gamma(t,r) &= Q_+(t) + O^\infty(r^{-1}), \\
(t,r) &= \widetilde{\Phi}_+(t) + O^\infty(r^{-1}),
\end{aligned}$$

where $\widetilde{N}_+(t) > 0$. $O^\infty(r^{-n})$ denotes a term that falls off at infinity as r^{-n}, and whose derivatives with respect to r fall off accordingly as $r^{-n-k}, k = 1,2,\ldots$. It is again straightforward to verify that these falloff conditions are consistent with the constraints and that they are preserved by the time evolution equations. Comparison with [83] shows that the metric is asymptotically anti-de Sitter, with the constant t hypersurfaces being asymptotic to hypersurfaces of constant Killing time, and $\widetilde{N}_+(t)$ gives the rate at which the Killing time evolves with respect to t at the infinity. Note that the lapse-function N diverges at the infinity for any nonzero value of $\widetilde{N}_+(t)$.For future use, we define the quantity

$$M_+(t) := \lambda(t) + 3\rho(t)$$

When the equations of motion hold, $M_+(t)$is independent of t, and it is equal to the mass parameter of the RNAdS metric.

Taken together, the falloff conditions achieve our aim. Every classical solution is an exterior region of a RNAdS spacetime with a nondegenerate event horizon, such that the constant t hypersurfaces begin at the horizon bifurcation two-sphere and reach the asymptotically anti-de Sitter infinity. In particular, the classical solutions satisfy $R_2 > 0$.

It is possible to replace

$$\begin{aligned}
\Lambda(t,r) &= r^{-1} - \frac{1}{2}\ell^3 r^{-3} + O^\infty(r^{-4}), \\
R(t,r) &= r + O^\infty(r^{-2}),
\end{aligned}$$

and then drop the assumption that the expansion proceed in integer powers of $(1/r)$ beyond the order shown, provided one makes more precise assumptions about what is meant by the symbol O^{∞}. Alternatively, it would be possible to strengthen the falloff conditions to read

$$\Lambda(t,r) = r^{-1} - \frac{1}{2}\ell^3 r^{-3} + \lambda(t)\ell^3 r^{-4} + O^{\infty}(r^{-4-\epsilon}),$$
$$R(t,r) = r + O^{\infty}(r^{-2-\epsilon})$$

where $0 < \epsilon \leq 1$, with similar changes in the rest. This would be analogous to the falloff conditions adopted for the asymptotically flat Schwarzschild case in [80], in that the value of the mass could then be read solely from the expansion of Λ. One might also consider writing the theory in terms of a lapse-function that has been rescaled by the factor $\Lambda^{-1}R'$: then, the falloff of the new lapse at $r \to \infty$ would then be independent of the canonical variables. For concreteness, we shall adhere to the theory as written above.

We can now write an action principle compatible with our falloff conditions. Consider the total action

$$S[\Lambda, R, \Gamma, P_\Lambda, P_R, P_\Gamma; N, N^r, \widetilde{\Phi}] = S_\Sigma[\Lambda, R, \Gamma, P_\Lambda, P_R, P_\Gamma; N, N^r, \widetilde{\Phi}] + S_{\partial\Sigma}[\Lambda, R, Q_0, Q_+; N, \widetilde{\Phi}_0, \widetilde{\Phi}_+]$$

where the boundary action is

$$S_{\partial\Sigma}[\Lambda R, Q_0, Q_+; N, \widetilde{\Phi}_0, \widetilde{\Phi}_+] = \int dt \left(\frac{1}{2}R_0^2 N_1 \Lambda_0^{-1} - \widetilde{N}_+ M_+ + \widetilde{\Phi}_0 Q_0 - \widetilde{\Phi}_+ Q_+\right)$$

The total action is clearly well-defined under our boundary conditions. Its variation contains a volume term proportional to the equations of motion, boundary terms from the initial and final hypersurfaces proportional to $\delta\Lambda, \delta R$and $\delta\Gamma$, and boundary terms from $r = 0$ and $r = \infty$ given by

$$\int dt \left(\frac{1}{2}R_0^2 \delta(N_1 \Lambda_0^{-1}) - M_+ \delta\widetilde{N}_+ + Q_0 \delta\widetilde{\Phi}_0 - Q_+ \delta\widetilde{\Phi}_+\right)$$

The variation thus gives the desired classical equations of motion provided we fix, in addition to the initial and final values of Λ, R, and Γ, also the quantities $N_1\Lambda_0^{-1}, \widetilde{N}_+\widetilde{\Phi}_+$, and $\widetilde{\Phi}_+$. On a classical solution all these quantities have a clear geometrical interpretation. $N_1\Lambda_0^{-1}$gives the evolution of the unit normal to the constant t hypersurface at the bifurcation two-sphere, and $\widetilde{N}_+$gives the evolution of the Killing time at the infinity. $\widetilde{\Phi}_0$and $\widetilde{\Phi}_+$ describe the electromagnetic gauge. Note that when a classical solution is written in coordinates that are regular at the bifurcation two- sphere, the electromagnetic potential will be regular at the bifurcation two-sphere only if $\tilde{\Phi}_+ = 0$.

6.4 Canonical transformation

In this section we perform a canonical transformation, which generalizes that given in [80] for the spherically symmetric vacuum Einstein theory. Following [80], we first examine how the variables appearing in the action carry the information about the geometry of the classical solution. We then use this information as a guide for finding the canonical transformation.

6.4.1 Reconstruction

Under our boundary conditions, every classical solution is an exterior region of a RNAdS spacetime with a nondegenerate Killing horizon (see Appendix B). We now assume that we are given the canonical data $(\Lambda, R, \Gamma, P_\Lambda, P_R, P_\Gamma)$ on a spacelike hypersurface embedded in such a RNAdS spacetime. We wish to recover from the canonical data the mass and charge parameters of the spacetime, the information about the embedding of the hypersurface in the spacetime, and the information about the electromagnetic gauge. Consider first the charge. The equations of motion imply that P_Γ is independent of both t and r. It is easily seen that in the curvature coordinates (see Appendix B), the value of P_Γ is just the charge Q. As P_Γ is unchanged by the gauge transformations generated by the constraints, it follows that in any gauge

$$Q = P_\Gamma$$

Consider then the mass. The reconstruction of the function F appearing in the metric proceeds exactly as in [80], with the result

$$F = \left(\frac{R'}{\Lambda}\right)^2 - \left(\frac{P_\Lambda}{R}\right)^2$$

We find for the mass the expression

$$M = \frac{R}{2}\left(\frac{R^2}{\ell^2} + 1 + \frac{P_\Gamma^2}{R^2} - F\right).$$

Consider next the embedding. By repeating the steps in [80], we obtain

$$-T' = R^{-1}F^{-1}\Lambda P_\Lambda$$

which determines the embedding up to an overall additive constant in T. To determine the value of the additive constant, one needs to know the value of T at one point on the hypersurface.

Consider finally the electromagnetic gauge. There exists a function

$(t, r\,)$such that

$$A = R^{-1}QdT + d\xi$$
$$= (R^{-1}QT' + \xi')dr + \left(R^{-1}QT' + \dot{\xi}\right)dt\,.$$

We then obtain

$$\xi' = \Gamma + R^{-2}F^{-1}\Lambda P_\Lambda P_\Gamma$$

which determines the value of ξ on the hypersurface up to an additive constant.

6.4.2 Transformation

We have seen that when the equations of motion hold, the quantities defined by the equations above have a transparent geometrical meaning. We now promote these equations into definitions of functions on the phase space, valid even when the equations of motion do not hold. Our aim is to complete the set of functions into a set that constitutes a canonical chart.

We shall from now on assume that the quantity R_2 is positive. This is always the case for our classical solutions.

The functions *M* and *Q* Poisson commute with each other. The function $-T'$ Poisson commutes with *Q* and is canonically conjugate to *M*. This suggests looking for a canonical transformation such that *M* and *Q* become two new coordinates, and $-T'$ becomes the momentum conjugate to *M*. As in the Schwarzschild case [80], the function $\boldsymbol{R} := R$ Poisson commutes with M, Q, and $-T'$, and provides therefore a candidate for a new canonical coordinate. The crucial issue then is whether one can find momenta conjugate to *Q* and R such that the transformation is canonical.

A necessary condition for the prospective new momenta arises from the observation that the supermomentum constraint generates spatial diffeomorphisms in all the variables. Since *M, Q,* and R are spatial scalars, the expression for H_r in the new variables must be $P_M M' + P_Q Q' + P_R R'$. Equating and substituting for *M* and $P_M = -T'$ gives only one equation for the two unknowns P_Q and $P'_{\square}$ but the structure of the equation as a linear combination of R' and $P'_{\square}$ suggests setting the coefficients of R' and $P'_{\square}$ individually to zero. These considerations suggest the transformation

$$M := \frac{1}{2}R(R^2\ell^{-2} + 1 + P_\Gamma^2 R^{-2} - F)\,,$$
$$P_M := R^{-1}F^{-1}\Lambda P_\Lambda\,,$$

$$R := R\,,$$
$$P_R := P_R - \tfrac{1}{2}R^{-1}\Lambda P_\Lambda - \tfrac{1}{2}R^{-1}F^{-1}\Lambda P_\Lambda$$
$$- R^{-1}\Lambda^{-2}F^{-1}\big((\Lambda P_\Lambda)'(RR') - (\Lambda P_\Lambda)(PP')'\big)$$
$$+ \tfrac{1}{2}R^{-1}F^{-1}\Lambda P_\Lambda(P_\Gamma^2 R^{-2} - 3R^3\ell^{-2})$$
$$Q := P_\Gamma\,,$$
$$P_Q := -\Gamma - R^{-2}F^{-1}\Lambda P_\Lambda P_\Gamma$$

The analogy between the pairs (M, P_M) and (Q, P_Q) becomes manifest by observing that on a classical solution, P_Q carries the information about the electromagnetic gauge via $P_Q = -\xi'$.

It is now straightforward to demonstrate that the transformation is indeed canonical. We begin with the identity

$$P_\Lambda\delta\Lambda + P_R\delta R + P_\Gamma\delta T - P_M\delta M - P_R\delta R - P_Q\delta Q$$
$$= \left(\frac{1}{2}R\delta R In\left|\frac{RR' + \Lambda P_\Lambda}{RR' - \Lambda P_\Lambda}\right|\right)' + \delta\left(\Gamma P_\Gamma + \Lambda P_\Lambda + \frac{1}{2}R\delta R In\left|\frac{RR' + \Lambda P_\Lambda}{RR' - \Lambda P_\Lambda}\right|\right)$$

and integrate both sides with respect to r from $r = 0$ to $r = \infty$. The first term on the right hand side gives substitution terms from $r = 0$ to $r = \infty$ that vanish by virtue of our falloff conditions, and we obtain

$$\int_0^\infty dr(P_\Lambda\delta + P_R\delta R + P_\Gamma\delta\Gamma)$$
$$-\int_0^\infty dr(P_M\delta M + P_R\delta Q) = \delta\omega[\Lambda, R, \Gamma, P_\Lambda, P_\Gamma]\,,$$

Where

$$\omega[\Lambda, R, \Gamma, P_\Lambda, P_\Gamma] = \int_0^\infty dr\left(\Gamma P_\Gamma + \Lambda P_\Lambda + \frac{1}{2}RR' In\left|\frac{RR' + \Lambda P_\Lambda}{RR' - \Lambda P_\Lambda}\right|\right)$$

The functional $\omega[\Lambda, R, \Gamma, P_\Lambda, P_\Gamma]$is well-defined by virtue of the falloff conditions. The Liouville forms of the old and new variables differ only by an exact form, and the transformation is thus canonical.

Let's now verify that the new variables have well-defined falloff properties at $r = 0$ and $r = \infty$. At r=0 we have:

$$F(t,r) = 4R_2^2\Lambda_0^{-2}r^2 + O(r^4)$$

And

$$M(t,r) = M_0(t) + M_2(t)r^2 + O(r^4)\,,$$
$$R(t,r) = R_0(t) + R_2(t)r^2 + O(r^4)\,,$$

$$Q(t,r) = Q_0(t) + Q_2(t)r^2 + O(r^4)\,,$$
$$P_M(t,r) = O(r),$$
$$P_R(t,r) = O(r),$$
$$P_Q(t,r) = O(r),$$

Where

$$M_0 = \frac{1}{2}R_0(R_0^2\ell^{-2} + 1 + Q_0^2R_0^{-2})\,,$$
$$M_2 = \frac{1}{2}R_2(3R_0^2\ell^{-2} + 1 - Q_0^2R_0^{-2} - 4R_0R_2\Lambda_0^{-2}) + Q_0Q_2R_0^{-1}\,,$$

At $r \to \infty$, we have

$$M(t,r) = M_+(t) + O^\infty(r^{-1})\,,$$
$$R(t,r) = r + \ell^2p(t)r^{-2} + O^\infty(r^{-3})\,,$$
$$Q(t,r) = Q_+(t) + O(r^{-1})\,,$$
$$P_M(t,r) = O^\infty(r^{-6}),$$
$$P_R(t,r) = O^\infty(r^{-4}),$$
$$P_Q(t,r) = O^\infty(r^{-2}),$$

The canonical transformation becomes singular when $F = 0$. Under our boundary conditions the classical solutions have $F > 0$ for $r > 0$. Our canonical transformation is therefore well-defined and differentiable near the classical solutions, and similarly the inverse transformation is well-defined and differentiable near the classical solutions. From now on we shall assume that we are always in a neighborhood of the classical solutions such that $F > 0$ holds for $r > 0$.

6.4.3 Action

It is possible to write an action in the new variables by simply re-expressing the constraints in terms of the new coordinates and momenta. A more transparent action can be found if we exercise the freedom to redefine the Lagrange multipliers.

The constraint terms in the bulk action take the form

$$NH + N^rH_r + \tilde{\Phi}G = N^MM' + N^RP_R + N^QQ'\ ,$$

where

$$N^M = -NF^{-1}\Lambda^{-1}R' + N^rR^{-1}F^{-1}\Lambda P_\Lambda\,,$$
$$N^R = -NR^{-1}P_\Lambda + N^rR'\ ,$$
$$N^Q = NR^{-1}F^{-1}\Lambda^{-1}R'P_\Gamma - N^r(\Gamma + R^{-2}F^{-1}\Lambda P_\Lambda P_\Gamma) - \tilde{\Phi}$$

When viewed as a linear transformation from $(N, N^r,)$ to (N^M, N^R, N^Q), The equations are nonsingular for $r > 0$. This suggests that we could take the constraint terms in the new bulk action to be those on the right hand side of the equations with N^M, N^R. and N^Q as independent Lagrange multipliers. At $r \to \infty$ this would be satisfactory since the asymptotic behavior gives

$$N^M(t,r) = -\tilde{N}_+(t) + O^\infty(r^{-5}),$$
$$N^R(t,r) = O^\infty(r^{-2}),$$
$$N^Q(t,r) = -\tilde{\Phi}_+(t) + O^\infty(r^{-5}),$$

and one could then fix $\tilde{N}_+(t)$ and $\Phi_+(t)$ after adding the boundary action

$$-\int dt\left(\tilde{N}_+ M_+ + \tilde{\Phi}_+ Q_+\right).$$

However, at $r = 0$ we have

$$N^M(t,r) = -\frac{1}{2} N_1 \Lambda_0 R_2^{-2} + O(r^2),$$
$$N^R(t,r) = O(r^2),$$
$$N^Q(t,r) = -\tilde{\Phi}_0(t) + \frac{1}{2} N_1 \Lambda_0 Q_0 R_2^{-1} + O(r^2),$$

which says that fixing N^M and N^Q at $r = 0$ to values that are independent of the canonical variables is not equivalent to fixing $N_1 \Lambda_0^{-1}$ and $\tilde{\Phi}_0$ to values that are independent of the canonical variables. We therefore need to redefine N^M and N^ near r = 0, without affecting their behavior at $r \to \infty$.

To proceed, we make two assumptions. First, we assume $M_0 > M_{crit}(Q_0)$, where the function M_{crit} is defined in Appendix B. Second, we regard the equations as defining R_o in terms of M_0 and Q_0 as $R_0 = R_{hor}(M_0, Q_0)$, where the function R_{hor} is defined in Appendix B. As discussed in Appendix B, these assumptions are always true for our classical solutions, and they therefore merely tighten the neighborhood of the classical solutions in which the held variables may take values. For future use, we note that these assumptions imply$3R_0^2 \ell^{-2} + 1 - Q_0^2 3R_0^{-2} > 0$, and the variation of R_0 takes the form

$$\delta R_0 = 2(3R_0^2\ell^{-2} + 1 - Q_0^2 R_0^{-2})^{-1}\left(\delta M_0 - R_0^{-1} Q_0 \delta Q_0\right).$$

Define now the quantities $\tilde{N}^M$ and $\tilde{N}^Q$ by

$$N^M = -\tilde{N}^M\left[(1-g) + 2_g R_0 (3R_0^2\ell^{-2} + 1 - Q_0^2 R_0^{-2})^{-1}\right],$$
$$N^Q = 2\tilde{N}^M g_{Q_0} (3R_0^2\ell^{-2} + 1 - Q_0^2 R_0^{-2})^{-1} - \tilde{N}^Q$$

where $g(r)$ is a smooth decreasing function that vanishes at $r \to \infty$ oc as $O^\infty(r^{-5})$, and approaches the value f at $r \to 0$ as $g(r) = 1 + O(r^2)$. Eqs. (2.40) then define a nonsingular linear transformation from $(N^M N^R)$

to $(\widetilde{N}^M, \widetilde{N}^Q)$. The asymptotic behavior at $r \rightarrow \infty$ is

$$\widetilde{N}^M(t,r) = \widetilde{N}_+(t) + O^\infty(r^{-5}),$$
$$\widetilde{N}^Q(t,r) = \widetilde{\Phi}_+(t) + O^\infty(r^{-1}),$$

and the asymptotic behavior at $r = 0$ is

$$\widetilde{N}^M(t,r) = \widetilde{N}_0^M(t) + O(r^2),$$
$$\widetilde{N}^Q(t,r) = \widetilde{\Phi}_0(t) + O(r^2),$$

where

$$\widetilde{N}_0^M = \frac{1}{4} N_1 \Lambda_0 R_0^{-1} R_2^{-1} (3R_0^2 \ell^{-2} + 1 - Q_0^2 R_0^{-2})$$

When the constraints $M' = 0$ and $Q' = 0$ hold. We then show that

$$\widetilde{N}_0^M = N_1 \Lambda_0^{-1},$$

where $Q' = 0$ and $M' = 0$. Thus, when the constraints hold, fixing $\widetilde{N}^M$ and $\widetilde{N}^Q$ at $r = 0$ is equivalent to fixing $N_1 \Lambda_0^{-1}$ and $\widetilde{\Phi}_0$. We therefore adopt $\widetilde{N}^M$. $\widetilde{N}^R$, and $\widetilde{N}^Q$ as a set of new independent Lagrange multipliers.

The bulk action takes the form

$$S[M, R, Q, P_M, P_R; P_Q; \widetilde{N}^M, N^R, \widetilde{N}^Q]$$
$$= \int dt \int_0^\infty dr \{P_M \dot{M} + P_R \dot{R} + P_Q \dot{Q} + \widetilde{N}^Q Q' - N^R P_R$$
$$+ \widetilde{N}^M [(1-g) 2_g (3R_0^2 \ell^{-2} + 1 - Q_0^2 R_0^{-2})^{-1} (R_O M'$$
$$- Q_0 Q')]\}$$

The total action is taken to be

$$S[M, R, Q, P_M, P_R; P_Q; \widetilde{N}^M, N^R, \widetilde{N}^Q] =$$
$$S_\Sigma[M, R, Q, P_M, P_R; P_Q; \widetilde{N}^M, N^R, \widetilde{N}^Q] + S_\partial[M_0, M_+, Q_0, Q_+; \widetilde{N}_+, \widetilde{\Phi}_0, \widetilde{\Phi}_+],$$

where

$$S_{\partial\Sigma}[M_0, M_+, Q_0, Q_+; \widetilde{N}_+, \widetilde{\Phi}_0, \widetilde{\Phi}_+]$$
$$= \int dt \left(\frac{1}{2} R_0^2 \widetilde{N}_0^M - \widetilde{N}_+ M_+ + \widetilde{\Phi}_0 Q_0 - \widetilde{\Phi}_+ Q_+\right)$$

The quantities to be varied independently are $M, R, Q, P_M, P_R, P_Q, \widetilde{N}^M, \widetilde{N}^R$, and $\widetilde{N}^Q$. The volume term in the variation of the action is proportional to the equations of motion

$$\dot{M} = 0$$
$$\dot{R} = N^R,$$
$$\dot{Q} = 0,$$
$$\dot{P}_R = 0,$$
$$\dot{P}_Q = (N^Q)',$$
$$M' = 0,$$

$$P_R = 0\,,$$
$$Q' = 0\,,$$

where $\widetilde{N}^M$ and N^Q are defined by the above. The boundary terms in the variation consist of terms proportional to $\delta M, \delta R,$ and δQ on the initial and final hypersurfaces, and terms from $r = 0$ and $r = \infty$ given by

$$\int dt\left(\frac{1}{2}R_0^2\delta\widetilde{N}_0^M - M_+\delta\widetilde{N}_+ + Q_0\delta\widetilde{\Phi}_0 - Q_+\delta\widetilde{\Phi}_+\right)$$

The action thus yields the equations of motion provided that we fix, in addition to the initial and final values of the new canonical coordinates, also the quantities $\widetilde{N}_0^M, \widetilde{N}_+, \widetilde{\Phi}_+$and $\widetilde{\Phi}_+$. These fixed quantities at the right and left ends have precisely the same interpretation in terms of the geometry of the classical solutions as the fixed quantities in the action.

6.5 Hamiltonian reduction

In this section we shall reduce the action to the true dynamical degrees of freedom by solving the constraints. The constraints imply that *M* and *Q* are independent of *r*. We can therefore write

$$M(t,r) = \boldsymbol{m}(t)$$
$$M(t,r) = \boldsymbol{m}(t)$$

substituting this and the constraint back yields the true Hamiltonian action

$$S[\boldsymbol{m},\boldsymbol{q},\boldsymbol{P_m},\boldsymbol{P_q};\widetilde{N}_0^M,\widetilde{N}_+,\widetilde{\boldsymbol{\Phi}}_+,\widetilde{\boldsymbol{\Phi}}_0] = \int dt(\boldsymbol{P_m}\dot{\boldsymbol{m}} + \boldsymbol{P_q}\dot{\boldsymbol{q}} - \boldsymbol{h})$$

where

$$\boldsymbol{P_m} = \int_0^\infty dr\, P_M\,,$$
$$\boldsymbol{P_q} = \int_0^\infty dr\, P_Q$$

The reduced Hamiltonian **h** is

$$\boldsymbol{h} = -\frac{1}{2}R_h^2\widetilde{N}_0^M + \widetilde{N}_+\boldsymbol{m} + (\widetilde{\boldsymbol{\Phi}}_+ - \boldsymbol{\Phi_0}\,)\boldsymbol{q}\,,$$

where $R_h := R_{hor}\,(\boldsymbol{m},\boldsymbol{q})$. The assumptions made in the previous section imply

$$\boldsymbol{m} > M_{crit}(\boldsymbol{q})\,,$$

and **h** is therefore well-defined. Note that **h** is, in general, explicitly time-dependent through the prescribed functions $\widetilde{N}_0^M(t)\widetilde{N}_+(t), \widetilde{\Phi}_+(t)$,and $\Phi_+(t)$.

The variational principle associated with the reduced action fixes the initial and final values of the coordinates **m** and **q.** The equations of motion are

$$\dot{m} = 0$$
$$\dot{q} = 0$$
$$\dot{P}_m = 2R_h(3R_0^2\ell^{-2} + 1 - q^2R_0^{-2})^{-1}\widetilde{N}_0^M - \widetilde{N}_+$$
$$\dot{P}_q = -2_q(3R_0^2\ell^{-2} + 1 - q^2R_0^{-2})^{-1}\widetilde{N}_0^M + \widetilde{\Phi}_0 - \widetilde{\Phi}_+$$

The equations above are readily understood in terms of the statement that on a classical solution **m** and **q** are respectively equal to the mass and charge parameters of the RNAdS solution. Also, recall that on a classical solution$P_M = -T'$, where T is the Killing time. Thus, $\boldsymbol{P_m} = T_0 — T_+$, where T_0 and T_+ are respectively the values of T at the left and right ends of the constant t hypersurface. As the constant t hypersurface evolves in the RNAdS spacetime, the first and second term on the right hand side are respectively equal to $\dot{T}_0$ and $—\dot{T}_+$. The interpretation of the last equation is analogous. On a classical solution we have $\boldsymbol{P_q} = \xi_0 — \xi_+$ where ξ is the function that specifies the electromagnetic gauge. The first two terms on the right hand side of the last equation give $\dot{\xi}_0$, and the last term gives $-\dot{\xi}_+$.

6.6 Quantum theory and the grand partition function

We shall now quantize the reduced Hamiltonian theory. Our aim is to construct the time evolution operator in the Hamiltonian quantum theory, and then to obtain a grand partition function via an analytic continuation of this operator.

6.6.1 Quantization

As is well known, the quantization of a given classical Hamiltonian theory requires input [87-89], and the questions of physically appropriate input for a quantum black hole remain largely open. For the purposes of the present analysis we shall be content to define the quantum theory in essence by fiat, following [68,80]. Our main physical conclusions will emerge from the semiclassical regime of the theory, and at this level one may reasonably hope the details of the quantization not to be crucial.

For convenience, the equation numbering in the section that follows is the same as in the thesis description given in [9].

We regard m and q as configuration variables. The wave functions are of the form (m; q), and the inner product is taken to be

$$(\psi, \chi) = \int_A \mu d\mathrm{m} d\mathrm{q} \, \overline{\psi} \chi \ , \tag{2.56}$$

where $A \subset \mathbb{R}^2$ is the domain and μ(m; q) is a smooth positive weight factor. The Hilbert space is thus $\mathcal{H} = L^2(A; \boldsymbol{dmdq})$. We assume that μ is a slowly varying function, in a sense to be made more precise later, but otherwise it will remain arbitrary.

The Hamiltonian operator $\hat{h}(t)$ is taken to act as pointwise multiplication by the function h(m; q; t): (m; q) → h(m; q; t) (m; q). $\hat{h}(t)$ is an unbounded essentially self-adjoint operator [90], and the corresponding unitary time evolution operator in $\mathcal{H}$ is

$$\hat{K}(t_2; t_1) = \exp\left[-i \int_{t_1}^{t_2} dt' \, \hat{\mathrm{h}}(t')\right] \ . \tag{2.57}$$

$\widehat{K}(t_2, t_1)$ acts in $\mathcal{H}$ by pointwise multiplication by the function

$$K(\mathrm{m}, \mathrm{q}; \mathcal{T}, \Xi_+, \Xi_0, \Theta) = \exp\left[-i\mathrm{m}\mathcal{T} - i\mathrm{q}\left(\Xi_+ - \Xi_0\right) + \tfrac{1}{2} i R_{\mathrm{h}}^2 \Theta\right] \ , \tag{2.58}$$

where

$$\mathcal{T} := \int_{t_1}^{t_2} dt \, \tilde{N}_+(t) \ , \tag{2.59a}$$

$$\Xi_+ := \int_{t_1}^{t_2} dt \, \tilde{\Phi}_+(t) \ , \tag{2.59b}$$

$$\Xi_0 := \int_{t_1}^{t_2} dt \, \tilde{\Phi}_0(t) \ , \tag{2.59c}$$

$$\Theta := \int_{t_1}^{t_2} dt \, \tilde{N}_0^M(t) \ . \tag{2.59d}$$

$\widehat{K}(t_2, t_1)$ therefore depends on t1 and t2 only through the quantities on the left hand side of (2.59), and we may write $\widehat{K}(t_2, t_1)$ as $\widehat{K}(\mathcal{T}, \Xi_+, \Xi_0, \Theta)$. The composition law,

$$\widehat{K}(t_3, t_2)\widehat{K}(t_2, t_1) = \widehat{K}(t_3, t_1),$$

amounts to independent addition in each of the four parameters in $\widehat{K}(\mathcal{T}, \Xi_+, \Xi_0, \Theta)$, and we may regard these four parameters as independent evolution parameters specified by the boundary conditions. $\mathcal{T}$ is the Killing time elapsed at the infinity, and Θ is the boost parameter elapsed at the bifurcation two-sphere. Ξ_+ and Ξ_0 can be computed from the line integral of the electromagnetic potential along the timelike curve of constant r and constant angular variables as this curve approaches respectively the infinity and the bifurcation two-sphere.

6.6.2 Grand partition function

We shall now construct a grand partition function by continuing the time evolution operator to imaginary time and taking the trace. We begin by discussing the boundary conditions for the relevant thermodynamical ensemble.

The envisaged semiclassical thermodynamical situation consists of a charged spherically symmetric black hole in asymptotically anti-de Sitter space, in thermal equilibrium with a bath of Hawking radiation. If the back-reaction from the radiation is neglected, the geometry is described by the RNAdS metric. Assuming that the local temperature is given in the usual manner in terms of the surface gravity and the redshift factor [54,71], we see that the local temperature is $F^{-1/2}\beta^{-1}$, where F is given in Appendix B and

$$\beta := 4\pi R_0\left(3\ell^{-2}R_0^2 + 1 - Q^2 R_0^{-2}\right)^{-1} . \tag{2.60}$$

At the infinity the local temperature vanishes as $\beta^{-1}\ell R^{-1}(1 + O^{\infty}(\ell^2 R^{-2}))$, and β^{-1} can thus be extracted from the asymptotic behavior as the coefficient of the leading order term ℓR^{-1}. We shall follow [54,71] and regard β^{-1} as a renormalized temperature at infinity.

The electromagnetic variable with thermodynamic interest for us is the electric potential difference between the horizon and the infinity, in the curvature coordinates (see Appendix B) and in an electromagnetic gauge that makes A invariant under the Killing time translations. We denote this quantity by ϕ. From Appendix B it is seen that on a classical solution $\phi = QR_0^{-1}$.

We shall consider a thermodynamical ensemble in which the fixed quantities are β and ϕ. This data can be interpreted as that for a grand canonical ensemble, with ϕ being analogous to the chemical potential [43,44,62]. Our aim is to obtain a grand partition function $Z(\beta, \phi)$ by continuing the time evolution operator of the Lorentzian Hamiltonian theory to imaginary time and taking the trace.

The continuation of $\mathcal{T}$ is straightforward: comparing the definition of β to the falloff of N in (2.13) and to the definition (2.59a), we are led to set $\mathcal{T} = -i\beta$. For the continuation of Θ we choose $\Theta = -2\pi i$, motivated by the regularity of the classical Euclidean solutions as in [68]. Previously, the regularity of the electromagnetic potential at the bifurcation two-sphere of the Lorentzian solutions requires $\widetilde{\Phi}_0 = 0$; similarly, requiring regularity of the electromagnetic potential at the horizon of the classical Euclidean solutions now leads us to set $\Xi_0 = 0$. Finally, recall that Ξ_+ gives the constant r line integral of the electromagnetic potential at the infinity. Comparing this to the definition of ϕ, we set $\Xi_+ = -\mathcal{T}\phi = i\beta\phi$. We are thus led to propose for the grand partition function the expression

$$\mathcal{Z}(\beta, \phi) = \mathrm{Tr}\left[\hat{K}(-i\beta, i\beta\phi, 0, -2\pi i)\right] \quad . \tag{2.61}$$

As it stands, the trace is divergent, but one can argue as in Refs. [68,70] that a suitable regularization and renormalization yields the result

$$\mathcal{Z}_{\mathrm{ren}}(\beta, \phi) = \mathcal{N} \int_A \mu d\mathrm{m} d\mathrm{q} \, \exp\left[-\beta(\mathrm{m} - \mathrm{q}\phi) + \pi R_{\mathrm{h}}^2\right] \quad , \tag{2.62}$$

where we have substituted for K the explicit expression (2.58). The normalization factor $\mathcal{N}$ may depend on ℓ, but we shall assume that it does not depend on β or ϕ.

Provided the weight factor μ is slowly varying compared with the exponential in (2.62), it is easy to verify, using the definition of R_h, that the integral in (2.62) is convergent. Equation (2.62) thus yields a well-defined grand partition function. Comparing with ordinary PVT systems [57], ϕ is now indeed seen to be analogous to the chemical potential, and the quantities m and q are respectively analogous to the energy and the particle number. We shall examine the thermodynamical properties of this grand partition function in the next section.

6.7 Thermodynamics in the grand canonical ensemble

It is useful to change the integration variables in (2.62) from the pair (m; q) to the pair (R_h; q). From (Appendix B) we obtain

$$\mathrm{m} = \tfrac{1}{2} R_{\mathrm{h}} \left(R_{\mathrm{h}}^2 \ell^{-2} + 1 + \mathrm{q}^2 R_{\mathrm{h}}^{-2}\right) \quad , \tag{2.63}$$

and the grand partition function takes the form

$$\mathcal{Z}_{\mathrm{ren}}(\beta, \phi) = \mathcal{N} \int_{A'} \tilde{\mu} \, dR_{\mathrm{h}} d\mathrm{q} \, \exp\left(-I_*\right) \quad , \tag{2.64}$$

where

$$I_*(R_h, q) := \tfrac{1}{2}\beta R_h\left(R_h^2\ell^{-2} + 1 + q^2 R_h^{-2}\right) - \beta\phi q - \pi R_h^2 \ . \tag{2.65}$$

One may view I_* as an effective action or a reduced action [60-62]. The integration domain A' is given by the inequalities

$$0 \leq R_h \tag{2.66a}$$

$$q^2 \leq R_h^2(1 + 3R_h^2\ell^{-2}) \tag{2.66b}$$

and the weight factor p is obtained from μ by including the Jacobian $|\partial(m,q)/\partial(/R_h,q)|$ Note that because of (2.66b), I_* remains finite as $R_h \to 0$.

As $\tilde{\mu}$ is assumed to be slowly varying, we can estimate $Z_{ren}(\beta,\phi)$ by the saddle point approximation to (2.64). For this, we need to find the critical points of I^* in the interior . When $\phi^2 < 1 - \frac{4}{3}\pi^2\ell^2\beta^{-2}$, R has no critical points. When $1 - \frac{4}{3}\pi^2\ell^2\beta^{-2} < \phi^2 < 1$, the two critical points of I_* are at

$$R_h = R_h^{\pm} := \frac{2\pi\ell^2}{3\beta}\left(1 \pm \sqrt{1 + \frac{3^2(\beta^2 - 1)}{4\pi^2\ell^2}}\right) \tag{2.66b}$$

The lower signs do not give a local extremum, but the upper signs give a local minimum. In the limiting case$1 - \frac{4}{3}\pi^2\ell^2\beta^{-2} = \phi^2$, the only critical point is (R_h^+, q^+), but it is not a local extremum. Finally, when$\phi^2 \geq 1$, the only critical point is (R_h^+, q^+), and it is a local minimum. Whenever the critical points exist, the value of I_*at these points can be written as

$$I_*(R_h^{\pm}, q^{\pm}) = \frac{\pi(R_h^{\pm})^2\left(1 - \phi^2 - \left(R_h^{\pm 2}\right)^2 \ell^{-2}\right)}{1 - \phi^2 + 3\left(R_h^{\pm 2}\right)^2 \ell^{-2}} \tag{2.66b}$$

As I_* grows without bound in the noncompact directions in A', the global minimum can be found by examining R at the critical points and on the boundary of A'. When $\phi^2 < 1 - \frac{4}{3}\pi^2\ell^2\beta^{-2}$, the global minimum is at the critical point (R_h^+, q^+),, and $I_*(R_h^+, q^+)$, is negative. When $\phi^2 < 1 - \frac{4}{3}\pi^2\ell^2\beta^{-2}$, the global minimum is at $R_h = 0 = q$, where I_* vanishes. In the limiting case $\phi^2 < 1 - \frac{4}{3}\pi^2\ell^2\beta^{-2}$, I_* vanishes at $(R_h^+, q^+)$and at $R_h = 0 = q$, and is positive everywhere else.

We thus see that for $\phi^2 > 1 - \pi^2\ell^2\beta^{-2}$, $\mathcal{Z}_{ren}$ can be approximated as

$$\mathcal{Z}_{\rm ren}(\beta, \phi) \approx P \exp[-I_*(R_h^+, q^+)] \quad , \tag{2.69}$$

where P is a slowly varying prefactor. The approximation becomes presumably progressively better with increasing $|I_*(R_h^+, q^+)|$. For $\phi^2 < 1 - \pi^2\ell^2\beta^{-2}$, the dominant contribution to $\mathcal{Z}_{ren}$ comes from the vicinity of $R_h = 0 = q$, and the behavior of $\mathcal{Z}_{ren}$ depends more sensitively on the weight factor $\tilde{\mu}$.

These results for $\mathcal{Z}_{ren}$ are consistent with what one would expect just from the existence of (Lorentzian) black hole solutions under fixing ϕ and the renormalized inverse Hawking temperature β (2.60). It can be verified that such solutions exist precisely at the critical points of I_*: the values of m and q at these critical points are just the mass and charge parameters of the black hole. Further, the value of I_* at a critical point is simply the Euclidean action of the corresponding Euclideanized black hole solution. When a unique classical solution exists, it dominates the grand partition function; when two distinct classical solutions exist, the grand partition function is dominated either by the larger mass classical solution or by no classical solutions. The situation is thus remarkably similar to that found in the absence of a cosmological constant when the boundary conditions are set on a finite size box [62].

Let us now consider the thermodynamical predictions from $\mathcal{Z}_{ren}$. Recall that the thermal expectation values of the energy and charge in the grand canonical ensemble are given by

$$\langle E \rangle = \left(-\frac{\partial}{\partial \beta} + \beta^{-1}\phi \frac{\partial}{\partial \phi} \right) (\ln \mathcal{Z}_{\rm ren}) \quad , \tag{2.70a}$$

$$\langle Q \rangle = \beta^{-1} \frac{\partial (\ln \mathcal{Z}_{\rm ren})}{\partial \phi} \quad . \tag{2.70b}$$

When $\mathcal{Z}_{ren}$ is dominated by the critical point (R_h^+, q^+), we find

$$\langle E \rangle \approx \mathrm{m}^+ \quad , \tag{2.71a}$$

$$\langle Q \rangle \approx \mathrm{q}^+ \quad , \tag{2.71b}$$

where m+ is obtained from (R_h^+, q^+) through (2.63). That is, the thermal expectation values of the energy and the charge are simply the mass and charge parameters of the dominant classical solution. In particular, there are no additional contributions to the mass from the gravitational binding energy associated with the thermal energy, or from the electrostatic binding energy associated with the charge. Such additional, finite size contributions were found to be present in the finite size ensembles of Refs. [50,60,62,70].

It is easily seen that $(\partial \boldsymbol{m}^{+}/\partial\beta) < 0$. This means that when the approximation (2.71a) is good, the (constant ϕ) heat capacity, $C_\phi = -\beta^2(\partial\langle E\rangle/\partial\beta)$, is positive. In the regime (2.71a), the system is thus stable under thermal fluctuations in the energy. Note that as $(\partial \boldsymbol{m}^{+}/\partial\beta) > 0$, a grand partition function dominated by the lower mass classical solution would be thermodynamically unstable [61]. This is analogous to what happens in the absence of a cosmological constant under the boxed boundary conditions considered in Refs. [50,60,62].

It is also easily seen that $(\partial \boldsymbol{q}^{+}/\partial\phi) > 0$. This shows that when the approximation (2.71b) is good, we have $(\partial\langle Q\rangle/\partial\phi) > 0$, and the system is stable under thermal fluctuations in the charge. More generally, one can show directly from the expressions (2.64), (2.65), and (2.70b) that $(\partial\langle Q\rangle/\partial\phi) > 0$ holds always, even when the approximation (2.71b) is not good.

The entropy in the grand canonical ensemble is given by

$$S = \left(1 - \beta\frac{\partial}{\partial\beta}\right)(\ln \mathcal{Z}_{\rm ren}) \quad . \tag{2.72}$$

When the approximation (2.69) is good, we have $S \approx \pi(R_h^{+})^2$, which means that the entropy is one quarter of the horizon area. This is the anticipated Bekenstein- Hawking result.

Finally, when $\mathcal{Z}_{ren}$ is not dominated by a critical point, the thermodynamical predictions become much more sensitive to the choice of the weight factor $\tilde{\mu}$. As in Refs. [60-62,70], one can view the transition in the qualitative behavior of $\mathcal{Z}_{ren}$ as evidence for a phase transition between a black hole sector and a topologically different sector of the theory; in the case at hand, the second sector might be referred to as “hot anti-de Sitter space.” On classical grounds one might have expected this transition to occur near $\phi^2 \approx 1 - \pi^2\ell^2\beta^{-2}$, where the classical solutions disappear. However, we saw that the transition in fact occurs near $\phi^2 \approx 1 - \pi^2\ell^2\beta^{-2}$, where the two classical solutions still exist. This is highly similar to what happens in four dimensions under boxed boundary conditions without a cosmological constant [60,62], but subtly different from what happens in two dimensions with Witten's dilatonic black hole [70].

6.8 The canonical ensemble

We have seen that the Hamiltonian formulation led into a thermodynamical grand canonical ensemble where the fixed quantities are the renormalized inverse temperature β at infinity and the electric potential difference ϕ between the horizon and the infinity. From the thermodynamical viewpoint, another natural ensemble for the charged black hole in asymptotically anti-de Sitter space is the canonical ensemble, where one allows fluctuations in ϕ but fixes instead the charge q. In this section we shall outline the recovery of the canonical ensemble from a Lorentzian Hamiltonian analysis, and briefly discuss the thermodynamical properties of the black hole in this ensemble.

As a starting point, we modify the boundary conditions of the Hamiltonian theory given previously by leaving $\tilde{\Phi}_0(t)$ and $\tilde{\Phi}_+(t)$ unspecified but fixing $Q_0(t)$ and $Q_+(t)$ to be prescribed functions of t. The action is obtained by omitting the terms $\int dt(\tilde{\Phi}_0 Q_0 - \tilde{\Phi}_+ Q_+)$. Clearly, classical solutions exist only when $Q_0(t)$ and $Q_+(t)$ are chosen independent of t and equal. We shall from now on assume that the boundary data is chosen in this manner.

One way to proceed is simply to push through the canonical transformation, noting that the new boundary conditions merely result into minor modifications. It is only when one subsequently performs a Hamiltonian reduction that the new boundary conditions give rise to important differences. Firstly, the boundary data for $Q_0(t)$ and $Q_+(t)$ implies that the quantity q(t) is a t-independent constant whose value is completely determined by the boundary conditions. Therefore, the Liouville term drops entirely out of the action. Secondly, because of the terms that were omitted from the boundary action, the term $(\tilde{\Phi}_+ - \tilde{\Phi}_0)\boldsymbol{q}$ drops out of the reduced Hamiltonian. This means that in the reduced Hamiltonian theory q has become an external parameter specified by the boundary conditions: it is not varied in the action, and it does not have a conjugate momentum. The new reduced action reads

$$S_C[\mathrm{m}, \mathrm{p_m}; \tilde{N}_0^M, \tilde{N}_+; \mathrm{q}] = \int dt\,(\mathrm{p_m}\dot{\mathrm{m}} - \mathrm{h}_C) \quad , \tag{2.73}$$

where

$$h_C = -\tfrac{1}{2}R_h^2\tilde{N}_0^M + \tilde{N}_+ m \quad . \tag{2.74}$$

Here $R_h = R_{hor}(\boldsymbol{m}, \boldsymbol{q})$ as before, and the assumptions made in the canonical transformation again holds.

An alternative way to proceed under the new boundary data is to partially reduce the action. One uses the constraint and the equation of motion to set P(t; r) equal to the constant specified in the boundary data, and substitutes this back in the action. The Liouville term $\int_0^\infty dr\, P_\Gamma \dot{\Gamma}$ then becomes a total time derivative and can be dropped. One thus obtains an action that no longer involves Γ or $\tilde{\Phi}$, involves P_Γ only as a prescribed constant, and correctly yields the equations of motion for the remaining variables. One can now perform a canonical transformation from the variables $(\Lambda, R, P_\Lambda, P_R)$ to the new variables $(M, \mathbf{R}, P_M, P_\mathbf{R})$, except that $P_\Gamma = Q$ is now regarded as a fixed external parameter. Finally, one can reduce the action by solving the constraints. The result is again the action given by (2.73) and (2.74).

Quantization of the reduced Hamiltonian theory proceeds as before. For the renormalized trace of the analytically continued time evolution operator, we obtain

$$Z_{\text{ren}}(\beta, q) = \int_{R_{\text{crit}}(q)}^{\infty} \tilde{\mu}\, dR_h \exp\left(-I_{C*}\right) \quad , \tag{2.75}$$

where the function R_{crit} is defined in Appendix B, the weight factor $\tilde{\mu}$ is a positive function of R_h (and possibly q), and

$$I_{C*}(R_h) := \tfrac{1}{2}\beta R_h \left(R_h^2 \ell^{-2} + 1 + q^2 R_h^{-2}\right) - \pi R_h^2 \quad . \tag{2.76}$$

Under the assumption that $\tilde{\mu}$ is slowly varying, the dominant contribution to Z_{ren} can be estimated by saddle point methods. The cases $q = 0$ and $q \neq 0$ merit each a separate analysis.

Consider first the special case $q = 0$. The lower limit of the integral in (2.75) is then at $R_h = 0$. The critical point structure of I_{C*} is identical to that of I_* (2.76) for $\phi = 0$, and the locations of the critical points and the values of the action at these points can simply be read off by setting $\phi = 0$.

Consider from now on the generic case $q \neq 0$. I_{C*} has one negative critical point, and from one to three positive critical points. The negative critical point is unphysical, but all the positive critical points lie in the physical domain $R_h > R_{crit}(q)$. As I_{C*} is decreasing at $R_h = R_{crit}(q)$ and tends to infinity as $R_h \to \infty$, the global minimum of I_{C*} in the domain $R_h > R_{crit}(q)$ is at a critical point. We can therefore concentrate on the positive critical points.

When $\beta^2 \geq \frac{3}{2}\pi^2\ell^2$, I_{C*} has only one positive critical point. When $\beta^2 < \frac{3}{2}\pi^2\ell^2$, the number of positive critical points is determined by the status of the double inequality

$$\frac{(1-3s)(1+s)}{36(1-s)^2} \leq \mathrm{q}^2\ell^{-2} \leq \frac{(1+3s)(1-s)}{36(1+s)^2} , \tag{2.77}$$

where

$$s := \sqrt{1 - \frac{2\beta^2}{3\pi^2\ell^2}} . \tag{2.78}$$

When (2.77) does not hold, there is only one positive critical point. When (2.77) holds as a genuine inequality, there are three positive critical points, and saturating the inequalities gives limiting cases where two of the three positive critical points merge. (Note that if $\beta^2 < \frac{4}{3}\pi^2\ell^2$, the leftmost expression in (2.77) is non-positive, and the left hand side inequality is then necessarily genuinely satisfied.) Now, when only one positive critical point exists, this critical point is the global minimum. On the other hand, when three positive critical points exist, they constitute a local maximum between two local minima, and the global minimum can be at either of the local minima depending on the values of the parameters. For example, when the right hand side inequality in (2.77) is close to being saturated, the global minimum is at the local minimum with the larger value of R_h.

The critical points can be examined further by parametrizing β and q as

$$4\pi\ell\beta^{-1} = \frac{(u-v)\left[3\left(u^2+v^2\right)+1\right]}{u^2+v^2-uv} , \tag{2.79a}$$

$$\mathrm{q}^2\ell^{-2} = \frac{u^2v^2(3uv+1)}{u^2+v^2-uv} , \tag{2.79b}$$

where the parameters u and v satisfy 0 < v < u. The negative, unphysical critical point is then at $R_h = -lv$, and $R_h = lu$ gives a positive critical point. The condition that only one positive critical point exist reads

$$\alpha(u) < v \ , \tag{2.80}$$

where $\alpha(u)$ is the unique solution to the equation

$$0 = 9u\alpha^3 - \left(6u^2 + 1\right)\alpha^2 + u\left(9u^2 + 2\right)\alpha - u^2 \tag{2.81}$$

in the interval 0 < α < u. In this case the parametrization (2.79) is unique. When the inequality in (2.80) is reversed and three positive critical points exist, the parametrization (2.79) can be made unique by imposing the conditions

$$u \ < \ \frac{1}{\sqrt{3}} \ , \tag{2.82a}$$

$$v \ < \ \frac{\sqrt{(1+6u^2)(1-3u^2)} - (1-3u^2)}{9u} \ , \tag{2.82b}$$

which make $R_h = lu$ the local maximum. The two local minima are then at the roots of the quadratic equation

$$0 = 3\left(u^2 + v^2 - uv\right)(R_\mathrm{h}/\ell)^2 - (u-v)(3uv+1)(R_\mathrm{h}/\ell) + uv(3uv+1) \ . \tag{2.83}$$

The global minimum is at the larger (smaller) local minimum when the inequality

$$0 < 12(6uv-1)\left(u^2+v^2-uv\right)^2 + (u-v)^2(3uv+1)\left[3\left(u^2+v^2\right)+1\right] \tag{2.84}$$

is satisfied (reversed).

It is of some interest to examine the behavior of the critical points in the limit $q^2 \to 0$ with fixed β. When $\beta^2 > \frac{4}{3}\pi^2\ell^2$, the above discussion shows that for sufficiently small q^2 there exists only one positive critical point, and in the limit $q^2 \to 0$ this critical point approaches zero as

$$R_\mathrm{h} = |\mathrm{q}|\left[1 + 2\pi\beta^{-1}|\mathrm{q}| + O\left(\mathrm{q}^2\ell^{-2}\right)\right] \ . \tag{2.85}$$

When $\beta^2 < \frac{4}{3}\pi^2\ell^2$, on the other hand, there exist three positive critical points for sufficiently small q^2. In the limit $q^2 \to 0$, the smallest positive critical point again approaches zero as (2.85), whereas the two larger ones approach the two critical points of the case $q = 0$. In the limiting case $\beta^2 = \frac{4}{3}\pi^2\ell^2$, the smallest of the three positive critical points once again approaches zero as (2.85), and the two larger ones merge into a $q = 0$ critical point that is not a local extremum. The limiting behavior is thus smooth, in spite of the changing number of critical points.

At any critical point, the value of the action can be written as

$$I^{c}_{C*} = -\frac{\pi R_{\rm h}^2\left(R_{\rm h}^2\ell^{-2} - 1 - 3{\rm q}^2 R_{\rm h}^{-2}\right)}{3R_{\rm h}^2\ell^{-2} + 1 - {\rm q}^2 R_{\rm h}^{-2}} \quad . \tag{2.86}$$

In the limit $q \to 0$, this agrees with the expression given in [71].

We thus see that for generic values of the parameters, $\mathcal{Z}_{ren}(\beta, q)$ can be approximated by exp h exp $[-I^{min}_{C*}]$, where I^{min}_{C*} stands for the value of $I^{\square}_{C*}$ at the critical point that is the global minimum. As this is consistent with what one would have expected just from the existence of (Lorentzian) black hole solutions under fixing the charge and the renormalized inverse Hawking temperature: such solutions exist precisely at the critical points of $I^{\square}_{C*}$, and the values of m and q at these critical points are just the mass and charge parameters of the black hole. One may view the shifting of the global minimum of $I^{\square}_{C*}$ from one local minimum to the other as a thermodynamical phase transition.

We end this section with some brief remarks on the thermodynamics in the canonical ensemble. Recall that the formulas for the thermal expectation values for the energy and the electric potential read

$$\langle E \rangle = -\frac{\partial(\ln Z_{\rm ren})}{\partial\beta} \quad , \tag{2.87a}$$

$$\langle \phi \rangle = -\beta^{-1}\frac{\partial(\ln Z_{\rm ren})}{\partial {\rm q}} \quad . \tag{2.87b}$$

When a critical point of $I^{\square}_{C*}$ dominates, we obtain

$$\langle E \rangle \approx {\rm m} \quad , \tag{2.88a}$$

$$\langle \phi \rangle \approx \frac{\rm q}{R_{\rm h}} \quad . \tag{2.88b}$$

These are, respectively, just the mass and the electric potential difference between the horizon and the infinity for the dominating classical solution. When the approximation (2.88a) is good, the positivity of the (constant q) heat capacity, $C_q = -\beta^2(\partial\langle E \rangle/\partial\beta)$, follows from the fact that the dominant critical point is a minimum of $I_{C*}^{\square}$ [60,61]. The positivity of C_q follows more generally, even when the saddle point approximation does not hold, by direct manipulations from the expression (2.75) and the assumption that $\tilde{\mu}$ is positive.

When the saddle point approximation is good, we have for the entropy the Bekenstein-Hawking result, $S = (1 - \beta\left(\frac{\partial}{\partial\beta}\right))(\ln Z_{red}) \approx \pi(R_h)^2$.

6.9 Conclusions and discussion

In this chapter we investigated the Hamiltonian dynamics and thermodynamics of spherically symmetric Einstein-Maxwell theory with a negative cosmological constant. We first set up a classical Lorentzian Hamiltonian theory in which the right end of the spacelike hypersurfaces is at the asymptotically anti-de Sitter infinity in an exterior region of a RNAdS black hole spacetime, and the left end of the hypersurfaces is at the bifurcation two-sphere of a nondegenerate Killing horizon. We then simplified the constraints by a canonical transformation, and we explicitly reduced the theory into an unconstrained Hamiltonian theory with two canonical pairs of degrees of freedom. The reduced theory was quantized by Hamiltonian methods, and a grand partition function for a thermodynamical grand canonical ensemble was obtained by analytically continuing the Schrodinger picture time evolution operator to imaginary time and taking the trace. The analytic continuation at the bifurcation two-sphere was done in a way motivated by the smoothness of Euclidean black hole geometries. A similar analysis with minor modifications to the boundary conditions led to a partition function for a thermodynamical canonical ensemble. Both the canonical ensemble and the grand canonical ensemble turned out to be well defined, and we were able to find the conditions under which the (grand) partition function is dominated by a classical Euclidean solution.

Both thermodynamical ensembles exhibited a phase transition. In the grand canonical ensemble the transition occurs when the grand partition function ceases to be dominated by any classical Euclidean black hole solution, in close analogy with what happens in the spherically symmetric vacuum canonical ensemble with a finite boundary. In the canonical ensemble this kind of a phase transition can occur only in the limit of a vanishing charge, whereas for nonvanishing charge there occurs a phase transition in which the dominating contribution to the partition function shifts from one classical Euclidean solution to another as the boundary data changes. In either ensemble, whenever the (grand) partition function is dominated by a classical solution, one recovers for the entropy the Bekenstein-Hawking value of one quarter of the horizon area.

The classical canonical transformation is a relatively straightforward generalization of the transformation that was found by Kuchar in the spherically symmetric vacuum Einstein theory under Kruskal-like boundary conditions. When the classical equations of motion hold, our new canonical coordinates M and Q are simply the mass and charge parameters of the RNAdS solution. By (generalized) Birkhoff's theorem, the spacetime is uniquely characterized by these two parameters and the cosmological constant. The conjugate momenta, P_M and P_Q, carry the information about the embedding of the spacelike hypersurface in the spacetime and the electromagnetic gauge. Upon elimination of the constraints, we saw in Section 2.4 that P_M and P_Q each give rise to one unconstrained momentum in the reduced Hamiltonian theory. These reduced momenta are global constructs with no local geometrical meaning, and they are associated with the anchoring of the spacelike hypersurfaces at the infinity and at the bifurcation two-sphere. The electromagnetic pair (Q; P_Q) is quite closely analogous to the gravitational pair (M; P_M). The third canonical pair, (R; P_R), is entirely gauge, and it completely disappears when the constraints are eliminated.

Although we have here discussed the canonical transformation only under boundary conditions motivated by our thermodynamical goal, it would appear possible to adapt this canonical transformation to boundary conditions under which the spacelike hypersurfaces extend from a left hand side asymptotically anti-de Sitter region to a right hand side asymptotically anti-de Sitter region, crossing the event horizons in arbitrary ways. The form taken by the constraints then suggests that, after introducing electromagnetic variables, it is possible to perform a canonical transformation that separates Q into the charge density Q_0 and the charge at the (say) left hand side infinity, in analogy with the transformation that separates M into the mass density M_0 and the mass at the left hand side infinity. Also, it appears possible to take the limit where the cosmological constant vanishes and the asymptotically anti-de Sitter regions are replaced by asymptotically flat regions. We have not investigated these issues in a systematic fashion; however, we shall outline in Appendix B how our canonical transformation can be adapted to the limit of a vanishing cosmological constant, under boundary conditions that still keep the left end of the hypersurfaces at the bifurcation two-sphere of a nondegenerate Killing horizon but replace the asymptotically anti-de Sitter falloff conditions at the right end by asymptotically at falloff conditions. In this case, each classical solution is the exterior region of a non-extremal Reissner-Nordstrom black hole.

The thermodynamical results show that the stabilizing effect of the negative cosmological constant is highly similar to the stabilizing effect of a finite "box" with fixed surface area and fixed local temperature. One important difference is, however, that in the asymptotically anti-de Sitter case various thermal expectation values are more directly related to the parameters of the dominant classical solutions. In the grand canonical ensemble, equations (2.71) show that the thermal expectation values of energy and charge are simply the mass and charge parameters of the dominant classical solution: there are no additional contributions to the mass from the gravitational binding energy associated with the thermal energy, or from the electrostatic binding energy associated with the charge. Such additional, finite size contributions were found to be present in oher finite size ensembles. In the canonical ensemble, the situation is similar with the thermal expectation values of the energy and the electric potential (2.88).

The stabilizing effect of the negative cosmological constant becomes fully apparent when one attempts to repeat the analysis with a vanishing cosmological constant, replacing the asymptotically anti-de Sitter infinity by an asymptotically flat infinity. We shall outline this analysis in Appendix B. While there is no difficulty in quantizing the reduced Hamiltonian theory, the trace of the analytically continued time evolution operator turns out to remain divergent even after renormalization. Neither the canonical ensemble nor the grand canonical ensemble exists. For the canonical ensemble this conclusion might be surprising in view of the observation that a Reissner-Nordstrom black hole in asymptotically at space is stable against Hawking evaporation when one fixes the charge and the temperature at the infinity, provided the mass and charge parameters of the hole satisfy the inequality$q^2 > \frac{3}{4}m^2$. However, as we shall see in Appendix B, the local stability of a classical solution is not sufficient to guarantee the existence of a full thermodynamical ensemble.

Finally, we recall that as the physical temperature of Hawking radiation is redshifted to zero at the anti-de Sitter infinity, we defined a renormalized temperature at infinity in terms of the rate at which the local Hawking temperature approaches zero. This definition led to physically reasonable conclusions; in particular, we recovered from the thermodynamical ensembles the Bekenstein-Hawking result for the black hole entropy. The definition can however be argued to have an ad hoc flavor, and one might wish to replace it by something that can be given a more immediate physical justification. What would be needed is a better understanding as to whether asymptotically anti-de Sitter infinity can in some sense be regarded as a physically realizable system, rather than just as a mathematically elegant set of boundary conditions.

Chapter 7. Lovelock Black Hole Thermodynamics

7.1 Introduction

In this chapter a description is given for the Hamiltonian dynamics and thermodynamics of spherically symmetric spacetimes within a one-parameter family of five-dimensional Lovelock theories [9,91]. We adopt boundary conditions that make every classical solution part of a black hole exterior region, with the spacelike hypersurfaces extending from the horizon bifurcation three-sphere to a timelike boundary with fixed intrinsic metric. The constraints are simplified by a Kuchar-type canonical transformation, and the theory is reduced to its true dynamical degrees of freedom. After quantization, the trace of the analytically continued Lorentzian time evolution operator is interpreted as the partition function of a thermodynamical canonical ensemble. Whenever the partition function is dominated by a Euclidean black hole solution, the entropy is given by the Lovelock analogue of the Bekenstein-Hawking entropy. The asymptotically flat space limit of the partition function does not exist. The results indicate qualitative robustness of the thermodynamics of five-dimensional Einstein theory upon the addition of the Lovelock parameter.

The analysis that follows parallels that of Chapter 6 closely, so will show less of the derivation (more placed in the Appendix C).

Background

A gravitational theory whose Lagrangian density consists of multiples of lower dimensional Euler densities has the property that the field equations are second order in the metric [92,93]. These theories, known as Lovelock theories, include Einstein's theory with a cosmological constant in all dimensions greater than two, and in five or more dimensions they provide genuine curvature squared generalizations of Einstein's theory. Among all curvature squared generalizations of Einstein's theory, Lovelock theories therefore have the special status that they preserve the number of degrees of freedom: a generic curvature squared action produces field equations that are fourth order in the metric, thus containing twice as many degrees of freedom as Einstein's theory. This has generated wide interest in Lovelock theories, especially in the contexts of cosmology and black hole physics [94-109].

The purpose of the chapter is to analyze the classical and quantum dynamics of spherically symmetric Lovelock gravity by the Hamiltonian methods recently developed by Kuchar [80]. These methods have previously been applied to spherically symmetric Einstein(-Maxwell) gravity in four dimensions [80,68,36,69,110], vacuum dilatonic gravity in two dimensions [69,111,112], and to related systems [113,114]; for related discussion, see [115-121]. At the classical level, we wish to find a canonical transformation that introduces the mass parameter of the spacetime as a new canonical variable, use this transformation to simplify the constraints, and reduce the theory to its true dynamical degrees of freedom. At the quantum level, we wish to derive from the quantum theory a partition function that describes the equilibrium thermodynamics of a Lovelock black hole in the canonical ensemble.

The issues of prime interest are twofold. First, although Lovelock theories have the same set of canonical variables as Einstein's theory, the Lovelock Hamiltonian is, in general, a multivalued function of the canonical variables [98,99]. One anticipates that this multivaluedness may introduce complications into the canonical formulation and Hamiltonian reduction, even though the Lovelock analogue of Birkhoff's theorem [96] strongly suggests that the local considerations should differ little from those in Einstein's theory. Second, certain Lovelock black holes have thermodynamical properties that differ qualitatively from those of Einstein black holes; in particular, a Lovelock black hole can be stable against Hawking evaporation in asymptotically flat space [102]. This raises the question whether Lovelock theories might admit quantum thermodynamical ensembles with boundary conditions that do not yield well-defined ensembles in Einstein's theory.

The number of possible Lovelock terms in the action increases with the dimension of the spacetime, and different choices for the coefficients give qualitatively different theories. In this paper we shall aim for concreteness at the expense of generality: we concentrate on a specific one-parameter family of Lovelock theories in which both of the above issues of interest are present.

We take the only bulk contributions to the action to be the Einstein-Hilbert term and the four-dimensional Euler density. In D spacetime dimensions, the action then reads [102]

$$S = \frac{1}{2\kappa}\int d^D x \sqrt{-g}\left[R + \frac{\lambda}{2}\left(R_{abcd}R^{abcd} - 4R_{ab}R^{ab} + R^2\right)\right]$$
$$+ \text{ (boundary terms)} \quad , \tag{3.1}$$

where κ is the D-dimensional gravitational constant and λ is the single Lovelock parameter. (We have set c = h = 1: the gravitational constant κ has the dimension of length to the D-2 power L^{D-2}, and the Lovelock parameter λ has the dimension of L^2.) For $D \geq 5$, the four-dimensional Euler density contributes to the equations of motion, and we obtain a one-parameter family of generalizations of Einstein's theory. In these theories, asymptotically flat black hole solutions that are stable against Hawking evaporation occur only when D = 5 and $\lambda > 0$ [102], and we shall therefore concentrate on this case. For the interest of comparison, we shall also include the limiting case of five-dimensional Einstein theory, $\lambda = 0$.

We shall formulate the spherically symmetric Hamiltonian theory with thermodynamically motivated boundary conditions similar to those introduced in [68]. On a classical solution, the left end of the spacelike hypersurfaces will be at the bifurcation three-sphere of a nondegenerate Killing horizon, and the right end will be on a timelike hypersurface in the right-hand-side exterior region of the spacetime. At the left end we fix the rate at which the spacelike hypersurfaces are boosted with respect to the coordinate time, and at the right end we fix the intrinsic metric on the timelike hypersurface. For $\lambda > 0$, the super-Hamiltonian H turns out to be a multivalued function of the canonical variables. However, our boundary conditions are sufficient to uniquely determine H near the left end of the spacelike hypersurfaces, and this solution for H can then be uniquely extended to the full spacelike hypersurfaces by continuity. Our boundary conditions at the horizon thus eliminate the potential difficulties due to the multivaluedness of the super-Hamiltonian.

We shall find that the theory admits a natural generalization of the canonical transformation of [67,68]. The constraints become simple, and a Hamiltonian reduction leads again to a single canonical pair of unconstrained degrees of freedom. On a classical solution, one member of the pair is the mass parameter, and its conjugate momentum is the difference of the Killing times at the left and right ends of the spacelike hypersurfaces.

After taking the curvature radius at the right end of the hypersurfaces to be time-independent, we quantize the reduced theory by Hamiltonian methods. Following [68], we analytically continue the time evolution operator to imaginary time and take the trace, and interpret the resulting object as the partition function of a thermodynamical canonical ensemble. This ensemble describes black hole spacetimes in a spherical \box" whose size and boundary temperature are fixed.

In the special case of Einstein's theory, $\lambda = 0$, we find that the thermodynamical properties of the system are highly similar to those of the corresponding system in four dimensions [68,50,60]. For high boundary temperatures, the partition function is dominated by a black hole that fills most of the box; for low boundary temperatures, there is no dominant classical solution, and one can argue that the behavior of the partition function suggests a topological phase transition from a black hole to "hot flat space" [50,60].

For $\lambda > 0$, the partition function displays several qualitatively different regions depending on the relative magnitudes of the box, the temperature, and λ. In the high temperature limit, with the other two parameters fixed, the partition function is again dominated by a classical black hole solution that fills most of the box. In the low temperature limit, with the other two parameters fixed, the partition function is now also dominated by a black hole solution: this black hole is small compared with the box, and it has no analogue in Einstein's theory. However, if λ is small compared with the size of the box, the existence of the new dominating solution at low temperatures only has a minor effect on the behavior of the partition function. In this sense, we can say that the qualitative thermodynamical behavior of the pure Einstein system is stable against the addition of the Lovelock parameter.

When the size of the box is taken to infinity, the partition function does not have a well-defined limit, neither for $\lambda = 0$ nor for $\lambda > 0$. For $\lambda = 0$ this is not surprising: just as in four dimensions, it reflects the fact that a Schwarzschild hole in asymptotically flat space is not stable against Hawking evaporation [50,60]. For $\lambda > 0$, on the other hand, the theory does admit asymptotically at black hole solutions that are stable against Hawking evaporation [102], and one might therefore have expected the infinite box limit to exist. The reason why this is not the case becomes apparent when one tries to repeat our analysis with boundary conditions that replace the right-hand-side timelike boundary by an asymptotically at infinity. The classical reduction and the construction of a quantum theory proceed without difficulty, but the effective Euclidean action of the system turns out to be unbounded below, and the formal integral expression for the partition function is divergent. The effective action has a local minimum, corresponding to the black hole that is stable against Hawking evaporation [61], but this is not sufficient to ensure the existence of the full canonical ensemble. Another system with locally stable classical solutions but no well-defined canonical ensemble is four-dimensional Einstein-Maxwell theory with fixed charge in asymptotically flat space [36].

Comparison of Reduced Quantization plan to Minisuperspace Quantization (Book 4 [4])

Overview of Minisuperspace Quantization for Dust Shell Collapse

(i) Reduced phase space formalism for spherical dust shell in spherically symmetric, asymptotically flat geometry is obtained.

(ii) For prescribed M_-:

$$S = \int dt\,(p\dot{\hat{r}} - h)$$

$$p = \sqrt{2M_-\hat{r}} - \sqrt{2M_+\hat{r}} + \hat{r}\ln\left(\frac{\hat{r} + \hat{p} + \sqrt{\hat{p}^2 + m^2} + \sqrt{2M_+\hat{r}}}{\hat{r} + \sqrt{2M_-\hat{r}}}\right)$$

$$\hat{p}: M_+ + M_- = \sqrt{\hat{p}^2 + m^2} + \frac{m^2}{2\hat{r}} - \hat{p}\sqrt{\frac{2M_+}{\hat{r}}}$$

$$h = M_+ + M_-$$

(iii) Time Reparameterization

Proper Time: $H(\hat{r}, P) = m\cosh\left({}^{P}\!/_{m}\right) - \frac{m^2}{2\hat{r}} + M_-\left[2 - \left(\frac{m}{\hat{r}}\right)e^{-P/m}\right]$

"Minkowski" Time: $\tilde{h}(\hat{r}, P) = \sqrt{P^2 + m^2} + \frac{m^2}{2\hat{r}} + M_-\left[2 - \frac{1}{\hat{r}}\left(\sqrt{P^2 + m^2} - P\right)\right]$

(iv) Quantization

- (a) Vacuum case (M_- dynamical)
- (b) $M_- = 0$
- (c) $M_- \neq 0$

Overview of Lovelock Black Hole Thermal Quantum Gravity

(i) Canonical formulation m metric variables

$$S = \frac{1}{2K}\int d^5 x\sqrt{-g}\left[{}^{(5)}R + \frac{\lambda}{2}\left({}^{(5)}R_{abcd}\,{}^{(5)}R^{abcd} - 4R_{ab}R^{ab} + R^2\right)\right]$$
$$+ \text{(boundary terms)}$$

(a) analysis of bulk term more involved, timelike boundary more involved also

(b) Hamiltonian bulk term:

$$S_\Sigma = \int dt \int_0^1 dr\,(P_\Lambda \Lambda + P_R \dot{R} - NH - N^r H_r)$$

$$H - y\left\{P_R + y\left[\Lambda R - \hat{\lambda}\left(\frac{R'}{\Lambda}\right)'\right]\right\} - \Lambda R\left[1 - \left(\frac{R'}{\Lambda}\right)^2\right]$$
$$+ \left(\frac{R'}{\Lambda}\right)'\left\{R^2 + \hat{\lambda}\left[1 - \left(\frac{R'}{\Lambda}\right)^2\right]\right\}$$

$$H_r = R' P_R - \Lambda P'_\Lambda \qquad \hat{\lambda} = \frac{1}{2}\lambda$$

$$0 = \frac{1}{3}\hat{\lambda} y^3 + y\left\{R^2 + \hat{\lambda}\left[1 - \left(\frac{R'}{\Lambda}\right)^2\right]\right\} + P_\Lambda$$

(ii)– (v) similar to previous → to get Partition Function:

$$Z(\beta; B, \hat{\lambda}) = N\int_0^1 \tilde{\mu}\, dx \exp(-I) \quad ; \quad x = \frac{R_0}{B}$$

$$I = \beta B^2(1 - \sqrt{F})\left[1 + \frac{\hat{\lambda}}{3B^2}(1 - \sqrt{F})(2 + \sqrt{F})\right]$$
$$- 2\pi B^2 x\left(\frac{1}{3}x^2 + \frac{\hat{\lambda}}{B^2}\right)$$

$$F = 1 + \frac{B^2}{\hat{\lambda}}\left(1 - \sqrt{1 + \frac{2x^2\hat{\lambda}}{B^2} + \frac{\hat{\lambda}^2}{B^4}}\right)$$

$$S = \left(1 - \beta\frac{\partial}{\partial\beta}\right)(\ln Z) \simeq 2\pi R_0\left(\frac{1}{3}R_0^2 + \hat{\lambda}\right)$$

in agreement with literature. We will now find, in an analysis that parallels that of Chapter 6:

(vi) For $\hat{\lambda} > 0$

(a) High temperature, large box, similar to $\hat{\lambda} = 0$, dominating Black Hole fills most of box
(b) Low temperature, dominating Black Hole (small) exits where none did for $\hat{\lambda} = 0$.
(c) As $B \rightarrow \infty$ partition function has no well-defined limit

(vii) Conjecture (reinforcing RNAdS result): Whenever global properties of Lovelock theory sufficiently similar to Einstein theory, then also the equilibrium thermodynamics will be qualitatively similar to that in Einstein's theory.

7.2 Metric formulation

In this section we introduce the model and present the Hamiltonian formulation in the metric variables. We begin with the general five-dimensional spherically symmetric Arnowitt- Deser-Misner (ADM) metric,

$$ds^2 = -N^2 dt^2 + \Lambda^2 (dr + N^r dt)^2 + R^2 d\Omega_3^2 \ . \tag{3.2}$$

Here $d\Omega_3^2$ is the metric on the unit three-sphere, and N, N^r, Λ, and R depend on the coordinates t and r only. The coordinate r has the range$0 \leq r \leq 1$; this is convenient in view of our boundary conditions, which will make the radial proper distance on the constant t hypersurfaces finite. Unless otherwise stated, we assume both the spatial metric and the spacetime metric to be nondegenerate. In particular, we take Λ, R, and N to be positive. For further details and the determination of the Lapse and Shift variables, see Appendix C. Note that equation numbering in the sections that follow are retained (from the authors thesis [9]) for convenience in the text references to those equation numbers that follow.

7.3 Canonical Formulation

Inserting the metric (3.2) in the Lovelock action (3.1) with D = 5, integrating over the three-sphere, and dropping a total derivative, we recover the action

$$S_{\Sigma}^{L} = \int dt \int_0^1 dr\, \mathcal{L} \ , \tag{3.3}$$

Where

$$\mathcal{L} = -\frac{[\dot{\Lambda}(N^r\Lambda)'](\dot{R} - N^r R')}{N}\left\{R^2 + \hat{\lambda}\left[1 - \left(\frac{R'}{\Lambda}\right)^2 + \frac{\left(\dot{R} - N^r R'\right)^2}{3N^2}\right]\right\}$$

$$-\frac{(\dot{R} - N^r R)^2}{N}\left[R - \lambda\left(\frac{R'}{\Lambda}\right)'\right]$$

$$+N\Lambda R\left[1 - \left(\frac{R'}{\Lambda}\right)^2\right] - N\left(\frac{R'}{\Lambda}\right)'\left\{R^2 + \lambda\left[1 - \left(\frac{R'}{\Lambda}\right)^2\right]\right\}$$

(3.4)

The overdot and the prime denote respectively $\partial/(\partial t)$and $\partial/(\partial r)$. We have written $\lambda = \frac{1}{2}\hat{\lambda}$, conforming to the notation of Ref. [102].

The Lagrangian equations of motion obtained from local variations of S_Σ^L (3.3) are equivalent to the full spherically symmetric Lovelock equations [94-96] derived from the action. The reduction of the action by spherical symmetry is therefore consistent with the equations of motion, and we can take S_Σ^L (3.3) as the starting point of the dynamical analysis. We shall address the boundary conditions and boundary terms within the Hamiltonian formulation below.

We take $\hat{\lambda} \geq 0$. For presentational simplicity, we shall assume $\hat{\hat{\lambda}} > 0$ until explicitly stated otherwise. In the limiting case of five-dimensional Einstein gravity, $\hat{\lambda} = 0$, the analysis would proceed in an entirely analogous manner, with the obvious technical simplifications.

The Hamiltonian form of the action (3.3) is

$$S_\Sigma \int dt \int_0^1 dr\left(P_\Lambda \dot{\Lambda} + P_R \dot{R} - NH - N^r H_r\right),$$

(3.5)

where the super-Hamiltonian constraint and the supermomentum constraint are given respectively by

$$H = y\left\{P_R + y\left[R - \hat{\lambda}\left(\frac{R'}{\Lambda}\right)'\right]\right\}$$

(3.6a)

$$-\Lambda R\left[1 - \left(\frac{R'}{\Lambda}\right)^2\right] + \left(\frac{R'}{\Lambda}\right)'\left\{R^2 + \hat{\lambda}\left[\left(\frac{R'}{\Lambda}\right)'\right]\right\}$$

$$H_r = R'^{P_R} - \Lambda P'_\Lambda$$

(3.6b)

The quantity y is determined in terms of the canonical variables by the cubic equation

$$0 = \frac{1}{3}\hat{\lambda}y^3 + y\left\{R^2 + \hat{\lambda}\left[1 - \left(\frac{R'}{\Lambda}\right)^2\right] + P_\Lambda\right\} \tag{3.7}$$

Note that the cubic (3.7) can have up to three real solutions for *y*, and the superHamiltonian is therefore potentially a multivalued function of the canonical variables. Such multivaluedness occurs generically in Lovelock theories [98,99], and we shall address it in more detail below.

Let us turn to the boundary conditions. From the Lovelock generalization of Birkhoff's theorem [96] it follows that the local properties of the classical solutions are completely characterized by a continuous, mass-like parameter, and a discrete parameter taking the values ± 1. The general solution is shown in curvature coordinates in Appendix C. We wish to concentrate on the black hole solutions, whose global structure is similar to that of the Kruskal manifold [102]. We further wish to attach the left end of our spacelike hypersurfaces at the bifurcation three-sphere, and to prescribe there the rate at which the hypersurfaces are boosted with respect to our coordinate time. The right end of the hypersurfaces will then be in the right-hand-side exterior region, and we wish to prescribe the metric on the timelike hypersurface that this end traces. We must now specify boundary conditions and boundary terms that achieve this.

Consider first the left end of the hypersurfaces. Following the analogous treatment in [68,69,36,70], we adopt at $r \to 0$ the falloff behavior

$$\begin{aligned}
\Lambda(t,r) &= \Lambda_0(t) + O(r^2)\,, \\
R(t,r) &= R_0(t) + R_2(t)r^2 + O(r^4)\,, \\
P_\Lambda(t,r) &= O(r^3)\,, \\
P_R(t,r) &= O(r)\,, \\
N(t,r) &= N_1(t)r + O(r^3)\,, \\
N^r(t,r) &= N_1^r(t)r + O(r^3)\,,
\end{aligned} \tag{3.8a-f}$$

where A_0 and R_0 are positive, and $N_1 \geq 0. O(r^n)$ stands for a term that falls off at $r \to 0$ as r^n, and whose derivatives fall off accordingly. As in [68,69,36,70], these conditions guarantee that the classical solutions have a bifurcate horizon, they put the left end of the spacelike hypersurfaces at the bifurcation sphere, and they are consistent with the constraints and preserved by the Hamiltonian evolution. They also ensure that the cubic (3.7) has a unique real solution for *y* near $r = 0$. On a classical solution, the future unit normal vector $n^a(t)$ to the spacelike hypersurfaces at $r = 0$ then evolves according to

$$n^a(t_1)n_a(t_2) = -\cosh\left(\int_{t_1}^{t_2} \Lambda_0^{-1}(t)N_1(t)dt\right)$$

(3.9)

Next, consider the boundary conditions in the variational principle. At $r = 0$, we follow [68,69,36,70] and make N_1/Λ_0 a prescribed function of *t*. By (3.9), this means fixing the rate at which the constant *t* hypersurfaces are boosted at $r = 0$. At $r = 1$, we make *R* and $-g_\mu = N^2 - (\Lambda N^r)^2$prescribed positive-valued functions of *t*. This means fixing the intrinsic metric on the three-surface $r = 1$, and in particular fixing this metric to be timelike.

Finally, we need the boundary terms to be added to the bulk action (3.5). As in [68], it can be verified that the appropriate term at r = 0 is

$$\int dt R_0\left(\frac{1}{3}R_0^2 + \hat{\lambda}\right)(N_1/\Lambda_0) \qquad (3.10)$$

and the appropriate term at $r = 1$ is the integral over *t* of

$$N^{-1}R^2R' - N^r\Lambda P_\Lambda - \frac{1}{2}\dot{R}\left(R^2 + \hat{\lambda}\right) In \left|\frac{N + \Lambda N^r}{N - \Lambda N^r}\right|$$
$$+ + \dot{\hat{\lambda}}N\left(\frac{R'}{\Lambda}\right)\left[1 - \frac{1}{3}\left(\frac{R'}{\Lambda}\right)^2 - \frac{\dot{R}(\dot{R} - N^rR')}{N^2}\right]$$
$$-\frac{\hat{\lambda}N^r}{3N}\left[\frac{\dot{R}^3}{N^2 - (\Lambda N^r)^2} + \frac{N^rR'^3}{\Lambda^2}\right]$$

We have therefore arrived at a variational principle with the desired boundary conditions. The Lovelock generalization of Birkhoff's theorem guarantees that classical solutions exist, and makes possible a complete description of the solutions.

Lovelock action

$S_\Sigma = \frac{1}{6}\int\sqrt{-g^{(5)}}\left[R + \frac{\lambda}{2}(R^2 - 4R_{ab}R^{ab} + R_{abcd}R^{abcd})\right]d^5x$

$\delta S_\Sigma = 0$ → gives metric equations of motion that are second order. As a tensorial relation this must be constructed from R_{ab} and g_{ab}.

Show that $\delta S_\Sigma = \int \delta(X)\, Y d^2x$, where *Y* is a second order equation of motion, as a check.

<u>Spherically symmetric 5-D minisuperspace:</u>

$ds^2 = -N^2dt^2 + \Lambda^2(dr + N^rdt)^2 + R^2d\Omega_3^2$

$d\Omega_3^2 = dx^2 + \sin^2 x\,(d\theta^2 + \sin^2\theta\, d\varphi)$

$0 \leq x \leq \pi$
$0 \leq \theta \leq \pi$
$0 \leq \varphi \leq \pi$
$\int dx \sin x d\theta \sin x \sin \theta d\varphi$
$N_r = g_{rk}N^k = g_{rr}N^r = L^2N^r$

Synopsis of Results (see Appendix C for details)
The Lagrangian bulk action is
$S_\Sigma = \int dt\, dr\, L$
$L = L_{kin} + L_{pot}$
$NL_{kin} = -(D\Lambda)(DR)\left\{R \div \hat{\lambda}\left[1 - \left(\frac{R'}{\Lambda}\right)^2 + \frac{(DR)^2}{3N^2}\right]\right\}$
$-(DR)^2\left\{\Lambda R - \hat{\lambda}\left(\frac{R'}{\Lambda}\right)'\right\}$
$N^{-1}L_{pot} = \Lambda R\left[1 - \left(\frac{R'}{\Lambda}\right)^2\right] - \left(\frac{R'}{\Lambda}\right)'\left\{R^2 + \hat{\lambda}\left[1 - \left(\frac{R'}{\Lambda}\right)^2\right]\right\}$
$DR := \dot{R} - sR'$
$D\Lambda := \dot{\Lambda} - (s\Lambda)'$
$s := N^r$
The Hamiltonian bulk action is
$S_\Sigma = \int dt\, dr\left(P_\Lambda\dot{\Lambda} + P_R\dot{R} - NH - sH_r\right)$
$H = H_{kin} + H_{pot}$
$H_r = R'^{P_R} - \Lambda P'_\Lambda$
$H_{pot} = -N^{-1}L_{pot}$
$H_{kin} = y\left\{P_R + y\left[\Lambda R - \hat{\lambda}\left(\frac{R'}{\Lambda}\right)'\right]\right\}$
Whereby y is solution of
$D = \frac{1}{3}\hat{\lambda}y^3 + y\left\{R^2 + \hat{\lambda}\left[1 - \left(\frac{R'}{\Lambda}\right)^2\right]\right\} + P_\Lambda$

Canonical Transformation Overview
Following the method of [80]:

$$M = \frac{1}{2}R^2(1-F) + \frac{1}{4}\hat{\lambda}(1-F)^3\,;\; M = \frac{1}{2}m + \frac{1}{4}\hat{\lambda}$$

$$T' = \frac{\Lambda y}{F}$$

Thus, $M, P_M = \frac{\Lambda y}{F} = T'$ form one canonical pair. Choose $R := R$ for config variable in new canonical pair and for conjugate momenta demand supermomentum constant remains in a form that generates spatial diffeomorphisms, so

$$R'^{P_R} - \Lambda P_\Lambda' = P_M M' + P_R R'$$

This suggests canonical transform of:

$M := \frac{1}{2}R^2(1-F) + \frac{1}{4}\hat{\lambda}(1-F)^2$

$P_M := -\Lambda y/F$

$R := R$

$P_R := F^{-1}(\Lambda^{-2}R'H_r - yH)$

Is the transformation canonical? Consider the difference in Louiville forms:

$P_\Lambda\delta\Lambda + P_R\delta R - P_M\delta M - P_R\delta R$

$= \delta\left[\Lambda P_\Lambda - \hat{\lambda}y\Lambda^{-1}(R')^2 + \frac{1}{2}R'^{(R^2+\hat{\lambda})}\ln\left|\frac{R'+y\Lambda}{R'-y\Lambda}\right|\right]$

$+\left\{\left[\hat{\lambda}y\Lambda^{-1} - \frac{1}{2}(R^2+\hat{\lambda})\ln\left|\frac{R'+y\Lambda}{R'-y\Lambda}\right|\right]\delta R\right\}$

Exact form is finite term.

Technical problem:

The expressions

$NH + N^+H_r = N^M M' + N^R P_R$

$N^M = -NF^{-1}\Lambda^{-1}R' - N^r F^{-1}\Lambda y,$

$N^R = Ny + N^R R'$

Suggest that N^M and N^R be taken as new Lagrange multipliers. However, fixing N^M at $r = 0$ to a unique independent is not the same as fixing $N_1\Lambda_o^{-1}$ to a value independent of canonical variables. This will be remedied by redefining the language multipliers near r=0 (see Appendix C).

Hamiltonian Reduction Overview

With constraints imposed in the new canonical theory the reduction is simple. The reduced action is then:

$$m' = 0 \rightarrow m = m(t)\ ;\ S_{red} = \int dt\,(p\dot{m} - h)$$

$$p := \int_0^1 dr\,P_M$$

$$h = -\hat{N}_0 R_0\left(\frac{1}{3}R_0^2 + \hat{\lambda}\right)$$
$$-\left(B^2+\hat{\lambda}\right)\left[\sqrt{Q^2F+\dot{B}^2} + \frac{1}{2}\dot{B}\ln\left(\frac{\sqrt{Q^2F+\dot{B}^2}-\dot{B}}{\sqrt{Q^2F+\dot{B}^2}+\dot{B}}\right)\right]$$
$$+\frac{1}{3}\hat{\lambda}Q^{-2}\left(Q^2F+\dot{B}^2\right)^{3/2}$$
$$\hat{N}_0 = N_1/\Lambda_0$$

$$R_0 = \sqrt{2m - \frac{1}{2}\hat{\lambda}}$$

B, Q^2, N_0 are prescribed functions of t, and $B > 0\,, Q^2 > 0\,, M_0 \geq 0$. Need $B^2 > \frac{1}{3}\hat{\lambda}$ for classical solutions to exist, so that is now assumed

7.4 Canonical Transformation

In this section we simplify the constraints by a canonical transformation and reduce the theory to unconstrained Hamiltonian variables. The treatment will closely follow Refs. [80,68,36].

To begin, suppose that we are given the canonical data $(\Lambda, R, P_\Lambda, P_R)$ on a spacelike hypersurface embedded in a classical solution. We wish to reconstruct from this data the spacetime and the location of the hypersurface.

The embedding of the hypersurface in the classical solution defines a unambiguous value of y: by the equation of motion obtained from the variation with respect to P_R, one finds that this value is $y_{true} = N^{-1}(\dot{R} - N^r R')$. To reconstruct y_{true} from the canonical data, one needs to solve the cubic (3.7), which may have up to three real solutions. Near r = 0, the falloff (3.8) guarantees that the cubic has a unique real solution, and this solution must therefore be equal to y_{true}. As r increases, two spurious real solutions may appear, but it is straightforward to verify that neither of the spurious real solutions can ever be equal to y_{true}. Therefore, y_{true} is recovered from (3.7) by choosing the unique real root near r = 0 and following this root by continuity to all r. We note that, generically, neither of the spurious roots for y satisfies the constraint H = 0.

After $y = y_{true}$ has been recovered, the reconstruction proceeds in full analogy with that in [80]. The function F is given by

$$F = \left(\frac{R'}{\Lambda}\right)^2 - y^2 \ , \qquad (3.12)$$

and from Appendix C one finds for the mass the expression

$$M = \frac{1}{2}R^2(1 - F) + \frac{1}{4}\hat{\lambda}(1 - F)^2 \qquad (3.13)$$

Finally, one finds

$$T' = \frac{\Lambda y}{F}$$

(3.14)

which specifies the location of the hypersurface up to translations in the Killing time. This completes the reconstruction.

Next, we wish to promote the reconstruction equations into a canonical transformation, valid even when the equations of motion do not hold. Provided we stay within a sufficiently narrow neighborhood of the classical solutions, y is again uniquely recovered as a function of the canonical data by taking the unique real root of (3.7) near $r = 0$ and continuously following this root as r increases. Computing the Poisson bracket between M and T' suggests that —T' could serve as the momentum conjugate to M; if this holds, the new momentum conjugate to $\mathbf{R} := R$ is fixed by the fact that the supermomentum constraint generates spatial diffeomorphisms in all the variables and must thus read $P_M M' + P_{\mathbf{R}} \mathbf{R}'$. These considerations suggest the transformation

$$M := \frac{1}{2}R^2(1-F)\frac{1}{4}\hat{\lambda}(1-F)^2$$

$$P_M := -\frac{\Lambda y}{F}$$

$$\mathbf{R} := R$$

$$P_{\mathbf{R}} := F^{-1}(\Lambda^{-2}R'^{H_r} - yH)$$

(3.15a-d)

We now need to examine whether this transformation is indeed canonical. To proceed, we arrange the difference of the integrands in the Liouville forms as

$$\begin{aligned} P_\Lambda \delta\Lambda \quad & + P_R\delta R - P_M\delta M - P_{\mathrm{R}}\delta \mathrm{R} \\ & = \delta\left[\Lambda P_\Lambda - \hat{\lambda} y \Lambda^{-1}(R')^2 + \tfrac{1}{2}R'\left(R^2+\hat{\lambda}\right)\ln\left|\frac{R'+y\Lambda}{R'-y\Lambda}\right|\right] \\ & + \left\{\left[\hat{\lambda} y \Lambda^{-1} R' - \tfrac{1}{2}\left(R^2+\hat{\lambda}\right)\ln\left|\frac{R'+y\Lambda}{R'-y\Lambda}\right|\right]\delta R\right\}' \quad . \qquad (3.16) \end{aligned}$$

Both terms on the right-hand side of (3.16) are well defined. Upon integration from r = 0 to r = 1, the second term only produces contributions from the two ends. The contribution from r = 0 vanishes because of the falloff (3.8). The contribution from r = 1 vanishes if δR vanishes there. As δ should in the context of the Liouville form be understood as a time derivative, this happens when the boundary conditions fix R to be independent of t at r = 1. If this is the case, we see that the difference of the Liouville forms is an exact form,

$$\int_0^1 dr\,(P_\Lambda \delta\Lambda + P_R \delta R) \quad - \int_0^1 dr\,(P_M \delta M + P_{\mathsf{R}} \delta \mathsf{R})$$
$$= \delta\left\{\int_0^1 dr \left[\Lambda P_\Lambda - \hat{\lambda} y \Lambda^{-1}(R')^2 + \tfrac{1}{2}R'\left(R^2 + \hat{\lambda}\right)\ln\left|\frac{R' + y\Lambda}{R' - y\Lambda}\right|\right]\right\} \quad ,(3.17)$$

and the transformation is canonical.

If, on the other hand, the boundary conditions fix R to be explicitly t-dependent at r = 1, one cannot similarly argue that δR would vanish at r = 1. The canonical variables at r = 1 do not cleanly split into "independent" degrees of freedom versus boundary data, and it us unclear what the proper attitude here should be. We shall, nevertheless, proceed to regard the transformation as canonical even when R is explicitly t-dependent at r = 1, as in [68,36], it will be seen that no apparent inconsistency will result. From the viewpoint of thermodynamics, the case of principal interest will in any case be the one where R is independent of t at r = 1.

By construction, our transformation is well defined in a sufficiently narrow neighborhood of the classical solutions. It also has a unique inverse. Equations (3.15a) and (3.15c), together with the falloff implied by (3.8), determine F uniquely in terms of M and R. Equations (3.12) and (3.15b), together with the fact that Λ is by assumption positive, then determine Λ and y. P_Λ is obtained from (3.7), and P_{R} finally from (3.15d).

To obtain the action in the new variables, we note that the constraint terms can be written as

$$NH + N^r H_r = N^M M' + N^{\mathsf{R}} P_{\mathsf{R}} \quad , \qquad (3.18)$$

where

$$N^M = -NF^{-1}\Lambda^{-1}R' - N^r F^{-1}\Lambda y \quad , \qquad (3.19a)$$

$$N^{\mathsf{R}} = Ny + N^r R' \quad . \qquad (3.19b)$$

This suggests that one could take N^M and N^R as the new independent Lagrange multipliers in the action. Examining the falloff at r = 0 reveals, however, that fixing N^M at r = 0 to a value that is independent of the canonical variables is not equivalent to fixing $N_1\Lambda_0^{-1}$ to a value that is independent of the canonical variables. This difficulty can be remedied by redefining the Lagrange multipliers near r = 0 as in [36], and the appropriate boundary terms at r = 0 and r = 1 can then be constructed as in [68,36]. After these steps, the constraints can be eliminated by a Hamiltonian reduction as in [80,68], and one recovers a reduced theory in a true Hamiltonian form. The steps follow the cited references so closely that we shall here omit the detail and proceed directly to the reduced action.

7.5 Hamiltonian Reduction

The reduced action reads

$$S_{\text{red}} = \int dt\,(\mathrm{p}\dot{\mathrm{m}} - \mathrm{h}) \quad . \tag{3.20}$$

The coordinate m is equal to the r-independent value that M takes when the constraint M' = 0 holds. The momentum p is related to the unreduced variables by

$$\mathrm{p} := \int_0^1 dr\, P_M \quad . \tag{3.21}$$

The Hamiltonian h is given by

$$\begin{aligned}
\mathrm{h} &= -\mathrm{N}_0 R_0 \left(\tfrac{1}{3}R_0^2 + \hat{\lambda}\right) \\
&\quad - \left(B^2 + \hat{\lambda}\right)\left[\sqrt{Q^2\mathrm{F} + \dot{B}^2} + \tfrac{1}{2}\dot{B}\ln\left(\frac{\sqrt{Q^2\mathrm{F} + \dot{B}^2} - \dot{B}}{\sqrt{Q^2\mathrm{F} + \dot{B}^2} + \dot{B}}\right)\right] \\
&\quad + \tfrac{1}{3}\hat{\lambda}Q^{-2}\left(Q^2\mathrm{F} + \dot{B}^2\right)^{3/2} \quad ,
\end{aligned} \tag{3.22}$$

where

$$N_0 := N_1/\Lambda_0 \tag{3.23}$$

$$R_0 := \sqrt{2m - \frac{1}{2}\hat{\lambda}} \tag{3.24}$$

$$F := 1 + \frac{B^2}{\hat{\lambda}}\left(1 - \sqrt{1 + \frac{4m\hat{\lambda}}{B^4}}\right) \tag{3.25}$$

and B and Q^2 are respectively the values of R $and\ -g_{tt}\ at\ r\ =$ 1. $B, Q^2, and\ N_0$ are considered prescribed functions of t, satisfying $B\ >$

$0, Q^2 > 0, and\ N_0 \geq 0$. For classical solutions to exist, we need $B^2 > \frac{1}{2}\hat{\lambda}$, and we shall from now on assume that this is the case. The range of **m** is $\frac{1}{4}\hat{\lambda} < \boldsymbol{m} < \frac{1}{2}B^2 + \frac{1}{4}\hat{\lambda}$, and the range of **p** is the full real axis.

The equation of motion for **m** implies that **m** is independent of *t:* the value of **m** is simply the mass parameter of the classical solution. The equation of motion for **p** reflects the fact that, by (3.14), (3.15b), and (3.21), **p** is equal to the difference in the Killing times at the two ends of the spacelike hypersurface.

7.6 Quantization and the partition function

In this section we quantize the reduced Hamiltonian theory and obtain a partition function as the trace of the analytically continued time evolution operator.

From now on, we take the boundary radius independent of time, $\dot{B} = 0$. We also subtract from the Hamiltonian (3.22) the value that the terms arising from r = 1 would take on flat spacetime. This subtraction does not affect the equations of motion, but it does renormalize the value of the action: it is analogous to subtracting the K_o term in Einstein's theory [48,49]. Writing $Q := \sqrt{Q^2} > 0$, the new Hamiltonian is given by

$$h = Q\left(1 - \sqrt{F}\right)\left[B^2 + \frac{1}{2}\hat{\lambda}\left(1 - \sqrt{F}\right)\left(2 + \sqrt{F}\right)\right] - N_0 R_0\left(\frac{1}{3}R_0^2 + \hat{\lambda}\right) \tag{3.26}$$

The first of the two terms in (3.26) is the Lovelock analogue of the quasilocal energy of Brown and York [50-52], The second term arises from the bifurcation three-sphere, and it will give rise to the black hole entropy.

Quantization proceeds exactly as in [68,36]. We take the wave functions $\psi$to be functions of the configuration variable **m,** with $\frac{1}{4}\hat{\lambda} < \boldsymbol{m} < \frac{1}{2}B^2 + \frac{1}{4}\hat{\lambda}$, and we introduce an inner product with some smooth and slowly varying weight factor. The Hamiltonian operator is taken to act by multiplication by the function **h** (3.26), $\psi(m) \mapsto h(m)\psi(m)$and the unitary time evolution operator is easily found. We then analytically continue the arguments of the time evolution operator to imaginary values: we set $\int Qdt = -i\beta$ interpreting $\beta > 0$ as the inverse temperature at the boundary, and $\int N_0 dt = -2\pi i$ motivated by the regularity of the classical Euclidean solutions. The trace of the analytically continued time evolution operator is divergent, but we can argue as in [68,36] that an acceptable renormalization is achieved by introducing a suitable regularization, dividing by the trace of the regularized identity operator, and finally eliminating the regulator. In this fashion, we obtain for the renormalized trace the manifestly well-defined expression

$$Z(\beta; B; \hat{\lambda}) = \left(\int_0^1 \tilde{\mu} dx\right)^{-1} \left[\int_0^1 \tilde{\mu} dx \exp -I_*\right] \tag{3.27}$$

where the effective action I^* is given by[4]

$$I_* = \beta B^2(1-\sqrt{F})\left[1+\frac{\hat{\lambda}}{3B^2}(1-\sqrt{F})\right] - 2B^3 x\left(\frac{1}{3}x^2+\frac{(\hat{\lambda})^2}{B^2}\right) \tag{3.28}$$

With

$$F := 1+\frac{B^2}{\hat{\lambda}}\left(1-\sqrt{\frac{2x^2\hat{\lambda}}{B^2}+\frac{(\hat{\lambda})^2}{B^2}}\right) \tag{3.29}$$

We have introduced the integration variable $x = R_0/B$, and the smooth and slowly varying positive function $\tilde{\mu}(x)$ arose from the choice of the inner product.

We now interpret the object $Z(\beta; B; \tilde{\lambda})$ (3.27) as the partition function of a thermodynamical canonical ensemble describing black holes in a spherical box with curvature radius B and inverse boundary temperature β. The thermodynamical properties of this ensemble will be analyzed in the next section.

7.7 Thermodynamics in the canonical ensemble

As noted above, the partition function $Z(\beta; B; \tilde{\lambda})$ (3.27) is manifestly well defined. Further, the form of the integral in (3.27) guarantees that the (constant volume) heat capacity, $C = \beta^2(\partial^2(\ln Z)/\partial\beta^2)$, is always positive (see, for example, section IV of Ref. [25]), and that the ensemble has a well-defined density of states [60-62]. These properties support the interpretation of the partition function in terms of a genuine thermodynamical equilibrium ensemble, in spite of the fact that we arrived at the partition function via an analytic continuation and not via direct statistical mechanics arguments.

To proceed, we shall estimate the integral in (3.27) by the saddle point approximation. We shall throughout assume $\tilde{\mu}(x)$ to be so slowly varying that its precise form will not affect the saddle point analysis. We shall also assume that the action is sufficiently rapidly varying to make the saddle point approximation is justified, without attempting to explicitly state the necessary conditions; typically, it will be throughout assumed that the system is "macroscopic," B >>1.

The critical points of I_* are at the roots of the equation

$$\frac{\beta x}{2\pi B} = \left(x^2 + \frac{\hat{\lambda}}{B^2}\right)\sqrt{F} \ . \tag{3.30}$$

The critical points give precisely the Lorentzian black hole solutions whose Hawking temperature at the boundary, calculated in the usual way from the surface gravity [102] and the blueshift factor, is equal to β. The mass of the hole is $m = \frac{1}{2}B^2x^2 + \frac{1}{4}\tilde{\lambda}$, and the value of I_* at a critical point equals the Euclidean action of the corresponding Euclideanized black hole solution. Whenever the partition function is dominated by a critical point, we recover for the thermal energy expectation value and the entropy the results

$$\langle E\rangle = -\frac{\partial(\ln Z)}{\partial\beta} \approx B^2\left(1-\sqrt{F}\right)\left[1+\frac{\hat{\lambda}}{3B^2}\left(1-\sqrt{F}\right)\left(2+\sqrt{F}\right)\right] \ , \tag{3.31a}$$

$$S = \left(1-\beta\frac{\partial}{\partial\beta}\right)(\ln Z) \approx 2\pi B^3 x\left(\tfrac{1}{3}x^2 + \frac{\hat{\lambda}}{B^2}\right) = 2\pi R_0\left(\tfrac{1}{3}R_0^2 + \hat{\lambda}\right) \tag{3.31b}$$

where x and F are evaluated at the critical point. The expression (3.31b) for the entropy agrees with the result first obtained by Euclidean methods [102].

We can now extract physical information by analyzing the critical point structure of I_* in various limits of interest in the three parameters β, B, and $\tilde{\lambda}$.

As a preliminary, consider the case $\tilde{\lambda} = 0$, in which our Lovelock theory reduces to Einstein's theory. Although we have for presentational simplicity assumed $\tilde{\lambda} > 0$, it is easy to see that the partition function for Einstein's theory is correctly recovered by taking the limit $\tilde{\lambda} \to 0$ in equations (3.27)-(3.29). In particular, (3.29) reduces to $F = 1 - x^2$, and the critical point equation (3.30) reduces to

$$\frac{\beta}{2\pi B} = x\sqrt{1-x^2} \quad . \tag{3.32}$$

The condition for critical points to exist is $\beta \leq \pi B$, and the critical points are then at

$x = x_{\pm} := 2^{-\frac{1}{2}}(1 \pm \sqrt{1 - \pi^{-2}B^{-2}\beta^2})^{1/2}$. When the critical points are distinct, x+ is a local minimum and x- a local maximum. When $\beta < \frac{3}{4}\pi B$, the partition function gets its dominant contribution from the global minimum at x = x+. When $\beta > \frac{3}{4}\pi B$, on the other hand, the partition function gets its dominant contribution from the vicinity of the global minimum at x = 0. The limiting case $\beta = \frac{3}{4}\pi B$ represents a phase transition where the dominant contribution shifts from x = x+ to x = 0 as β increases. When the saddle point dominates, the thermal energy and entropy (3.31) take the form

$$\langle E \rangle \approx B^2 \left(1 - \sqrt{1 - x_+^2}\right) \quad , \tag{3.33a}$$

$$S \approx \tfrac{2}{3}\pi B^3 x_+^3 = \tfrac{2}{3}\pi R_0^3 \quad , \tag{3.33b}$$

and the relation between the thermal energy and the mass can be written as

$$\mathrm{m} \approx \langle E \rangle - \frac{\langle E \rangle^2}{2B^2} \quad . \tag{3.34}$$

Equation (3.34) displays explicitly how the mass gets a contribution both from the thermal energy and from the gravitational binding energy associated with the thermal energy. Expectedly, the situation is closely similar to that in four dimensional Einstein theory [61].

We now turn to the case $\tilde{\lambda} > 0$, in which I_ always has at least one critical point.

Consider first the limit of small $\tilde{\lambda}$ with fixed B and β. The situation differs from that in the case $\tilde{\lambda} = 0$ only in that there is now one new critical point, a local minimum, at $x = 2\pi\tilde{\lambda}B^{-1}\beta^{-1} + \mathcal{O}(\tilde{\lambda}^2)$. At the new critical point, $I_* = \frac{1}{4}\tilde{\lambda}\beta + \mathcal{O}(\tilde{\lambda}^2)$. Therefore, as $\tilde{\lambda} \to 0$, the partition function smoothly approaches that of Einstein's theory. In particular, when $\beta < \frac{3}{4}\pi B$, it would be straightforward to compute the first order correction in $\tilde{\lambda}$ to the thermal energy and the entropy (3.33), assuming that the corrections to the saddle point approximation are small.

Consider next the small β limit with fixed B and $\tilde{\lambda}$. There is only one critical point, at $x = 1 - \frac{1}{8}\pi^{-2}(B^2 + \tilde{\lambda})\beta^2 + \mathcal{O}(\beta^4)$, and this critical point is the global minimum of I_*. One can think of this critical point as the counterpart of the larger of the two critical points of the case $\tilde{\lambda} = 0$: the black hole fills almost all of the box. The disappearance of the smaller critical point of the case $\tilde{\lambda} = 0$ is related to the fact that, for fixed $\tilde{\lambda}$, the Hawking temperature of the Lovelock hole in asymptotically at space is bounded below by $\frac{1}{4}\pi^{-1}\tilde{\lambda}^{-1/2}$ [102]. If the saddle point approximation to the partition function remains good, the thermal energy and the entropy are given by

$$\langle E\rangle \approx B^2 + \tfrac{2}{3}\hat{\lambda} - \tfrac{1}{2}\pi^{-1}B\beta + O(\beta^2) \ , \tag{3.35a}$$

$$S \approx 2\pi B^3\left(\frac{1}{3} + \frac{\hat{\lambda}}{B^2} - \frac{\beta^2}{8\pi^2 B^2}\right) + O(\beta^4) \ . \tag{3.35b}$$

Consider next the large β limit with fixed B and $\tilde{\lambda}$. There is again only one critical point, at $x = 2\pi\tilde{\lambda}B^{-1}\beta^{-1}F_0^{1/2} + \mathcal{O}(\beta^{-3})$, where $F_0 = 1 + B^2\tilde{\lambda}^{-1}(1 - \sqrt{1 + \tilde{\lambda}^2 B^{-4}})$. This critical point is the global minimum, and it has no counterpart in the Einstein theory: it corresponds to a small, "purely Lovelock," black hole. If the saddle point approximation to the partition function remains good, the thermal energy and the entropy are easily read off from (3.31) as

$$\langle E\rangle \approx B^2\left(1 - \mathsf{F}_0^{1/2}\right)\left[1 + \frac{\hat{\lambda}}{3B^2}\left(1 - \mathsf{F}_0^{1/2}\right)\left(2 + \mathsf{F}_0^{1/2}\right)\right] + O(\beta^{-2}) \ , \tag{3.36a}$$

$$S \approx \frac{4\pi^2\hat{\lambda}^2\mathsf{F}_0^{1/2}}{\beta} + O(\beta^{-3}) \ . \tag{3.36b}$$

Consider then the large $\tilde{\lambda}$ limit with fixed B and β. There is again only one critical point, at $x = 1 - \frac{1}{8}\pi^{-2}\tilde{\lambda}^{-1}\beta^2 + \mathcal{O}(\tilde{\lambda}^{-2})$, and this critical point is the global minimum. The hole is again "purely Lovelock," but it now fills almost all of the box.

Finally, consider the large B limit with fixed $\tilde{\lambda}$ and β. One critical point is at $x = 1 - \frac{1}{8}\pi^{-2}B^{-2}\beta^2 + \mathcal{O}(B^{-4})$. This critical point is the global minimum, and it can be regarded as the counterpart of the larger of the two critical points of the case $\tilde{\lambda} = 0$. If $\beta > 4\pi\tilde{\lambda}^{1/2}$, there are in addition two other critical points, at $x = \frac{1}{4}\pi^{-1}B^{-1}\beta(1 \pm \sqrt{1 - 16\pi^2\tilde{\lambda}\beta^{-2}}) + \mathcal{O}(B^{-2})$. The fact that the two small critical points exist only for $\beta > 4\pi\tilde{\lambda}^{1/2}$ is related to the above-mentioned phenomenon that the Hawking temperature of our Lovelock hole in asymptotically at space is bounded below by $\frac{1}{4}\pi^{-1}\tilde{\lambda}^{-1/2}$ [11]. If the saddle point approximation is good, the thermal energy and the entropy are obtained by replacing both O-terms in (3.35) by $\mathcal{O}(B^{-1})$.

We therefore see that for $\tilde{\lambda} > 0$, the partition function is always dominated by a black hole solution in the limits that we have considered. In the high temperature limit and in the large box limit, the situation is very similar to that for $\tilde{\lambda} = 0$, in that the dominating black hole solution fills most of the box. In the low temperature limit, on the other hand, the Lovelock theory does exhibit a dominating black hole solution where none existed in the case $\tilde{\lambda} = 0$. For a macroscopic box and $\tilde{\lambda} \ll B^2$, however, the presence of the new dominating solution does not appear to make the qualitative thermodynamical behavior substantially different from that in the case $\tilde{\lambda} = 0$. One can read these results as evidence for stability of the qualitative thermodynamical behavior of Einstein's theory upon the addition of the Lovelock parameter.

It should be emphasized that the partition function has no well-defined limit as $B \to \infty$ with fixed β and $\tilde{\lambda}$, neither for $\tilde{\lambda} = 0$ nor for $\tilde{\lambda} > 0$. As with Einstein's theory in four dimensions [60], this reflects the fact that the thermodynamical canonical ensemble is not well defined in asymptotically at space. We shall give a more detailed comparison of the boxed Lovelock theory to Lovelock theory in asymptotically at space in appendix C.

7.8 Summary and discussion

In this chapter we investigated the Hamiltonian dynamics and thermodynamics of five-dimensional spherically symmetric Lovelock theories in which the only contributions to the Lagrangian density are the Einstein-Hilbert term and the four-dimensional Euler density. We adopted boundary conditions that enforce every classical solution to be part of the exterior region of a black hole, with the spacelike hypersurfaces extending from the horizon bifurcation three-sphere to a timelike boundary with fixed intrinsic metric. We simplified the constraints by a canonical transformation that generalizes the one introduced by Kuchar in four dimensional spherically symmetric Einstein theory, and we reduced the theory classically to its true dynamical degrees of freedom.

After Hamiltonian quantization, we interpreted the trace of the analytically continued time evolution operator as the partition function of a thermodynamical canonical ensemble, describing black holes in a spherical box whose size and boundary temperature are fixed. In the special case where the Lovelock parameter $\tilde{\lambda}$ vanishes and the theory reduces to Einstein's theory, we found that the thermodynamics is highly similar to that of the corresponding system in four-dimensional Einstein theory: in particular, for high boundary temperatures the partition function is dominated by a classical black hole solution that fills most of the box. When $\tilde{\lambda} > 0$, the situation was more versatile. In the high temperature limit, with $\tilde{\lambda}$ and the box size fixed, the partition function is again dominated by a black hole that fills most of the box. In the low temperature limit, on the other hand, the partition function is now also dominated by a black hole solution; this black hole is small, and it has no analogue in Einstein theory. Nevertheless, if $\tilde{\lambda}$ is small compared with the size of the box, the new dominating solution has little qualitative effect on the thermodynamical properties. In this sense, the qualitative thermodynamical behavior of the Einstein system is stable upon the addition of the Lovelock parameter.

When the box size is taken to infinity, we found that the partition function has no well-defined limit, neither for $\tilde{\lambda} = 0$ nor for $\tilde{\lambda} > 0$. While this is not surprising for Einstein's theory, in view of the similar phenomenon in four dimensions, one might have hoped the theory with $\tilde{\lambda} > 0$ to fare better on the grounds that this theory admits asymptotically at black hole solutions that are stable against Hawking evaporation. However, even though a classical solution that dominates a well-defined partition function must be stable against Hawking evaporation, our Lovelock theory in asymptotically flat space provides an example where the mere existence of such a locally stable classical solution does not imply the existence of well-defined canonical ensemble. Another such example occurs in four-dimensional Einstein-Maxwell theory in asymptotically flat space.

In the classical theory with $\tilde{\lambda} > 0$, we saw that the super-Hamiltonian emerges as a multivalued function of the canonical variables, as is generically the case in Lovelock theories. Nevertheless, our thermodynamically motivated boundary conditions were sufficient to uniquely specify the super-Hamiltonian near the horizon, and the uniqueness could be extended to the full spacelike hypersurfaces by continuity. Another boundary condition that would uniquely specify the super- Hamiltonian in this fashion is the asymptotically flat falloff discussed in Appendix C. However, one expects there to exist boundary conditions of interest for which such uniqueness does not occur, and in such cases one would need to seek other criteria for specifying the super-Hamiltonian. If one regards the Lovelock theory as a perturbation to Einstein's theory, or as a toy model for semiclassical gravity with back-reaction, one possible criterion of this kind is perturbative expandability of the solutions in $\tilde{\lambda}$.

In conclusion, our results provide evidence for robustness of the classical Hamiltonian structure and the qualitative thermodynamical structure of spherically symmetric Einstein gravity in five dimensions upon the addition of the Lovelock parameter. To put this conclusion in proper perspective, one should remember that both our particular Lovelock theory and our boundary conditions were hand-picked so that the global aspects of the problem remained virtually identical to those in pure Einstein gravity. One might conjecture that whenever the global properties of a Lovelock theory are sufficiently similar to those of Einstein's theory, then also the equilibrium thermodynamics, with finite or infinite boundary conditions, will be qualitatively similar to that in Einstein's theory. Another example supporting such a conjecture is found in the asymptotically anti-de Sitter Lovelock theories, which include as a special case Einstein gravity in three and four dimensions with a negative cosmological constant. However, to elevate the content of the conjecture substantially beyond tautology, one would need a more systematic understanding of the possible global structures that the various Lovelock theories may have.

Chapter 8. Analytic Time and Thermal Quantum Gravity

8.1 The Quantum Mechanics Propagator and the Statistical Mechanics Partition Function

Indications of a connection between Quantum Mechanics and Statistical Mechanics (likewise, quantum field theory and statistical field theory) have occurred in prior examples, let's now explore this in detail. Time consisting of a real parameter that labels causal events is familiar and is part of the dynamical description in a special role from the earliest classical mechanical descriptions. From special relativity we learn time has different forms (parametrizations) and in the context of spacetime, is simply another coordinate variable (albeit with differently signed signature). Once demoted to being a coordinate variable, even part of the time (no pun intended), we can ask about complex time in that context. The metric signature will change from Lorentzian to Euclidean if we switch to pure imaginary time, for example, where all of the Euclidean-based sums will be well-defined. In essence, we will find that the connection between Quantum Mechanics and Statistical Mechanics is that they share the same analytic time, one referencing the real-part of time in a standard dynamical context (Quantum Mechanics) while the other references the imaginary part of time in a standard equilibrium thermodynamics description of the system. The analyticity of time overall will be maximally extended in the manner that gives rise to the Feynman propagator. (which will embed the proper causality into the theory).

As usual, when a convenient mathematical connection is used a lot (Planck's original introduction of his constant and quantization [122]; or use of Vector potential in electromagnetism, proven to be real by the Aharonov-Bohm Effect [123]) we should consider what it means if analytic time really 'exists'. And it probably does, a supporting reason for this is that bound state descriptions, such as the solution for the hydrogen atom, so easy in standard Schrodinger or Heisenberg analysis, if attempted with the path integral formalism that works everywhere else (when other methods don't), is found to only work describing bound states if in curved spacetime with analytic time. So analytic time is needed to fully utilize the path integral formalism. What does it mean to have complexified time? From the Unruh analysis of the accelerated observer [5], one possibility is that this indicates a universal thermality according to acceleration. In the analysis, imaginary time is periodic and that periodicity defines the inverse temperature of the system. what results for the accelerated observer (in a standard quantum vacuum) is the appearance of a thermal flux from direction accelerating away from, with thermal spectrum at temperature:

$$T = \frac{\hbar a}{2\pi c k_B},$$

where $\hbar$ is Planck's constant, a is the acceleration, c is the speed of light, and k_B is Boltzmann's constant. For an acceleration of $1 m/s^2$ we have a temperature $T = 4.06 \times 10^{-21} K$. The strongest acceleration we usually feel is due to Earth, 9.8 m/s^2, so we are bathed in a thermal bath at temperature $T = 3.98 \times 10^{-20} K$ from this effect (so drowned out by the CMBR). As explored in Chapters 6 and 7, in cosmological and Black Hole analysis complexified time provides a bridge to the thermodynamics of the system, which allows a thermal quantum gravity solution to be described (but not a quantum gravity solution, especially since such a solution might not exist). Once we've adjusted to the notion of analytic time, its natural to ask for the maximal analytic extension and if there is more than one choice (there is) the choice consistent with other aspects of the physics (such as causality) can be adopted at this juncture, and that is what is done with the aforementioned Feynman propagator. Analytic time provides a more complete and interconnected physical description, but it also constrains time and other aspects even more than before. So we now need to know why we have a theory with analytic time in addition to the odd local gauge group that is a product group of U(1), SU(2), and SU(3), and the set of 19 (or 22?) constants, etc. To have a deeper understanding of time, and the rest, requires an enveloping formalism to the structured formalism currently seen, and that is what is explored in Book 7 [7], with a synopsis provided in Appendix D.

8.2 The Propagator with Complexified Time

From the first appearance of the Schrodinger equation it was noted that changing time to imaginary time, $t \to -i\tau$, would give the diffusion equation. Let's now consider complexification of time when working with the propagator.

8.2.1 Direct Substitution

Recall that the propagator is based on the unitary evolution operator that is defined by the Hamiltonian for the system, or by the Action on paths:

$$K(t,q,q') = \langle q|e^{itH/\hbar}|q'\rangle = \int_{\chi(0)=q'}^{\chi(t)=q} \mathcal{D}\chi e^{iS[\chi]/\hbar},$$

with paths parametrized by $\chi(t)$, that start at q' and end at q. Let's now shift from $t \to i\tau$ in this context:

$$K_E(\tau,q,q') = \langle q|e^{-\tau H/\hbar}|q'\rangle = \int_{\chi(0)=q'}^{\chi(\tau)=q} \mathcal{D}\chi e^{-S_E[\chi]/\hbar},$$

where now the integrand is real and well-defined. The partition function is then simply given by:

$$\int dq K_E(\tau,q,q') = tr[e^{-\tau H/\hbar}]$$

where the temperature is $T = \hbar/\tau k_B = 1/\beta$. In standard notation ($k_B = 1$) using for partition function $Z(\beta)$ and free energy $F(\beta)$ we then have:

$$Z(\beta) = tr\left[e^{-\beta H}\right] = e^{-\beta F} = \oint_{\chi(0)=\chi(\tau)} \mathcal{D}\chi e^{-S_E[\chi]/\hbar}$$

Note that to have the correspondence with the definition of partition function we identified the ends of the paths, or equivalently, we've shifted to integration on periodic paths with period $\tau = \hbar\beta$. Let's now apply this to the harmonic oscillator fundamental case to see if it makes sense. For the Euclideanized harmonic oscillator, we have:

$$K_E(\hbar\beta,q,q) = \sqrt{\frac{m\omega}{2\pi\hbar\sinh(\hbar\omega\beta)}}\exp\left[-\frac{2m\omega q^2}{\hbar}\frac{\sinh^2(\hbar\omega\beta/2)}{\sinh(\hbar\omega\beta)}\right]$$

$$Z(\beta) = \int dq K_E(\tau,q,q') = \frac{1}{2\sinh\left(\frac{\hbar\omega\beta}{2}\right)} = \sum_n \exp\{-\beta E_n\},$$

where

$$E_n = \hbar\omega\left(n+\frac{1}{2}\right).$$

For low temperature, $\beta \to \infty$, and $F(\beta) \to E_0 = \frac{1}{2}\hbar\omega$, as expected. This simple substitution works out, so let's now consider complex time in more detail. In particular, is time analytic?

8.2.2 Full analyticity – Wick rotation

We've seen that the direct substitution $t \to -i\tau$ provides interesting connections. If we arrive at this change more formally in terms of analytic time, we will have better understanding. First, as regards a Lorentzian spacetime with integrations (the Action) described as integrals in real time, it was observed that complexified time being analytic allowed for a change in the contour of integration, effectively a rotation from the real axis by 90 degrees about the origin to turn the integration into an integral along the imaginary axis, i.e., we've achieved $t \to -i\tau$ by way of a "Wick rotation". There are different ways to do this in the context of the quantum propagator, however, but only one encodes the causal structure consistently, the Feynman propagator.

Before continuing with analysis of the Feynman propagator, consider the significance of analyticity and the Wick rotation is not only that a 1-dimensional time and (N-1)-dimensional spatial dynamics problem can be turned into a N-dimensional statics problem, allowing for Euclideanized path integrals that are convergent and well-defined to obtain system integral solutions (that can be analytically carried back to the Lorentzian system representation, thereby making the path integrals well-defined via analyticity). We can also take this in the other direction, an intractable N-dimensional statics problem might be more easily solvable as a (N-1)-dimensional dynamics problem.

8.2.2.1 Green's Function – Feynman Propagator – Choice of analytic extension

Let's follow the discussion of analyticity given by [45] and consider the Feynman propagator given in their notation by:

$$G_F\left(x^\alpha, t; x'^\alpha, t'\right) = \left\{ \begin{array}{l} i\sum_n \dfrac{\psi_n(x^\alpha)\psi_n\left(x'^\alpha\right)}{2\omega_n} \exp[-i\omega_n(t-t')], \quad t > t' \\ i\sum_n \dfrac{\psi_n(x^\alpha)\psi_n\left(x'^\alpha\right)}{2\omega_n} \exp[i\omega_n(t-t')], \quad t < t' \end{array} \right\}$$

To effect the substitution by $t \to -i\tau$ by analytically extending off the real line, there must be a rotation as indicated in Figure 1, which requires analyticity in quadrants Two and Four as shown in shade.

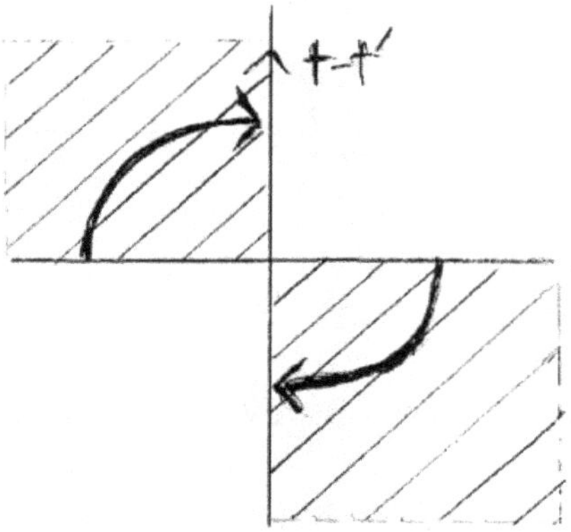

Here's the new Euclideanized Green's function G_E (or Euclideanized Feynman propagator):

$$G_E(x^\alpha, t; x'^\alpha, t') = \begin{cases} i\sum_n \frac{\psi_n(x^\alpha)\psi_n(x'^\alpha)}{2\omega_n} \exp[-\omega_n(\tau - \tau')], & \tau > \tau' \\ i\sum_n \frac{\psi_n(x^\alpha)\psi_n(x'^\alpha)}{2\omega_n} \exp[\omega_n(\tau - \tau')], & \tau < \tau' \end{cases}$$

which is well-defined and unique given the fall-off condition to zero for large$|\tau - \tau'|$.

The choice of shaded regions allows the "Wick rotation" shown, and this convention for analytic extension is usually captured in the mathematics (e.g., not diagrammatically as shown) by introduction of a small imaginary part to ω_n, or in momentum representation (taking a Fourier Transform), to $p^2 - m^2$. Let's shift to the standard Feynman propagator form with this in mind. Staring with the standard Green's function to the Klein Gordon equation we have the standard solution

$$G(x,y) = \frac{1}{(2\pi)^4}\int d^4p \frac{e^{-ip(x-y)}}{p^2 - m^2 \pm i\varepsilon}$$

where the $\pm i\varepsilon$ denotes various choice of deformation to the integration contour to have a well-defined propagator, and these solutions are different. We want the Feynman propagator convention allowing the "Wick rotation" indicated, thus, we want:

This is equivalent to the definition in terms of the limit as the $i\varepsilon$ contour deformation goes to zero:

$$G_F(x,y) = \lim_{\varepsilon \to 0} \frac{1}{(2\pi)^4} \int d^4p \frac{e^{-ip(x-y)}}{p^2 - m^2 + i\varepsilon}.$$

8.2.2.2 Thermal Green's Function

Let's now consider a system at a temperature $T = 1/\beta$ and get the associated Thermal Green's Function. If we have a temperature and using the grand canonical ensemble formalism we can write the expectation value of any operator A as:

$$\langle A \rangle_\beta = \frac{Tr[\exp(-\beta H)\, A]}{Tr[\exp(-\beta H)]}$$

The thermal Green's function would then be:

$$G_T(x,y) = i\langle T\varphi(x)\varphi(y)\rangle_\beta.$$

The solution has the same terms as before, but now has a Bose-Einstein statistics contribution:

$$G_T(x^\alpha, t; x'^\alpha, t') = \left\{ \begin{array}{ll} i\sum_n \frac{\psi_n(x^\alpha)\psi_n(x'^\alpha)}{2\omega_n} \left\{ \begin{array}{l} (1+n_B)\exp[-i\omega_n(t-t')] \\ +n_B \exp[i\omega_n(t-t')] \end{array} \right\}, & t > t' \\ i\sum_n \frac{\psi_n(x^\alpha)\psi_n(x'^\alpha)}{2\omega_n} \left\{ \begin{array}{l} (1+n_B)\exp[i\omega_n(t-t')] \\ +n_B \exp[-i\omega_n(t-t')] \end{array} \right\}, & t < t' \end{array} \right\}$$

where

$$n_B = \frac{1}{(\exp(\omega_n \beta) - 1)}$$

As before, we can analytically continue to obtain the more manageable Euclideanized form, and this will correspond to the analytic continuation under conditions requiring periodicity in imaginary time with period β:

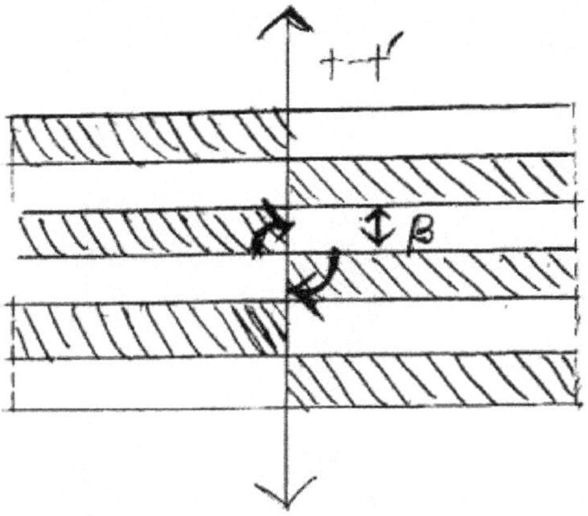

8.3. Thermal Quantum Field Theory

8.3.1 Complexification of time arrives at thermal state and partition function

If we have a quantum system that is in equilibrium with its environment, and where nontrivial exchange is possible in energy and particles with that environment, the 'Euclideanization' of the spacetime described above brings us to a Grand-Canonical Ensemble formulation for the thermal state given by:

$$\Phi_{\beta,\mu} = \frac{e^{-\beta(H-\mu N)}}{Z}, \quad Z = Tr\left[e^{-\beta(H-\mu N)}\right]$$

where H is the Hamiltonian operator of the system, N is the number operator, μ is the chemical potential, and Z is the partition function (from which the entire system thermodynamics can be derived).

8.3.2 Non-interacting particles: Bose-Einstein and Fermi-Dirac statistics

Suppose we have non-interacting particles with energy and number density given by (harmonic oscillator form):

$$H = \sum_n \varepsilon_n {a_n}^\dagger a_n, \quad N = \sum_n {a_n}^\dagger a_n$$

where $\{{a_n}^\dagger, a_n\}$ are the standard creation and annihilation operators, which for Bosons will obey the standard commutation relations:

$$\left[a_j, {a_k}^\dagger\right] = \delta_{jk}, \qquad \left[a_j, a_k\right] = 0, \qquad \left[{a_j}^\dagger, {a_k}^\dagger\right] = 0$$

The mean number of Bosons molecules in mode n is then n_B:

$$n_B = Tr\left[{a_n}^\dagger a_n \Phi_{\beta,\mu}\right] = \frac{1}{\exp(\beta[\varepsilon_n - \mu]) - 1},$$

where $\varepsilon_n > \mu$ is required to have a positive mode number.

For Fermions the commutator relations above get replaced with anticommutators, and we now have:

$$n_F = Tr\left[{a_n}^\dagger a_n \Phi_{\beta,\mu}\right] = \frac{1}{\exp(\beta[\varepsilon_n - \mu]) + 1},$$

there is no longer a problem $\varepsilon_n < \mu$ in this situation to stay consistent with positive occupation numbers (we have antimatter).

8.3.3 The Laws of Thermodynamics are recovered

8.3.3.1 Definition of Heat Bath

Let's consider a total system Hamiltonian H consisting of the system Hamiltonian H_S and of a heat bath Hamiltonian H_B, and an interaction Hamiltonian between System and Heat Bath, H_{SB}. Thus:

$$H = H_S + H_B + H_{SB}.$$

The Hamiltonian gives the unitary time evolution operator as

$$U(t) = T\{\exp -i \int_0^t dt' H(t')\}$$
$$\equiv \lim_{\delta t \to 0} e^{-i\delta t H(t)} e^{-i\delta t H(t-\delta t)} \dots e^{-i\delta t H(\delta t)} e^{-i\delta t H(0)}$$

(the starting point form many path integral formulations). Instead of a wavefunction, with mixing, and with Euclideanization, we have a system density matrix that is acted on by the unitary evolution operator:

$$\rho(t) = U(t)\rho(0)U^\dagger(t), \quad \frac{\partial \rho(t)}{\partial t} = -i[H(t), \rho(t)].$$

Let's consider the mean energy change in the heat reservoir:

$$\frac{\partial \langle H_B(t) \rangle}{\partial t} = \frac{\partial}{\partial t} \langle H_B(t) - \mu N \rangle + \mu \frac{\partial}{\partial t} \langle N \rangle = F + P,$$

where $F = \frac{\partial}{\partial t} \langle H_B(t) - \mu N \rangle$ is the heat flow that enters the reservoir and $P = \mu \frac{\partial}{\partial t} \langle N \rangle$ is the power that enters the reservoir. The separation into F and P parts is indicated by the definition of entropy in the next section.

8.3.3.2 Entropy

Recall the Boltzmann entropy formula in terms of system probabilities (Shannon Entropy) is:

$$S = -k_B p \ln p,$$

here we take $k_B = 1$ and switch to the density matrix formulation, to get the standard von Neumann entropy form relevant to our situation:

$$S = -Tr\{\rho \ln \rho\}$$

and we then have:

$$\partial_t S = -\partial_t Tr\{\rho \ln \rho\} = 0$$

under unitary evolution. Now suppose we split the density matrix into a non-thermal system part (non-system parts traced out) and a thermal system part $\Phi_{\beta,\mu}$ as indicated previously (thereby ignoring reservoir information). Denote the effective description by $\rho_S \otimes \Phi_{\beta,\mu}$. The entropy difference (divergence) that describes the difference between the full density matrix and the non-thermal state density matrix, the information loss by such an approximation, is then $(\rho || \rho_S) \otimes \Phi_{\beta,\mu}$ and the entropy of this can be written

$$S[(\rho || \rho_S) \otimes \Phi_{\beta,\mu}] = S[\rho_S] - S[\rho] + \beta \langle H_B - \mu N \rangle + \ln Z$$

Thus (showing k_B):

$$\partial_t S[(\rho || \rho_S) \otimes \Phi_{\beta,\mu}] = \partial_t S[\rho_S] + \frac{F}{k_B T}.$$

This allows us to write:

$$\partial_t S[\Phi_{\beta,\mu}] = \beta \partial_t \langle H_B - \mu N \rangle = \frac{F}{k_B T}$$

8.3.3.3 First Law of Thermodynamics

Let's start with the implications of number conservation. Global number conservation gives:

$$[N_S + N_B, H] = 0,$$

while local number conservation gives:

$$[N_B, H_B] = 0, \;\; [N_S, H_S] = 0.$$

Taken together we then know that the thermal coupling terms satisfies:

$$[N_S + N_B, H_{SB}] = 0,$$

which means that particles are either in the system or in the bath (the coupling does not change their total number). If there were a coupling number, this would be the same as the change in that number being zero. There is a coupling energy term, however, so let's make change in the coupling energy over time equal to zero. This means that:

$$\partial_t \langle H_{SB} \rangle = i \langle [H_S + H_B, H_{SB}] \rangle = 0.$$

In terms of the F, P variables introduced earlier:

$$F = i\langle [H_S - \mu N_S, H_{SB}] \rangle, \qquad P = i\mu \langle [N_S, H_{SB}] \rangle,$$

and

$$\partial_t \langle H_S \rangle = \langle \partial_t H_S \rangle - (F + P),$$

where $\langle \partial_t H_S \rangle$ is power entering the system, and F, P describe energy flows leaving the system. Thus, the first law of thermodynamics is shown for the system.

8.3.3.4 Second Law of Thermodynamics

Let's return to the product notation for the effective density of states given our (typical) missing information of a system: $\rho_S \otimes \Phi_{\beta,\mu}$. Let's now consider the real system to have precisely this tensor product form as initial state:

$$\rho(0) = \rho_S(0) \otimes \Phi_{\beta,\mu}.$$

We can now write change in entropy as:

$$\Delta S(t) = S\left[(\rho(t) || \rho_S(t)) \otimes \Phi_{\beta,\mu}\right] = S[\rho_S(t)] - S[\rho(t)] + \frac{Q}{k_B T}$$

where heat is

$$Q = Tr\{(H_B - \mu N)\rho(t)\} - Tr\{(H_B - \mu N)\rho(0)\}.$$

Note that the expression above for change in entropy is in terms of a relative entropy, which is always positive of zero. Thus,

$$\Delta S(t) \geq 0,$$

the Second Law of Thermodynamics.

8.4 Wightman axioms and the Osterwalder-Schrader reconstruction theorem

For quantum theory there are a variety of prescriptions for making the path integral formulation well-defined (see [124]). The one most compact mathematically is to work with Euclideanization with use of analytic time. In this process, however, we are speaking of a quantum description, where an integral description is given, for which we Wick rotate to arrive at statistical description where the integral is well-defined and can be solved, and then Wick rotating back with our solution. How general is this procedure? How general is the Wick rotation? What does this suggest?

The Wightman axioms [124] answer how general such a construction can be is very general as long as the key axiom is satisfied: "If the supports of two fields are spacelike separated, then the fields either commute or anticommute". Note that this burdens the commutator relations with imposing the causality relation unlike in local quantum field theory. The Wightman axioms require the standard spin statistics relations. A true spinor representation is incompatible with the Euclideanization maneuvers used in the Wightman axioms, so not directly generalizable in that sense (but suggested of a variety of ways to make some consistent for fermions. The Wightman framework also can't, technically, work with gauge theories. It makes sense there would be a complications for Non-Abelian, but there's even a complication for Abelian gauge theories because gauge theories allow an indefinite norm (incompatible with the positive definite norm requirement for a Hilbert space). This is trivially overcome with appropriate choice of gauge (Coulomb), thus falling under the purview of the Wightman axioms in that form.

The Field Theories that can be 'Wick' rotated appear to be a large enough family to cover many cases of interest (if not all approximately) but this requires that we look at the analytic continuation Wick rotation maneuver more closely. In doing so we arrive at the Osterwalder-Schrader reconstruction theorem [125], where we find a set of rules for when a (well-defined) Euclidean quantum field theory can be Wick-rotated not a Wightman quantum field theory. Equivalence of the technical conditions for the Osterwalder-Schrader theories and Wightman quantum field theory axioms was shown in 1994.

8.5 Conclusion

The time variable in a quantum theory, upon complexation into pure imaginary, gives rise to the partition-function based statistical mechanical theory for the same state system (as described in the quantum state system). This association is not surprising given time and energy paired in the unitary quantum evolution operator (that transforms into the partition function upon complexation). Time complexation isn't merely a convenience, however, of simultaneous representation of the quantum unitary or thermodynamic partition function (for the same system of states). The time complexation appears to be fully analytic, as evinced by the use of the standard Wick rotation to rotate the contour integral for Feynman integrals. Together with the pole convention of the Feynman propagator (that captures causal information corresponding to the n-point function of a time-ordered product) we see a propagator form not only for analytic time, but for how to extend to maximal analyticity.

Application of analytic time to thermal quantum gravity began with the complexified Schwarzschild analysis by Hartle and Hawking in 1976 [38]. They found that complexified Schwarzschild had an effective particle bath with thermal spectrum. This was followed a year later by Gibbons and Hawking, 1977 [48] where a more general description of how complexation of any Lorentzian system would give rise to a partition function, and require positive heat capacity for the ensemble to be deemed to even exist (not merely have stable solution).

Having identified the problem with requiring a positive heat capacity, it was clear that the standard Schwarzschild black hole in asymptotically flat spacetime would never suffice since the known Hawking radiation would clearly lead to an unstable situation (e.g., negative heat capacity). (This was proven in a full general relativistic analysis in 1995 by Louko and Whiting [68].) To avoid the instability due to asymptotically flat spacetime, in 1983 Hawking and Page [71] considered the Schwarzschild black hole in a spacetime that was asymptotically Anti-deSitter. As hoped, it exhibited positive heat capacity, such that stable states could be analyzed. The approach did not work with full general relativity formalism, however, so behavior of boundary-term fall-off and other aspects were not resolved. In 1996 the full general relativity solution for the Reissner-Nordstrom Anti-deSitter spacetime weas considered (Louko and Winters-Hilt [36]). This spacetime family includes Schwarzschild and includes the presence of an electromagnetic field, and since asymptotically Anti-deSitter, it has positive heat capacity, with surface terms now fully determined (this required significant work, to is mainly placed in the Appendix). In 1997 the full general relativity solution was done for the family of Lovelock spacetimes [91], again showing the existence of stable solutions.

In chapters 6 and 7, that use analytic time, we see that due to required periodic boundary conditions for well-definedness, the pure imaginary time part is proportional to the inverse temperature of the thermal system. In these thermal quantum gravity systems there are shown to exist stable solutions (positive heat capacity). We, thus, see how time complexation can inform us about stable solutions in such systems.

Thermodynamics is thus interwoven with the definition of analytic time. Thermodynamics is the original phenomenological theory and appears to manifest statistical laws in the form of a Martingale process for which equilibrium state always exists, and system evolution tends to that equilibrium. Thermodynamics also exhibits global structure emergence. Thus, while thermodynamics is the oldest physics discipline, it also the youngest, with much yet to be understood. With the advent of generalized artificial intelligence and large-scale computing, it is likely that significant advances in thermodynamics, statistical mechanics and applied areas of condensed matter physics and solid state physics, will see significant growth in the decades ahead.

Appendix

Appendix A. Schwarzschild Black Hole Thermodynamics

A.1 Schwarzschild Black Hole with Asymptotically Flat ∞
Concerning fall-off conditions at bifurcation 2-sphere
At $r \to 0$, need to consider Gaussian Normal coordinates in the vicinity of the "$r = 0,$" the Horizon surface, this leads to consideration of a metric regular there, that is also compatible with the ADM formalization – i.e. Kruskal. This permits a perturbative analysis and also indicates that there will be no 1st order deviation terms (from the properties of the Gaussian Normal coordinates):
$R = R_0 + r^2 R_2 + \cdots$
$\Lambda = \Lambda_0 + O(r^2) + \cdots$

Derivation of fall-off conditions at bifurcation 2-sphere for spherically symmetric
The Schwarzschild line element in curvature coordinates (T,R):

$$ds^2 = -f(R)dT^2 + f^{-1}(R)dR^2 + R^2 d\Omega^2 \ , \qquad f = 1 - 2M/R$$

Substitute
$T = T(t, r)$
$R = R(t, r) = 2M + 2M.p(t, r)$

Which describes a general foliation (not necessarily a Schwarzschild slicing) and shifts the new $p(t, r)$ variable in a manner convenient to perturbation analysis in small $R - 2M = 2Mp$:

$$R = 2M(1 + p)$$

$$\left(1 - \frac{2M}{R}\right) = 1 - \frac{1}{1 + p} = \left(\frac{p}{1 + p}\right)$$

$$dR = 2Mdp$$

So,

$$ds^2 = -\left(\frac{p}{1 + p}\right) dT^2 + \left(\frac{1 + p}{p}\right)(2Mdp)^2 + R^2 d\Omega^2$$

If, to leading order in p, the fall-off conditions on the metrical variables at the bifurcation 2-sphere are required to be of the same form for all slices, including the Schwarzschild slices, then

$$p = p_1(t)r + p_2(t)r^2 + \cdots$$

This is because the r label is taken to approach zero at $R = 2M$, for convenience in the perturbative analysis, and $R = 2M$ is the location of the bifurcation 2-sphere on the Schwartzschild slices.
Expanding

$$dT = \dot{T}dt + T'dr$$
$$dp = \dot{p}_1 r dt + p_1 dr$$

Let's only carry p_1 for now, if isn't compatible with later conditions the analysis returns here and takes the $p_2(t)$ term as leading order.

$$ds^2 = -\left(\frac{p_1 r}{1 + p_1 r}\right)\left(\dot{T}^2 dt^2 + 2\dot{T}T' dt dr + T'^2 dr^2\right) + \left(\frac{1 + p_1 r}{p_1 r}\right)(2M)^2 dp^2 + R^2 d\Omega$$
$$= \left[-\dot{T}^2\left(\frac{p_1 r}{1+p_1 r}\right) + 4M^2\left(\frac{1+p_1 r}{p_1 r}\right)(\dot{p}_1 r)^2\right]dt^2 + \left[-2\dot{T}T'\left(\frac{p_1 r}{1+p_1 r}\right) + 8M^2\left(\frac{1+p_1 r}{p_1 r}\right)(p_1 r)\dot{p}_1\right]drdr$$
$$+\left[-T'^2\left(\frac{p_1 r}{1+p_1 r}\right) + 4M^2\left(\frac{1+p_1 r}{p_1 r}\right){p_1}^2\right]dr^2 + R^2 d\Omega^2$$

Now to compare with the spherically symmetric from for the ADM metric:

$$ds^2 = -N^2 dt^2 + \Lambda^2(dr + N^r dt)^2 + R^2 d\Omega^2$$

Note:

(1) That as $r \to 0$, Λ becomes singular unless $p_1 = 0$
(2) With $t, p, \dot{p}$ as independent parameters the freedom in T' is redundant – so take T=T(t).

As $r \to 0$, $\Lambda = O\left(r^{-1/2}\right)$, $N^r \Lambda^2 = O(1)$, so $N^r = O(r^{-1})$, will this be consistent?
As $r \to 0$,

$$N = O\left(r^{1/2}\right) \to 0$$
$$N^r = O(r^1) \to 0$$

If the bifurcation 2-sphere is to remain the boundary then $N \ and \ N^r$ mus approach zero there, so this works so far…

$$ds^2 = -\left(\frac{p_1 r}{1+p_1 r}\right)\dot{T}^2 dt^2 + 4M^2\left(\frac{1+p_1 r}{p_1 r}\right)p_1^2\left(dr + \frac{\dot{p}_1}{p_1} r dt\right)^2 + R^2 d\Omega^2$$
$$N = \dot{T}\sqrt{\frac{p_1 r}{1 + p_1 r}} = O\left(r^{1/2}\right)$$

$$\Lambda = 2Mp_1\sqrt{\frac{1+p_1r}{p_1r}} = O\left(r^{-1/2}\right)$$

$$N^r = \frac{\dot{p}_1 r}{p_1} = O(r)$$

$$R = 2M(1+p_1r) = O(1)$$

$$P_\Lambda = -N^{-1}R\left(\dot{R} - R'N^r\right) = -\frac{2M(1+p_1)^{3/2}}{\dot{T}\sqrt{p_1r}}\left(2Mr\dot{p}_1 - 2Mp_1\left(\frac{\dot{p}_1}{p_1}\right)r\right)$$

$$= 0$$

$$P_\Lambda = 0$$

$$P_R = \frac{\Lambda P_\Lambda}{R'} = 0$$

Is this consistent with the super-Hamiltonian constant?

$0? = \Lambda^{-1}\, RR'' - \Lambda^{-2}RR'\Lambda' + \frac{1}{2}\Lambda^{-1}R'^2 - \frac{1}{2}\Lambda$

The singular terms cancel at lowest order:

$$0 \cong -(2Mp_1)^{-2}\left(\frac{p_1r}{1+p_1r}\right)2M(1$$

$$+ p_1r)2Mp_1(2Mp_1)\frac{1}{2}\left(\frac{1+p_1r}{p_1r}\right)^{-1/2}\left(\frac{p_1}{p_1r} + \frac{1+p_1r}{p_1r}\right)$$

$$-\frac{1}{2}(2Mp_1)\sqrt{\frac{1+p_1r}{p_1r}}$$

Now try the mass relation to double check:

$M = \frac{1}{2}R^{-1}P_\Lambda^2 - \frac{1}{2}\Lambda^{-2}RR'^2 + \frac{1}{2}R$

$= -\frac{1}{2}\left[2Mp_1\sqrt{\frac{1+p_1r}{p_1r}}\right]^{-2} 2M(1+p_1r)(2Mp_1)^2 + \frac{1}{2}2M(1+p_1r)$

$= -M\left(\frac{p_1r}{1+p_1r}\right)(1+p_1r) + M + Mp_1r = M$

so consistent.

What of the equations of motion:

$\dot{\Lambda} = -NR^{-2}(RP_R - \Lambda P_\Lambda) + (\Lambda N^r)'$

$\dot{R} = -NR^{-2}P_\Lambda + R'N^r$

$\dot{P}_\Lambda = \frac{1}{2}N\left[1 - \Lambda^{-2}R'^2 - R^{-2}P_\Lambda^2\right] - \Lambda^{-2}RR'N' + N^rP_\Lambda'$

$\dot{P}_R = N\left[R^{-3}\Lambda {P_\Lambda}^2 - R^{-2}P_\Lambda P_R\right] + (N^rP_R)' - N\left(\frac{R'}{\Lambda}\right)' - \left(\frac{N'}{\Lambda}R\right)'$

Consider $\dot{R}$:

$2M\dot{p}_1 \;?= 2Mp_1\left(\frac{\dot{p}_1}{p_1}r\right)$

which checks.

Consider $\dot{\Lambda}$:

$2Mp_1\sqrt{\frac{1+p_1r}{p_1r}}+\frac{2Mp_1}{2}\sqrt{\frac{p_1r}{1+p_1r}}\left[\frac{r\dot{p}_1}{rp_1}-\frac{(1+p_1r)\dot{p}_1r}{(rp_1)^2}\right]=\frac{(2M\dot{p}_1)}{\sqrt{p_1}}\left(\sqrt{r+p_1r^2}\right)'=$
$\frac{M\dot{p}_1}{\sqrt{p_1}}\frac{(1+2p_1r)}{\sqrt{r+p_1r^2}}$

which checks.

Consider $\dot{P}_\Lambda$:

$\dot{P}_\Lambda=\frac{1}{2}N\left[1-\Lambda^{-2}R'^2\right]-\Lambda^{-2}RR'N^r$

Does

$$0?=\frac{1}{2}\dot{T}\sqrt{\frac{p_1r}{1+p_1r}}\left[1-\left(\frac{p_1r}{1+p_1r}\right)\right]$$
$$-\left[\left(\frac{p_1r}{1+p_1r}\right)\frac{(2M)^2p_1}{(2\mu p_1)^2}(1+p_1r)\dot{T}\left(\sqrt{\frac{p_1r}{1+p_1r}}\right)\right]$$
$$=\left[1-\left(\frac{p_1r}{1+p_1r}\right)\right]-\left[\frac{(1+p_1r)}{p_1}\left\{\frac{p_1r}{1+p_1r}\right\}'\right]$$
$$=-\left(\frac{p_1r}{1+p_1r}\right)-\frac{(1+p_1r)}{p_1}\frac{(p_1r)[-(p_1)]}{1+p_1r}=0$$

so also checks.

Now, consider $\dot{P}_R$

$$\dot{P}_R=-N\left(\frac{R'}{\Lambda}\right)'-\left(\frac{N'}{\Lambda}R\right)'$$

$0?=-\dot{T}\sqrt{\frac{p_1r}{1+p_1r}}\left(\sqrt{\frac{p_1r}{1+p_1r}}\right)'-\left(\dot{T}\left(\sqrt{\frac{p_1r}{1+p_1r}}\right)'\frac{1}{2Mp_1}\sqrt{\frac{p_1r}{1+p_1r}}2M(1+p_1r)\right)'$

$0?=\left(\frac{p_1r}{1+p_1r}\right)'-\left[\left(\frac{p_1r}{1+p_1r}\right)'\frac{1+p_1r}{p_1}\right]'$

$0\neq-\left(\frac{p_1r}{1+p_1r}\right)''\frac{(1+p_1r)}{p_1}=\frac{p_1}{(1+p_1r)^2}\cdots$

$\left(\frac{p_1r}{1+p_1r}\right)'=\frac{p_1r}{1+p_1r}-\frac{(p_1r)p_1}{(1+p_1r)^2}=\frac{p_1}{(1+p_1r)^2}$

$\left(\frac{p_1r}{1+p_1r}\right)''=\frac{-2p_1^2}{(1+p_1r)^3}$

Thus, there is an inconsistency in the $\hat{P}_R$ term, and the assumed first order expansion fails (on the very last verification step). the expansion must proceed with lowest order second-order form.

Aside from the problem getting $\dot{P}_R$ to work – which we will find it does at higher order – there is the problem with $\Lambda = O\left(r^{-1/2}\right)$. This precludes the possibility of a perturbative analysis in the neighborhood of the bifurcation point. This isn't surprising given the singular behavior in the Schwarzschild solution at the bifurcation point. What is desired is a perturbative analysis, however, so regular behavior at the bifurcation point is needed. Thus the solution will not be couched in terms of the Schwarzschild coordinates and p_1 must be taken as zero in hopes that the p_2 term will yield regular behaviour. This is plausible since $R'|_{r=0} = 0$ in that instance, instead of p_1, indicating a perturbative form for the metrical variables near the horizon.

Now consider the theory with

$p = p_2^2(t)r^2 + O(?)$

$dp = 2p_2\dot{p}_2 r^2 dt + 2rp_2^2 dr$

$ds^2 = -\left(\frac{p_2^2 r^2}{1+p_2^2 r^2}\right)\dot{T}^2 dt^2 + 4M^2\left(\frac{1+p_2^2 r^2}{p_2^2 r^2}\right)(2rp_2^2)^2\left(dr + r\frac{\dot{p}_2}{p_2}dt\right)^2 +$
$R^2 d\Omega^2$

$$N = \frac{\dot{T}p_2}{\sqrt{1+(p_2 r)^2}}r$$
$$\Lambda = 4Mp_2\sqrt{1+(p_2 r)^2}$$
$$N^r = \frac{\dot{p}_2}{p_2}r$$
$$R = 2M(1+(p_2 r)^2)$$
$$P_\Lambda = -N^{-1}R\left(\dot{R} - R'N^r\right)$$

and since

$$2M\cdot 2\cdot p_2\dot{p}_2 r^2 - 2M\cdot 2\cdot p_2^2 r\left(\frac{\dot{p}_2}{p_2}\right)r = 0$$

we get

$$P_\Lambda = 0$$
$$P_R = 0$$

With this form $N, N^r \sim O(r)$, so go to zero a bifurcation pt. as desired. Λ is no longer singular as $r \to \infty$. Let's verify that this is actually a solution to the constraints and equations of motion:

$M? = \frac{1}{2}R^{-1}P_\Lambda^2 - \frac{1}{3}\Lambda^{-2}RR'^2 + \frac{1}{2}R$
$?= -\frac{1}{2}(4Mp_2)^{-2}(1+(p_2r)^2)\,(2M)^3(1+(p_\theta r)^2)(2p_2^2r)^2 +$
$\frac{1}{2}(2\mu)(1+(p_2r)^2)$
$= -M(p_2r^2) + M + M(p_2r)^2 = M$

The Super-momentum solved trivially.

Does $\dot{\Lambda} = (\Lambda N^r)'$?

$$4M\dot{p}_2\left(\sqrt{1+(p_2r)^2} + \frac{(2p_2\dot{p}_2r^2)\frac{1}{2}}{\sqrt{1+(p_2r)^2}}\right) ?= \dot{p}_2\left(4Mr\sqrt{1+(p_2r)^2}\right)$$

$$4M\dot{p}_2\left(\sqrt{1+(p_2r)^2} + \frac{p_2r^2}{\sqrt{1+(p_2r)^2}}\right) = 4M\dot{p}_2\left(r\sqrt{1+(p_2r)^2}\right)'$$

Does $\dot{R} = R'N^r$?

$$4\dot{p}_2p_2r^2 = 2Mp_2^2 2r\left(\frac{\dot{p}_2}{p_2}\right)r$$

$$\dot{P}_\Lambda = \frac{1}{2}\left(\frac{\dot{T}p_2}{\sqrt{1+(p_2r)^2}}r\right)\left[1 - \left(\frac{2MP_2^2r\cdot 2}{4Mp_2\sqrt{1+(p_2r)^2}}\right)^2\right] -$$

$$\frac{(2m)^2(1+(p_2r)^2)(2p_2^2r)}{(4Mp_2)^2((1+p_2r)^2)}\dot{T}p_2\left(\frac{r}{1+(p_2r)^2}\right)^2 ?= 0$$

$$0 = \frac{r}{1+(p_2r)^2}\left[1 - \left(\frac{rp_3}{1+(p_2r)^2}\right)\right] - r\left(\frac{r}{1+(p_2r)^2}\right)'$$

$$\dot{P}_R = -N\left(\frac{R'}{\Lambda}\right)' - \left(\frac{N'}{\Lambda}R\right)' ?= 0$$

$$0? = -\left(\frac{\dot{T}p_2r}{\sqrt{1+(p_2r)^2}}\right)\left(\frac{4Mp_2^2r}{4Mp_2\sqrt{1+(p_2r)^2}}\right)' - \left(\left(\frac{\dot{T}p_2r}{\sqrt{1+(p_2r)^2}}\right)'\frac{2m(1+(p_2r)^2)}{4Mp_2\sqrt{1+(p_2r)^2}}\right)'$$

$$0 = -\left(\frac{p_2r}{\sqrt{1+(p_2r)^2}}\right)^2 - \left(\left(\frac{r}{\sqrt{1+(p_2r)^2}}\right)\sqrt{(1+(p_2r)^2)}\right)'$$

So, everything works.

We now have a consistent theory with the form $p = p_2^2(t)r^2 + O(r^3)$. Now to determine the general fall-off conditions. To leading order in r the fall-off conditions on the metrical variables at the bifurcation 2-sphere are required to be of the same form for all slices (that have perturbative expressions). Thus, the solution from the Schwarzschild slicing indicates:
$N = N_1(t)r + O_\geq(r^2)$
"$O_\geq(r^k)$" means order greater or possible equal to r^k

$\Lambda = \Lambda_0(t) + \Lambda_2(t)r^2 + O_{\geq}(r^2)$
$N^r = N_1^r(t)r + O_{\geq}(r^2)$
$R = R_0(t) + R_2(t)r^2 + O_{\geq}(r^3)$
$P_\Lambda = O_{\geq}(r)\ , P_R = O_{\geq}(r)$

Consistency with the constraint equations and the dynamical equations will now dictate the form of the $O_{\geq}(r^k)$ terms.

Consider the super-momentum: $H_r = P_R R' - \Lambda P_\Lambda$, with $H_r = 0$ constraint.
$P_\Lambda = -N^{-1}R(\dot{R} - R'N^r)$ (from $\dot{R}$ dynamical equation), So:

$$P_\Lambda = -\underbrace{\left(R_0(t) + R_2(t)r^2 + O_{\geq}(r^3)\right)}_{N_1(t)r+O_{\geq}(r^2)}\left(\dot{R}_0 + \dot{R}_2 r^2 + O_{\geq}(r^3) - \underbrace{[2rR_2 + O(r^2)]\left(N_4^r r + O_{\geq}(r^3)\right)}_{O(r^2)}\right)$$

$$= -\left(\frac{R_0\dot{R}_0}{N_1 r}\right)r^{-1} - \left(\frac{R_0}{N_1}[R_2 - 2N_1^r R_2] + R_2\dot{R}_0\right)r/N_1 + O(r^2)$$

$$= -\left(\frac{R_0\dot{R}_0}{N_1 r}\right)r^{-1} - \left(\frac{R_0}{N_1}\left[\dot{R}_2 - 2N_1^r R_2\right] + \frac{R_2}{N_1}\dot{R}_0\right)r + O(r^2)$$

In order to maintain a perturbative theory the r^{-1} term must be eliminated, this is only possible if $\boxed{\dot{R}_0 = 0}$, taking R_0 = constant observe that the "r" term is unconstrained. So, P_Λ *is* $O(r)$ at least.

Consider information on P_Λ from the mass equation:

$$M = \frac{1}{2}R^{-1}P_\Lambda^2 - \frac{1}{2}\Lambda^{-2}RR'^2 + \frac{1}{2}R$$

$$= \frac{\frac{1}{2}P_\Lambda^2}{(R_0 + R_2r^2 + O_{\geq}(r^3))} - \frac{1}{2}\frac{\left(R_0 + R_2r^2 + O_{\geq}(r^3)\right)\left(2R_2r + O_{\geq}(r^2)\right)}{\left(\Lambda_0 + \Lambda_2r^2 + O_{\geq}(r^3)\right)^2} + \frac{1}{2}\left(R_0 + R_2r^2 + O_{\geq}(r^3)\right)$$

$$(2M - R_0) = \frac{P_\Lambda^2}{R_0} - \left[\frac{R_0R_0^2}{\Lambda_0^2} - R_2\right]r^2 + O_{\geq}(r^3)$$

This indicates $R_0 = 2M$ and $P_\Lambda = O(r)$ unless

$$(P_\Lambda)_1 = \frac{4R_0R_0^2}{\Lambda_0^2} - R_0R_2 ?= 0$$

So, does $4R_0R_2 = \Lambda_0^2$?

Consider the super-momentum:

$$0 = H_r = P_R R' - \Lambda P_\Lambda' \rightarrow P_R = \frac{\Lambda P_\Lambda'}{R'}$$

Now

$$P_R = (P_R)_1 r + (P_R)_2 r^2 + (P_R)_3 r^3 + \cdots \qquad (P_R)_0 = 0$$
$$P_\Lambda = (P_\Lambda)_1 r + (P_\Lambda)_2 r^2 + (P_\Lambda)_3 r^3 + \cdots \qquad (P_\Lambda)_0 = 0$$

Now, to have consistency:

$$P_R = \underbrace{\left(\Lambda_0 + \Lambda_2 r^2 + O_\geq(r^3)\right)}_{(2R_2 r + O_\geq(r^3))} [(P_\Lambda)_\Lambda + 2(P_\Lambda)_2 r + 3(P_\Lambda)_3 r^2]$$
$$= \frac{1}{2R_2 r}\left(\Lambda_0 + \Lambda_2 r^2 + O_\geq(r^3)\right)[(P_\Lambda)_\Lambda + 2(P_\Lambda)_2 r + 3(P_\Lambda)_3 r^2 + \cdots]$$
$$= \frac{1}{2R_2 r}(P_\Lambda)_\Lambda r^{-1} + \frac{\Lambda_0 (P_\Lambda)_2}{R_2} r^0 + \left(\frac{3}{2}\frac{(P_\Lambda)_3 \Lambda_0}{R_2} + \frac{\Lambda_2 (P_\Lambda)_1}{2R_2}\right) r + \cdots$$

Since $(P_R)_{-1}, (P_R)_0$ are zero this indicates:

$$(P_\Lambda)_1 = 0\,, (P_\Lambda)_2 = 0$$

So, $P_\Lambda = O(r^3)$, at least, and $P_R = O(r)$ at least. Furthermore, we have $4R_0 R_2 = \Lambda_0^2$ from the boxed equation as well as $R_0 = 2M$. This indicates that R_2 is positive for the $M > 0$ class of solutions considered.

So, far:

$$N = N_1 r + O_\geq(r^2)$$
$$N^r = N_1^r r + O_\geq(r^2)$$

$$\Lambda = \Lambda_0 + \Lambda_2 r^2 + O_\geq(r^3)$$
$$R = R_0 + R_2 r^2 + O_\geq(r^3)$$

Thus

$$P_\Lambda = O_\geq(r^3)$$
$$P_R - O_\geq(r)$$

as indicated by constraint equations and $\dot{R}$ dynamical equation (as well as $R_0 = 2M\,, \Lambda_0^2 = 4R_0 R_2$). Now to examine the $\dot{\Lambda}, \dot{P}_\Lambda$ *and* $\dot{P}_R$ equations:

$$\dot{\Lambda} = -NR^{-2}(RP_R - \Lambda P_\Lambda) + (\Lambda N^r)'$$
$$\dot{P}_\Lambda = \frac{1}{2}N\left[1 - \Lambda^{-2}R'^2 - R^{-2}P_\Lambda^2\right] - \Lambda^{-2}RR'N' + N^r P_\Lambda'$$
$$\dot{P}_R = N\left[R^{-3}\Lambda {P_\Lambda}^2 - R^{-2}P_\Lambda P_R\right] + (N^r P_R)' - N\left(\frac{R'}{\Lambda}\right)' - \left(\frac{N^r}{\Lambda}R\right)'$$

First $\dot{\Lambda}$

$$\dot{\Lambda} = \left(\dot{\Lambda}_0 + \dot{\Lambda}_2 r^2 + O_\geq(r^3)\right) = -\frac{\left(N_1 r + O_\geq(r^3)\right)}{(R_0 + R_2 r^2 + O_\geq(r^3))}\left(R_0 (P_R)_1 r + O(r^3)\right) + (\Lambda N^r)'$$

$$\dot{\Lambda}_0 + \dot{\Lambda}_2 r^2 + O_{\geq}(r^3) = -\left(\frac{N_1}{R_0} r + O_{\geq}(r^2) - \frac{N_1 R_2}{{R_0}^2} r^3 + O_{\geq}(r^4)\right)\left(R_0 (P_R)_1 r + O(r^3)\right)$$

$$+\left[\left(\Lambda_0 + \Lambda_2 r^2 + O_{\geq}(r^3)\right)\left(N_1^r r + O_{\geq}(r^2)\right)\right]'$$

$$= -N_1 (P_R)_1 r^2 + O_{\geq}(r^3) + \left(\Lambda_0 N^r + \Lambda_0 O_{\geq}(r) + 3\Lambda_2 N_1^r r^2 + O_{\geq}(r^3)\right)$$

There is no $O(r)$ term in Λ, so $N_2^r = 0$, i.e.

$$N^r = N_1^r r + O_{\geq}(r^3)$$

Now consider $\dot{P}_\Lambda$:

$$\dot{P}_\Lambda = O_{\geq}(r^3) = \frac{1}{2}\left(N_1 r + O_{\geq}(r^2)\right)\left[1 - \frac{\left(2R_2 r + O_{\geq}(r^2)\right)^2}{\left(\Lambda_0 + \Lambda_2 r^2 + O_{\geq}(r^3)\right)^2} - O_{\geq}(r^6)\right]$$

$$-\frac{\left(R_0 + R_2 r^2 + O_{\geq}(r^3)\right)\left(2R_2 r + O_{\geq}(r^2)\right)\left(N_1 + O_{\geq}(r)\right)}{\left(\Lambda_0 + \Lambda_2 r^2 + O_{\geq}(r^3)\right)^2} + \left(N_1^r + O_{\geq}(r^3)\right) O_{\geq}(r^3)$$

$O(r)$ *terms* : $\frac{1}{2} N_1 r - \frac{2R_0 R_2}{\Lambda_0^2} N_1 r = 0$ from the $\Lambda_0^2 = 4R_0 R_2$ relation

$O(r^2)$*terms* : $-\frac{2R_0 R_2}{\Lambda_0} r O_{\geq}(r) + \frac{1}{2} O_{\geq}(r^2) = 0$

This indicates nothing new so far, but now:

$O(r^3)$ terms:

$$O_{\geq}(r^3) = \frac{1}{2} O_{\geq}(r^2) - \frac{1}{2} N_1 \left(\frac{2R_2}{\Lambda_0}\right)^2 r^3 - \begin{bmatrix} \frac{2R_0 R_2}{\Lambda_0^2} O_{\geq}(r^2) r - \frac{2R_0 R_2}{\Lambda_0^2} N_1 (2\Lambda_0 \Lambda_2) r^3 \\ + \frac{R_0}{\Lambda_0^2} O_{\geq}(r^2) O_{\geq}(r) + \frac{2R_2^2 N_1 r^3}{\Lambda_0^2} \end{bmatrix} + N_1^r O_{\geq}(r^2) r$$

$$= \left[-4\left(\frac{R_2}{\Lambda_0}\right)^2 N_1 + \Lambda_0 \Lambda_2 N_1\right] r^3 + O_{\geq}(r^3)$$

This indicates that the $O_{\geq}(r^3)$ term in P_Λ is indeed $O(r^3)$ (unless the r^3 term is cancelled by contributions from the $O_{\geq}(r^3)$ on the RHS).

Consider $\dot{P}_R$:

$$\dot{P}_R = \left(N_1 r + O_{\geq}(r^2)\right)\left[O_{\geq}(r^6) - O_{\geq}(r^4)\right] + \left[\left(N_1^r r + O_{\geq}(r^2)\right) O_{\geq}(r)\right]'$$

$$-\left(N_1 r + O_{\geq}(r^2)\right)\left[\frac{2R_2 + O_{\geq}(r^2)}{(\Lambda_0 + \Lambda_2 r^2)}\right]' - \left[\frac{N_1 + O_{\geq}(r)}{(\Lambda_0 + \Lambda_2 r^2)}(R_1 + R_2 r^2)\right]'$$

$$O_{\geq}(r) = \left[N_1^r r O_{\geq}(r^0) + N_1^r O_{\geq}(r)\right] - \frac{N_1 (2R_2)}{N_0} r - \frac{R_0}{\Lambda_0} N_1^r r O_{\geq}(r^0) - \frac{2N_1 R_2}{\Lambda_0} r + \frac{2N_1 R_2}{{\Lambda_0}^2}$$

$= \frac{R_0}{\Lambda_0} O_{\geq}(r^0) + \left[-4N_1\left(\frac{R_2}{\Lambda_0}\right) + 2N_1\frac{\Lambda_2}{\Lambda_0}\left(\frac{R_2}{\Lambda_0}\right)\right]r - O_{\geq}(r)$

The first term cannot be $O_{\geq}(r^0)$, and the order $O_{\geq}$ term came from N. Thus $N_2 = 0$, i.e., we have:

$$N = N_1 r + O_{\geq}(r^3)$$

This remaining terms indicate, similarly to P_Λ, that $P_R = O(r)$ is consistent.

Consider the P_Λ (from $\dot{R}$) equation with $P_R = O(r)$, $P_\Lambda = O(r^3)$

$P_\Lambda = -N^{-1}R(\dot{R} - R'N^r)$

$O(r^3) = -\frac{\left(R_0+R_2r^2+O_{\geq}(r^3)\right)}{\left(N_1^r+O(r^3)\right)}\left[\left(\dot{R}_0 + R_2r^2 + O_{\geq}(r^3)\right) - \left(R_2(2r) + \right.\right.$

$\left.\left.O_{\geq}(r^2)\right)\left(N_1^r + O(r^3)\right)\right]$

$= -\left(R_0 + R_2r^2 + O_{\geq}(r^3)\right)\left(\frac{1}{N_1r} - \frac{1}{N_1}O(r)\right)\left[\left(\dot{R}_2 - 2R_2N_1^r\right)r^2 + \right.$

$\left.O_{\geq}(r^3) + O(r^4)\right]$

The $\left(\dot{R}_2 - 2R_2N_1\right)$ terms to give rise to a $O(r)$ term, so

$\boxed{\dot{R}_2 - 2R_2N_1^r = 0}$

The $O_{\geq}(r^3)$ term yields a $\frac{1}{r}O_{\geq}(r^3)$ term so $O_{\geq}(r^3) \neq O(r^3)$ in R, so take the $O(r^4)$ term:

$$\boxed{R = R_0 + R_2r^2 + O(r^4)}$$

Consider again $M = \left(to\ O(r^3)\right)$ now:

$M = O(r^6) - \frac{1}{2}\frac{\left(R_0+R_2r^2+O_{\geq}(r^4)\right)\left(2R_2r+O(r^3)\right)^2}{\left(\Lambda_0+\Lambda_2r^2+O_{\geq}(r^3)\right)^2} + \frac{1}{2}\left(R_0 + R_2r^2 + O(r^4)\right)$

$0 = \frac{1}{2}R_2r^2 + O(r^4) - 2\frac{R_0R_2^2}{\Lambda_0^2}r^2 + O(r^4)$

The above is consistent with $O_{\geq}(r^3)$ term in Λ and other order terms.

Consider the super-momentum again:

$0 = O(r)\left(2R_2r + O(r^3)\right) - \left(\Lambda_0 + \Lambda_2r^2 + +O_{\geq}(r^3)\right)O(r^2)$

This is consistent with $O_{\geq}(r^3)$, etc., and the $\dot{\Lambda}$ eqn. remains consistent.

Consider $\dot{P}_\Lambda$:

$$O(r^3) = \frac{1}{2}(N_1 r + O(r^3))\left[1 - \frac{(2R_2 r + O(r^3))^2}{(\Lambda_0 + \Lambda_2 r^2 + O_{\geq}(r^3))^2} - O(r^6)\right]$$
$$-\frac{(R_0 + R_2 r^2 + O_{\geq}(r^4))(2R_2 r + O(r^3))^2(N_1 + O(r^2))}{(\Lambda_0 + \Lambda_2 r^2 + O_{\geq}(r^3))^2} + (N_1 r + O(r^3))O(r^2)$$
$$= r\left\{\frac{1}{2}N_1 - N_1\left(\frac{2R_0 R_2^2}{\Lambda_0^2}\right)\right\} + r^2\{0\} + O(r^3)$$
which is consistent.

$\dot{P}_R$ will offer nothing new since $P_R = O(r)$ already works consistently, where the $O(r^3)$, or $O(r^4)$, terms remain undetermined.

Double check Λ in the following

$N = N_1 r + O(r^3)$

$N^r = N_1^r r + O(r^3)$

$\Lambda = \Lambda_0 + \Lambda_2 r^2 + O(r^3) \rightarrow \Lambda = \Lambda_0 + O(r^2)$

$R = R_0 + R_2 r^2 + O(r^4)$

$P_\Lambda = O(r^3)$

$P_R = O(r)$

The constraints:

$0 = H = -R^{-1}P_R P_\Lambda + \frac{1}{2}R^{-2}\Lambda P_\Lambda^2 + R\left(\frac{R'}{\Lambda}\right)' + \frac{1}{2}\Lambda^{-1}R'^2 - \frac{1}{2}\Lambda$

$0 = H_r = P_R R' - \Lambda P_\Lambda'$

Or grouped for the mass term:

$M = \frac{1}{2}R^{-1}P_\Lambda^2 - \frac{1}{2}\Lambda^{-2}RR'^2 + \frac{1}{2}R$

The EOM's :

$\dot{R} = -NR^{-1}P_\Lambda + N^r R'$

$\dot{\Lambda} = N[R^{-2}\Lambda P_\Lambda - R^{-1}P_R] + (\Lambda N^r)'$

$\dot{P}_\Lambda = \frac{1}{2}N[1 - \Lambda^{-2}R'^2 - R^2 P_\Lambda^2] - \Lambda^2 RR'N' + N^r P_\Lambda'$

$P_R = N[R^{-3}\Lambda P_\Lambda^2 - R^{-2}P_\Lambda P_R] + (N^r P_R)' - N\left(\frac{R'}{\Lambda}\right)' - \left(\frac{N'}{\Lambda}R\right)'$

The only equation not explicitly checked for consistency in the above choice is the super-Hamiltonian, this is because the mass equation was used instead. The difficulty in checking the $O_{\geq}(r^3)$ term in Λ has been in isolating the Λ term – clearly this should have been done using the super-Hamiltonian, something I'll amend right now … it is found that the $O(r^4)$ terms from R obscure any resolution via the R' and R'' contributions. Viewing the other eqn.'s in this manner it is realized that the $O(r^3)$ *or* $O(r^4)$ term in Λ is unconstrained. Thus, the above form is consistent for the metrical variables, with, $\Lambda = \Lambda_0 + \Lambda_2 r^2 + O(r^3) =$

$\Lambda_0 + O(r^2)$ consistent, and this completes the analysis to obtain a set of fall-off conditions.

A.2 Schwarzschild Black Hole with Asymptotic Anti-deSitter

$$F(R) = 1 - \frac{2M}{R} + \frac{R^2}{l^2} \ ; \ F = \left(\frac{R'}{\Lambda}\right)^2 - \left(\frac{P_\Lambda}{R}\right)^2$$

$M = \frac{1}{2}R\left(1 + \frac{R^2}{l^2} - F\right)$
$P_M = R^{-1}F^{-1}\Lambda P_\Lambda$
$\mathfrak{R} = R$
$P_{\mathfrak{R}} = F^{-1}(R^{-1}P_\Lambda H + R'\Lambda^{-2}H_r)$
$H = -R^{-1}P_R P_\Lambda + \frac{1}{2}R^{-2}\Lambda P_\Lambda^2 + \Lambda^{-1}RR'' - \Lambda^{-2}RR'\Lambda' + \frac{1}{2}\Lambda^{-1}R'^2 -$
$\frac{1}{2}\Lambda - \frac{1}{2}\left(\frac{3}{l^2}\right)\Lambda R$
$H_r = P_R R' - \Lambda P_\Lambda'$
$P_{\mathfrak{R}} = P_R - \frac{1}{2}R^{-1}\Lambda P_\Lambda - \frac{1}{2}R^{-1}F^{-1}\Lambda P_\Lambda - R^{-1}\Lambda^{-2}F^{-1}[(\Lambda P_\Lambda)'(RR') -$
$(\Lambda P_\Lambda)(RR')]$
$-\frac{1}{2}\Lambda R P_\Lambda F^{-1}\left(\frac{3}{l^2}\right)$

Now to regroup:

$S_\Sigma = \int dtdr\,\{P_M\dot{M} + P_{\mathfrak{R}}\dot{\mathfrak{R}} - NH - N^r H_r\} = \int dtdr\,\{P_M\dot{M} + P_{\mathfrak{R}}\dot{\mathfrak{R}} -$
$N^M M' - N^{\mathfrak{R}} P_{\mathfrak{R}}\}$

where use is made of the relation $M' = -\frac{R'}{\Lambda}\left[H + \frac{P_\Lambda}{RR'}H_r\right]$.

So,

$$N^M\left\{-\frac{R'}{\Lambda}\left(H + \frac{P_\Lambda}{RR'}H_r\right)\right\} + N^{\mathfrak{R}}\{F^{-1}(R^{-1}P_\Lambda H + R'\Lambda^{-2}H_r)\}$$
$$= NH + N^r H_r$$

$$H\left\{-N^M\left(\frac{R'}{\Lambda}\right) + R^{\mathfrak{R}}(F^{-1}R^{-1})\right\} + H_r\left\{-N^M\left(\frac{P_\Lambda}{\Lambda R}\right) + N^{\mathfrak{R}}(F^{-1}R'\Lambda^{-2})\right\}$$
$$= NH + N^r H_r$$

$N = -N^M\left(\frac{R'}{\Lambda}\right) + N^{\mathfrak{R}}(F^{-1}R^{-1}P_\Lambda)$
$N^r = -N^M\left(\frac{P_\Lambda}{\Lambda R}\right) + N^{\mathfrak{R}}(F^{-1}R'\Lambda^{-2})$
$(NR') - (N^r R^{-1}\Lambda^2 P_\Lambda) = -N^M\left[\left(\frac{R'}{\Lambda}\right)R' - \frac{\Lambda}{R^2}P_\Lambda^2\right] = -N^M \Lambda F$
$N^M = N(-F^{-1}\Lambda^{-1}R') + N^r(R^{-1}F^{-1}\Lambda P_\Lambda)$
$N \equiv -(4M)^{-1}N^M = (4M)^{-1}(NF^{-1}\Lambda^{-1}R' - N^r R^{-1}F^{-1}\Lambda P_\Lambda)$

and

$(P_\Lambda R^{-1})N - (R')N^r = N^{\mathfrak{R}}F^{-1}\left(\left(\frac{P_\Lambda}{R}\right)^2 - \left(\frac{R'}{\Lambda}\right)^2\right) = -N^{\mathfrak{R}}$

So,

$N^{\mathfrak{R}} = N^r R' - NR^{-1}P_\Lambda$

With the above change in Lagrange multipliers:

$$S_\Sigma = \int dtdr\left\{P_M\dot{M} + P_{\mathfrak{R}}\dot{R} + N(4MM') - N^{\mathfrak{R}}P_{\mathfrak{R}}\right\},$$

as before, the change in Lagrange multipliers is permitted by the general properties. The grouping N was useful for the Schwarzschild spacetime bifurcation two sphere because:

$$\int N\,(2M^2)'dr \rightarrow \int [N\delta(2M^2)]'dr \rightarrow \int [N\delta 2M^2]'dr$$
$$\rightarrow \int (N)2M^2|_0^\infty dr \rightarrow 0$$

For $S - Ads$ (with or without electromagnetic field) goal is to take "$N^M M'$" and regroup so as to obtain the form:

$$n^M M' = \begin{cases} \tilde{N}\left(-\frac{1}{2}X^2\right)' & as\, r \rightarrow 0\,, \tilde{N}_0 = \left(N_1\Lambda_0^{-1}\right) \\ -\tilde{N}_t M' & as\, r \rightarrow \infty\,, \tilde{N}_t = \tilde{N}_t(t)\left(1 + O(r^{-4})\right) \end{cases}$$

$N^M M'|_{r\rightarrow 0} =?$

$$N^M = F^{-1}\left[-N\Lambda^{-1}R' + N^r N^{-1}\Lambda P_\Lambda\right];\;\; 2M = R + \frac{R^3}{l^2} - RF;\;\; 2M'$$
$$= R' + \frac{3R'^{R^2}}{l^2} - R'F - RF'$$

and recall $\left(1 + \frac{3R_0^2}{\Lambda_0}\right) = \frac{4R_0R_2}{\Lambda_0^2}$:

<u>$r \rightarrow 0$</u>

$R = R_0 + R_2r^2 + O(r^4)$

$\Lambda = \Lambda_0 + O(r^2)$

$N = N_1r + O(r^3)$

$N^r = N_1^r + O(r^3)$

$P_\Lambda = O(r^3)$

$P_R = O(r)$

$F = \left(\frac{R'}{\Lambda}\right)^2 - \left(\frac{P_\Lambda}{R}\right)^2 = \left(\frac{2R_1r+O(r^3)}{\Lambda_0+O(r^2)}\right)^2 - \left(\frac{O(r^3)}{R_0+R_2r^2}\right) = \frac{4R_2^2}{\Lambda_0^2}r^2 + O(r^4)$

$2M = R_0 + R_2r^2O(r^4) + \frac{(\ldots)^3}{l^2} - (\ldots)F$

$$2M = \left(R_0 + \frac{R_0^2}{l^2}\right) + \left(R_2 + \frac{3R_0^2 R_2}{\Lambda_0^2}\right) r^2 + O(r^4) = \left(R_0 + \frac{R_0^2}{l^2}\right) + O(r^4)$$

Expanding to lowest order:

$$N^M = \left[\begin{array}{c} -(N_1 r + O(r^3))(2R_2 r + O(r^3))\left(\Lambda_0^{-1} + O(r^2)\right) \\ +(N_1^r r + O(r^3))\left(R_0^{-1} - \frac{R_2}{R_0^2} r^2 + O(r^4)\right)(\Lambda_0 + O(r^2))\left(O(r^2)(O(r^3)\right) \end{array}\right.$$

$$\times\left(\frac{\Lambda_0^2}{4R_2^2} r^{-2} + O(r^0)\right)$$

$$N^M = -N_1 \Lambda_0^{-1}(2R_2)\left(\frac{\Lambda_0^2}{4R_2}\right) + O(r^2) = -N_1 \Lambda_0^1 \left(\frac{\Lambda_0^2}{4R_2}\right)^2 + O(r^2)$$

$$= -N_1 \Lambda_0^1 \left[-2R_0\left(1 + \frac{3R_0{}^2}{l^2}\right)^{-1}\right] + O(r^2)$$

$$\equiv \left(N_1 \Lambda_0^1\right)[f(M)|_{r\to 0}] + O(r^2)$$

$N^M M' = \left(N_1 \Lambda_0^{-1}\right)(f(M)M')_{r\to 0} + O(r^2) = \left(N_1 \Lambda_0^{-1}\right)(2X^2)'_{r\to 0}$

$4XX'|_{r\to 0} = (f(M)M')|_{r\to 0} = f(M_0)[M']_{r\to 0}$

$X|_{r\to 0} = X(M_0)$ and we choose $X = X(M)$

$X'|_{r\to 0} = \frac{\partial X}{\partial M} M'|_{r\to 0} = \left[\frac{\partial X}{\partial M}(M_0)\right][M']_{r\to 0}$

$4X\frac{\partial X}{\partial M}|_{r\to 0} = f(M_0)$

$\frac{\partial}{\partial M}(2X^2)|_{r\to 0} = f(M_0)$

$f(M)M'|_{r\to 0} = -2R_0\left(1 + \frac{3R_0^2}{l^2}\right)^2 M'|_{r\to 0}$

Guess $2M = X + X^3 l^{-2}$

$2M' = X'(1 + 3X^2 l^{-2})$

$(2X^2)' = 4XX' = 4X\left(1 + \frac{3X^2}{l^2}\right)^{-1}(2M')$

$(2X^2)'|_{r\to 0} = 4\left[-2X_0\left(1 + \frac{3X^2}{l^2}\right)^{-1} M'\right]$

$= -4[f(M)M']_{r\to 0}$

So,

$$N^M M'|_{r\to 0} = \left\{\frac{N}{\Lambda}\frac{R''}{R'}\left(-\frac{1}{2}X^2\right)'\right\}|_{r\to 0}$$

$R = R_0 + R_2 r^2 + O(r^4)$

$R' = 2R_2 r + O(r^3)$

$R'' = 2R_2 + O(r^2)$
$R''/R' = r^{-1} + O(r^1)$
Now want to use the freedom in the Lagrange multiplier N^M such that $N^M \rightarrow N^M g$ where g is a function of the canonical variables that $N^M M' \rightarrow \left(\frac{N}{\Lambda}\frac{R''}{R'}\right)\left(-\frac{1}{2}X^2\right)' = (N^M g)M'$
This can be done with $g = g_1 g_2$ where

$$g_1 = -\frac{F}{R'}\left(\frac{R''}{R'}\right)\left(-\frac{1}{2}X^2\right)'\frac{1}{M'}$$

and g_2 is a function of the canonical variables that is unitary at the bifurcation 2-sphere and sufficiently dampened for $r > 0$ that g is well behaved.

$$N^M M' = F^{-1}\left[-\frac{N}{\Lambda}R' + N^r\frac{\Lambda}{R}P_\Lambda\right]\left(-\frac{F}{R'}\frac{R''}{R'}\left(-\frac{1}{2}X^2\right)'\frac{1}{M'}\right)M' g_2$$
$$= g_2\left(\frac{N}{\Lambda}\frac{R''}{R'}\right)\left(-\frac{1}{2}X^2\right)' + g_2 N^r\left(\frac{\Lambda P_\Lambda}{RR'}\right)\frac{R''}{R'}\left(-\frac{1}{2}X^2\right)'$$
$$= \delta\left[g_2\left(\frac{N}{\Lambda}\frac{R''}{R'}\right)\left(-\frac{1}{2}X^2\right)|_{r=0}\right] + \delta\left[g_2\left(\frac{N}{\Lambda}\frac{R''}{R'}\right)\left(-\frac{1}{2}X^2\right)\right]|_{r=0}$$
$$= \delta\left(\frac{N_1}{\Lambda_0}\right)\left(-\frac{1}{2}X^2\right) + \delta(O(r^2))O(r^0) = 0 + 0 = 0$$

The boundary term that arises from this is

$$N^M M' g = \tilde{N}\left(-\frac{1}{2}X^2\right)' + \left(g_2 N^r\left(\frac{\Lambda P_\Lambda}{RR'}\right)\frac{R''}{R'}\left(-\frac{1}{2}X^2\right)'\right),$$
$$\tilde{N} = g_2\left(\frac{N}{\Lambda}\frac{R''}{R'}\right)$$

Boundary term needed: $\int\left[\tilde{N}\left(\frac{1}{2}X^2\right)\right]_{r=0} dt = S_\Sigma|_{r=0}$

Now consider N^m as $r \rightarrow \infty$:

$\Lambda = \frac{l}{r} - \frac{1}{2}\left(\frac{l}{r}\right)^3 + r^{-4}\lambda + O(r^{-5})$
$R = r + r^{-2}p_2 + O(r^{-3})$
$N_r = O(r^{-2})$
$N = \frac{r}{l} + \frac{1}{2}\frac{l}{r} + \eta r^{-2} + O(r^{-3})$
$P_\Lambda = O(r^{-2})$
$P_R = O(r^{-4})$

$$F = \left(\frac{R'}{\Lambda}\right)^2 - \left(\frac{P_\Lambda}{R}\right)^2 = \left(\frac{1-2r^{-3}p_2+O(r^{-4})}{\left(\frac{l}{r}\right)\left[1-\frac{1}{2}\left(\frac{l}{r}\right)^2+r^{-3}\lambda l^{-1}+O(r^4)\right]}\right)^2 - \left(\frac{O(r^{-2})}{r}\right)^2$$

$$= [1 - 4r^{-3}p_2 + (r^{-4})]\left(\frac{l}{r}\right)^2\left[1 + \left(\frac{l}{r}\right)^2 - 2r^{-3}\lambda l^{-1} + O(r^4)\right]$$
$$= \left(\frac{l}{r}\right)^2\left[1 + \left(\frac{l}{r}\right)^2 - r^{-3}(4p_2 + 2\lambda l^{-1}) + O(r^4)\right]$$
$$N\Lambda^{-1}R' = \left(\frac{l}{r} + \frac{1}{2}\frac{l}{r} + \eta r^{-2} + O(r^3)\right)\left(\frac{l}{r}\right)^{-1}$$
$$\times\left[1 + \frac{1}{2}\left(\frac{l}{r}\right)^2 - r^{-3}\lambda l^{-1} + O(r^4)\right](1 - 2r^{-3}p_2 O(r^4))$$
$$= \left[\left(\frac{r}{l} + \frac{1}{2}\frac{l}{r}\right)^2 - \left(\frac{r}{l} + \frac{1}{2}\frac{l}{r}\right)r^{-2}\lambda l^{-2} + O(r^{-2}) + \eta l^{-1}r^{-1}\right](1 - 2r^{-3}p_2 + O(r^4))$$
$$= \left(\frac{r}{l}\right)^2 + 1 + \frac{1}{4}\left(\frac{r}{l}\right)^2 - r^{-1}(\lambda l_{-3} - \eta l^{-1} + 2p_2 l^{-2}) + O(r^{-2})$$
$$N^r R^{-1}\Lambda P_\Lambda = O(r^{-5})$$
$$N^M = \frac{-\left(\frac{r}{l}\right)^2\left[1+\left(\frac{r}{l}\right)^2 - r^{-3}(\lambda l^{-1} - \eta l + 2p_2) + O(r^{-4})\right]}{\left(\frac{r}{l}\right)^2\left[1+\left(\frac{l}{r}\right)^2 - r^3(4p_2 + 2\lambda l^{-1}) + O(r^{-4})\right]}$$
$$= -\left[1 + \left(\frac{l}{r}\right)^2 - r^{-3}(\lambda l^{-1} - \eta l + 2p_2) + O(r^{-4})\right]\left[1 - \left(\frac{l}{r}\right)^2 - r^{-3}(4p_2 + 2\lambda l^{-1}) + O(r^{-4})\right]$$
$$= 1 - \left(\frac{l}{r}\right)^2 - r^{-3}(4p_2 + 2\lambda l^{-1}) + O(r^{-4}) + \left(\frac{l}{r}\right)^2 - r^{-3}(\lambda l^{-1} - \eta l + 2p_2)$$
$$= -1 - r^3(4p_2 + 2\lambda l^{-1} - \lambda l^{-1} + \eta l - 2p_2) + O(r^{-4})$$
$$= -1 - r^3(2p_2 + \lambda l^{-1} + \eta l) + O(r^{-4})$$

There is the condition $\eta l^2 + \lambda l^{-4} + 2p_2 l^{-3} = 0$, so we get:

$$N^M = -r + O(r^{-4})$$

So $N^M M'|_{r=\infty} = -M'$, and if the rescaling freedom in the lapse is used, $N^M \to N^M \tilde{N}_+(t)$, the desired form is obtained.

$$S_\Sigma|_{r=\infty} = \int dtdr\,[\tilde{N}_t M']$$
$$(\delta S_\Sigma)|_{r=\infty} = \int dt\,[\tilde{N}_t \delta M]_{r=\infty}$$
$$S_{\partial\Sigma}|_{r=\infty} = -\int dt\,[\tilde{N}_t M]_{r=\infty}$$
$$\delta(S_\Sigma + S_{\delta\Sigma})|_{eom} = \int dt\,[\delta\tilde{N}_t]M \to \delta\tilde{N}_t = 0$$

The family of geometries considered is restricted so that all have the same boost parameter at the bifurcation 2-sphere. Thus, the slicing of the 4-geometry is fixed at the bifurcation 2-sphere. A further condition on the slicing is that $\int(N_1\Lambda_0^{-1})d\tau \propto 2\pi$ in order that a Euclidean correspondence be possible (without angle deficit singularities). So, $N_1 \Lambda_0^{-1}$ need not be constant, but it must integrate appropriately over the S^1 complex time path. The slicing of the 4-geometry is also fixed at spatial infinity, so that the family of geometries considered is restricted to have the same Lapse function as $r \to \infty$. So,

$$S_{\partial\Sigma} = \int dt \left[\tilde{N}\left(\frac{1}{2}X^2\right)\right]_{r=0} - \int dt \left[\tilde{N}_t M\right]_{r=\infty}$$

A.3 Reduced Hamiltonian and Partition Function

From the constraint $M' = 0 \to M(t,r) = m(t)$, together with $P_{\Re} = 0$:

$$S = \int dt\, \dot{m} \int_0^\infty dr\, P_M + S_{\partial\Sigma}, \quad P_M = R^{-1}F^{-1}\Lambda P_\Lambda$$

$r \to 0 \quad P_M = (R_0 + R_2 r^2)^{-1}\left(\frac{4R_2^2}{\Lambda_0} r^2 + O(r^4)\right)^{-1} (\Lambda_0 + O(r^2))O(r^3) = O(r)$

$r \to \infty \quad P_M = (r + r^{-2}p_2)^{-1}\left(\frac{r}{l}\right)^{-2}\left(1 + \left(\frac{l}{r}\right)^{-2}\right)\left(\frac{l}{r} - O(r^{-3})\right)O(r^{-2}) = O(r^{-6})$

Let $P = \int_0^\infty dr\, P_M$, $S_{\partial\Sigma} = \int dt(-h)$, then

$$S = \int dt\,(p\dot{m} - h), \quad h = \tilde{N}_t M_t - \tilde{N}\left(\frac{1}{2}X^2\right)$$

Quantization $h \to \hat{h}$ is then appropriately defined to arrive at a unitary time evolution operator:

$$\tilde{K}(t_2, t_1) = \mathcal{P}\exp\left[-i\int_{t_1}^{t_2} dt'\, \hat{h}(t)\right]$$

h commutes with itself, so $\mathcal{P}$ trivial. Define $\Theta = \int_{t_1}^{t_2} dt\tilde{N}$ and $T_B = \int_{t_1}^{t_2} dt\tilde{N}_t$ and using the fact that the equations of motion give $\dot{m} = 0$:

$$\tilde{K} = \exp\left[-i\left(mT_B - \frac{1}{2}X^2\Theta\right)\right]$$

From the unitary evolution operator we arrive at a partition function by an analytic continuation to complex time with periodic boundary conditions (e.g., Wick rotation, Euclideanization), the period is related to the inverse temperature of the thermodynamic system that is indicated by the partition function obtained.

Partition function

$$Ƶ = (\beta) = T_R\left[\widetilde{K}(-i\beta i; -2\pi i)\right]$$

$$Ƶ = (\beta) = \eta \int_0^{\infty} dm \exp[-m\beta + \pi X^2] = \exp[-m_s\beta + \pi X_s]$$

$$\sigma = \left(1 - \beta\frac{\partial}{\partial\beta}\right) \ln Ƶ = (-m_s\beta + \pi X_s^2) - \beta(-m_s) - \beta\left[-\beta\frac{\partial}{\partial\beta} m_s + \pi\frac{\partial}{\partial\beta}(X_s^2)\right]$$

$$= \pi X_s^2 = \frac{1}{2}(Area)$$

$$2m = X + \frac{X^3}{l^2} \rightarrow 2 = \frac{\partial X}{\partial m}\left(1 + \frac{3X^2}{l^2}\right)$$

$$f(m) = -m\beta + \pi X^2$$

$$\frac{\partial f}{\partial m} = -\beta + 2\pi X\frac{\partial X}{\partial m} \quad = -\beta + 2\pi X_s \frac{2}{\left(1+\frac{3X_s^2}{l^2}\right)} = 0$$

$$-\beta\left(1 + \frac{3X_s^2}{l^2}\right) + 4\pi X_s = 0$$

$$\left(-\frac{3\beta}{l^2}\right)X_s^2 + 4\pi X_s = 0$$

$$X_s = \frac{-4\pi \pm \sqrt{(4\pi)^2 - 4\left(\frac{3\beta}{l^2}\right)\beta}}{2\left(-\frac{3\beta}{l^2}\right)}$$

B. Reissner-Nordstrom Anti-deSitter Spacetime Thermodynamics

B.1 Background

In this appendix we recall some relevant properties of the Reissner-Nordstrom anti-deSitter (RNAdS) metric. We concentrate on the case where a nondegenerate event horizon exists, and on the region exterior to this horizon. Note that the equation numbering used in what follows from my PhD Thesis [9].

In the curvature coordinates (T;R), the RNAdS metric is given by

$$ds^2 = -F dT^2 + F^{-1} dR^2 + R^2 d\Omega^2 \quad , \tag{A1a}$$

where d2 is the metric on the unit two-sphere and

$$F := \frac{R^2}{\ell^2} + 1 - \frac{2M}{R} + \frac{Q^2}{R^2} \quad . \tag{A1b}$$

T and R are called respectively the Killing time and the curvature radius. The parameter ` is positive, and we take the parameters M and Q to be real. Together with the electromagnetic potential

$$A = \frac{Q}{R} dT \quad , \tag{A2}$$

the metric (A1) is a solution to the Einstein-Maxwell equations with the cosmological constant $-3l^{-2}$. The parameters M and Q are referred to respectively as the mass and the (electric) charge. The case Q = 0 yields the Schwarzschild anti- de Sitter metric, and the case Q = M = 0 yields the metric on (the universal covering space of) anti-de Sitter space.

The metric (A1) has an asymptotically anti-de Sitter infinity at $R \to \infty$ for all values of the parameters. We wish to restrict the parameters so that the metric describes the exterior of a black hole with a nondegenerate horizon. This happens when the quartic polynomial $R^2 F(R)$ has a simple positive root $R = R_0$, such that F is positive for $R > R_0$. The necessary and sufficient condition is M > $M_{crit}(Q)$, where

$$M_{\text{crit}}(Q) := \frac{\ell}{3\sqrt{6}} \left(\sqrt{1 + 12(Q/\ell)^2} + 2\right) \left(\sqrt{1 + 12(Q/\ell)^2} - 1\right)^{1/2} \quad . \tag{A3}$$

Note that M is then necessarily positive. R0 can now be determined uniquely as the function $R_{hor}(M; Q)$ of M and Q: for Q = 0, $R_{hor}(M; Q)$ is defined as the unique positive solution to the equation F = 0; for Q 6= 0, $R_{hor}(M; Q)$ is defined as the larger of the two positive solutions. In either case, if Q is considered fixed, $R_{hor}(M; Q)$ is a monotonically increasing function of M that takes the values $R_{crit}(Q) < R_{hor}(M; Q) < 1$ as $M_{crit}(Q) < M < 1$, where

$$R_{\rm crit}(Q) := \frac{\ell}{\sqrt{6}}\left(\sqrt{1+12(Q/\ell)^2}-1\right)^{1/2} . \tag{A4}$$

The metric can thus be uniquely parametrized by Q and R_0. The only restriction for these parameters is

$$R_0 > R_{\rm crit}(Q) \quad , \tag{A5}$$

and the mass is then given by

$$M = \frac{R_0}{2}\left(\frac{R_0^2}{\ell^2}+1+\frac{Q^2}{R_0^2}\right) . \tag{A6}$$

With $R_0 < R < 1$, the metric (A1a) covers the region from the horizon to the asymptotically anti-de Sitter infinity.

In the main text we took the cosmological constant to be strictly negative. In this appendix we shall outline the corresponding classical and quantum mechanical analysis in the case where the cosmological constant vanishes. The classical solutions are then not asymptotically anti-de Sitter but asymptotically flat, and the falloff conditions at r → 1 must be modified to reflect this fact. Thus, we retain the falloff conditions at r → 0, but at r → 1 we introduce the new falloff conditions:

$$R(t,r) = r + O^{\infty}(r^{-\epsilon})$$
$$P_{\Lambda}(t,r) = r + O^{\infty}(r^{-\epsilon})$$
$$P_R(t,r) = O^{\infty}(r^{-1-\epsilon})$$
$$N(t,r) = N_+(t) + O^{\infty}(r^{-\epsilon})$$
$$N^r(t,r) = O^{\infty}(r^{-\epsilon})$$
$$\Gamma(t,r) = O^{\infty}(r^{-1-\epsilon})$$
$$P_{\Gamma}(t,r) = Q_+(t) + O^{\infty}(r^{-\epsilon})$$
$$\widetilde{\Phi}(t,r) = \widetilde{\Phi}_+(t) + O^{\infty}(r^{-\epsilon})$$

where $0 < \epsilon \leq 1$. For the metric quantities these conditions are precisely those used in Ref. [80], ensuring asymptotic flatness. These conditions make the bulk action well-defined, and they are preserved under the time evolution. Adding the boundary action

$$\int dt\left(\frac{1}{2}R_0^2 N_1\Lambda_0^{-1} - N_+M_+ + \widetilde{\Phi}_0 Q_0 - \widetilde{\Phi}_+ Q_+\right)$$

yields an action for a variational principle in which $N_+, N_1\Lambda_0^{-1}, \widetilde{\Phi}_+$and $\widetilde{\Phi}_0$ are prescribed functions of *t.* Dropping the last two terms yields an action for a variational principle in which $\widetilde{\Phi}_+$ and $\widetilde{\Phi}_0$are free but Q_+ and Q_0 are prescribed.

The canonical transformation of the main text can now be adapted to the present boundary conditions by simply taking the limit $\ell \to \infty$. A new action can be constructed, the new falloff conditions only giving rise to minor technical modifications to the redefinition of the Lagrange multipliers. The theory that prescribes $\widetilde{\Phi}_+$ and $\widetilde{\Phi}_0$ elimination of the constraints then yields the reduced action

$$S\left[m, q, P_m, P_q; \widetilde{N}_0^M, N_+, \widetilde{\Phi}_+, \widetilde{\Phi}_0\right] = \int dt\left(P_m\dot{m} + P_q\dot{q} - h\right)$$

where the reduced Hamiltonian is given by

$$h = -\frac{1}{2}R_h^2\widetilde{N}_0^M + N_+m + \left(\widetilde{\Phi}_+ - \widetilde{\Phi}_0\right)q$$

with $R_h := m + \sqrt{m^2 - q^2}$. The range of the variables is $0 < \boldsymbol{m},\ \boldsymbol{q}^2 < \boldsymbol{m}^2$. In the theory that prescribes Q_+ and Q_+, one proceeds to obtain the reduced action

$$S_c\left[m, P_m; \widetilde{N}_0^M, N_+; q\right] = \int dt(P_m\dot{m} - h_c)$$

where q is now regarded as an external parameter and

$$h_c = -\frac{1}{2}R_h^2\widetilde{N}_0^M + N_+m$$

Quantization of the two reduced theories proceeds as in the main text. For the renormalized trace of the analytically continued time evolution operator, we obtain formally

$$\mathcal{Z}_{ren}(\beta, \phi) = N \int_{R_h > |q|} \tilde{\mu} dR_h dq\ \exp(-I_*)$$

$$z_{ren}(\beta, q) = \int_{R_h > |q|}^{\infty} \tilde{\mu} dR_h dq\ \exp(-I_c)$$

where $I_*$and $I_{C*}$are respectively given by dropping the term proportional to ℓ^{-2}. β is now interpreted as the inverse Hawking temperature at the infinity, with no renormalization. However, both integrals are divergent because of the behavior of $I_*$and I_{C*} at large R_h. Thus, neither the canonical ensemble nor the grand canonical ensemble exists under the asymptotically flat boundary conditions. In this respect, the inclusion of the charge has therefore not made a qualitative difference from the asymptotically flat vacuum case [68],

The critical points of $I_*$and I_{C*} give again the (Lorentzian) classical solutions that have the inverse Hawking temperature at infinity and the prescribed value of respectively ϕ or q. The condition that I_* possess critical points is $|\phi| < 1$: when this condition is satisfied there exists exactly one critical point, but this critical point is not a local extremum. This reproduces the observations made by Davies in [43,44] about charged black hole equilibria with fixed ϕ, and reflects. In particular, the fact that a semiclassical charged black hole under these boundary conditions is not stable against Hawking evaporation.

The condition that I_{C*} possess critical points is $\beta/|q| \geq 6\pi\sqrt{3}$, and when the inequality is genuine, there exist two critical points. The critical point with the smaller (larger) value of R_h is a local minimum (maximum, respectively). The local minimum satisfies $q^2 > \frac{3}{4}m^2$, and it corresponds to the classical solution that Davies [43] showed to be stable against Hawking evaporation under these boundary conditions (see also Refs. [99-102]). While the thermodynamical stability of this semiclassical solution is reflected in its being a local minimum of our I_{C*} [61], the divergence of the integral demonstrates that this local stability is not sufficient to guarantee the existence of a full thermodynamical canonical ensemble.

B.2 Metric and Canonical Formulation

The spacetimes considered are spherically symmetric solutions to Einstein's equations with negative cosmological constant. Such solutions are locally isometric to the Schwarzschild-anti-deSitter line element.

$$ds^2 = -F(R)dT^2 + F^{-1}(R)dR^2 + R^2 d\Omega^2$$

Written in the curvature coordinates (T, R) in order to follow the analysis of LVK (an the notation as well):

$$F(R) = (1 - 2M/R + R^2/l^2)\,,$$

where $G = C = 1$ and the cosmological constant is set to: $\Lambda = -3/l^2$ (eliminates consequent notational clash as well as being convenient). Since $d\Omega^2$ is the line element on the unit sphere, the curvature coordinate R is invariably defined through $4\pi R^2$ being the area of the 2-sphere.

Now to reconstruct Mass and Time from the canonical data:
(1) First insert the hypersurface (spacetime) coordinates $T(t,r)$ and $R(t,r)$ into the line element.

(2) Then compare with the ADM form of the line element, solving for lapse and shift. This requires appropriate choice of radial label in region I-IV so that lapse N is positive.

(3) Substitute the relations for N and N^r into the momenta derived from the ADM action $(P_\Lambda \;\; and \; P_R)$. This yields the information:

$$-T' = R^{-1}F^{-1}\Lambda P_\Lambda$$

$$F = \left(\frac{R'}{\Lambda}\right)^2 - \left(\frac{P_\Lambda}{R}\right)^2$$

Since there are no derivatives on F, this part of the derivation remains general. Together this expresses T' in terms of the canonical data as well as the Schwarzschild mass (where the cosmological constant is taken to be a fixed parameter of the theory).

So,

$$\left(1 - \frac{2M}{R} + \frac{R^2}{\Lambda^2}\right) = \left(\frac{R'}{\Lambda}\right)^2 - \left(\frac{P_\Lambda}{R}\right)^2$$

and matching up:

$$M = \left(\frac{R}{2}\right)\left\{R^{-2}P_\Lambda^2 - \Lambda^{-2}R'^2 + 1 + \frac{R^2}{l^2}\right\}$$

$$= \frac{1}{2}R^{-1}P_\Lambda^2 - \frac{1}{2}\Lambda^{-2}RR'^2 + \frac{1}{2}R + \frac{1}{2l^2}R^3$$

Now to take the Mass function and Time Gradient as canonical variables:
(1) Promote $M(r)$ and $-T'^{(r)}$ to definitions of two sets of dynamical variables on the phase space. If they are to be conjugate: M(r) is the spatial scalar while $-T'^{(r)}$ is a scalar density, thus $-T'^{(r)}$ is chosen to be the conjugate momentum $P_M(r)$.
(2) $M(r)$ and $-T'(r)$ have vanishing Poisson bracket with $R(r)$ so that is still a good canonical variable, not so with $P_R(r)$, however, In order to

define a new momenta conjugate to $R(r)$ that also has vanishing Poisson bracket with $M(r)$ and $-T'(r)$ there is the freedom to add a term that has no dependence on P_R:

$$P_{\bar{R}} = P_R + \theta(r; R, \Lambda, P_\Lambda)$$

At this juncture [80] we could require that the new canonical variables should also respect the properly that the canonical coordinates are spatial scalar while the momenta are scalar densities. This is done to determine the form of the supermomentum by the requirement that $H_r(r)$ generate Diff R. This enforces a relation between the supermomenta:

$$R_R R' - \Lambda P'_\Lambda = +P_M M' + P_{\bar{R}} R'$$

where $H_r := P_R R' - \Lambda P'_\Lambda$ (the sign is chosen so that H_r generates Diff R on the canonical variables). Let's test that the indicated H_r generates diffeomorphisms (Diff R) starting with the invariance of the spatial line element:

$$dl^2 = \Lambda^2(r)dr^2 + R^2(r)d\Omega^2 = \Lambda^2(\tilde{r})d\tilde{r}^2 + \tilde{R}(r)d\Omega^2$$

So, $R(r) = \tilde{R}(\tilde{r})$ and is thus a scalar. Consider the diffeomorphism generated by the infinitesimal radial shift: $\tilde{r} + \varepsilon(\tilde{r})$:

$$R\big(\tilde{r} + \varepsilon(\tilde{r})\big) = R(\tilde{r}) + \varepsilon(\tilde{r})R'(\tilde{r}) + \cdots$$

$$\delta R = \tilde{R}(\tilde{r}) - R(\tilde{r}) = \varepsilon R'$$

Thus

$$\delta R = \varepsilon R.$$

Similarly, with $\tilde{\Lambda}(\tilde{r})d\tilde{r} = \Lambda(r)dr$ a density we have:

$$\Lambda\big(\tilde{r} + \varepsilon(\tilde{r})\big)\big(d\tilde{r} + \varepsilon'(\tilde{r})d\tilde{r}\big) = \tilde{\Lambda}(\tilde{r})d\tilde{r}$$

$$\Lambda(\tilde{r})d\tilde{r} + \varepsilon(\tilde{r})\Lambda'(\tilde{r}) + \varepsilon'(\tilde{r})\Lambda(\tilde{r}) + O(\varepsilon^2) = \tilde{\Lambda}(\tilde{r})d\tilde{r}$$

$$\delta\Lambda(\tilde{r}) = \tilde{\Lambda}(\tilde{r}) - \Lambda(\tilde{r}) = \left[\varepsilon\,(\tilde{r})\tilde{\Lambda}(\tilde{r})\right]'$$

Note: the integrand of the action must be a density, so from the grouping $P_\Lambda\dot{\Lambda} + P_R R$, if R is a scalar, P_R is a density, and if Λ is a density P_R is a scalar. Now to check:

$$\left\{R, \int N^r H_r\right\} = \left\{R, \int N^r P_R R'\right\} = N^r(r)R'(r) = \delta R$$

$$\left\{P_R, \int N^r H_r\right\} = \left\{P_R \int N^r P_R R'\right\} = \left(N^r(r)P_R(r)\right)' = \delta P_R$$

where N^r is taken to be the diffeomorphism parameter, like $\varepsilon(r)$ above.

Similarly

$$\left\{\Lambda, \int N^r H_r\right\} = \left\{\Lambda, \int N^r(-\Lambda P'_\Lambda)\right\} = \left(N^r(r)\Lambda(r)\right)' = \delta\Lambda$$

$$\left\{P_\Lambda, \int N^r H_r\right\} = \left\{P_\Lambda, \int N^r(-\Lambda P'_\Lambda)\right\} = N^r(r)P_\Lambda(r) = \delta P_\Lambda$$

So, H_r as chosen does generate diffeomorphisms.

Consider the new canonical variables:

$$R_{new} = R_{old}$$

where R_{New} is a scalar, and

$$P_{R_{new}} = P_{R_{old}} + \theta(r; R, \Lambda, P_\Lambda)$$

where $P_{R_{new}}$ is a density. So, the sign and placement of derivative in the new P_R, R term in H_r is as before. Consider $P_M = R^{-1}F^{-1}\Lambda P_\Lambda$. The R,F and P_Λ terms are scalars – the product of which is still a scalar. The Λ term is a density. The overall product of scalar and density is a density, so P_m is a density. From this it can be concluded that M is a scalar – since $P_M\dot{M}$ must be a density. If the canonical variable M is a scalar, the canonical pair enters the H_r expression in the same way as the R, $P_{\bar{R}}$. This enforces the relation:

$$P_R R' - \Lambda P_\Lambda' = P_M M' + P_{\bar{R}} R'$$

Now to calculate $P_{\bar{R}}$ from the relation on H_r:

$$P_{\bar{R}} = P_R - \frac{1}{R'}(\Lambda P_\Lambda' + P_M M') = \frac{1}{R'}H_r - \frac{1}{R'}P_M M'$$

$$P_M = R^{-1}F^{-1}\Lambda P_\Lambda$$

$$M = \frac{1}{2}R^{-1}P_\Lambda^2 - \frac{1}{2}\Lambda^{-2}RR'^2 + \frac{1}{2}R + \left(\frac{1}{2l^2}R^3\right)$$

$$M' = \left(\frac{1}{2}R^{-2}R' + R^{-1}P_\Lambda P_\Lambda'\right) + \left(\Lambda^{-3}\Lambda' RR'^2 - \frac{1}{2}\Lambda^{-2}(R')^3 - \Lambda^{-2}RR'R''\right) + \frac{1}{2}R' + \left(\frac{3}{2l^2}R^2R'\right)$$

Let's now compare the $\frac{1}{R'}P_M M'$ part to $H_t (\equiv H)$:

$$H = -R^{-1}P_R P_\Lambda + \frac{1}{2}R^{-2}\Lambda P_\Lambda^2 + \Lambda^{-1}RR'' - \Lambda' + \frac{1}{2}\Lambda^{-1} + X$$

To determine the possible form of X consider the contribution if a cosmological constant is present:

$$\frac{1}{6\pi}\int d^4x\sqrt{-g}\,(-2\Lambda_{cosm}) = \frac{1}{2}\int drdt\,[-N^t LR^2\Lambda_{cosm}]$$

which suggests the form

$$X = \frac{1}{2}\left(-\frac{3}{l^2}\right)(\Lambda R^2)$$

$$H = -R^{-1}\{(R')^{-1}H_r + (R')^{-1}\Lambda P_\Lambda{}'\}P_\Lambda + \frac{1}{2}R^{-2}\Lambda P_\Lambda{}^2 + \Lambda^{-1}RR''$$

$$-\Lambda^{-2}RR'\Lambda' + \frac{1}{2}\Lambda^{-1}R'^2 - \frac{1}{2}\Lambda - \frac{3}{2l^2}\Lambda R^2$$

We can now show that

$$M' = -\frac{R'}{\Lambda}\left[H + \frac{P_\Lambda}{RR'}H_r\right]$$

So,

$$P_{\bar{R}} = \frac{1}{R'}H_r - \frac{1}{R'}(R^{-1}F^{-1}\Lambda P_\Lambda)\left(-\frac{R'}{\Lambda}\right)\left(H + \frac{1}{RR'}H_r\right)$$

$$= \frac{1}{R'}\left(1 + \frac{P_\Lambda}{R^2F}\right)H_r + \frac{P_\Lambda}{RF}H$$

Recall that $F + R^{-2}P_\Lambda^2 = \left(\frac{R'}{\Lambda}\right)^2$, so:

$$P_{\bar{R}} = F^{-1}\{R^{-1}P_\Lambda H + R'\Lambda^{-2}H_r\}$$

(the same as in the zero cosmological constraint analysis).

Let's now examine if the difference in Liouville forms is an exact form:

$$\int_{-\infty}^{\infty} dr\,\{P_\Lambda(r)\delta\Lambda(r) + P_R(r)\delta R(r) - P_M(r)\delta M(r) - P_{\bar{R}}(r)\delta R(r)\}$$

$$= \delta w[\Lambda, P_\Lambda, R, P_R]$$

$$P_m = R^{-1}F^{-1}\Lambda P_\Lambda$$

$$P_{\bar{R}} = (R')^{-1}(P_R R' - \Lambda P_\Lambda) - (R')^{-1}P_m M'$$

$$\delta M = -\frac{1}{2}R^{-2}P_\Lambda^2\delta R + R^{-1}P_\Lambda\delta P_\Lambda + \Lambda^{-3}\delta\Lambda RR'^2 - \frac{1}{2}\Lambda^{-2}(R')^2\delta R$$

$$-\Lambda^{-2}RR'\delta(R') + \frac{1}{2}\delta R - \frac{3}{2l^2}R^2\delta R$$

Thus

$$P_R\delta R - P_{\bar{R}}\delta R = (R')^{-1}(\Lambda P_m^{m'})\delta R$$

Consider the $\frac{M'}{R'}$ part:

$$\frac{M'}{R'} = -\frac{1}{2}R^{-2}P_\Lambda^2 + (R')^{-1}R^{-1}P_\Lambda P_\Lambda' + \Lambda^{-3}\Lambda' RR' - \frac{1}{2}\Lambda^{-2}(R')^2 - \Lambda^{-2}RR'' + \frac{1}{2} - \frac{3}{2l^2}R^2$$

which we can use with

$$P_R\delta R - P_{\bar{R}}\delta R - P_m\delta M = (R')^{-1}\Lambda P_\Lambda'\delta R + P_M\left(\frac{M'}{R'}\delta R - \delta M\right)$$

$$= (R')^{-1} \Lambda P'_\Lambda \delta R$$

$$+ P_M \left[\begin{array}{c} R^{-1} P_\Lambda \left(\frac{P_\Lambda}{R'} \delta R - \delta P_\Lambda \right) + \Lambda^{-3} R R'^2 \left(\frac{\Lambda'}{R'} \delta R - \delta \Lambda \right) \\ - \Lambda^{-2} R R' \left(\frac{R''}{R'} \delta R - \delta R' \right) \end{array} \right]$$

Putting everything together:

$$P_R \delta R + P_\Lambda \delta \Lambda - P_{\bar{R}} \delta R - P_m \delta M = P_\Lambda \delta \Lambda + (R')' \Lambda P'_\Lambda \delta R$$

$$+ \left\{ \frac{(R^{-1} \Lambda P_\Lambda)}{\left(\frac{R'}{\Lambda}\right)^2 - \left(\frac{P_\Lambda}{R}\right)^2} \right\} \left\{ \left(\frac{P_\Lambda}{R} \right) \left(\frac{P'_\Lambda}{R'} \delta R - \delta P_\Lambda \right) \right.$$

$$\left. + (R \Lambda^{-1}) \left(\frac{R'}{\Lambda} \right)^2 \left(\frac{\Lambda'}{R'} \delta R - \delta \Lambda \right) \right\}$$

$$- (R\Lambda) \left(\frac{R'}{\Lambda} \right) \left(\frac{R''}{R'} \delta R - \delta R \right)$$

$$= P_\Lambda \delta \Lambda + \left\{ \left(\frac{R'}{\Lambda} \right)^2 - \left(\frac{P_\Lambda}{R} \right)^2 \right\}^{-1} \left\{ \left(\frac{P'_\Lambda}{R'} \right) \Lambda \delta R \left[\left(\frac{R'}{\Lambda} \right)^2 - \left(\frac{P'_\Lambda}{R'} \right)^2 \right] \right.$$

$$\left. + \left(\frac{P'_\Lambda}{R'} \right) \Lambda \delta R \left(\frac{P'_\Lambda}{R'} \right)^2 \right\}$$

$$- \left(\frac{P'_\Lambda}{R'} \right)^2 \Lambda \delta R - \left(\frac{R'}{\Lambda} \right)^2 P \delta \Lambda + P_\Lambda \left(\frac{R'}{\Lambda} \right) \delta R' - P_\Lambda \left(\frac{R'}{\Lambda} \right) \frac{R''}{R'} \delta R$$

$$+ \left(\frac{R'}{\Lambda} \right) \frac{P_R \Lambda'}{R'}$$

$$= P_\Lambda \delta \Lambda + F^{-1} \left\{ \left(\frac{R'}{\Lambda} \right)^2 \Lambda \left(\frac{P'_\Lambda}{R'} \right) \delta R + \left(\frac{R'}{\Lambda} \right) P_\Lambda \left[\delta R' - \frac{R''}{R'} \delta R \right] \right\}$$

$$- \left[\left(\frac{P_\Lambda}{R} \right)^2 \Lambda \delta P_\Lambda + \left(\frac{R'}{\Lambda} \right) P_\Lambda \delta \Lambda \right] + \left(\frac{R'}{\Lambda} \right)^2 \frac{P_R \Lambda'}{R'}$$

$$= P_\Lambda \delta \Lambda + \frac{1}{[RR' - \Lambda P_\Lambda][RR' + \Lambda P_\Lambda]} \left\{ R^2 \Lambda R' P'_\Lambda \delta R \right.$$

$$+ R^2 R' P_\Lambda \left[\delta R' - \frac{R''}{R'} \delta R \right] - (\Lambda P_\Lambda)^2 \Lambda \delta P_\Lambda \, R^2 R'^{P_\Lambda \Lambda'} \delta R$$

$$\left. - (RR')^2 P_\Lambda \delta \Lambda \right\}$$

$$= P_\Lambda \delta\Lambda + \frac{RR'}{(RR')^2 - (\Lambda P_\Lambda)}\left\{\Lambda P_\Lambda' \delta\Lambda + (\Lambda P_\Lambda) R\left[\delta R' - \frac{R''}{R'}\delta R\right] \right.$$
$$\left. - (RR')[P_\Lambda \delta\Lambda + \Lambda \delta P_\Lambda]\right\}$$
$$+ \frac{1}{(RR')^2 - (\Lambda P_\Lambda)^2}\{(RR')^2(+\Lambda P_\Lambda) - (\Lambda P_\Lambda)^2\} + \frac{R^2 R' P_\Lambda \Lambda' \delta R}{(RR')^2 - (\Lambda P_\Lambda)^2}$$
$$= \delta(\Lambda P_\Lambda) + \frac{RR'}{[RR' - \Lambda P_\Lambda][RR' + \Lambda P_\Lambda]}\left\{\Lambda P_\Lambda' R f R - (\Lambda P_\Lambda)\frac{RR''}{R'}\delta R \right.$$
$$\left. - (RR')\delta(\Lambda P_\Lambda)\right\}$$
$$+ \frac{R^2 R'(\Lambda P_\Lambda)}{(RR')^2 - (\Lambda P_\Lambda)^2}\delta R' + \frac{R P_\Lambda \Lambda' \delta R}{(RR')^2 - (\Lambda P_\Lambda)^2}$$

$$P_R \delta R + P_\Lambda \delta\Lambda - P_{\bar{R}} \delta R - P_m \delta M =$$
$$\delta(\Lambda P_\Lambda)$$
$$+ \frac{RR'}{(RR')^2 - (\Lambda P_\Lambda)^2}\left\{\begin{matrix} [(\Lambda P_\Lambda)\delta(RR') - (\Lambda P_\Lambda) R' \delta R] - (RR')\delta(\Lambda P_\Lambda) \\ + \Lambda P_\Lambda' R \delta R - (\Lambda P_\Lambda)\frac{RR''}{R'}\delta R + (P_\Lambda \Lambda') R \delta R \end{matrix}\right\}$$
$$= \delta(\Lambda P_\Lambda) + \frac{1}{2} RR' \delta \ln\left|\frac{RR' - \Lambda P_\Lambda}{RR' + \Lambda P_\Lambda}\right| + \frac{RR\prime}{(RR\prime)^2 - (\Lambda P_\Lambda)^2}\left\{(\Lambda P_\Lambda' + P_\Lambda \Lambda') R \delta R - (\Lambda P_\Lambda)\left[R' + \frac{RR\prime}{R\prime}\right]\delta R\right\}$$
$$= \delta(\Lambda P_\Lambda) + \frac{1}{2} RR' \delta \ln\left|\frac{RR' - \Lambda P_\Lambda}{RR' + \Lambda P_\Lambda}\right| + \frac{R\delta R}{(RR\prime)^2 - (\Lambda P_\Lambda)^2}\{(RR')(\Lambda P_\Lambda)' - (\Lambda P_\Lambda)(RR')'\}$$
$$= \delta(\Lambda P_\Lambda) + \frac{1}{2} RR' \delta\left\{\ln\left|\frac{RR' - \Lambda P_\Lambda}{RR' + \Lambda P_\Lambda}\right|\right\} + \frac{1}{2} R\delta R\left\{\left|\frac{RR' + \Lambda P_\Lambda}{RR' - \Lambda P_\Lambda}\right|\right\}'$$
$$= \delta(\Lambda P_\Lambda) + \delta\left(\frac{1}{2} RR' \ln\left|\frac{RR' - \Lambda P_\Lambda}{RR' + \Lambda P_\Lambda}\right|\right) + \left(\frac{1}{2} R\delta R \ln\left|\frac{RR' + \Lambda P_\Lambda}{RR' - \Lambda P_\Lambda}\right|\right)'$$
$$= \delta\left(\Lambda P_\Lambda + \frac{1}{2} RR' \ln\left|\frac{RR' - \Lambda P_\Lambda}{RR' + \Lambda P_\Lambda}\right|\right) + \left(\frac{1}{2} R\delta R \ln\left|\frac{RR' + \Lambda P_\Lambda}{RR' - \Lambda P_\Lambda}\right|\right)'$$

To obtain the exact form the integration of the total divergence must vanish, i.e., the boundary terms:

$$\left(\frac{1}{2} R\delta R \ln\left|\frac{RR' + \Lambda P_\Lambda}{RR' - \Lambda P_\Lambda}\right|\right)$$

must vanish.

To determine if the term vanishes at the boundary at asymptotic spatial infinity, we must examine the fall-off conditions for the various parameters at spatial infinity…

B.3 Fall-off as $r \to \infty$

Let's now consider the derivation of fall-off conditions at spatial infinity for spherically symmetric, asymptotically Anti-deSitter (AdS) spacetimes. The boundary conditions at spatial infinity are devised so as to satisfy the following three requirements (following [83]):
(i) They should contain the Schwarzschild – AdS metric (a reasonable starting point since all isolated systems in spherically symmetric Ads spacetime will approach such a metric at infinity [77]).
(ii) They should be invariant under the Ads group O(3,2).
(iii) They should make the surface integrals associated with the generators of O(3,2) finite.

Here are the steps in determining the fall-off conditions:
(1) Express the Schwarzschild–AdS metric in terms of its leading order r^{-1} deviation from the Ads metric.
(2) The asymptotic symmetry group is shown by [83] to be the AdS group. Act on the asymptotic structure of the Schwarzschild–AdS metric with O(3,2) in all possible ways to determine the general expression for the asymptotic structure.
(3) Take the generalized metric deviations from AdS and calculate leading order terms in the extrinsic curvature and, consequently, the conjugate momenta.
(4) Later, consider evaluation of the general metric deviation by use of the equations of motion and self-consistency.

The Ads metric:

$$ds_0^2 = -\left(1+\left(\frac{r}{l}\right)^2\right)dt^2 + \left(1+\left(\frac{r}{l}\right)^2\right)^{-1} dr^2 + r^2 d\Omega^2$$

The Schwarzschild–Ads metric:

$$ds^2 = -\left(1-\frac{2M}{r}+\left(\frac{r}{l}\right)^2\right)dt^2 + \left(1-\frac{2M}{r}+\left(\frac{r}{l}\right)^2\right)^{-1} dr^2 + r^2 d\Omega^2$$

We want $ds^2 = ds_0^2 + h_{\mu v}dx^\mu dx^v$ with $h_{\mu v}$ at leading order:

$$h_{tt} = \left(\frac{2M}{r}\right), h_{rr} = \left(1-\frac{2M}{r}+\left(\frac{r}{l}\right)^2\right)^{-1} - \left(1+\left(\frac{r}{l}\right)^2\right)^{-1}$$

with no other deviation terms. The leading order term in h_{rr} is:

$$h_{rr} = \left(\frac{2M}{r}\right)\left(1-\frac{2M}{r}+\left(\frac{r}{l}\right)^2\right)^{-1}\left(1+\left(\frac{r}{l}\right)^2\right)^{-1}$$

$$= \left(\frac{2M}{r}\right)\left(\frac{l}{r}\right)^4\left(1+\left(\frac{l}{r}\right)^2\left(1-\frac{2M}{r}\right)\right)\left(1+\left(\frac{l}{r}\right)^2\right)^{-1}$$

$$= \left(\frac{2ml^4}{r^5}\right) + O(r^{-7})$$

So, $h_{tt} = \left(\frac{2M}{r}\right)$ and $h_{rr} = \left(\frac{2ml^4}{r^5}\right) + O(r^{-7})$ the rest zero. Now to act on this with O(3,2) mod O(3) (this should eliminate the rotations so as to maintain the restriction to spherical symmetry). The AdS Killing vectors are given in Book 5 [5] and in [Appendix A of [83]]. Denote the Killing vectors by $\mathfrak{z}_a^\mu$ where a is an index ranging from 1 to 10 for the 10 independent AdS Killing vectors (μ is a concrete spacetime index).

One acts on the perturbations $h_{\mu\nu}$ with O(3,2) by Lie dragging them along the Killing vector fields to obtain the most general form. To be certain of this process, lets first confirm that the Kerr-AdS deviation terms give rise the general form.

Kerr-AdS leading order deviations:

$$ds^2 = -f dt^2 + f^{-1} dr^2 + r^2 d\Omega^2, f = \left(1+\left(\frac{r}{l}\right)^2\right)$$

$h_{tt} = \frac{2m}{r}(1-\alpha^2\sin^2\theta)^{-5/2} + O(r^{-3}) = r^{-1} f_{tt}^1(\theta) + O(r^{-3})$, $(\alpha = a/l)$,

with 'a' related to angular momentum.

$h_{t\varphi} = -\frac{2am\sin^2\theta}{r}(1-\alpha^2\sin^2\theta)^{-5/2} + O(r^{-3}) = r^{-1} f_{t\varphi}^1(\theta) + O(r^{-3})$

$h_{\varphi\varphi} = \frac{2am\sin^2\theta}{r}(1-\alpha^2\sin^2\theta)^{-5/2} + O(r^{-3}) = r^{-1} f_{\varphi\varphi}^1(\theta) + O(r^{-3})$

$h_{rr} = \frac{2ml^4}{r^4}(1-\alpha^2\sin^2\theta)^{-3/2} + O(r^{-7}) = r^{-5} f_{rr}^1(\theta) + O(r^{-7})$

$h_{\theta r} = \frac{2ml^2a^2}{r^4}(1-\alpha^2\sin^2\theta)^{-5/2}\sin\theta\cos\theta + O(r^{-6}) = r^{-4} f_{\theta r}^1(\theta) + O(r^{-6})$

$h_{\theta\theta} = \frac{2ma^4}{r^3}(1-\alpha^2\sin^2\theta)^{-7/2}\sin\theta\cos\theta + O(r^{-5}) = r^{-3} f_{\theta\theta}^1(\theta) + O(r^{-5})$

Now consider the asymptotic form of the Killing vectors

$$\mathfrak{z}_1 = l\frac{\partial}{\partial t} = \lim_{r\to\infty}\mathfrak{z}_1 = \mathfrak{z}_1^\infty$$

$$\mathfrak{z}_2 = -r\sin\left(\frac{t}{l}\right)\sin\theta\cos\varphi\left(1+\left(\frac{r}{l}\right)^2\right)^{-1/2}\frac{\partial}{\partial t}$$

$$+\left(1+\left(\frac{r}{l}\right)^2\right)^{1/2}\cos\left(\frac{t}{l}\right)\sin\theta\cos\varphi\, l\frac{\partial}{\partial r}$$

$$+\left(\frac{r}{l}\right)^{-1}\left(1+\left(\frac{r}{l}\right)^2\right)^{1/2}\cos\left(\frac{t}{l}\right)\left(\cos\theta\cos\varphi\frac{\partial}{\partial\theta}-\frac{\sin\varphi}{\sin\theta}\frac{\partial}{\partial\varphi}\right)$$

$$\mathfrak{z}_2^\infty = \lim_{r\to\infty}\mathfrak{z}_1^\infty = -\sin\left(\frac{r}{l}\right)\sin\theta\cos\varphi\, l\frac{\partial}{\partial t}+\cos\left(\frac{t}{l}\right)\sin\theta\cos\varphi\, r\frac{\partial}{\partial r}$$

$$+\cos\left(\frac{t}{l}\right)\left(\cos\theta\cos\varphi\frac{\partial}{\partial t}-\frac{\sin\varphi}{\sin\theta}\frac{\partial}{\partial t}\right)$$

$$\mathfrak{z}_3 = -r\sin\left(\frac{t}{l}\right)\sin\theta\sin\varphi\left(1+\left(\frac{r}{l}\right)^2\right)^{-1/2}\frac{\partial}{\partial t}$$

$$+\left(1+\left(\frac{r}{l}\right)^2\right)^{1/2}\cos\left(\frac{t}{l}\right)\sin\theta\sin\varphi\, l\frac{\partial}{\partial r}$$

$$+\left(\frac{r}{l}\right)^{-1}\left(1+\left(\frac{r}{l}\right)^2\right)^{1/2}\cos\left(\frac{r}{l}\right)\left(\cos\theta\sin\varphi\frac{\partial}{\partial\theta}+\frac{\cos\varphi}{\sin\theta}\frac{\partial}{\partial\varphi}\right)$$

$$\mathfrak{z}_3{}^\infty = -\sin\left(\frac{r}{l}\right)\sin\theta\sin\varphi\, l\frac{\partial}{\partial t}+\cos\left(\frac{1}{l}\right)\sin\theta\sin\varphi\, r\frac{\partial}{\partial r}$$

$$+\cos\left(\frac{1}{l}\right)\left(\cos\theta\cos\varphi\frac{\partial}{\partial\theta}-\frac{\sin\varphi}{\sin\theta}\frac{\partial}{\partial\varphi}\right)$$

$$\mathfrak{z}_4{}^{\square} = -r\sin\left(\frac{1}{l}\right)\cos\left(1+\left(\frac{r}{l}\right)^2\right)^{-1/2}\frac{\partial}{\partial t}$$

$$+\left(1+\left(\frac{r}{l}\right)^2\right)^{1/2}\cos\left(\frac{t}{l}\right)\cos\theta\, l\frac{\partial}{\partial r}$$

$$-\frac{l}{r}\left(1+\left(\frac{r}{l}\right)^2\right)^{1/2}\cos\left(\frac{1}{r}\right)\sin\theta$$

$$\mathfrak{z}_4{}^\infty = -\sin\left(\frac{t}{l}\right)\cos\theta\, l\frac{\partial}{\partial t}+\cos\left(\frac{t}{l}\right)\cos\theta\, r\frac{\partial}{\partial r}-\cos\left(\frac{1}{l}\right)\sin\theta\frac{\partial}{\partial\theta}$$

$$\mathfrak{z}_5^{\square} = r\cos\left(\frac{t}{l}\right)\cos\theta\left(1+\left(\frac{r}{l}\right)^2\right)^{-1/2}\frac{\partial}{\partial t}$$

$$+\left(1+\left(\frac{r}{l}\right)^2\right)^{1/2}\sin\left(\frac{1}{l}\right)\cos\theta\, l\frac{\partial}{\partial r}$$

$$-\frac{l}{r}\left(1+\left(\frac{r}{l}\right)^2\right)^{1/2}\sin\left(\frac{1}{l}\right)\sin\theta$$

$$\mathfrak{z}_5^\infty = \cos\left(\frac{1}{l}\right)\cos\theta\, l\frac{\partial}{\partial t}+\sin\left(\frac{t}{l}\right)\cos\theta\, r\frac{\partial}{\partial r}-\sin\left(\frac{t}{l}\right)\sin\theta\frac{\partial}{\partial\theta}$$

$$\mathfrak{z}_6 = r\cos\left(\frac{t}{l}\right)\sin\theta\cos\varphi\left(1+\left(\frac{r}{l}\right)^2\right)^{-1/2}\frac{\partial}{\partial t}$$
$$+\left(1+\left(\frac{r}{l}\right)^2\right)^{1/2}\sin\left(\frac{t}{l}\right)\sin\theta\frac{\partial}{\partial r}$$
$$+\left(\frac{r}{l}\right)^{-1}\left(1+\left(\frac{r}{l}\right)^2\right)^{1/2}\sin\left(\frac{t}{l}\right)\left(\cos\theta\cos\varphi\frac{\partial}{\partial\theta}-\frac{\sin\varphi}{\sin\theta}\frac{\partial}{\partial\varphi}\right)$$
$$\mathfrak{z}_6^{\infty} = \cos\left(\frac{t}{l}\right)\sin\theta\cos\varphi\, l\frac{\partial}{\partial t}+\sin\left(\frac{t}{l}\right)\sin\theta\cos\varphi\, r\frac{\partial}{\partial r}$$
$$+\sin\left(\frac{t}{l}\right)\left(\cos\theta\cos\varphi\frac{\partial}{\partial\theta}-\frac{\sin\varphi}{\sin\theta}\frac{\partial}{\partial\varphi}\right)$$
$$\mathfrak{z}_7 = rcos\left(\frac{t}{l}\right)\sin\theta\sin\varphi\left(1+\left(\frac{r}{l}\right)^2\right)^{-1/2}\frac{\partial}{\partial t}$$
$$+\left(1+\left(\frac{r}{l}\right)^2\right)^{1/2}\sin\left(\frac{t}{l}\right)\sin\theta\sin\varphi\, l\frac{\partial}{\partial r}$$
$$+\left(\frac{r}{l}\right)^{-1}\left(1+\left(\frac{r}{l}\right)^2\right)^{1/2}\sin\left(\frac{t}{l}\right)\left(\cos\theta\sin\varphi\frac{\partial}{\partial\theta}+\frac{\cos\varphi}{\sin\theta}\frac{\partial}{\partial\varphi}\right)$$
$$\mathfrak{z}_7^{\infty} = \cos\left(\frac{t}{l}\right)\sin\theta\sin\varphi\, l\frac{\partial}{\partial t}+\sin\left(\frac{t}{l}\right)\sin\theta\sin\varphi\, r\frac{\partial}{\partial r}$$
$$+\sin\left(\frac{t}{l}\right)\left(\cos\theta\sin\varphi\frac{\partial}{\partial\theta}+\frac{\cos\varphi}{\sin\theta}\frac{\partial}{\partial\varphi}\right)$$
$$\mathfrak{z}_8^{\infty} = \mathfrak{z}_8 = -\sin\varphi\frac{\partial}{\partial\theta}-\cot\theta\cos\varphi\frac{\partial}{\partial\varphi}$$
$$\mathfrak{z}_9^{\infty} = \mathfrak{z}_9 = \cos\varphi\frac{\partial}{\partial\theta}-\cot\theta\sin\varphi\frac{\partial}{\partial\varphi}$$
$$\mathfrak{z}_{10}^{\infty} = \mathfrak{z}_{10} = \frac{\partial}{\partial\varphi}$$

Consider

$$L_{\mathfrak{z}_a}(g_{\mu v}) = \mathfrak{z}_a^{\alpha}\nabla_{\alpha}g_{\mu v}+g_{\mu\alpha}\nabla_v\mathfrak{z}_a^{\alpha}+g_{\alpha v}\nabla_{\mu}\mathfrak{z}_a^{\alpha}$$

where $g_{\mu v} = (AdS)g_{\mu v}+h_{\mu v}$ and $\mathfrak{z}_a$ is a Killing vector for the Kerr-AdS $g_{\mu v}$ metric, and the asymptotic form of the $\mathfrak{z}_a^{\prime s}$ are used to obtain leading order effects.

$h_{\mu v}^{(g)} \equiv$ the general form for the deviations

$$h_{\mu v}^{(g)} = \Sigma_a\left[\underbrace{\mathfrak{z}_a^{\infty}\cdot\nabla\left(h_{\mu v}^{(k-Ads)}\right)}_{I}+\underbrace{h_{\mu v}^{(k-Ads)}\nabla_v\mathfrak{z}_a^{\alpha}}_{II}+\underbrace{h_{\mu v}^{(k-Ads)}\nabla_{\mu}\mathfrak{z}_a^{\alpha}}_{III}\right]\omega^a$$

The nonzero $h_{\mu\nu}^{(k-Ads)}$ are $h_{tt}, h_{t\varphi}, h_{\varphi\varphi}, h_{rr}, h_{\theta r}, h_{\theta\theta}$. Term I directly generalizes these expressions while terms I and III ensure mutual consistency. Terms II and III also lead to new nonzero $h_{\mu\nu}^{\prime s}$.

Consider $h_{tt}^{(g)}$, in the derivation the order of r^{-1} dependence is important as well as the possible dependence on the θ, φ, t variables. This is all the information that need be maintained.

$h_{tt}^{(g)} = \sum_a w^a \{\mathfrak{z}_a^{\infty} \cdot \nabla(r^{-1} f_{tt}^1(\theta) + O(r^{-3})) + h_{ta} \nabla_t \mathfrak{z}_a^{\alpha,\infty} + h_{\alpha t} \nabla_t \mathfrak{z}_a^{\alpha,\infty}\}$

Break down the contribution in terms of Killing vectors $\mathfrak{z}_a$, $a = \{1 \dots 10\}$:

$(a = 1)$ No contribution

$(a = 2)$ $I : \cos\left(\frac{t}{l}\right) \sin\theta \cos\varphi \, r\left(-r^{-2} f_H^1(\theta) + O(r^{-4})\right)$

$+ \cos\left(\frac{t}{l}\right) \cos\theta \sin\varphi \left(r^{-1} \frac{\partial}{\partial\theta} f_H^1(\theta)\right)$

$+\mathfrak{z}_2^{\mu} \Gamma_{\mu t}^{\nu} h_{\nu t}$

II: ……

This becomes much too cumbersome, need to extract only the information necessary from the Killing vector fields.

Expressing the asymptotic Killing vectors in terms of functional dependencies and leading order r^{-1} dependencies:

$\mathfrak{z}_1^{\infty} = l \frac{\partial}{\partial t}$

$\mathfrak{z}_2^{\infty} = (f_{2t}(t,\theta,\varphi) + O(r^{-2})) \frac{\partial}{\partial t} + (f_2(t,\theta,\varphi) + O(r^{-1})) \frac{\partial}{\partial r}$

$+(f_{2\theta}(t,\theta,\varphi) + O(r^{-2})) \frac{\partial}{\partial\theta} + \left(f_{2\varphi}(t,\theta,\varphi) + O(r^{-2})\right) \frac{\partial}{\partial\varphi}$

$\mathfrak{z}_3^{\infty} \to \mathfrak{z}_2^{\infty}$ (has same form)

$\mathfrak{z}_4^{\infty} = (f_{4t}(t,\theta) + O(r^{-2})) \frac{\partial}{\partial t} + (f_{4r}(t,\theta) \cdot r + O(r^{-1})) \frac{\partial}{\partial r}$

$+(f_{4\theta}(t,\theta) + O(r^{-2})) \frac{\partial}{\partial\theta} + \left(f_{4\varphi}(t,\theta) + O(r^{-2})\right) \frac{\partial}{\partial\varphi}$

(also same form as $\mathfrak{z}_2^{\infty}$ but no φ dependence).

$\mathfrak{z}_5^{\infty} \to$ same form as $\mathfrak{z}_4^{\infty}$

$\mathfrak{z}_6^{\infty}, \mathfrak{z}_7^{\infty} \to$ same functional dependencies as $\mathfrak{z}_{2,3}^{\infty}$

$\mathfrak{z}_8^{\infty} = f_{8\theta}(\varphi) \frac{\partial}{\partial\theta} + f_{8\varphi}(\theta,\varphi) \frac{\partial}{\partial\varphi}$

$\mathfrak{z}_9^{\infty} \to \mathfrak{z}_8^{\infty}$ (same form)

$\mathfrak{z}_{10}^{\infty} = \frac{\partial}{\partial\varphi}$

So, evaluating the general perturbations it is only necessary to consider the form:

$$\mathfrak{z}_\infty = [f_t(t,\theta,\varphi) + O(r^{-2})]\frac{\partial}{\partial t} + [f_r(t,\theta,\varphi) + O(r^{-2})]r\cdot\frac{\partial}{\partial r}$$
$$+\left(f_\theta(t,\theta,\varphi) + O(r^{-2})\right)\frac{\partial}{\partial\theta} + \left(f_\varphi(t,\theta,\varphi) + O(r^{-2})\right)\frac{\partial}{\partial\varphi}$$

Now its possible to write:

$$h_{\mu v}^{(g)} = \mathfrak{z}_\infty^\alpha \nabla_\alpha\left(h_{\mu v}^{(k-Ads)}\right) + h_{\mu\alpha}^{(k-Ads)}\nabla_v \mathfrak{z}_\infty^\alpha + h_{\alpha v}^{(k-Ads)}\nabla_\mu \mathfrak{z}_\infty^\alpha$$
$$= \mathfrak{z}^\alpha\partial_\alpha\left(h_{\mu v}\right) + \mathfrak{z}^\alpha\left(\Gamma_{\alpha\mu}^\beta h_{\beta v} + \Gamma_{\alpha v}^\beta h_{\mu\beta}\right) + h_{\mu\alpha}\partial_v\mathfrak{z}^l - h_{\mu\alpha}\Gamma_{v\beta}^\alpha\mathfrak{z}^\beta +$$
$$h_{\alpha v}\partial_\mu\mathfrak{z}^\alpha - h_{\alpha v}\Gamma_{v\beta}^\alpha\mathfrak{z}^\beta$$
$$h_\mu^{(g)} = \mathfrak{z}^\alpha\partial_\alpha\left(h_{\mu v}\right) + h_{\mu\alpha}\partial_v\mathfrak{z}^\alpha + h_{\alpha v}\partial_\mu\mathfrak{z}^\alpha$$
$$h_{tt}^{(g)} = \mathfrak{z}^\alpha\partial_\alpha\left(r^{-1}f_{tt}^1(\theta) + O(r^{-3})\right) + h_{t\alpha}\partial_t\mathfrak{z}^\alpha + h_{\alpha t}\partial_t\mathfrak{z}^\alpha$$
$$= \left(f_\theta - f_r + O(r^{-2})\right)\left(r^{-1}f_{tt}^1(\theta) + O(r^3)\right) + 2h_{tt}\partial_t\mathfrak{z}^\alpha + 2h_{t\varphi}\partial_t\mathfrak{z}^\alpha$$
$$= \left(r^{-1}g(t,\theta,\varphi) + O(r^{-3})\right) + \left(r^{-1}f_{tt}{}^1(\theta) + O(r^{-3})\right)\left(\frac{\partial}{\partial t}f_t\right) +$$
$$\left(r^{-1}f_{t\varphi}^1(\theta) + O(r^{-3})\left(\frac{\partial}{\partial t}f_\varphi\right)\right)$$
$$\boxed{h_{tt}^{(g)} = r^{-1}f_{tt}{}^{\square}(t,\theta,\varphi) + O(r^{-3})}$$

$$h_{t\varphi}^{(g)} = \mathfrak{z}^\alpha\partial_\alpha\left(r^{-1}f_{t\varphi}'(\theta)O(r^{-3})\right) + h_{t\alpha}\partial_\varphi\mathfrak{z}^\alpha + h_{\alpha\varphi}\partial_t\mathfrak{z}^\alpha$$
$$= \left(r^{-1}g(t,\theta,\varphi) + O(r^{-3})\right) +$$
$$\underbrace{h_{tt}\partial_\varphi\mathfrak{z}^t + h_{\varphi\varphi}\partial_\varphi\mathfrak{z}^\varphi + h_{t\varphi}\partial_t\mathfrak{z}^\alpha + h_{tt}\partial_\varphi\mathfrak{z}^t}_{\left(r^{-1}k(t,\theta,\varphi)+O(r^{-3})\right)\left(j(t,\theta\varphi)+O(r^2)\right)}$$
$$\boxed{h_{t\varphi}^{(g)} = r^{-1}f_{t\varphi}{}^{\square}(t,\theta,\varphi) + O(r^{-3})}$$

$$h_{t\varphi}^{(g)} = h_{t\alpha}\partial_\theta\mathfrak{z}^\alpha + h_{\alpha\theta}\partial_t\mathfrak{z}^\alpha$$
$$= \underbrace{h_{tt}\partial_\theta\mathfrak{z}^t + h_{1\varphi}\partial_\theta\mathfrak{z}^\varphi}_{\left(r^{-1}h(t,\theta,\varphi)+O(r^{-3})\right)} +$$
$$\underbrace{h_{r\theta}\partial_t\mathfrak{z}^r + h_{\theta\varphi}\partial_\theta\mathfrak{z}^\varphi}$$
$$\left(r^{-4}f_{r\theta}^1+O(r^{-6})\right)\left(\frac{\partial}{\partial t}f_r(t,\theta,\varphi)+O(r)\right)+\left(r^{-3}f_{\theta\theta}^1-O(r^{-5})\right)\left(\frac{\partial}{\partial t}f_\theta\right)$$
$$\boxed{h_{t\varphi}^{(g)} = r^{-1}f_{t\varphi}{}^{\square}(t,\theta,\varphi) + O(r^{-3})}$$

$$h_{t\varphi}^{(g)} = \mathfrak{z}^\alpha\partial_\alpha(h_{tr}) + h_{tr}\partial_r\mathfrak{z}^\alpha + h_{\alpha r}\partial_t\mathfrak{z}^\alpha$$
$$h_{tt}\partial_r\mathfrak{z}^t + h_{t\varphi}\partial_r\mathfrak{z}^\varphi + h_{\theta r}\partial_r\mathfrak{z}^\theta + h_{rr}\partial_{rr}\partial_t\mathfrak{z}^r$$
$$= \left(r^{-1}f_{tt}'(\theta)\right) + \left(O(r^{-3})\right) + \left(r^{-1}\left[f_{\varphi\varphi}^1 + f_{\theta\theta}^1\right] + \left(O(r^{-3})\right)\right)$$

$$+\left(r^{-4}f_{\theta r}{}^{1}(t,\theta,\varphi)+O(r^{-6})\right)\left(\frac{\partial f^{6}}{\partial t}\right)+\left(r^{-5}f_{rr}{}^{1}(t,\theta,\varphi)+O(r^{-7})\right)\left(\frac{\partial}{\partial t}[rf_{r}]\right)$$

$$\boxed{h_{tr}^{(g)}=r^{-4}f_{tr}{}^{1}(t,\theta,\varphi)+O(r^{-6})}$$

$$h_{rr}^{(g)}=ʒ^{\alpha}\partial_{\alpha}\left(r^{-5}f_{rr}{}^{1}(\theta)+O(r^{-7})\right)+h_{r\alpha}\partial_{r}ʒ^{\alpha}+h_{\alpha r}\partial_{r}ʒ^{\alpha}$$

$$\left(r^{-5}g(t,\theta,\varphi)+O(r^{-7})\right)+\left(r^{-5}f_{rr}{}^{1}+O(r^{-7})\right)\left(f_{r}+O(r^{2})\right)+\left(r^{-5}f_{r\theta}{}^{1}(\theta)+O(r^{-6})\right)O(r^{-3})$$

$$\boxed{h_{rr}^{(g)}=r^{-5}f_{tr}{}^{1}(t,\theta,\varphi)+O(r^{-7})}$$

$$h_{rr}^{(g)}=ʒ^{\alpha}\partial_{\alpha}(h_{r\theta})+h_{r\alpha}\partial_{\theta}ʒ^{\alpha}+h_{\alpha\theta}\partial_{r}ʒ^{\alpha}$$

$$h_{r\theta}\partial_{\theta}ʒ^{\theta}+h_{rr}\partial_{\theta}ʒ^{r}+h_{r\theta}\partial_{r}ʒ^{r}+h_{\theta\theta}\partial_{r}ʒ^{\theta}$$

$$\boxed{h_{r\theta}^{(g)}=r^{-4}f_{r\theta}(t,\theta,\varphi)+O(r^{-6})}$$

$$\boxed{h_{r\varphi}^{(g)}=r^{-4}f_{r\varphi}(t,\theta,\varphi)+O(r^{-6})}$$

$$h_{\theta\varphi}^{(g)}=ʒ^{\alpha}\partial_{\alpha}\left(h_{\varphi\varphi}\right)+2h_{\varphi\alpha}\partial_{\varphi}ʒ^{\alpha}$$

$h_{\varphi\alpha}\partial_{\varphi}ʒ^{\varphi}$ - same leading order structure $h_{\varphi t}\partial_{\varphi}ʒ^{t}$

$$\boxed{h_{\varphi\varphi}^{(g)}=r^{-1}f_{r\theta}(t,\theta,\varphi)+O(r^{-3})}$$

$$h_{\varphi\varphi}^{(g)}=h_{\theta\alpha}\partial_{\varphi}ʒ^{\alpha}+h_{\alpha\varphi}\partial_{\theta}ʒ^{\alpha}$$

$$=\frac{h_{\theta\theta}\partial_{\varphi}ʒ^{\theta}}{O(r^{-3})}+\frac{h_{\theta r}\partial_{\varphi}ʒ^{r}}{O(r^{-3})}+\frac{h_{\theta t}\partial_{\varphi}ʒ^{t}}{O(r^{-1})}+\frac{h_{\varphi\varphi}\partial_{\varphi}ʒ^{\varphi}}{O(r^{-1})}+\frac{h_{t\varphi}\partial_{\theta}ʒ^{t}}{O(r^{-1})}+\frac{h_{r\varphi}\partial_{\theta}ʒ^{r}}{O(r^{-3})}$$

$$\boxed{h_{\varphi\varphi}^{(g)}=r^{-1}f_{\theta\theta}(t,\theta,\varphi)+O(r^{-3})}$$

$$h_{\varphi\varphi}^{(g)}=ʒ^{\alpha}(h_{\theta\theta})+2h_{\theta\alpha}\partial_{\theta}ʒ^{\alpha}$$

$$=\frac{ʒ^{\theta}\partial_{\alpha}h_{\theta\theta}}{O(r^{-3})}+\frac{2h_{\theta\varphi}\partial_{\varphi}ʒ^{\varphi}}{O(r^{-1})}+\frac{2h_{\theta\theta}\partial_{\theta}ʒ^{\theta}}{O(r^{-3})}+\frac{2h_{\theta t}\partial_{\theta}ʒ^{t}}{O(r^{-1})}+\frac{2h_{\theta r}\partial_{\theta}ʒ^{r}}{O(r^{-3})}$$

$$\boxed{h_{\theta\theta}^{(g)}=r^{-1}f_{\theta\theta}(t,\theta,\varphi)+O(r^{-3})}$$

Now it is finally possible to consider the Schwarzschild –Ads problem. The S-AdS leading order derivations from AdS $ds_0^2 - fdt^2 + f^{-1}dr^{-1} + r^2 d\Omega^2, f = \left(1 + \left(\frac{r}{l}\right)^2\right)$ are simply:

$h_{tt}^{S-Ads} = \left(\frac{2M}{r}\right)$ (exact)

$h_{rr}^{S-Ads} = \left(\frac{2Ml^4}{r^{-5}}\right) + O(r^{-7})$

As before, the general from for the asymptotic Killing vector field is:

$$\mathfrak{z}_\infty = [f_t(t,\theta,\varphi) + O(r^{-2})]\frac{\partial}{\partial t} + [f_r(t,\theta,\varphi) + O(r^{-2})]r \cdot \frac{\partial}{\partial r}$$
$$+[f_\theta(t,\theta,\varphi) + O(r^{-2})]\frac{\partial}{\partial \theta} + [f_\varphi(t,\theta,\varphi) + O(r^{-2})]\frac{\partial}{\partial \varphi}$$

However, now that the spacetime is restricted to be spherically symmetric we are only interested in the asymptotic Killing vector fields that respect this when Lie dragging $h_{\mu\nu}^{ads}$ to obtain the most general form. Consequently, the most general allowable $\mathfrak{z}_\infty$ is that for which the functions f_t, f_r, f_θ *and* f_φ do not depend on the angular variables θ, φ:

$$\mathfrak{z}_\infty = [f_t(t) + O(r^{-2})]\frac{\partial}{\partial t} + [f_r(t) + O(r^{-2})]r \cdot \frac{\partial}{\partial r}$$
$$+[f_\theta(t) + O(r^{-2})]\frac{\partial}{\partial \theta} + [f_\varphi(t) + O(r^{-2})]\frac{\partial}{\partial \varphi}$$

As before, the general metric is:

$$h_{\mu\nu}^{(g)} = L_{\mathfrak{z}_\infty}\left(h_{\mu\nu}^{S-Ads}\right) = \mathfrak{z}^\alpha \partial_\alpha\left(h_{\mu\nu}^{S-Ads}\right) + h_{\mu\alpha}\partial_v \mathfrak{z}^\alpha + h_{\alpha v}\partial_v \mathfrak{z}^\alpha$$
$$= [f_r(t) + O(r^{-2})]h_{\mu\nu}^{S-Ads} + h_{\mu\alpha}\partial_v \mathfrak{z}^\alpha + h_{\alpha v}\partial_v \mathfrak{z}^\alpha$$

$h_{tt}^{(g)} = [f_r(t) + O(r^{-2})]\left(\frac{2M}{r}\right)\left(\frac{\partial f_t}{\partial t}\right) + O(r^{-2})$

$\boxed{h_{tt}^{(g)} = r^{-1}f_{tt}(t) + O(r^{-3})}$

$h_{tr}^{(g)} = h_{tt}\partial_r \mathfrak{z}^t + h_{rr}\partial_r \mathfrak{z}^r = [rf_{tt} + O(r^{-3})] + \left(r^{-5}f_{rr} + O(r^{-7})\right)\left[\frac{\partial f_r}{\partial t} + O(r^{-2})\right]$

$\boxed{h_{tr}^{(g)} = r^{-1}f_{tr}(t) + O(r^{-6})}$

$h_{rr}^{(g)} = \left(f_t(r) + O(r^{-2})\right)\left[\frac{2Ml^4}{r^5} + O(r^{-7})\right] + \left[\frac{2Ml^4}{r^5} + O(r^{-7})\right]\left[\left(\frac{\partial f_r}{\partial t} + O(r^{-3})\right)\right]$

$\boxed{h_{rr}^{(g)} = r^{-5}f_{rr}(t) + O(r^{-7})}$

$h_{t\varphi}^{(g)} = h_{t\alpha}\left(\partial_\varphi \mathfrak{z}^\alpha\right) + \left(h_{\alpha\varphi}\right)\partial_t \mathfrak{z}^\alpha = 0$ similarly for $h_{t\theta}^{(g)}$

$h_{r\varphi}^{(g)} = h_{r\alpha}(\partial_\varphi \mathfrak{z}^\alpha) + (h_{\alpha\varphi})\partial_r \mathfrak{z}^\alpha = 0$ similarly for $h_{r\theta}^{(g)}$

$h_{r\varphi}^{(g)} = 0$

$h_{\varphi\varphi}^{(g)} = 0$

$h_{\theta\varphi}^{(g)} = 0$

This is not right, expect $R(r,t)$ to have non-trivial fall-off, …, the repression/simplification to no angular variables has been done too soon – so backtrack and reintroduce the angular variables until this step is done.

Again, we have:

$h_{tt}^{s-Ads} = \left(\frac{2M}{r}\right)$ (exact)

$h_{rr}^{s-Ads} = \left(\frac{2Ml^4}{r^5}\right) + O(r^{-7})$

but now keep the general form for asymptotic Killing vector field:

$$\mathfrak{z}_\infty = [f_t(t,\theta,\varphi) + O(r^{-2})]\frac{\partial}{\partial t} + [f_r(t,\theta,\varphi) + O(r^{-2})]r \cdot \frac{\partial}{\partial r}$$
$$+[f_\theta(t,\theta,\varphi) + O(r^{-2})]\frac{\partial}{\partial \theta} + [f_\varphi(t,\theta,\varphi) + O(r^{-2})]\frac{\partial}{\partial \varphi}$$

As before:

$h_{\mu\nu}^{(g)} = L_{\mathfrak{z}_\infty}(h_{\mu\nu}^{s-Ads}) = \mathfrak{z}^\alpha \partial_\alpha (h_{\mu\nu}^{s-Ads}) + h_{\mu\alpha}\partial_\nu \mathfrak{z}^\alpha + h_{\alpha\nu}\partial_\nu \mathfrak{z}^\alpha$

$h_{tt}^{(g)} = [-f_r + O(r^{-2})]\left(\frac{2M}{r}\right) 2h_{tt}\partial_t \mathfrak{z}^t \qquad (\partial_t f_t + O(r^{-2}))$

$\boxed{h_{tt}^{(g)} = r^{-1} f_{tt}(t,\theta,\varphi) + O(r^{-3})}$

$h_{rr}^{(g)} = (-gf_r + O(r^{-2}))\left[\left(\frac{2Ml^4}{r^5}\right) + O(r^{-7})\right] + 2h_{rr}\partial_r \mathfrak{z}^r \quad (f_r + O(r^{-2}))$

$\boxed{h_{rr}^{(g)} = r^{-5} f_{tt}(t,\theta,\varphi) + O(r^{-7})}$

All other terms have no contribution from the $\partial_r \mathfrak{z}^r (h_{\mu\nu})\ term$:

$h_{t\theta}^{(g)} = h_{t\alpha}\partial_\varphi \mathfrak{z}^\alpha + h_{\alpha\varphi}\partial_t \mathfrak{z}^\alpha = h_{tt}\partial_\varphi \mathfrak{z}^t$

$\boxed{h_{t\varphi}^{(g)} = r^{-1} f_{t\varphi}(t,\theta,\varphi) + O(r^{-3})}$

$\boxed{h_{t\theta}^{(g)} = r^{-1} f_{t\theta}(t,\theta,\varphi) + O(r^{-3})}$

$h_{tr}^{(g)} = h_{t\alpha}\partial_r \mathfrak{z}^\alpha + h_{\alpha r}\partial_t \mathfrak{z}^\alpha = h_{tt}\partial_r \mathfrak{z}^t + h_{rr}\partial_t \mathfrak{z}^r$

$= \left(\frac{2M}{r}\right)[O(r^{-3})] + (r^{-5}[2Ml^4] + O(r^{-7}))r\left(\frac{\partial f_r(t,\theta,\varphi)}{\partial t}\right)$

$\boxed{h_{tr}^{(g)} = r^{-4} f_{tr}(t,\theta,\varphi) + O(r^{-6})}$

$h^{(g)}_{r\theta} = h_{r\alpha}\partial_\theta \mathfrak{z}^\alpha + h_{\alpha\theta}\partial_t \mathfrak{z}^\alpha = h_{rr}\partial_\theta \mathfrak{z}^t$

$\left[\left(\frac{2Ml^4}{r^5}\right) O(r^{-7})\right]\left(r\frac{\partial}{\partial\theta}\right) f_r(t,\theta,\varphi)$

$\boxed{h^{(g)}_{r\theta} = r^{-4} f_{r\theta}(t,\theta,\varphi) + O(r^{-6})}$ (no longer have zero error)

$\boxed{h^{(g)}_{r\varphi} = r^{-4} f_{r\varphi}(t,\theta,\varphi) + O(r^{-6})}$ (no longer have zero error)

$h^{(g)}_{\theta\theta} = h_{\theta\alpha}\partial_\theta \mathfrak{z}^\alpha = 0$

$h^{(g)}_{\theta\theta} = 0$

$h^{(g)}_{\varphi\varphi} = 0$ still zero

$h^{(g)}_{\varphi\theta} = 0$

Now, drop the θ, φ functional dependencies due to spherical symmetry and there results: (drop g for "general" superscript also)

$h_{tt} = r^{-1} f_{tt} + O(r^{-3})$

$h_{rr} = r^{-5} f_{tt} + O(r^{-7})$

$h_{t\varphi} = r^{-1} f_{t\varphi} + O(r^{-3})$

$h_{t\theta} = r^{-1} f_{t\theta} + O(r^{-3})$

$h_{t\varphi} = r^{-4} f_{t\varphi} + O(r^{-3})$

$h_{r\theta} = r^{-4} f_{r\theta} + O(r^{-6})$

$h_{tr} = r^{-4} f_{tr} + O(r^{-6})$

Other $h'_{\mu\nu}s$ zero.

Let's restate the solution obtained. Starting from

$ds^2 = ds_0^2 + h_{\mu\nu}dx^\mu dx^\nu$

$ds_0^2 = -f dt^2 + f^{-1}dr^2 + r^2 d\Omega^2 \quad , f = \left(1 + \left(\frac{r}{l}\right)^2\right)$

we have:

$h^{(g)}_{tt} = r^{-1} f_{tt}(t) + O(r^{-3})$; $h^{(g)}_{t\theta} = r^{-1} f_{t\theta}(t) + O(r^{-3})$; $h^{(g)}_{t\varphi} = r^{-1} f_{t\varphi}(t) + O(r^{-3})$

$h^{(g)}_{tr} = r^{-4} f_{tr}(t) + O(r^{-6})$; $h^{(g)}_{r\theta} = r^{-4} f_{r\theta}(t) + O(r^{-6})$; $h^{(g)}_{r\varphi} = r^{-1} f_{r\varphi}(t) + O(r^{-6})$

$h^{(g)}_{tr} = r^{-5} f_{rr}(t) + O(r^{-7})$; the rest zero.

Compare this to the ADM form for the metric:

$$ds^2 = -N^2 dt^2 + \Lambda^2 (dr + N^r dt)^2 + l^2 d\Omega^2$$

Again R(r,t) has trivial fall-off conditions, perhaps the most general fall-off conditions are arrived at by considering the most general solution,

Kerr-AdS, obtaining the fall-off of metrical components, and then imposing spherical symmetry via restricting functional dependencies in the metrical components and their deviations (this is equivalent to setting angular momentum to zero so that the background goes from Kerr-Ads to Schwarzschild-Ads). If this were the case, then a similar analysis of Schwarzschild would indicate its deviation starting from Kerr. Perhaps this is the best way to address the supertranslations problem. Recall that for Schwarzschild stronger boundary conditions were required than were obtained from requiring that the asymptotic symmetry group contain the Poincare transformations. The stronger conditions can be imposed via parity conditions on the canonical metrical variables or via asymptotic conditions on the magnetic part of the Weyl tensor-both methods may relate to the modified analysis associated with Kerr. In what follows I'll simply take the Kerr-Ads deviations and drop all angular dependencies from the functions in the deviations and metric (a=0). So,
$h_{tt}, h_{t\theta}, h_{t\varphi}, h_{\theta\theta}, h_{\theta\varphi}, h_{\varphi\varphi}$ have the form: $r^{-1}f(t) + O(r^{-3})$
$h_{tr}, h_{\theta r}, h_{\theta r}$ have the form: $r^{-4}f(t) + O(r^{-6})$
h_{rr} has the form $r^{-5}f(t) + O(r^{-7})$
where

$$ds_0^2 = -fdt^2 + f^{-1}dr^2 + r^2 d\Omega^2 \ , \ f = \left(1 + \left(\frac{r}{l}\right)^2\right)$$
$$ds^2 = ds_0^2 + h_{\mu\nu}dx^\mu dx^\nu$$
$$ds^2 = -N^2 dt^2 + \Lambda^2(dr + N^r t)^2 + R^2 d\Omega^2$$

Imposing spherical symmetry more involved than just dropping θ, φ dependencies. We have spherically symmetric if isometry contains a subgroup isomorphic to the group SO(3) and the orbits of this subgroup are two-dimensional spheres. So must have angular terms enter the metric only as a multiple of the metric of a 2-sphere. This restriction generally eliminates the θ, φ dependencies in the metric and its deviations – except in the $g_{\varphi\varphi}$ term which has a $\sin^2\theta$ term relative to the $g_{\theta\theta}$ term (also for $h_{\varphi\varphi}$ versus $h_{\theta\theta}$). Thus

$$\boxed{\begin{array}{c} h_{tt}, h_{\theta\theta}, \sin^{-2}\theta\, h_{\varphi\varphi} \ \rightarrow \ r^{-1}f(t) + O(r^{-3}) \\ h_{tr} \ \rightarrow \ r^{-4}f(t) + O(r^{-6}) \\ h_{rr} \ \rightarrow \ r^{-5}f(t) + O(r^{-7}) \end{array}}$$

Comparing with the ADM form:

$$ds^2 = \left(-N^2 + \Lambda^2 N^{r2}\right)dt^2 + 2\Lambda^2 dr dt + \Lambda^2 dr^2 + R^2 d\Omega^2$$
$$\cong \left(-f + r^{-1}f_{tt}(t) + O(r^{-3})\right)dt^2 + 2\left(r^{-4}f_{tr}(t) + O(r^{-6})\right)drdt$$

$$+\left(f^{-1}+r^{-5}f_{rr}(t)+O(r^{-7})\right)dr^2+\left(r^2+r^{-1}f_\Omega(t)+O(r^{-3})\right)d\Omega^2$$
$$R^2\approx r^2+r^{-1}f_\Omega(t)+O(r^{-3})=r^2\left(1+r^{-3}f_\Omega(t)+O(r^{-3})\right)$$
$$R=r\left(1+\frac{1}{2}r^{-3}f_\Omega(t)+\frac{1}{2}O(r^{-3})\right)$$
$$\boxed{R\approx r+r^{-2}p(t)+O(r^{-4})}$$
$$\Lambda^2\approx\left[\left(1+\left(\frac{r}{l}\right)^2\right)^{-1}+r^{-5}f_{rr}(t)+O(r^{-7})\right]\approx\left(1+\left(\frac{r}{l}\right)^2\right)^{-1}\left[1+\right.$$
$$\left.r^{-3}\frac{1}{l^2}f_{rr}(t)+O(r^{-5})\right]$$
$$\boxed{\Lambda\approx\left(1+\left(\frac{r}{l}\right)^2\right)^{-1/2}+r^{-4}\lambda(t)+O(r^{-6})}$$
$$2\Lambda^2N^r=2\left(r^{-4}f_{tr}(t)+O(r^{-6})\right)$$
$$N^r=\left(r^{-4}f_{tr}(t)+O(r^{-6})\right)\left(1+\left(\frac{r}{l}\right)^2\right)$$
$$\boxed{N^r=r^{-2}v(t)+O(r^{-4})}$$
$$-N^2+\Lambda^2N^{r2}=\left(-\left(1+\left(\frac{r}{l}\right)^2\right)-1r^{-1}f_{tt}(t)+O(r^{-3})\right)$$
$$N^2=\left(1+\left(\frac{r}{l}\right)^2\right)-r^{-1}f_{tt}(t)+O(r^{-3})=\left(\frac{r}{l}\right)^2 1+\left(\frac{l}{r}\right)^2+r^{-3}f_{tt}+$$
$$O(r^{-1})$$
$$\boxed{N\approx\frac{r}{l}+\frac{1}{2}\left(\frac{l}{r}\right)+r^{-2}\mathcal{N}(t)+O(r^{-3})}$$

$$K_{ij}=(2N)^{-1}\left(-g_{ij,t}+N_{(a|b)}\right)$$
$$K_{rr}=-N^{-1}\Lambda\left(\dot{\Lambda}-(\Lambda N^r)'\right)$$
$$K_{\theta\theta}=-N^{-1}R\left(\dot{R}-R'N^r\right)$$
$$K_{\varphi\varphi}=\sin^2\theta\,K_{\theta\theta}$$
(the other K's being zero from spherical symmetry condition)
$$K_{rr}\cong-\left(\frac{r}{l}+\frac{1}{2}\left(\frac{l}{r}\right)+r^{-2}\mathcal{N}(t)\right)^{-1}\left[\left(1+\left(\frac{r}{l}\right)^2\right)^{-1/2}+r^{-4}\lambda(t)\right]$$
$$\times\left(r^{-4}\dot{\lambda}(t)-\left\{\left[\left(1+\left(\frac{r}{l}\right)^2\right)^{-1/2}+r^{-4}\lambda\right][r^{-2}v(t)+O(r^{-4})]\right\}'\right)$$
$$\cong-\left(\frac{l}{r}\right)^2\left(1-\frac{1}{2}\left(\frac{l}{r}\right)^2\right)^2\left(-r^{-3}v(t)\right)\cong r^{-5}K_r(t)+O(r^{-7})$$

$$\pi^{ij}=|g|^{1/2}\left(Kg^{ij}-K^{ij}\right)$$

$$K_{\theta\theta} = -\left(\frac{l}{r}\right)\left(1-\frac{1}{2}\left(\frac{l}{r}\right)^2\right)\left(r + r^{-2}p(t)\right)\left(r^{-2}\dot{p} - (1-2r^{-3}p)r^{-2}v(t)\right)$$
$$= r^{-2}K_\theta(t) + O(r^{-4})$$
$$K = K_{\mu\nu}g^{\mu\nu} = \left(r^{-5}K_r(t) + O(r^{-7})\right) + 2\left(r^{-2}K_\theta(t) + O(r^{-5})\right)r^{-2}$$
$$= r^{-4}K_1(t) + r^{-5}K_2(t) + O(r^{-6})$$
$$\pi^{rr} = (r^3)2K_{\theta\theta}g^{\theta\theta}g^{rr}$$
$$\pi^{\theta\theta} = r^3\left(K_{rr}g^{rr} + K_{\varphi\varphi}g^{\varphi\varphi}\right)g^{\theta\theta}$$
$$\pi^{rr} = r^{-1}p^{rr}(t) + O(r^{-3})$$
$$\pi^{\theta\theta} = r^{-3}p^{\theta\theta}(t) + O(r^{-4})$$

$$P_\Lambda = R^2K^\theta_\theta$$
$$P_R = R\Lambda\left(K^r_r + K^\theta_\theta\right)$$
$$P_\Lambda = \left(r^2 + O(r^{-1})\right)\left(r^{-2}K_\theta(t) + O(r^{-4})\right)r^{-2}$$
$$\boxed{P_\Lambda = r^{-2}p_\lambda(t) + O(r^{-4})}$$
$$P_R = r\left(\frac{l}{r} - \frac{1}{2}\left(\frac{l}{r}\right)^3 \dots\right)\left(\left(r^{-5}k_r(t) + O(r^{-7})\right) + \left(r^{-4}k_\theta(t) + O(r^{-6})\right)\right)$$
$$\boxed{P_R = r^{-4}p_R(t) + O(r^{-5})}$$

Finally, it's possible to examine the boundary term from the analysis of the Louisville forms:

$$X = \frac{1}{2}R\delta R \ln\left|\frac{RR' + \Lambda P_\Lambda}{RR' - \Lambda P_\Lambda}\right|$$

$$R \to r \ : R' \to 1$$
$$\delta R = O(r^{-2})$$
$$\Lambda \to \frac{l}{r} - \frac{1}{2}\left(\frac{l}{r}\right)^3$$
$$P_\Lambda \to r^{-2}p\lambda(t) + O(r^{-4})$$
$$X = rO(r^{-2}) \ln\left|\frac{r + r^{-3}p_\lambda}{r - r^{-3}p_\lambda}\right| = O(r^{-1}) \ln\left|\frac{r + r^{-4}p_\lambda}{r - r^{-4}p_\lambda}\right|$$

Recall $\frac{1}{2}\ln\left(\frac{1+x}{1-x}\right) = x + \frac{x^3}{3} + \cdots$, so

$$X = O(r^{-1})O(r^{-4}) = O(r^{-5})$$

at leading order. For consideration of the w-form

$$\frac{1}{2}RR' \ln\left|\frac{RR' - \Lambda P_\Lambda}{RR' - \Lambda P_\Lambda}\right| = RR'\left(\left(\frac{\Lambda P_\Lambda}{RR'}\right) + \frac{1}{3}\left(\frac{\Lambda P_\Lambda}{RR'}\right)^3 + \cdots\right) = -\Lambda P_\Lambda + O(r^{-11})$$

which easily avoids the logarithmic singularity.

B.4 Fall-off as $r \to \infty$, Alternate derivation using Ker-AdS
The derivation begins by first considering the more general case of Kerr-Anti de Sitter (K-Ads). This analysis is discussed in [83].

The general form of the fall-off conditions analysis starts with the AdS metric:

$$ds_0^2 = -\left(1+\left(\frac{r}{l}\right)^2\right)dt^2 + \left(1+\left(\frac{r}{l}\right)^2\right)^{-1} dr^2 + r^2 d\Omega^2$$

Leading order r^{-1} deviations (general) have the form:

$$h_{\mu v} = r^{-k} f_{\mu v}(t,\theta,\varphi) + O(r^{-k-2})$$

Where the $f_{\mu v}$ are arbitrary functions and for $h_{tt}, h_{t\theta}, h_{t\varphi}, h_{\theta\theta}, h_{\theta\varphi}, h_{\varphi\varphi}$: $k = 1$, for $h_{tr}, h_{\theta r}, h_{\varphi r}$ $k = 4$, and for h_{rr}, $k = 5$.

The spherically symmetric subclass of the above general form for the fall-off conditions is obtained after two steps:

(1) Angular terms in the metric are eliminated-excepting those that enter as a multiple of the 2-sphere (spherically symmetric isometry group contains a sub-group isomorphic to SO(3) for which the orbits are two-dimensional spheres). Thus, the functional dependencies are reduced from $f_{\mu v}(t,\theta,\varphi)$ to $f_{\mu v}(t)$. Furthermore, all mixed angular index $h'_{\mu v}s$ are zero while $h_{\theta\theta} = \sin^{-2} h_{\varphi\varphi}$. Thus, the nonzero deviations are reduced to the following list: $h_{tt}, h_{\theta\theta}, \sin^{-2} h_{\varphi\varphi}$ (with $k = 1$), h_{tr} (with $k = 4$), and h_{rr} (with $k = 5$).
(2) The perturbative expansion of the AdS metric should be done in a manner that respects the spherical symmetry. Thus, if Λ_0 for ds_0^2 has a non-trivial expansion then so will R_0 via $\Lambda^2 = \left(1 - \frac{2M}{R} + \left(\frac{R}{l}\right)^2\right)^{-1}$. Furthermore, the expansions should be taken to the order of the unspecified terms in the deviations.

$$\Lambda = r^{-1}l - r^{-3}\left(\frac{1}{2}l^3\right) + r^{-4}\lambda + r^{-5}\left(\frac{3}{8}l^5\right) + O(r^{-6})$$
$$\Lambda^{-1} = rl^{-1} + r^{-1}\left(\frac{1}{2}l\right) - r^{-2}(\lambda l^{-2}) - r^{-3}\left(\frac{1}{8}l^3\right) + O(r^{-4})$$
$$\Lambda^{-1} = r^2 l^{-2} + 1 - r^{-1}(2\lambda l^{-3}) + O(r^{-3})$$

$$R = r + r^{-2}p_2 + O(r^{-4})$$
$$N_r = r^{-2}v(t) + O(r^{-4})$$

$N = rl^{-1} + r^{-1}\left(\frac{1}{2}l\right) + r^{-2}\eta(t) - \frac{1}{8}r^{-3}l^3 + O(r^{-4})$
$P_\Lambda = O(r^{-2})$
$P_R = O(r^{-4})$
The constraints:
$0 = H = -R^{-1}P_R P_\Lambda + \frac{1}{2}R^{-2}\Lambda P_\Lambda^2 + \Lambda^{-1}RR'' - \Lambda^{-2}RR'\Lambda' + \frac{1}{2}\Lambda - \frac{3}{2l^2}(\Lambda R^2)$
$0 = H_r = P_R R' - \Lambda P_\Lambda'$
Mass equation:
$M = \frac{1}{2}R^{-1}P_\Lambda^2 - \frac{1}{2}\Lambda^{-2}RR'^2 + \frac{1}{2}R + \frac{1}{2l^2}R^3$
The Equations of motion:
$\dot{P}_\Lambda = \frac{1}{2}N\left[1 - \Lambda^{-2}R'^2 - R^2P_\Lambda^2\right] - \Lambda^{-2}RR'N' + N^r P_\Lambda' + \frac{3}{2l^2}NR^2$
$\dot{P}_R = N[r^{-3}\Lambda P_\Lambda^2 - R^{-2}P_\Lambda P_R] + (N^r P_R)' - N\left(\frac{R'}{\Lambda}\right)' - \left(\frac{N'}{\Lambda}R\right)' + \frac{3}{l^2}N\Lambda R$
$\dot{R} = -NR^{-1}P_\Lambda + N^r R'$
$\dot{\Lambda} = N[R^{-2}\Lambda P_\Lambda - R^{-1}P_R] + (\Lambda N^r)'$

Consider $\dot{P}_\Lambda$:
$\dot{P}_\Lambda = O(r^{-2})$
$\left[1 - \Lambda^{-2}R'^2 - R^{-2}P_\Lambda^2\right] = 1 - \left(r^2l^{-2} + 1 - r^{-1}(2\lambda l^{-3}) + O(r^{-3})\right)\left(1 - 2r^{-3}p_2 + O(r^{-5})\right)^2$
$= -r^2l^{-2} + r^{-1}(2\lambda l^{-3} + 4p_2 l^{-2}) + O(r^{-3})$

$\frac{1}{2}N[\ldots] = -\frac{1}{2}r^3l^{-3} - \frac{1}{4}rl^{-1} + \left(\lambda l^{-4} + 2p_2l^{-3} - \frac{1}{2}\eta l^{-2}\right) + \frac{1}{16}r^{-1}l + O(r^{-2})$

$(RN') = rl^{-1} - r^{-1}\left(\frac{1}{2}l\right) + r^{-2}(-2\eta + p_2l^{-1}) + r^{-3}\left(\frac{3}{8}l^3\right) + O(r^{-4})$
$R'\Lambda^{-2}(RN') = r^3l^{-3} - r\left(\frac{1}{2}l^{-1}\right) + (-2\eta l^{-2} + p_2l^{-3}) + \frac{3}{8}r^{-1}l + O(r^{-2})$
$= r^3l^{-3} + r\left(\frac{1}{2}l^{-1}\right) + (-2\eta l^{-2} - 2\lambda l^{-4} + p_2l^{-3}) + r^{-1}\left(-\frac{1}{8}l\right) + O(r^{-2})$

$N^r P_\Lambda' = O(r^{-5})$
$\frac{3}{2l^2}NR^2 = \frac{3}{2l^2}\left(r^3l^{-1} + r\left(\frac{1}{2}l\right) + \eta - \frac{1}{8}r^{-1}l^3 + O(r^{-2}) + 2l^{-1}p_2\right)$
$\frac{3}{2}r^3l^{-3} + \frac{3}{4}rl^{-1} + \left(\frac{3}{2}\eta l^{-2} + 3p_2l^{-3}\right) - \frac{3}{16}r^{-1}l + O(r^{-2})$

$$\dot{P}_R = r^3\left(-\tfrac{1}{2}-1+\tfrac{3}{2}\right)l^{-3} + r\left(-\tfrac{1}{4}-\tfrac{1}{2}+\tfrac{3}{4}\right)l^{-1} + \left(2\eta - \tfrac{1}{2}\eta + \tfrac{3}{2}\eta\right)l^{-2}$$
$$+(\lambda + 2\lambda)l^{-4} + (2p_2 + p_2 + 3p_2)l^{-3} + r^{-1}\left(\frac{1}{16}+\frac{1}{8}-\frac{3}{16}\right)l + O(r^{-2})$$
$$= O(r^{-2})$$
if $\boxed{\eta l^{-2} + \lambda l^{-4} + 2p_2 l^{-3} = 0}$

Consider $\dot{P}_R$:
$\dot{P}_R = O(r^{-4})$
$NR^{-3}\Lambda P_\Lambda^2 = O(r^{-7})\,; NR^{-2}P_\Lambda P_R = O(r^{-7})\,;\; (N^r P_R)' = O(r^{-7})$

$$\left(\frac{R'}{\Lambda}\right)' = \Big[\left(1 - 2r^{-3}p_2 + O(r^{-5})\right)\Big(rl^{-1} + r^{-1}\left(\tfrac{1}{2}l\right) - r^{-2}(\lambda l^{-2}) -$$
$$r^{-3}\left(\tfrac{1}{8}l^3\right) + O(r^{-4})\Big)\Big]'$$
$$= \left[rl^{-1} + r^{-1}\left(\tfrac{1}{2}l\right) - r^{-2}(\lambda l^{-2} + 2p_2 l^{-1}) - r^{-3}\left(\tfrac{1}{8}l^3\right) + O(r^{-4})\right]$$
$$= l^{-1} - r^{-1}\left(\tfrac{1}{2}l\right) - r^{-2}(\lambda l^{-2} + 4p_2 l^{-1})r^{-3} - r^{-4}\left(\tfrac{3}{8}l^3\right) + O(r^{-4})$$

$$N\left(\tfrac{R'}{\Lambda}\right)\Lambda^{-1} = r^{-2} + r^{-2}(2\lambda l^{-2} + 4p_2 l^{-2} + \eta l^{-1}) + O(r^{-4})$$
$$(N'R)\Lambda^{-1} = r^2 l^{-2} - r^{-1}(2\eta l^{-1} - p_2 l^{-2} + \lambda l^{-3}) + O(r^{-3})$$
$$\left(N'R\Lambda^{-1}\right) = r(2l^{-2}) + r^{-2}(2\eta l^{-1} - pl^2 + \lambda l^{-3}) + O(r^{-4})$$
$$\Lambda N = 1 + r^{-3}(\eta l + \lambda l^{-1}) + O(r^{-5})$$

$$\tfrac{3}{l^2}\Lambda NR = r(3l^{-2}) + r^{-2}(3\eta l^{-1} + 3\lambda l^{-3} + 3p_2 l^{-2}) + O(r^{-4})$$
$$\dot{P}_R = r(-1-2+3)l^{-2} + r^{-2}\begin{bmatrix}(3\eta - 2\eta - r\eta)l^{-1} + (3\lambda - \lambda - 2\lambda)l^{-3} \\ +(3p_2 + p_2 - 4p_2)l^{-1}\end{bmatrix}$$
$$+ O(r^{-4})$$
$$= O(r^{-4})$$

Consider H:
$$H = O(r^{-7}) + O(r^{-7}) + \left(\Lambda^{-1}\right)\left(r + r^{-2}p_2 + O(r^{-4})\right)\left(6p_2 r^{-4} +\right.$$
$$\left.O(r^{-6})\right)$$
$$+\left(l^{-1} - r^{-2}\left(\frac{1}{2}l\right) + r^{-3}(2\lambda l^{-2}) + r^{-4}\left(\frac{3}{8}l^3\right) + O(r^{-5})\right)$$
$$\times\left(r + r^{-2}p_2 + O(r^{-4})\right)\left(1 + 2r^{-3}p_2 + O(r^{-4})\right)$$

$$+\frac{1}{2}\Bigg(rl^{-1}+r^{-1}\left(\frac{1}{2}l\right)-r^{-2}(\lambda l^{-2})-r^{-3}\left(\frac{1}{8}l^3\right)$$
$$+O(r^{-4})\Bigg)\left(1-2r^{-3}p_2+O(r^{-5})\right)^2$$
$$-\frac{1}{2}\left(rl^{-1}-r^{-3}\left(\frac{1}{2}l^3\right)+r^{-4}\lambda+r^{-5}\left(\frac{3}{8}l^5\right)+O(r^{-6})\right)$$
$$-\frac{3}{2}l^2\Bigg(r^{-1}-r^{-3}\left(\frac{1}{2}l^2\right)+r^{-4}\lambda+r^{-5}\left(\frac{3}{8}l^5\right)$$
$$+O(r^{-6})\Bigg)\left(r+r^{-2}p_2+O(r^{-4})\right)^2$$
$$=r\left(l^{-1}+\tfrac{1}{2}l^{-1}-\tfrac{3}{2}l^{-1}\right)+r^{-1}l\left(-\tfrac{1}{2}+\tfrac{1}{4}+\tfrac{3}{4}\right)+r^{-2}\left[2\lambda-\tfrac{1}{2}\lambda-\tfrac{3}{2}\lambda\right]$$
$$+l^{-1}(6p_2+p_2-2p_2-2p_2-3p_2)]+r^{-3}\left(\tfrac{3}{8}-\tfrac{1}{16}+\tfrac{1}{4}-\tfrac{7}{16}\right)+O(r^{-4})$$

So,

$$H=O(r^{-4})$$
$$H_r=O(r^{-4})$$

Consider $\dot{R}$:
$\dot{R}=O(r^{-2})$
$-NR^{-1}P_\Lambda=O(r^{-2})$; $N^rR'=O(r^{-2})$; so $\dot{R}=O(r^{-2})$
Consider $\dot{\Lambda}$
$\dot{\Lambda}=O(r^{-4})$
$NR^{-2}\Lambda P_\Lambda=O(r^{-4})\,; NR^{-1}P_R=O(r^{-4})$; $(NV^r)'=O(r^{-4})\to\dot{\Lambda}=O(r^{-4})$

Mass equation: $M=\frac{1}{2}R^{-1}P_\Lambda^2-\frac{1}{2}\Lambda^{-2}RR'^2+\frac{1}{2}R+\frac{1}{2l^2}R^3$

$$M=O(r^{-5})-\frac{1}{2}\left(r^2l^2+1-r^{-1}(2\lambda l^{-3})+O(r^{-3})\right)\big(r+r^{-2}p_2$$
$$+O(r^{-4})\big)\left(1-2r^{-3}p_2+O(r^{-5})\right)^2$$
$$+\frac{1}{2}\left(r+r^{-2}p_2+O(r^{-4})\right)+\frac{1}{2l^2}\left(r+r^{-2}p_2+O(r^{-4})\right)^3$$
$$=r^{-3}\left(-\frac{1}{2}l^{-2}+\frac{1}{2}l^{-2}\right)+r\left(-\frac{1}{2}+\frac{1}{2}\right)+\left(\lambda l^{-3}+\frac{3}{2}p_2l^{-2}+\frac{3}{2}l^{-2}\right)$$
$$+O(r^{-2})$$

Thus

$$\boxed{Ml^3=\lambda+3p_2l}$$

Is there consistency with $\Lambda^2 = \left(1 - \frac{2M}{R} + \left(\frac{R}{l}\right)^2\right)^{-1}$?

$\Lambda^2 - 1 - R^2 l^{-2} = -2MR^{-1}$

$\left(r^2 l^{-2} + 1 - r^{-1}(2\lambda l^{-3}) + O(r^{-3})\right) - 1 - \left(r + r^{-2}p_2 + O(r^{-4})\right)^2 l^{-2} = \frac{2M}{r}(1 + r^{-3}p_2)^{-1}$

$r^{-1}(-2\lambda l^{-3} - 2p_2 l^{-2}) = r^{-1}(2M)$

$$\boxed{\lambda + p_2 l = Ml^3}$$

which is known as "Schwarzschild Gauge". If the Schwarzschild Gauge is desired, compatibility with the relation from the mass equation requires that $p_2 = 0$.

N can be arbitrarily scaled by $N_\infty(t)$. This is because such changes do not alter the consistency analysis on the fall-off conditions. N(t) is only expressed in the EOM's, consider those: for $\dot{R}, \dot{\Lambda}$, N_∞ merely multiplies order factors that were ignored. For $\dot{P}_R, \dot{P}_\Lambda, N$ it multiplies all the nontrivial terms linearly. Thus, N(t) can be scaled. If a form $N_\infty(t,r)$ is considered, the leading order r dependence that is consistent with the fall-off analysis is $O(r^{-5})$ - which is expected from the leading non-fixed or constrained order term being $O(r^{-4})$ in N itself.

Consider $\left(\frac{R'}{\Lambda}\right)$ and its relation to N:

$\left(\frac{R'}{\Lambda}\right) = \left(1 + r^{-3}(-2p_2) + O(r^{-5})\right)\left(rl^{-1} + r^{-1}\left(\frac{1}{2}l\right) - r^{-2}(\lambda l^{-2}) - r^{-3}\left(\frac{1}{8}l^3\right) + O(r^{-5})\right)$

$= rl^{-1} + r^{-1}\left(\frac{1}{2}l\right) - r^{-2}(\lambda l^{-2} + 2p_2 l^{-1}) - r^{-3}\left(\frac{1}{8}l^3\right) + O(r^{-4})$

$N = rl^{-1} + r^{-1}\left(\frac{1}{2}l\right) + r^{-2}(\eta) - r^{-3}\left(\frac{1}{8}l^3\right) + O(r^{-4})$

$N = \left(\frac{R'}{\Lambda}\right) + O(r^{-4})$ if $\eta = -(\lambda l^{-2} + 2p_2 l^{-1})$

From the relation required for the consistency of $\dot{P}_\Lambda's$ fall off:

$$\eta = -(\lambda l^{-2} + 2p_2 l^{-1})$$

So, N does indeed have the form indicated and can be "rescaled" by

$$N_\infty(t,r) = N_\infty(t) + O(r^{-5})$$

to get:

$$N = N_\infty(t,r)\left(\frac{R'}{\Lambda}\right) + O(r^{-4}) = N_\infty(t)\left(\frac{R'}{\Lambda}\right) + O(r^{-4})$$

Now that the fall-off conditions are known, the correct Hamiltonian, accounting for surface terms, may be calculated. (The canonically transformed boundary terms are also checked.)We start with:

$H = \int_0^\infty dr\, \{N\mathcal{H} + N^r\mathcal{H}_r\}$

$\mathcal{H} = -R^{-1}P_R P_\Lambda + \frac{1}{2}R^{-2}\Lambda P_\Lambda^2 + \Lambda^{-1}RR'' - \Lambda^{-2}RR'\Lambda' RR'\Lambda' + \frac{1}{2}\Lambda^{-1}R'^2 - \frac{1}{2}\Lambda - \frac{3}{2l^2}(\Lambda \dot{R})$

$\mathcal{H}_r = P_R R' - \Lambda P'_\Lambda$

$\delta H =$

$$\int_0^\infty dr \begin{cases} \delta(-NR^{-1}P_R P_\Lambda) + \delta\left(\frac{1}{2}NR^{-2}\Lambda P_\Lambda^3\right) + \delta(N\Lambda^{-1}R)R'' - \delta(N\Lambda^{-2}R)R'\Lambda \\ +\delta(N\Lambda^{-1})R'^2 - \delta\left(\frac{1}{2}\Lambda\right) + \delta(N_r P_R)R' - \delta\Lambda P'_\Lambda - \delta\left(N\frac{3}{2l^2}\Lambda R^2\right) \end{cases}$$

$$+\int_0^\infty dr\,\{(N\Lambda^{-1}R)\delta R'' - (N\Lambda^{-2}R)\delta(R'\Lambda') - (N\Lambda^{-1}R)\delta R' + (N^r P_R)\delta R' - (N^r\Lambda)\delta P'_\Lambda\}$$

$$\int_0^\infty dr\,(N\Lambda^{-1}R)\delta R'' = \int_0^\infty dr\left\{[N\Lambda^{-1}R\delta R']' - (N\Lambda^{-1}R)'\delta R'\right\}$$

$$= \int_0^\infty dr\left\{[N\Lambda^{-1}R\delta R']' - \left[(N\Lambda^{-1}R)'\delta R\right]' + (N\Lambda^{-2}R\Lambda')''\delta R\right\}$$

$$\int_0^\infty dr\,(N\Lambda^{-2}R)\delta(R'\Lambda')$$

$$= \int_0^\infty dr\left\{[(N\Lambda^{-1}R)(\Lambda'\delta R + R'\delta\Lambda)]' - (N\Lambda^{-2}R\Lambda')'\delta R - (N\Lambda^{-2}RR')\delta\Lambda\right\}$$

$$\int_0^\infty dr\,(N\Lambda' R')\delta R' = \int_0^\infty dr\left\{[(N\Lambda^{-1}R')\delta R]' - (N\Lambda' R')'\delta R'\right\}$$

$$\int_0^\infty dr\,\{[(N^r P_R)\delta R]' - (N^r P_R)'\delta R\}$$

$$\int_0^\infty dr\,(N^r \Lambda)\delta P'_\Lambda = \int_0^\infty dr\,\{[(N^r \Lambda)\delta P_\Lambda]' - (N^r \Lambda)'\delta P_\Lambda\}$$

We can now evaluate:

$$\delta H = \int_0^\infty dr\,\{P_\Lambda \delta \Lambda + P_R \delta R - \dot{\Lambda}\delta P_\Lambda - \dot{R}\delta P_\Lambda\} + S(r)|_0^\infty$$

S is the surface term, where:

$$S = N\Lambda^{-1}R\delta R' - \left(N\Lambda^{-1}R\right)'\delta R - \left(N\Lambda^{-2}R\right)(\Lambda'\delta R)$$
$$-\left(N\Lambda^{-2}R\right)(R'\delta\Lambda) + \left(N\Lambda^{-1}R'\right)\delta R + (N^r P_R)\delta R - N^r \Lambda \delta P_\Lambda$$

(the S terms above, seven in number, will be referenced sequentially as (i)-(vii) in what follows).

Now, as r – 0:

$$R = R_0 + R_2 r^2 + O(r^4)$$

from constraints $2M = R_0 + R_0^3 l^{-2}$ and $R_2 = \frac{1}{4}\Lambda_0^2 R_0^{-1}(1 + 3R_0^2 l^{-2})$, but these re not imposed a priori, so $\delta R_0 \neq 0$, etc. So,

$$R' = R_0 r + O(r^3)$$
$$\delta R = O(r^0)$$
$$\delta R' = O(r^1)$$
$$\Lambda = \Lambda_0 + O(r^2)$$
$$\Lambda' = O(r)$$
$$\delta\Lambda = O(r^0)$$

$$N = N_1 r + O(r^3)$$
$$N^r = N^r{}_1 r + O(r^3)$$
$$P_\Lambda = O(r^3)$$
$$\delta P_\Lambda = O(r^3)$$
$$P_R = O(r)$$

Now, evaluating the terms that were indicated above with labels (i)-(vii):

(i) $= O(r^2)$
(ii) $= O(r^0)$
(iii) $= O(r^2)$
(iv) $= O(r^2)$
(v) $= O(r^2)$
(vi) $= O(r^2)$
(vii) $= O(r^4)$

The part labeled (ii) gives the surface term, so let's examine that term in detail:

$$(N\Lambda^{-1}R)'\delta R = N'(\Lambda^{-1}R)\delta R + N(\Lambda^{-1}R)'\delta R$$

$\lim_{r\to 0} S(r) = \{-N'(\Lambda^{-1}R)\delta R\}|_0 = Y|_0$
$Y = -N'\Lambda' R\delta R$
$= (N_1 + O(r^2))\Lambda_0^{-1}(1 + O(r^2))(R_0 + R_2 r^2 + O(r^4))(\delta R_0 + \delta R_2 r^2 + O(r^4))$
$= -\frac{N_1 R_0}{\Lambda_0}\delta R_0 + O(r^2)$
$= \frac{N_1}{\Lambda_0}(\delta R_0^2) + O(r^2)$
$= 0 + O(r^2)$ upon fixing $R_0 = \alpha$ by the constraint due to the mass equation.

How to express $Y = \delta Z$? If $\delta Z = \delta\left(N'^{\Lambda^{-1}\left(\frac{1}{2}r^2\right)}\right) = N'\Lambda^{-1}\delta\left(\frac{1}{2}r^2\right) + \frac{1}{2}R^2\delta\left(N'^{\Lambda^{-1}}\right)$, then $\delta Z = Y$ in the limit of $r \to 0$ if $N_1\Lambda_0^{-1}$ is fixed. Thus, the surface term for $r \to 0$ may be expressed by:

$$S_{\partial\Sigma}^{(0)} = \frac{1}{2}\int dt\, [R^2 N'\Lambda^{-1}]_{r=0}$$

The variation of $S_{\partial\Sigma}^{(0)}$ is

$$\delta S_{\partial\Sigma}^{(0)} = \frac{1}{2}\int dt\, [\delta(R^2)N'\Lambda^{-1}]_{r=0} + \frac{1}{2}\int dt\, [R^2\delta(N'\Lambda^{-1})]_{r=0}$$

where the first term on the right is zero when the gravitational constraints are satisfied, and the second term is zero when the slicing constraint is satisfied.

Consider $X = \frac{1}{2}R\delta R \ln\left|\frac{RR' + \Lambda P_\Lambda}{RR' - \Lambda P_\Lambda}\right|$ at the bifurcation of the boubdary

$\lim_{r\to 0} X = \frac{1}{2}R_0\delta R_0 \ln\left|\frac{1 + \frac{\Lambda_0 O(r^3)}{2R_0R_2r}}{1 - \frac{\Lambda_0 O(r^3)}{2R_0R_2r}}\right| + O(r^2) = \frac{1}{2}R_0(\delta R_0)\left(\frac{\Lambda_0 O(r^3)}{2R_0R_2r}\right) + O(r^2) = O(r^2) \to 0$

In order that the exact form "w", be well defined it must be shown to be finite:

$w = \int_0^\infty dr\left(\Lambda P_\Lambda + \frac{1}{2}RR'\ln\left|\frac{1 - \left(\frac{\Lambda P_\Lambda}{RR'}\right)}{1 + \left(\frac{\Lambda P_\Lambda}{RR'}\right)}\right|\right)$

The integrand is nonsingular, so only contributions bear the boundaries need be considered. For the bifurcation boundary

$$\frac{1}{2}RR'\ln\left|\frac{1-\left(\frac{\Lambda P_\Lambda}{RR'}\right)}{1+\left(\frac{\Lambda P_\Lambda}{RR'}\right)}\right| = -RR'\left[\left(\frac{\Lambda P_\Lambda}{RR'}\right)+\frac{1}{3}\left(\frac{\Lambda P_\Lambda}{RR'}\right)^3+\cdots\right]$$
$$= -\Lambda P_\Lambda + \frac{1}{3}\Lambda P_\Lambda\left(\frac{\Lambda P_\Lambda}{RR'}\right)^2 + \text{ higher orders}$$
$$= -\Lambda P_\Lambda + O(r^7)$$
So, even the logarithmic singularity is easily avoided.

As $r \to \infty$:

$$\Lambda = r^{-1}l - r^{-3}\left(\frac{1}{2}l^3\right) + r^{-4}\lambda + r^{-5}\left(\frac{3}{8}l^5\right) + O(r^{-6})$$
$$R = r + r^{-2}p_2 + O(r^{-4})$$
$$N_r = r^{-2}v(t) + O(r^{-4})$$
$$N = rl^{-1} + r^{-1}\left(\frac{1}{2}l\right) + r^{-2}\eta(t) - \frac{1}{8}r^{-3}l^3 + O(r^{-4})$$
$$P_\Lambda = O(r^{-2})$$
$$P_R = O(r^{-4})$$

$$\delta R = O(r^{-2})$$
$$\delta R' = O(r^{-3})$$
$$\delta\Lambda = O(r^{-4})$$
$$\delta P_\Lambda = O(r^{-2})$$
When the evaluate the surface terms labeled (i)-(vii) as before, we now find that terms (i)-(v) remain finite:

(i) $= O(r^0)$
(ii) $= O(r^0)$
(iii) $= O(r^0)$
(iv) $= O(r^0)$
(v) $= O(r^0)$
(vi) $= O(r^{-8})$
(vii) $= O(r^{-5})$

Let's evaluate the surface terms explicitly:

$$(ii)+(iii)+(v) = \left[-N'^{(\Lambda^{-1}R)}\delta R + N\Lambda^{-2}\Lambda' R\delta R + N\Lambda^{-1}R'\delta R\right]$$
$$+\left[-N\Lambda^{-2}\Lambda' R\delta R\right] + \left[N\Lambda^{-1}R'\delta R\right]$$
$$= -N'\left(\Lambda^{-1}R\right)\delta R$$

$$(i)+(iv) = NR\left(\Lambda^{-1}R\delta R'\right) + NR\left(-\Lambda^{-2}\delta\Lambda R'\right) = NR\delta\left(\Lambda^{-1}R\right)$$

So,

$$\lim_{r\to\infty} S(r) = \left\{NR\delta\left(\Lambda^{-1}R'\right) - N'\Lambda^{-1}R\delta R\right\}\Big|^\infty = X\Big|^\infty$$

$X = NR\delta(\Lambda^{-1}R') - N'\Lambda^{-1}R\delta R$
$NR = r^2 l^{-1} + \frac{1}{2}l + r^{-1}\eta - \frac{1}{8}r^{-2}l^3 + O(r^3)$
$\Lambda^{-1}R' = rl^{-1} + r^{-1}\left(\frac{1}{2}l\right) - r^{-2}(\lambda l^{-2}) - r^{-3}\left(\frac{1}{8}l^3\right) + O(r^{-4}) +$
$r^{-2}(-2p_2 l^{-1})$
$\delta(\Lambda^{-1}R') = r^{-2}(-\delta\lambda l^{-2} - 2\delta p_2 l^{-1}) + O(r^{-4})$
$NR\delta(\Lambda^{-1}R') = \delta(-\lambda l^{-3} - 2\delta p_2 l^{-2}) + O(r^{-2})$
$N'R = rl^{-1} - r^{-1}\left(\frac{1}{2}l\right) - 2r^{-2}\eta + \frac{3}{8}r^{-3}l^3 + O(r^{-4})$
$\Lambda^{-1}(N'R) = r^2 l^{-2} - r^{-2}\left(\frac{1}{4}l^2\right) + r^{-1}(-2\eta + p_2 l^{-1}) + \frac{3}{8}r^{-2}l^2 +$
$O(r^{-3})$
$\delta R = r^{-2}\delta p_2$
$\Lambda^{-1}(N^1 R)\delta R = \delta(p_2 l^{-2}) + O(r^{-3})$
$X = \delta(-\lambda l^3 - 3p_2 l^{-2}) = \delta(-M)$

$$\boxed{\lim_{r\to\infty} S(r) = \delta(-M)}$$

Consider $X = \frac{1}{2}R\delta R \ln\left|\frac{RR'+\Lambda P_\Lambda}{RR'+\Lambda P_\Lambda}\right|$ as $r \to \infty$:

$$\lim_{r\to\infty} X = (r)O(r^{-2})\left(\frac{(r^{-1}l)O(r^{-2})}{r(-2r^{-3}p_2)}\right) + \text{higher orders} = O(r^{-2}) \to 0$$

As for the $r \to \infty$ portion of the integral expressing the exact form w:

$$\frac{1}{2}RR'\ln\left|\frac{1-\left(\frac{\Lambda P_\Lambda}{RR'}\right)}{1+\left(\frac{\Lambda P_\Lambda}{RR'}\right)}\right| = -\Lambda P_\Lambda + \frac{1}{3}\Lambda P_\Lambda\left(\frac{\Lambda P_\Lambda}{RR'}\right)^2 + \text{higher orders}$$

$$= -\Lambda P_\Lambda + (r^{-1})O(r^{-2})\left(\frac{O(r^{-3})}{r}\right)^2$$

$$= -\Lambda P_\Lambda + (r^{-11})$$

B.5 Fall-off as $r \to 0$ at bifurcate 2-sphere

In the following is a derivation of fall-off conditions at the bifurcation 2-sphere for spherically symmetric asymptotically Anti-deSitter spacetimes.The Schwarzschild–AdS metric:

$$ds^2 = -f(R)dT^2 + f^{-1}(R)dR^2 + R^2 d\Omega^2 \;;\; f = 1 - \frac{2M}{R} + \left(\frac{R}{l}\right)^2$$

As before, consider $R = 2M(1 + p^2)$

$$f = 1 - \frac{2M}{2M(1+p^2)} + \left(\frac{2M}{l}\right)^2 (1 + p^2)^2$$

$$\boxed{f = \frac{p^2}{1+p^2} + X^2(1 + p^2)^2 \quad X = \left(\frac{2M}{l}\right)}$$

Consider $p = rp_2(t)$
$dR = 4M(rp_2^2 dr + r^2 p_2 \dot{p}_2 dt)$
$dR = -\left[\left(\frac{(p_2 r)^2}{1+(p_2 r)^2}\right) + (1 + (p_2 r)^2)^2\right] \dot{T}^2 dt^2$
$+\left[\left(\frac{(p_2 r)^2}{1+(p_2 r)^2}\right) + (1 + (p_2 r)^2)^2\right]^{-1} (4Mrp_2^2)^2 \left(dr + r\frac{\dot{p}_2}{p_2} dt\right)^2 + R^2 d\Omega^2$
$N = \dot{T}\left[\left(\frac{(p_2 r)^2}{1+(p_2 r)^2}\right) + X^2(1 + (p_2 r)^2)^2\right]^{1/2}$

This indicates problems, as $r \to 0,\ N \to X$. Consider for replacement $R = \alpha(1 + p^2)$ such that

$$F(R)_\alpha \to 0 \text{ as } p \to 0$$

for some choice of α. What is sought is a perturbative analysis about the bifurcation 2-sphere at $R = \alpha$:

$R = \alpha(1 + p^2)$
$f = 1 - \frac{2M}{\alpha}\frac{1}{(1+p^2)} + \left(\frac{\alpha}{l}\right)^2 (1 + p^2)^2$
As $p \to 0$
$f \cong 1 - \frac{2M}{\alpha}(1 - p^2) + \left(\frac{\alpha}{l}\right)^2 (1 + p^2)$
$\cong \underbrace{\left(1 - \frac{2M}{\alpha} + \left(\frac{\alpha}{l}\right)^2\right)}_{want\ zero} + \left[\left(\frac{DM}{\alpha}\right) + 2\left(\frac{\alpha}{l}\right)^2\right] p^2$
$a\alpha^3 + b\alpha + c = 0 \quad a = l^{-2}\ , b = 1\ , c = -2M$
From Math Handbook:
$x^3 + a_1 x^2 + a_2 x + a_3 = 0 \qquad (here\ a_1 = 0\,, a_2 = l^2\ ,\ a_3 = -2M)$
$Q = \frac{3a_2 - a_1^2}{9} = \left(\frac{l^2}{3}\right)$
$R = \frac{9a_1 a_2 - 27a_3 - 2a_1^3}{54} = Ml^2$

$$S = \sqrt[3]{R + \sqrt{Q^3 + R^2}} = \sqrt[3]{M + \sqrt{\left(\frac{l^2}{3}\right)^3 + M^2 l^4}}$$

$$T = \sqrt[3]{R - \sqrt{Q^3 + R^2}} = \sqrt[3]{M - \sqrt{\left(\frac{l^2}{3}\right)^3 + M^2 l^4}}$$

$D = \sqrt{Q^3 + R^2}$ is the discriminant. If D>0 (as is the case) then there is only one real root. This is convenient. The solution is clearer if reanalyzed however. ($l \to \infty$ limit not meaningful here)

Consider $f = \left(1 - \frac{2M}{\alpha} + \left(\frac{\alpha}{l}\right)^2\right)$ in the new form:

$$l = \lambda^{-1}$$
$$\alpha = \beta^{-1}$$

so have $\lambda \to 0$ good.

$1 - 2M\beta + \lambda^2\beta^2 = 0$

$-2M\beta^2 + \beta^2 + \lambda^2 = 0$

$\beta^3 - \left(\frac{1}{2M}\right)\beta^2 - \frac{\lambda^2}{2M} = 0, \ \lambda = 0 \to \beta = 2M^{-1} \to \alpha = 2M$

$a_M = -\frac{1}{2M}, a_2 = 0, a_3 = -\frac{\lambda^2}{2M}$

$Q = \frac{3a_2 - a_1^2}{9} = -\frac{1}{(6M)^2}$

$R = \frac{9a_1a_2 - 27a_3 - 2a_1^3}{54} = \left(\frac{\lambda}{2}\right)^2 \frac{1}{M} + \left(\frac{1}{6M}\right)^3$

$D = Q^3 + R^2 = -\frac{1}{(6M)^6} + \left(\frac{\lambda}{2}\right)^4 \frac{1}{M^2} + \frac{2}{M}\left(\frac{\lambda}{2}\right)^2\left(\frac{1}{6M}\right)^3 + \frac{1}{(6M)^6}$

$= \left(\frac{\lambda}{2M}\right)^2 \left(\left(\left(\frac{\lambda}{2}\right)\right) + \frac{2}{6^3}\frac{1}{m^2}\right) > 0$

$S = \sqrt[3]{R + \sqrt{Q^3 + R^2}} \quad , T = \sqrt[3]{R + \sqrt{Q^3 + R^2}}$

$$\beta_1 = S + T - 3a$$

Solutions: $\beta_2 = -\frac{1}{2}(S + T) - \frac{1}{3}a_1 + \frac{1}{2}i\sqrt{3}(S - T)$

$$\beta_3 = -\frac{1}{2}(S + T) - \frac{1}{3}a_1 + \frac{1}{2}i\sqrt{3}(S - T)$$

Consider $\lambda = 0 \ \to D = 0 \quad \beta_1 = -2\left(\frac{a_1}{3}\right) - \left(\frac{a_1}{3}\right) = -a_1 = \frac{1}{2M}$

$\beta_2 = \beta_3 = 0$

For D>0 there is one real root, β_1 .

Take α to be one real solution to $0 = 1 - \frac{2M}{\alpha} + \left(\frac{\alpha}{l}\right)^2$:

$$\alpha =$$

$$\left[\sqrt[3]{\left[\left(\frac{\lambda}{2}\right)^2 \frac{1}{M} + \left(\frac{1}{6M}\right)^3\right] + \frac{\lambda}{2}\sqrt{\left(\frac{\lambda}{2M}\right)^2 + \frac{2}{M}\left(\frac{1}{6M}\right)^3}} + \sqrt{[\dots] - \frac{\lambda}{2}\sqrt{\dots}} + \frac{1}{6M}\right]^{-1}$$

$R = \alpha(1 + p^2)$

$f = 1 - \frac{2M}{\alpha(1+p^2)} + \left(\frac{\alpha}{l}\right)^2 (1 + p^2)^2 = 1 - \frac{\left(1+\left(\frac{\alpha}{l}\right)^2\right)}{(1+p^2)} + \left(\frac{\alpha}{l}\right)^2 (1 + p^2)^2$

$$\boxed{f = \left(\frac{p^2}{1 + p^2}\right) + \left(\frac{\alpha}{l}\right)^2 \left[(1 + p^2) - \frac{1}{(1 + p^2)}\right]}$$

Again, consider $p = r^2 p_2(t)$, $T = T(t)$

$dR = 2\alpha(r^2 p_2 \dot{p} dt + r p_2^2 dr)$

$ds^2 = -f\dot{T}^2 dt + f^{-1}(2\alpha r p_2^2)^2 \left(dr + r\frac{\dot{p}_2}{p_2}dt\right)^2 + R^2 d\Omega^2$

$$N = f^{1/2}\dot{T}$$

$$\Lambda = (2\alpha r p_2^2) f^{-1/2}$$

$$N^r = r\frac{\dot{p}_2}{p_2}$$

$$R = \alpha(1 + (p_2 r)^2)$$

$P_\Lambda = -N^{-1}R(\dot{R} - R'^{N^r})$; $(\dot{R} - R'^{N^r}) = 2\alpha p_2 \dot{p}_2 r^2 - 2\alpha p_2^2 r \left(r\frac{\dot{p}_2}{p_2}\right) = 0,$

and using $P_R = \frac{\Lambda P'_\Lambda}{R\prime}$:

$$\boxed{\begin{array}{l} P_\Lambda = 0 \\ P_R = 0 \end{array}}$$

The leading order expression for the various terms:

$f = \frac{(p_2 r)^2}{1+(p_2 r)^2} + \left(\frac{\alpha}{l}\right)^2 \left[(1 + p^2) - \frac{1}{(1+p^2)}\right]$

$f = (p_2 r)^2(1 - (p_2 r)^2 + O(r^4)) + \left(\frac{\alpha}{l}\right)^2 [1 + 2p^2 + p^4 - (1 - p^2 + p^4 + O(r^6))]$

$= \left(p_2^2 + 3\left(\frac{\alpha}{l}\right)^2\right) r^2 + O(r^4)$

$f^{1/2} = \sqrt{p_2^2 + 3\left(\frac{\alpha}{l}\right)^2}\, r + O(r^3)$

$f^{1/2} = \frac{r^{-1}}{\sqrt{p_2^2 + 3f^{1/2}}} + O(r)$

$$\boxed{\begin{array}{c} N = f^{1/2}\dot{T} = \left(\dfrac{\dot{T}}{\sqrt{p_2^2 + 3f^{1/2}}}\right) + O(r^3) \\ N^r = \dfrac{\dot{p}_2}{p_2} r \\ \Lambda = \left(\dfrac{2\alpha p_2^2}{\sqrt{p_2^2 + 3f^{1/2}}}\right) + O(r^2) \\ R = \alpha + (\alpha p_2^2) r^2 \end{array}}$$

This form is essentially the same as for Schwarzschild, all else being equal, the results would be similar as well. However, the super-Hamiltonian is different, and consequently $P_{R,\Lambda}$, so there may possibly be changes.

The constraints:
$0 = H = -R^{-1}P_R P_\Lambda + \frac{1}{2}R^{-2}\Lambda P_\Lambda^2 + \Lambda^{-1}RR'' - \Lambda^{-2}RR'\Lambda' + \frac{1}{2}\Lambda^{-1}R'^2 - \frac{1}{2}\left(\frac{3}{l^2}\right)(\Lambda R^2)$
$0 = H_r = P_R R' - \Lambda P_\Lambda'$
$M = \frac{1}{2}R^{-1}P_\Lambda^2 - \frac{1}{2}\Lambda^{-2}RR'^2 + \frac{1}{2}R + \left(\frac{1}{2l^2}R^3\right)$
The EOM's:
$\dot{R} = -NR^{-1}P_\Lambda + N^r R'$
$\dot{\Lambda} = N[R^{-2}\Lambda P_\Lambda - R^{-1}P_R] + (\Lambda N^r)'$
$\dot{P}_\Lambda = -\frac{1}{2}N\left[R^{-2}P_\Lambda^2 - \Lambda^{-2}\left(2RR'' + R'^2 - 2(RR')\right) - 1\right] - \Lambda^{-2}RR'N' + N^r P_\Lambda' + \frac{1}{2}\left(\frac{3}{l^2}\right)R^2 N$
$= \frac{1}{2}N\left[1 - \Lambda^{-2}R'^2 - R^2 P_\Lambda^2\right] - \Lambda^{-2}RR'N^r + N^r P_\Lambda' + \frac{1}{2}N\left(\frac{3}{l^2}\right)R^2$
$\dot{P}_R = N[R^{-3}\Lambda P_\Lambda - R^{-2}P_\Lambda P_R] - N\left(\frac{R''}{\Lambda}\right) - \left(N\frac{R}{\Lambda}\right)'' + NR\left(\frac{1}{\Lambda}\right)' - \left(\left(\frac{1}{\Lambda}\right)' NRR'\right)' + \left(\Lambda^{-1}R'N\right)' + N\left(\frac{3}{l^2}\right)(\Lambda R) + (N_r P_R)'$
$= N[R^{-3}\Lambda P_\Lambda^2 - R^{-2}P_\Lambda P_R] + (N^r P_R)' - N\left(\frac{R'}{\Lambda}\right)' - \left(\frac{N'}{\Lambda}R\right)' + N\left(\frac{3}{l^2}\right)(\Lambda R)$
This is known about the general slice (regular at the bifurcation point):
$N = N_1(t)r + O_\geq(r^2)$
$N^r = N_1^r(t)r + O_\geq(r^2)$
$\Lambda = \Lambda_0(t) + O_\geq(r^2)$
$R = R_0(t) + R_2(t)r^2 + O_\geq(r^3)$
$P_\Lambda = O_\geq(r)$
$P_R = O_\geq(r)$
As before, the super-momentum constraint requires $O(P_\Lambda) = r^2 O(P_R)$
So try $P_\Lambda = O(r^3), P_R = O(r)$ again.

From $\dot{R}$ eqn.:
$\left(\dot{R}_0 + \dot{R}_2 + O_\geq(r^3)\right) = \frac{\left(N_1 r + O_\geq(r^2)\right)}{\left(R_0 + R_2 r^2 + O_\geq(r^2)\right)}O(r^3) + \left(N_1^r r + O_\geq(r^2)\right)\left(2R_2 r + O_\geq(r^2)\right)$

$= O(r^4) + 2R_2 N_1^r r^2 + O_{\geq}(r^3)$

Indicating $\boxed{\dot{R}_0 = 0}$ and $\dot{R}_2 = 2R_2 N_1^r$.

From the $\dot{\Lambda}$ eqn.:

$\left(\dot{\Lambda}_0 + O_{\geq}(r^2)\right) = \left(N_1 r + O_{\geq}(r^2)\right)\left[O(r^3) - \frac{O(r)}{R}\right] + \Big((\Lambda_0 + O_{\geq}(r^2))(N_1^r r + O_{\geq}(r^2))\Big)'$

$= O(r^2) + N_1^r \Lambda_0 + O_{\geq}(r)$

indicating $\dot{\Lambda}_0 = N_1^r \Lambda_0$ and $N^r = N_1^r(t) r + O(r^3)$.

Consider the mass equation:

$M = \frac{1}{2}R^{-1}P_\Lambda^2 - \frac{1}{2}\Lambda^{-2}RR'^2 + \frac{1}{2}R + \left(\frac{1}{2l^2}R^3\right)$

$$2M = O(r^6) - \frac{\left(R_0 + R_2 r^2 + O_{\geq}(r^3)\right)\left(2R_2 r + O_{\geq}(r^2)\right)^2}{\left(\Lambda_0 + O_{\geq}(r^2)\right)^2} + \left(R_0 + R_2 r^2 + O_{\geq}(r^3)\right) + \frac{1}{l^2}\left(R_0 + R_2 r^2 + O_{\geq}(r^3)\right)$$

Indicating $2M = R_0 + \left(\frac{{R_0}^3}{l^2}\right)$; $R_2\left(1 + \frac{3R_0^2}{l^2}\right) = \frac{4R_0 R_2^2}{\Lambda_0^2}$

Recall: $2M = \alpha + \frac{\alpha^3}{2l^2} \rightarrow \boxed{R_0 = \alpha}$

Now consider $\dot{P}_\Lambda$:

$$O(r^3) = \frac{1}{2}\left(N_1 r + O_{\geq}(r^2)\right)\left[1 - \left(2R_2 r + O_{\geq}(r^2)\right)^2\left(\Lambda_0 + O_{\geq}(r^2)\right)^{-2} - O(r^3)\right]$$
$$- \frac{\left(R_0 + R_2 r^2 + O_{\geq}(r^3)\right)}{\left(\Lambda_0 + O_{\geq}(r^2)\right)^2}\left(2R_2 r + O_{\geq}(r^2)\right)\left(N_1 + O_{\geq}(r)\right)$$
$$+ \left(N_1 r + O(r^3)\right)O(r^2) + \frac{1}{2}\left(N_1 r + O_{\geq}(r^2)\right)\left(\frac{3}{l^2}\right)\left(R_0 + R_2 r^2 + O_{\geq}(r^3)\right)$$
$$= r\left[\frac{1}{2}N_1 - \frac{2R_0 R_2}{\Lambda_0^2}N_1 + \frac{1}{2}N_1\left(\frac{3}{l^2}\right)R_0\right]$$

$$= -\underbrace{rO_{\geq}(r)}_{from\ N}\left(\frac{2R_0R_2}{\Lambda_0^2}\right) - \underbrace{O_{\geq}(r^2)}_{from\ R}\left(\frac{N_1R_0}{\Lambda_0^2}\right) + \underbrace{\frac{1}{2}O_{\geq}(r^2)}_{from\ N} + O(r^3)$$

This line indicates:

$$\boxed{\begin{array}{c} N = N_1 r + O(r^3) \\ R = R_0 + R_2 r^2 + O(r^4) \end{array}}$$

The preceding line indicates:

$$\left(1 - \frac{4R_0R_2}{\Lambda_0^2}\right) + R_0\left(\frac{3}{l^2}\right) = 0$$

indicating $R_2 > 0$ as before (from the mass equation there was the same relation).

Now consider $\dot{P}_R$:

$$\dot{P}_R = (N_1 r + O(r^2))[O(r^6) - O(r^4)] + [(N_1 r + O(r^3))O(r)]'$$
$$-(N_1 r + O(r^3))\left[\frac{2R_2 r + O(r^3)}{\Lambda_0 + O_{\geq}(r^3)}\right]'$$
$$-\left[\frac{N_1 + O(r^2)}{\Lambda_0 + O_{\geq}(r^3)}(R_0 + R_2 r^2 + O_{\geq}(r^4))\right]'$$
$$+(N_1 r + O(r^3))\left(\frac{3}{l^2}\right)(\Lambda_0 + O_{\geq}(r^2))(R_0 + R_2 r^2 + O(r^4))$$

$O(r) = O(r)$ no further restriction on $O_{\geq}(r^2)$ of Λ, so can take

$$\Lambda = \Lambda_0 + O(r^2)$$

This completes the fall-off condition analysis.

B.6 Saddle-point evaluation

In this section we show the saddle-point evaluation for thermodynamic analysis of the RNAdS Black Hole system.

The domain of horizon radii for given parameter values:

$ds^2 = -FdT^2 + F^{-1}dR^2 + R^2 d\Omega^2$

$F = \ell^{-2}R^2 + 1 - 2MR^{-1} + Q^2R^{-2}$

Write $R = \ell x, M = \ell \propto, Q = \ell \gamma,$ and $f(x) = x^2 F$

$f(x) = x^4 + x^2 - 2 \propto x + \gamma^2$

The horizon occurs at $F = 0 \Rightarrow f = 0.$

$f'(x) = 4x^3 + 2x - 2 \propto$

$f''(x) = 12x^2 + 2$

There is only one real (positive) root for $f'(x) = 0$, and that root is a minimum. The minimum satisfies $\underline{f'(r) = 0}$: $\underline{\alpha = 2r^3 + r}$.

In order that a horizon exist the curvature radius must satisfy $f(R_0) = 0$, and this is only possible if $\underline{f(r) < 0}$. Of the two possible horizon solutions, the larger is the horizon for the class of spacetimes considered here.

$$f(r) < 0 \Rightarrow \gamma^2 < 3r^4 + r^2 \longrightarrow \boxed{|Q| < Q_{crit}\, , Q_{crit}\ell^{-1} = \sqrt{3r^4 + r^2}}$$

$$\boxed{\begin{array}{l} R > R_0 > r > R_{crit}\, , R_{crit} = \dfrac{\ell}{\sqrt{6}}\left(\sqrt{12(Q/\ell)^2 + 1} - 1\right)^{1/2} \\ M > M_{crit} = \dfrac{\ell}{3\sqrt{6}}\left(\sqrt{12(Q/\ell)^2 + 1} - 1\right)^{1/2}\left(\sqrt{12(Q/\ell)^2 + 1} + 2\right) \end{array}}$$

From the Hamiltonian analysis we can get the partition function ($\delta\Phi = 0$ on boundaries):

$$S\left[m, q, p_m, p_q; \widetilde{N}_0^M, \widetilde{N}_t, \widetilde{\Phi}_t, \widetilde{\Phi}_0\right] = \int dt\, (p_m\dot{m} + p_q\dot{q} - h)$$

$$h = -\frac{1}{2}R_0^2\widetilde{N}_0^M + \widetilde{N}_t + (\widetilde{\Phi}_t - \widetilde{\Phi}_0)q$$

$\widehat{K}(t_2, t_1) = \exp\left\{-i \int_{t_1}^{t_2} dt'\, \hat{h}(t')\right\}$

$K(m, q; \Theta_H, T_\infty) = \exp\left\{-i\left(-\frac{1}{2}R_0^2\left\{\int_{t_1}^{t_2} \widetilde{N}_0^m\, dt\right\} + m\left\{\int_{t_1}^{t_2} \widetilde{N}_t\, dt\right\} + q\left\{\int_{t_1}^{t_2}(\widetilde{\Phi}_t - \widetilde{\Phi}_0)\, dt\right\}\right)\right\}$

$= \exp\left\{-i\left(-\frac{1}{2}R_0^2\Theta_H + mT_\infty + q\phi\right)\right\}$

The description is Euclideanized to obtain the partition function:

(1) $t \to i\tau, \tau$ an angular coordinate

(2) $T_\infty = \int_{t_1}^{t_2} \widetilde{N}_t\, dt = -i\beta$

(3) $\Theta_H = \int_{t_1}^{t_2} \widetilde{N}_0^m\, dt = -i2\pi$

(4) $\phi = \int_{t_1}^{t_2}(\widetilde{\Phi}_t - \widetilde{\Phi}_0)\, dt = -i\Phi\beta$

$K_E(m, q; -i2\pi, -i\beta\Phi) = \exp\{-m\beta + \pi R_0^2 - q\Phi\beta\}$

$Z = \int dmdq\, K_E(m, q) = \int dmdq \exp[-m(R_0, q)\beta + \pi R_0^2 - q\Phi\beta]$

(the partition function is defined up to "slowly" varying measure terms).

$$m(R_0, q) = \frac{1}{2} R_0 (\ell^{-2} R_0^2 + 1 + q^2 R_0^{-2})$$

m has a monotonically increasing dependence on R_0 in the domain of allowed horizon radil:

$$\frac{\partial m}{\partial R_0} = \frac{3}{2\ell^2} R_0^2 + \frac{1}{2} - \frac{1}{2}\frac{q^2}{R_0^2} \rightarrow \text{monotonic if } \frac{1}{2}\left(\frac{q}{R_0}\right)^2 < \frac{3}{2}\left(\frac{R_0}{\ell}\right)^2 + \frac{1}{2}$$

$(q = Q)$

$$Q^2 < Q_{crit}^2 = \ell^2(3r^4 + r^2) < \ell^2\left(3\left(\frac{R_0}{\ell}\right)^4 + \left(\frac{R_0}{\ell}\right)^2\right) \rightarrow \frac{1}{2}\left(\frac{Q}{R_0}\right)^2 < \frac{3}{2}\left(\frac{R_0}{\ell}\right)^2 + \frac{1}{2}$$

So, ignoring measure factors, it is possible to change to R_0 as integration variable:

$$Z = \int_0^\infty d\,R_0 \int_{-Q_{crit}}^{+Q_{crit}} dQ \exp[-m(R_0, q)\beta + \pi R_0^2 - Q\Phi\beta] = \int dR_0 dQ e^{-I}$$

If δQ is fixed at the boundaries, instead of $\delta\Phi$, a different partition function results from the Hamiltonian theory. With $\delta Q = 0$ at the boundaries the boundary action simplifies to:

$$S_{\partial\Sigma} = \int dt \left(\frac{1}{2} R_0^2 \tilde{N}_0^M - \tilde{N}_t M_t\right)$$

The reduced Hamiltonian is now just

$$h = -\frac{1}{2} R_0^2 \tilde{N}_0^M - \tilde{N}_t M$$

In this theory it is consistent to view Q as merely a parameter – no longer a dynamical variable – in which case the partition function is:

$$Z_{\delta\phi} = \int_0^\infty dR_0 \exp[-m(R_0, Q)\beta + \pi R_0^2]$$

Are these partition functions well defined (finite)? Consider the Q integration for $Z(\delta\Phi = 0)$:

$$A_Q = \int_{-Q_{crit}}^{+Q_{crit}} dQ \exp\left[-\frac{\beta}{2R_0}Q^2 - \Phi\beta Q\right]$$

$$< \int_{-\infty}^{\infty} dQ \exp\left[-\frac{\beta}{2R_0}Q^2 - \Phi\beta Q\right] \equiv A_Q^{\infty}$$

$$A_Q^{\infty} = \sqrt{\frac{\pi R_0}{2\beta}} \exp\left(\frac{R_0 \Phi^2 \beta}{2}\right)$$

Thus $Z_{\delta\Phi}$ is bounded by Z_B (as is $Z_{\delta Q}$ when $Q < Q_{crit}$ is accounted for):

$$Z_B = \int_0^{\infty} dR_0 \exp\left[-\frac{\beta}{2\ell^2}R_0^3 + \pi R_0^2 - \frac{\beta}{2}R_0 + \frac{1}{2}\Phi^2 \beta R_0\right]$$

Z_B, in turn, is finite due to the R_0^3 term that arises from the asymptotic AdS constraint on geometries. If the spacetime were asymptotically flat the positive R_0^2 term would dominate in a similar expression bounding from below, leading to a divergent Z.

<u>Configurations that dominate the ($\delta\Phi = 0$ at boundaries) partition function:</u>

$$Z = \int d\,R_0 dQ e^{-I}$$

$$I - \left(\frac{\beta}{2\ell^2}\right)R_0^3 - \pi R_0^2 - \frac{\beta}{2}R_0 + Q\Phi\beta + \frac{1}{2}\beta Q^2 R_0^{-1}$$

$$\frac{\partial I}{\partial R_0} = \frac{3\beta}{2\ell^2}R_0^2 - 2\pi R_0 + \frac{\beta}{2} - \frac{1}{2}\beta Q^2 R_0^{-2}$$

$$\frac{\partial I}{\partial Q} = \Phi\beta + \beta Q R_0^{-1}$$

$$\left.\begin{aligned} \frac{\partial I}{\partial R_0} = 0 &\Rightarrow \left(\frac{Q}{R_0}\right)^2 = \left(\frac{3}{\ell^2}R_0^2 - \frac{4\pi}{\beta}R_0 + 1\right) \\ \frac{\partial I}{\partial Q} = 0 &\Rightarrow Q = -\Phi R_0 \end{aligned}\right\} \Rightarrow$$

$$R_0^2 - \left(\frac{4\pi\ell^2}{3\beta}\right)R_0 - \frac{\ell^2}{3}(\Phi^2 - 1) = 0$$

$$R_0^{(\pm)} = \left(\frac{2\pi\ell^2}{3\beta}\right) \pm \sqrt{\left(\frac{2\pi\ell^2}{3\beta}\right)^2 + \frac{\ell^2}{3}(\Phi^2 - 1)}$$

$$\frac{\partial^2 I}{\partial R_0^2} = \frac{3\beta}{\ell^2} R_0 - 3\pi\pi + \beta Q^2 R_0^{-3}$$

$$\frac{\partial^2 I}{\partial Q^2} = \beta R_0^{-1} > 0$$

$$X = \det \begin{vmatrix} I_{,R_0 R_0} & I_{,R_0 Q} \\ I_{,Q R_0} & I_{,QQ} \end{vmatrix} = 3\beta^2 \ell^{-2} \left(1 - \frac{2\pi\ell^2}{3\beta} R_0^{-1}\right)$$

If $X = \left(R_0^{(\pm)}\right) > 0$ (with $I_{,QQ} > 0$), then $R_0^{(\pm)}$ is a minimum of I.
If $X < 0$, then the horizon root represents a saddle point.

$X > 0$ if $R_0^{(\pm)} > \left(\frac{2\pi\ell^2}{3\beta}\right)$, this is true for R_0^{+} but not R_0^{-}. So, $R_0^{(+)}$ is a minimium and R_0^{-} is a saddle point. The extremal roots $R_0^{(\pm)}$ only represent horizon solutions when real and positive. Consequently:

$1 < \Phi^2$ permits only one root: $R_0^{(+)}$

$\left(1 - \left[\frac{2\pi}{\beta}\right]^2 \frac{\ell^2}{3}\right) \leq \Phi^2 \leq 1$: both roots allowed

$\Phi^2 < \left(1 - \left[\frac{2\pi}{\beta}\right]^2 \frac{\ell^2}{3}\right)$ neither root exists (complex)

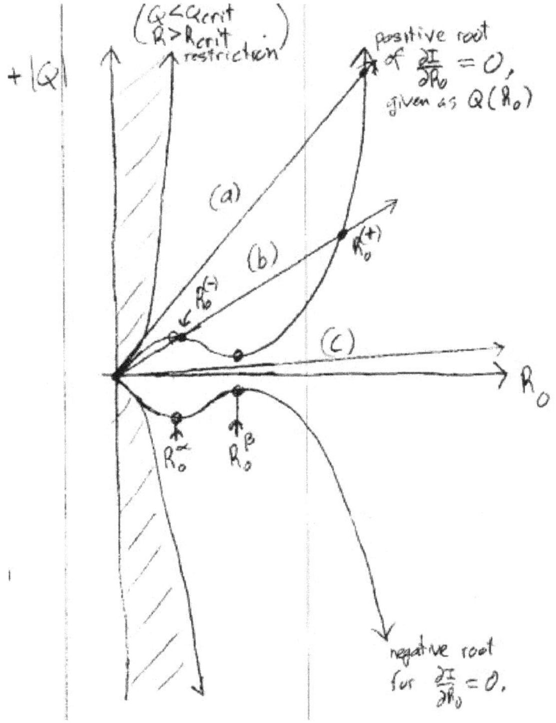

(a) $\frac{\partial I}{\partial Q} = 0$ solution with $\Phi < -1$

(b) $\frac{\partial I}{\partial Q} = 0$ solution with Φ negative and $\left(1 - \left(\frac{2\pi}{\beta}\right)^2 \frac{\ell^2}{3}\right) \leq \Phi^2 \leq 1$

(c) $\frac{\partial I}{\partial Q} = 0$ solution with negative Φ and $\Phi^2 < \left(1 - \left(\frac{2\pi}{\beta}\right)^2 \frac{\ell^2}{3}\right)$

Intercepts between $\frac{\partial I}{\partial R_0} = 0$ and $\frac{\partial I}{\partial Q} = 0$ also exist for the negative root of $\frac{\partial I}{\partial R_0} = 0$ with positive Φ in $\frac{\partial I}{\partial Q} = 0$. Consider the curve $Q(R_0)$ for the positive root of $\frac{\partial I}{\partial R_0} = 0$:

$$Q = R_0 \sqrt{\frac{3}{\ell^2} R_0^2 - \frac{4\pi}{\beta} R_0 + 1}$$

$$\frac{\partial Q}{\partial R_0} = \frac{1}{\sqrt{\frac{3}{\ell^2} R_0^2 - \frac{4\pi}{\beta} R_0 + 1}} \left\{ \left(\frac{3}{\ell^2} R_0^2 - \frac{4\pi}{\beta} R_0 + 1 \right) + \frac{1}{2} R_0 \left(\frac{3}{\ell^2} R_0 - \frac{4\pi}{\beta} \right) \right\}$$

$$= \frac{1}{\sqrt{\frac{3}{\ell^2} R_0^2 - \frac{4\pi}{\beta} R_0 + 1}} \left\{ \frac{6}{\ell^2} R_0^2 - \frac{6\pi}{\beta} R_0 + 1 \right\} = \frac{R_0}{Q} \left\{ \frac{6}{\ell^2} R_0^2 - \frac{6\pi}{\beta} R_0 + 1 \right\}$$

$$= \frac{R_0^2}{Q\beta} \left\{ \left. \frac{\partial^2 I}{\partial R_0^2} \right|_{\frac{\partial I}{\partial R_0} = 0} \right\}$$

Since:

$$\frac{\partial^2 I}{\partial R_0^2} = \frac{3\beta}{\ell^2} R_0 - 2\pi + \beta Q^2 R_0^{-3}; \qquad \left. \frac{\partial^2 I}{\partial R_0^2} \right|_{\frac{\partial I}{\partial R_0} = 0} = \frac{6\beta}{\ell^2} R_0 - 6\pi + \beta R_0^{-1}$$

So,

$$\frac{\partial Q}{\partial R_0} > 0 \Rightarrow \left. \frac{\partial^2 I}{\partial R_0^2} \right|_{\frac{\partial I}{\partial R_0} = 0} > 0,$$

a minima for variation confined to R_0, and likewise $\frac{\partial Q}{\partial R_0} < 0 \Rightarrow$ maxima. This will prove useful in reading off the stable points from the overlay of plots on the $Q - R_0$ plane. The restriction on the parameter space due to $|Q| < Q_{crit}$ and $R > R_{crit}$ is given by the shaded region in the sketch. The border of the shaded region is found from

$$\left(\frac{Q}{R_0}\right) = \frac{3}{\ell^2}R_0^2 + 1$$

As this border curve never intersects the $\frac{\partial I}{\partial R_0}$ curve, it introduces no further complication in determining stable points. In fact, the parameter space restriction now makes the $R_0 \longrightarrow 0, Q \longrightarrow 0$ limit of $I(R_0, Q)$ well defined since $(Q/R_0)^2 \geq 1$.

The sketch shows a graph (in the real Q, R_0) for $\left(\frac{\partial I}{\partial R_0}\right) = 0$ in terms of $Q(R_0)$. Such a graph is not possible for all values of $\{\ell, \beta\}$:

$$\left(\frac{Q}{R_0}\right)^2 = \left(\frac{3}{\ell^2}R_0^2 - \frac{4\pi}{\beta}R_0 + 1\right) = g(R_0)$$

For the graph to exist, $g(R_0) > 0$:

$g(x) = 0$: $x^2 - \frac{4\pi\ell^2}{3\beta}x + \frac{\ell^2}{3} = 0$ (same as $\Phi = 0$)

$$x^{(\pm)} = \frac{2\pi\ell^2}{3\beta} \pm \sqrt{\left(\frac{2\pi\ell^2}{3\beta}\right)^2 - \frac{\ell^2}{3}}$$

Want zero or one real intercept: $\frac{\ell^2}{3} \geq \left(\frac{2\pi\ell^2}{3\beta}\right)^2$

$$1 \geq \left(\frac{2\pi}{\beta}\right)^2 \frac{\ell^2}{3}$$

The graph of $g(R_0)$ with one intercept $\left(x_0 = \left(\frac{2\pi}{\beta}\right)\frac{\ell^2}{3} = \frac{\beta}{2\pi}\right)$ is possible.

For $\left(\frac{2\pi}{\beta}\right)^2 \frac{\ell^2}{3} = 1$ the graph of $Q(R_0)$ becomes:

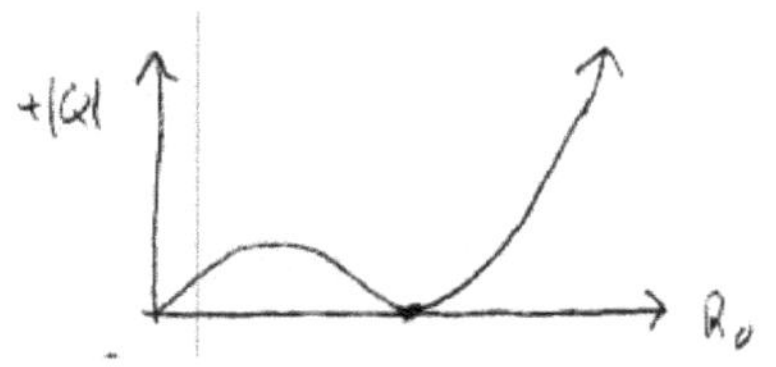

For $\left(\frac{2\pi}{\beta}\right)^2 \frac{\ell^2}{3} < 1$ the graph of real $Q(R_0)$ solutions is:

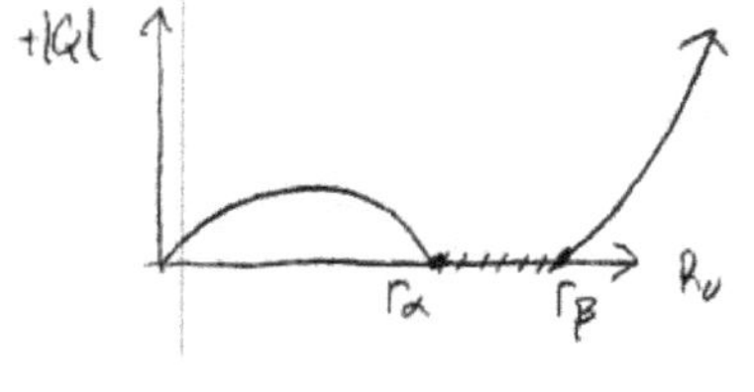

Where the graph of $Q(R_0)$ between $r_\propto$ and r_β does not extend into negative Q but imaginary Q instead – values obviously excluded for the spacetimes studies. Now to examine the stable minima for $I(R_0, Q)$ on the allowed parameter domain:

<u>$|\Phi| \geq 1$</u>

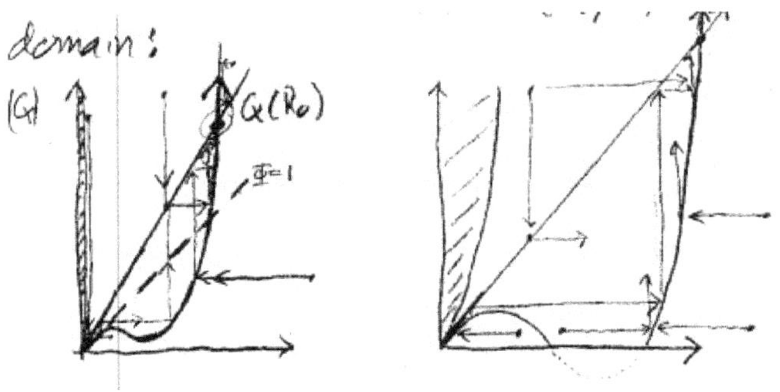

There is only one local minimum for $|\Phi| \geq 1, R_0^{(+)}$, there being no additional minima from the parameter boundaries.

$$\left(1 - \left[\frac{2\pi}{\beta}\right]^2 \frac{\ell^2}{3}\right) \leq \Phi^2 \leq 1:$$

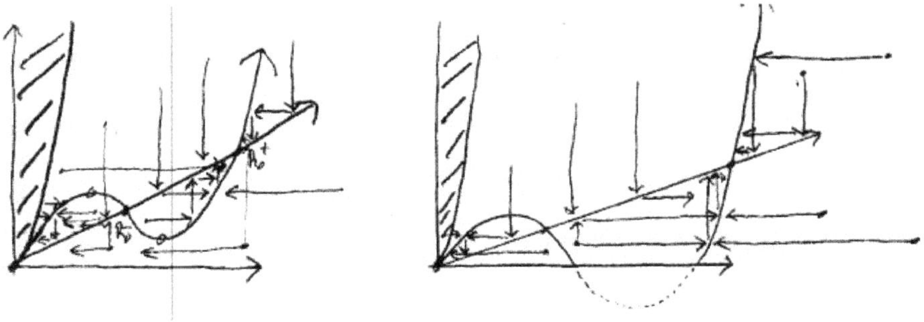

In this instance there are two local minima: R_0^+ and the $R_0 - Q$ origin. The extrema at R_0^- is a saddle point, as can be seen graphically above or via analysis using the X term described earlier. To determine which minima is the global minimum:

$I(\text{origin}; QR_0^- = -\Phi) = 0$

$$I\left(R_0^{(+)}\right) = \frac{R_0}{3}\left\{2\pi R_0 - \frac{\beta}{2} + \beta\frac{Q^2}{2}R_0^{-2}\right\} - \pi R_0^2 + \frac{\beta}{2}R_0 + Q\Phi\beta$$
$$+ \frac{1}{2}\beta Q^2 R_0^{-1}$$
$$= \frac{1}{3}\pi R_0^{(+)^2} + \frac{\beta}{3}R_0 + Q\Phi\beta + \frac{2}{3}\beta Q^2 R_0^{-1}$$

$$= \frac{1}{3}\pi R_0^{(+)^2} + \frac{\beta}{3}R_0^{(+)} + \frac{1}{3}Q\Phi\beta = \frac{1}{3}\left(-\pi\left(R_0^{(+)^2}\right) + \beta R_0^{(+)} - \Phi^2\beta R_0^{(+)}\right)$$
$$= \frac{\beta}{3}R_0^{(+)}\left\{-\frac{\pi}{\beta}R_0^{(+)} + 1 - \Phi^2\right\} \qquad 0 \leq 1 - \Phi^2 \leq \left(\frac{2\pi}{\beta}\right)^2\frac{\ell^2}{3}$$

$$I((R_0^+)_{min}) = \frac{\beta}{3}\left(R_0^{(+)}\right)_m\left\{-\frac{\pi}{\beta}\left(R_0^{(+)}\right)_m + \left(\frac{2\pi}{\beta}\right)^2\frac{\ell^2}{3}\right\}; \quad (R_0^+)_{min}$$
$$= \left(\frac{2\pi\ell^2}{3\beta}\right)$$
$$= +\frac{\beta}{3}(R_0^+)_{min}^2$$

$$\beta \to 0, R_0^+ \to (R_0^+)_{max} = \left(\frac{4\pi\ell^2}{3\beta}\right) \text{ and } (1 - \Phi^2) \approx 0$$
$$I((R_0^+)_{max}) = -\frac{\pi}{3}(R_0^+)_{max}^2$$

So, the global minimum depends on the parameters.

$$\Phi^2 < \left(1 - \left(\frac{2\pi}{\beta}\right)^2\frac{\ell^2}{3}\right):$$

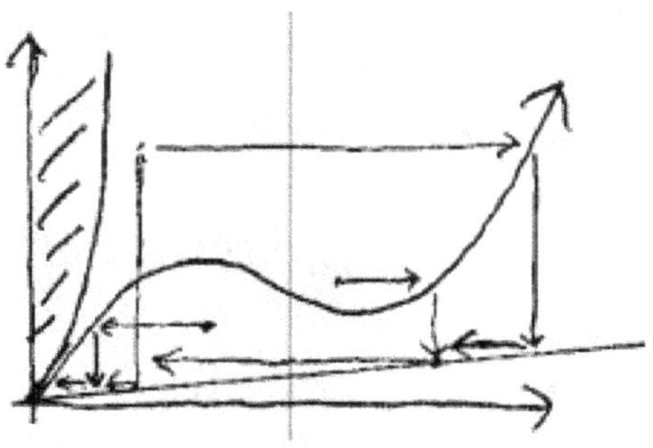

There is one minima now, and it is at the origin.

Summary of minima:

$$\Phi^2 \geq 1 \rightarrow \; R_0^{(+)} = \left(\frac{2\pi\ell^2}{3\beta}\right) + \sqrt{\left(\frac{2\pi\ell^2}{3\beta}\right)^2 + \frac{\ell^2}{3}(\Phi^2 - 1)}$$

$$\left(1 - \left(\frac{2\pi}{\beta}\right)^2 + \frac{\ell^2}{3}\right) \leq \Phi^2 < 1 \rightarrow \; R_0^{(+)} \text{ and } R_0 - Q \text{ origin}$$

$$\Phi^2 < \left(1 - \left(\frac{2\pi}{\beta}\right)^2 + \frac{\ell^2}{3}\right) \rightarrow \; R_0 - Q \text{ origin}$$

B.7 Thermodynamics from heuristics

In this section we describe the thermodynamics for RNAdS – Heuristic derivation based on Euclideanization.

$$ds^2 = -FdT^2 + F^{-1}dR^2 + R^2 d\Omega^2 \;\; (R_h = R_0)$$
$$F = R^2\ell^2 + 1 - 2MR^{-1} + Q^2R^{-2}$$

$F(R_h) = 0$ defines the horizon solution R_h:

$$M = \frac{R_h}{2}\left(\frac{R_h^2}{\ell^2} + 1 + \frac{Q^2}{R_h^2}\right)$$

Euclideanization defines temperature:

Write $R = R_h(1 + \rho^2), for\ \rho \ll 1$, i.e., near the horizon

$$F = \underline{R_h^2\ell^2} + R_h^2(2\rho^2)\ell^{-2} + \underline{1} - 2MR_h^{-1}(\underline{1} - \rho^2) + Q^2R_h^{-2}(\underline{1} - 2\rho^2) + O(\rho^4)$$
$$= \rho^2(2R_h^2\ell^{-2} + 2MR_h^{-1} - 2Q^2R_h^{-2}) + O(\rho^4)$$
$$= F(R_h)_0 + F'(R_h)(R - R_h) + O(R - R_h)^2 = \rho^2 R_h F'(R_h) + O(\rho^4)$$

Using $dF = R_h(2\rho d\rho)$, ds^2 near the horizon is:

$$ds^2 = -\rho^2 R_h F'(R_h) dT^2 + \frac{R_h^2 4\rho^2 d\rho^2}{R_h \rho^2 F'(R_h)} + R^2 d\Omega^2 + O(\rho^4)$$
$$= \frac{4R_h}{F'(R_h)} d\rho^2 - R_h F'(R_h)\rho^2 dT^2 + R^2 d\Omega^2 + O(\rho^4)$$

Upon Euclideanization, $T \rightarrow i\tau$, and τ is taken to be an angular coordinate with periodicity such that the 2-plane described by coordinates $\{\rho, \tau\}$ has no conical singularities.

$$(ds^2)_E = \frac{4R_h}{F'(R_h)}\left[d\rho^2 + \rho^2\left(\frac{F'(R_h)}{2}\right)^2 d\tau^2\right] + R^2 d\Omega^2 + O(\rho^4)$$

So, $\frac{F'(R_h)}{2}\tau$ is periodic with period 2π, the period in τ is taken to be inverse temperature, denoted by β:

$$\beta = \frac{4\pi}{F'(R_h)}$$

$$\beta(R_h^2\ell^{-2} + MR_h^{-1} - Q^2R_h^{-2}) = 2\pi R_h$$

Use M relation to get:

$$\beta\left(\frac{3}{2}\frac{1}{\ell^2}R_h^2 + \frac{1}{2} - \frac{1}{2}Q^2R_h^{-2}\right) = 2\pi R_h$$

This is precisely the equation that results from $\frac{\partial I}{\partial R_0} = 0$.

If Q and R_h are restricted to varying in a co-dependent manner such that

$$\Phi = -QR_h^{-1}$$

is a fixed quantity, then the relation $\frac{\partial I}{\partial Q} = 0$ is effectively imposed. The analysis then reveals the same behavior as before, except the minimum at the origin is not clearly exhibited and unless further assumption are made to obtain "I", there is not a simple comparative basis between stable minima to see which dominates.

If there is one minimum that clearly dominates, with solution M_0, R_0:

$$Z(\beta) \simeq \exp[-M_0\beta + \pi R_0^2 - Q\Phi\beta]$$

$$S = \left(1 - \beta\frac{\partial}{\partial\beta}\right)\ln Z\,\{-M_0\beta + \pi R_0^2 - Q\Phi\beta\} - (-M_0\beta - Q\Phi\beta)$$

$$+\beta\frac{\partial R_0}{\partial\beta}\frac{\partial}{\partial R_0}(M_0\beta - \pi R_0)$$

$$S = \pi R_0^2$$

The global properties that can yield minima at the boundaries (i.e., origin thus far) do not play a role in the $\delta Q = 0|_{boundary}$ problem. This will be confirmed in a concluding section but for now the thermodynamic analysis will be performed starting from the Euclideanization results.

For the $\delta Q = 0$ theory, a thermodynamic analysis is performed starting directly from the Euclideanization:

$$T_\infty = \beta^{-1} = \frac{1}{2\pi R_h}\left(\frac{3}{2\ell^2}R_h^2 + \frac{1}{2} - \frac{1}{2}\frac{Q^2}{R_h^2}\right)$$

(the $\frac{\partial I}{\partial R_h}$ equation from Hamiltonian theory). And

$$M = \frac{R_h}{2}\left(\frac{R_h^2}{\ell^2} + 1 + \frac{Q^2}{R_h^2}\right)$$

With the same substitutions as previously:

$$f(x) = \left(\frac{2}{3\beta}\right)\left(\frac{R_h}{\ell}\right)^2\left(\frac{\partial I}{\partial R_h}\right) = 0$$

$$f(x) = x^4 - \left(\frac{4\pi\ell}{3\beta}\right)x^3 + \frac{1}{3}x^2 - \frac{1}{3}\left(\frac{Q}{\ell}\right)^2$$

$g(x) = 3f(x) = 3x^4 - \underbrace{\left(\frac{4\pi\ell}{\beta}\right)}_{\gamma}x^3 + x^2 - \left(\frac{Q}{\ell}\right)^2$ (form studied by Jorma)

$g'(x) = 12x^3 - 3\gamma x^2 + 2x = x(12x^2 - 3\gamma x + 2)$

Zeroes: $\qquad x = 0, \quad x^2 - \frac{1}{4}\gamma x + \frac{1}{6} = 0$

$$r_\pm = \left(\frac{\pi\ell}{2\beta}\right) = \sqrt{\left(\frac{\pi\ell}{2\beta}\right)^2 - \frac{1}{\gamma}}$$

<u>$\frac{\gamma^2}{64} < \frac{1}{6}$:</u>

One root for $\left|\left(\frac{Q}{\ell}\right)\right| \neq 0$

The root for $\left|\left(\frac{Q}{\ell}\right)\right| = 0$ does not directly relate to one for $g(x)$:

$$\frac{\partial I}{\partial R_h} = 0 = \frac{3\beta}{2\ell^2}R_h^2 - 2\pi R_h + \frac{\beta}{2}$$

$$R_h^{+,-} = \frac{2\pi \pm \sqrt{(2\pi)^2 - 2\beta\left(\frac{3\beta}{2\ell^2}\right)}}{2\left(\frac{3\beta}{2\ell^2}\right)}$$

So, for $R_h^{+,-}$ to be real: $(2\pi)^2 > \left(\frac{3\beta^2}{\ell^2}\right)$ and

$$\frac{1}{6} > \left(\frac{\pi\ell}{2\beta}\right)^2 > \frac{3}{16} \to \frac{1}{6} \ngtr \frac{3}{16},$$

So, $\left|\left(\frac{Q}{\ell}\right)\right| = 0$ does not describe a root for $\left(\frac{\partial I}{\partial R_h}\right)$.

If $Q = 0$, there is a second minima if $\left(\frac{\pi\ell}{2\beta}\right)^2 > \frac{3}{16}$. Consider the $Q = 0$ subcase of $\delta Q = 0$ in detail:

$$0 = \frac{\partial I}{\partial R_h} = \frac{3\beta}{2\ell^2} R_h^2 - 2\pi R_h + \frac{\beta}{2}$$

$$R_h^{(\pm)} = \frac{2\pi \pm \sqrt{(2\pi)^2 - 4\left(\frac{3\beta}{2\ell^2}\right)\left(\frac{\beta}{2}\right)}}{2\left(\frac{3\beta}{2\ell^2}\right)} = \left(\frac{2\pi\ell^2}{3\beta}\right) \pm \sqrt{\left(\frac{2\pi\ell^2}{3\beta}\right)^2 - \frac{\ell^2}{3}}$$

There are minima as long as $\left(\frac{2\pi\ell^2}{3\beta}\right)^2 > \frac{\ell^2}{3} \Rightarrow \left(\frac{2\pi}{\beta}\right)^2 - \frac{\ell^2}{3} > 1$

$$\frac{\partial^2 I}{\partial R_h} = \frac{3\beta}{\ell^2} R_h - 2\pi = 2\pi\left(\frac{3\beta}{2\pi\ell^2} R_h - 1\right)$$

and $2\left(\frac{2\pi\ell^2}{3\beta}\right) > R_h^{(+)} > \left(\frac{2\pi\ell^2}{3\beta}\right)$, when it exists, while $R_h^{(-)} < \left(\frac{2\pi\ell^2}{3\beta}\right)$. So, for $\left(\frac{2\pi}{\beta}\right)^2 \frac{\ell^2}{3} > 1$ there is a minimum.

$R_h^{(+)}$ and the origin are local minima. Which is the global minimum?

$$I\left(R_h^{(+)}\right) = (R_h^+)\left[\left(\frac{\beta}{2\ell^2}\right)(R_h^+)^2 - \pi R_h^+ + \frac{\beta}{2}\right]$$

$$= (R_h^+)\left[\left(\frac{2}{3}\pi R_h^+ - \frac{\beta}{6}\right) - \pi R_h^+ + \frac{\beta}{2}\right] = R_h^+\left[-\frac{\pi}{3} R_h^+ + \frac{1}{3}\beta\right]$$

$R_h^{(+)}$ is the global minimum if $R_h^+ > \frac{\beta}{\pi}$.

Consider heat capacity at constant Q:

$$\left(\frac{\partial M}{\partial T_\infty}\right)_Q = ?$$

$$\left(\frac{\partial M}{\partial T_\infty}\right)_Q = \left(\frac{\partial M}{\partial R_h}\right)_Q \left(\frac{\partial R_h}{\partial T_\infty}\right)_Q$$

$$= \left(\frac{3}{2\ell^2} R_h^2 + \frac{1}{2} - \frac{1}{2}\frac{Q^2}{R_h^{-2}}\right)\left(\frac{3}{4\pi\ell^2} - \frac{1}{4\pi} R_h^{-2} + \frac{3}{4\pi} Q^2 R_h^{-4}\right)^{-1}$$

$$= 2\pi R_h^2 \frac{\left(\frac{3R_h^2}{\ell^2} + 1 - \frac{Q^2}{R_h^2}\right)}{\left(\frac{3R_h^2}{\ell^2} - 1 + \frac{3Q^2}{R_h^2}\right)}$$

Recall $Q^2 < R_h^2 \left(\frac{3R_h^2}{\ell^2} + 1\right) \Rightarrow$ so numerator positive for stability $\left(\frac{\partial M}{\partial T_\infty}\right)_Q > 0$:

$$\left(\frac{3R_h^2}{\ell^2} - 1 + \frac{3Q^2}{R_h^2}\right) > 0$$

$$3x^2 - 1 - 3\left(\frac{Q^2}{\ell^2}\right)x^{-2} > 0$$

Compare this with $\frac{\partial^2 I}{\partial R_h^2} = 3\left(\frac{\beta}{\ell^2}\right)R_h - 2\pi + \beta Q^2 R_h^{-3} = \beta R_h^{-1}\left(3x^2 - \left(\frac{2\pi\ell}{\beta}\right)x + \left(\frac{Q}{\ell}\right)^2 x^{-2}\right)$

For $Q = 0$: Thermodynamic stability: $3x^2 - 1 > 0 \rightarrow \underline{x^2 > \frac{1}{3}} \rightarrow x > \frac{+1}{\sqrt{3}}$

Hamiltonian Z minima: $3x - \left(\frac{2\pi\ell}{\beta}\right) > 0 \rightarrow x > \left(\frac{2\pi\ell}{\beta}\right) \rightarrow$
$\begin{pmatrix} R_h^{(+)} & minima \\ R_h^{(-)} & maxima \end{pmatrix}$

$$R_h^{(\pm)} = \ell\left\{\left(\frac{2\pi\ell}{3\beta}\right) \pm \sqrt{\left(\frac{2\pi\ell}{3\beta}\right)^2 - \frac{1}{3}}\right\} \rightarrow \left(\frac{2\pi\ell}{3\beta}\right)^2 > \frac{1}{3}$$

$$\rightarrow \begin{matrix} R_h^+ > \dfrac{1}{\sqrt{3}} & minima \\ R_h^- < \dfrac{1}{\sqrt{3}} & maxima \end{matrix}$$

So, the criteria for stability are equivalent.

Consider $\delta Q = 0$ theory with $\ell^2 = \infty$:

$$T_\infty = \frac{1}{2\pi R_h}\left\{\frac{1}{2} - \frac{1}{2}\frac{Q^2}{R_h^2}\right\} \quad ; \quad M = \frac{R_h}{2}\left(1 + \left(\frac{Q}{R_h}\right)^2\right)$$

$$I = -\pi R_h^2 + \frac{\beta}{2}R_h + \frac{1}{2}\beta Q^2 R_h^{-1}$$

$$\frac{\partial I}{\partial R_h} = -2\pi R_h + \frac{\beta}{2} - \frac{1}{2}\beta Q^2 R_h^{-2}$$

In order for a solution R_h to exist in the T_∞ eqn.:

$$T_\infty = \frac{1}{4\pi Q} x(1-x^2) \ ; \ x = \frac{Q}{R_h}$$
$$f(x) = x(1-x^2)$$
$$f'(x) = 1 - 3x^2 \rightarrow \text{extrema at } x = \frac{1}{\sqrt{3}}$$
$$f''(x) = -6x$$

So, $T_\infty(4\pi|Q|) < \frac{2}{3\sqrt{3}}$

Stability $\left(\text{using } \frac{\partial M}{\partial T_\infty}\right) : \left(\frac{Q}{R_h}\right)^2 > \frac{1}{3}$
$$3Q^2 > R_h^2 = \left(M + \sqrt{M^2 - Q^2}\right)^2$$
$$|Q| > \frac{\sqrt{3}}{2} M$$

Consider the thermodynamic analysis with Φ and T_∞ fixed:
$$\Phi = \frac{\Phi}{R_h}$$
$$T_\infty = \frac{1}{4\pi R_h}\left[\frac{3R_h^2}{\ell^2} + 1 - \Phi^2\right]$$
$$M = \frac{R_h}{2}\left[\frac{R_h^2}{\ell^2} + 1 + \Phi^2\right]$$

Stability is governed by $\left(\frac{\partial T_\infty}{\partial R_h}\right)_\Phi$ contribution to $\left(\frac{\partial M}{\partial T_\infty}\right)_\Phi$:
$$\frac{3R_h^2}{\ell^2} + \Phi^2 - 1 > 0$$
for stability (compare with $X > 0 \Rightarrow R_h > \left(\frac{2\pi\ell^2}{3\beta}\right)$ from Ham.-Z analysis.)

Solutions for T_∞ can be extracted from Ham.-Z analysis:

$$1 - \left(\frac{\beta_0}{\beta}\right)^2 < \begin{array}{lll} |\Phi|^2 \geq 1 & \Rightarrow & \text{one solution} \\ |\Phi|^2 < 1 & \longrightarrow & \text{three solutions} \\ |\Phi|^2 < 1 - \left(\frac{\beta_0}{\beta}\right)^2 & \longrightarrow & \text{one solution} \end{array}$$

Stability: $\frac{3R_h^2}{\ell^2} > (1-|\Phi|^2) > \left(\frac{\beta_0}{\beta}\right)^2$

$R_h > \frac{1}{\sqrt{3}}\left(\frac{\ell\beta_0}{\beta}\right)$; $\beta_0 = \ell\frac{2\pi}{\sqrt{3}}$

$R_h > \frac{2\pi\ell^2}{3\beta}$ (the stabilit criterian is the same for the roots).

<u>For the $\delta Q = 0$ theory, Hamiltonian thermodynamic analysis</u>:

$$Z = \int d\,R_h e^{-I}$$

$$I = \left(\frac{\beta}{2\ell^2}\right)R_h^3 - \pi R_h^2 + \frac{\beta}{2}R_h + \frac{1}{2}\beta Q^2 R_h^{-1}$$

Z is dominated by the contributions near the minima of I. The extrema for I are given by:

$$0 = \frac{\partial I}{\partial R_h} = \frac{3\beta}{2\ell^2}R_h^2 - 2\pi R_h + \frac{\beta}{2} - \frac{1}{2}\beta Q^2 R_h^{-2}$$

The extrema satisfy: $\left(\frac{2}{3\beta}\right)\left(\frac{R_h}{\ell}\right)^2\left(\frac{\partial I}{\partial R_h}\right) = f\left(\frac{R_h}{\ell}\right) = f(x) = 0, x \leq \frac{R_h}{\ell}$

$$f(x) = x^4 - \left(\frac{4\pi\ell}{3\beta}\right)x^3 + \frac{1}{3}x^2 - \frac{1}{3}\left(\frac{Q}{\ell}\right)^2$$

$f(0) = -\frac{1}{3}\left(\frac{Q}{\ell}\right)^2 < 0$, while for large x, $f(x)$ is dominated by $+x^4$. Clearly there is at least one positive real root for $f(x)$. When is there more than one extremum? That can be determined by examining the behavior of $f(x)$:

$$f'(x) = x\left(4x^2 - 4\pi\frac{\ell}{\beta}x + \frac{2}{3}\right)$$

Write $g(x) = \frac{1}{x}f'(x)$

$$g(r_\pm) = 0 \Rightarrow r_\pm = \left(\frac{\pi\ell}{2\beta}\right) \pm \sqrt{\left(\frac{\pi\ell}{2\beta}\right)^2 - \frac{1}{6}}$$

For $g(x)$ to have real roots: $\left(\frac{\pi\ell}{2\beta}\right)^2 > \frac{1}{6}$. There are six types of behavior for $f(x)$ depending upon the values of $f(x)$ at $r_\pm$.

The allowed horizon radii, R_h, are constrained such that $R_h > R_{crit}(Q)$, and R_{crit} satisfies:

$$\left(\frac{R_c}{\ell}\right)^4 + \frac{1}{3}\left(\frac{R_c}{\ell}\right)^2 - \frac{1}{3}\left(\frac{Q}{\ell}\right)^2 = 0$$

Comparison with the equation for the extrema, $I_{,R_h} = 0$, reveals that the entire extremal curve lies in the domain of allowed parameters. So, the parameter restriction does not effect the analysis of the minima.

$$f(x) = \left(\frac{2}{3\beta}\right)\left(\frac{R_h}{\ell}\right)^2\left(\frac{\partial I}{\partial R_h}\right)$$

$$f'(x) = \left(\frac{2}{3\beta}\right)2\left(\frac{R_h}{\ell}\right)\left(\frac{\partial I}{\partial R_h}\right) + \left(\frac{2\ell}{3\beta}\right)\left(\frac{R_h}{\ell}\right)^2\left(\frac{\partial^2 I}{\partial R_h^2}\right)$$

At the zero's of $f(x)$:
$f(r) = 0$

$$f'(r) = \left(\frac{2\ell}{3\beta}\right)r^2\left(\frac{\partial^2 I}{\partial R_h^2}\right)$$

So, at the zero's of $f(x)$ there is a direct relation between $f'(r)$ and $\left(\frac{\partial^2 I}{\partial R_h^2}\right)(r)$. The zero's with positive slope in $f(x), f'(r) > 0$, have positive $\left(\frac{\partial^2 I}{\partial R_h^2}\right)$, and thus represent minima; zero slope → inflection and negative slope → maxima. Thus, there is always a minima. A second minima arises when:

$$\left(\frac{\pi\ell}{2\beta}\right)^2 > \frac{1}{6}$$

$$f(r_+) < 0 < f(r_-)$$

where $r_\pm = \left(\frac{\pi\ell}{2\beta}\right) \pm \sqrt{\left(\frac{\pi\ell}{2\beta}\right)^2 - \frac{1}{6}}$ (satisfies: $4x^2 - 4\pi\frac{\ell}{\beta}x + \frac{2}{3} = 0$):

$f(r) = (r^2)^2 - \left(\frac{4\pi\ell}{3\beta}\right)r(r^2) + \frac{1}{3}r^2 - \frac{1}{3}\left(\frac{Q}{\ell}\right)^2 = r^2\left\{r^2 - \left(\frac{4\pi\ell}{3\beta}\right)r + \frac{1}{3}\right\} - \frac{1}{3}\left(\frac{Q}{\ell}\right)^2$

$$= \left(\frac{\pi\ell}{\beta}r - \frac{1}{6}\right)\left\{\left(\frac{\pi\ell}{\beta} - \frac{4\pi\ell}{3\beta}\right)r + \underbrace{\frac{1}{3} - \frac{1}{6}}_{\frac{1}{6}}\right\} - \frac{1}{3}\left(\frac{Q}{\ell}\right)^2$$

$$= \left(\frac{\pi\ell}{\beta}r - \frac{1}{6}\right)\left(-\frac{\pi\ell}{\beta}r + \frac{1}{6}\right)^{-\frac{\pi\ell}{3\beta}} - \frac{1}{3}\left(\frac{Q}{\ell}\right)^2$$

$$= -\frac{1}{3}\left(\frac{\pi\ell}{\beta}\right)^2 r^2 + \frac{1}{6}\left(\frac{4\pi\ell}{3\beta}\right)r - \frac{1}{36} - \frac{1}{3}\left(\frac{Q}{\ell}\right)^2$$

$$= \left\{-\frac{1}{3}\left(\frac{\pi\ell}{\beta}\right)^3 + \frac{2}{9}\left(\frac{\pi\ell}{\beta}\right)\right\}r + \frac{1}{18}\left(\frac{\pi\ell}{\beta}\right)^2 - \frac{1}{36} - \frac{1}{3}\left(\frac{Q}{\ell}\right)^2$$

If $f(r_+) < 0 < f(r_-)$ then $\underline{f(r_-) - f(r_+) > 0}$, thus $\left\{-\frac{1}{3}\left(\frac{\pi\ell}{\beta}\right)^3 + \frac{2}{9}\left(\frac{\pi\ell}{\beta}\right)\right\}(r_- - r_+) > 0$, must be true. This is only possible if

$$\left\{-\frac{1}{3}\left(\frac{\pi\ell}{\beta}\right)^3 + \frac{2}{9}\left(\frac{\pi\ell}{\beta}\right)\right\} < 0 \Rightarrow \left\{-\frac{1}{3}\left(\frac{\pi\ell}{\beta}\right)^2 + \frac{2}{9}\right\} < 0,$$

Is this compatible with $\left(\frac{\pi\ell}{2\beta}\right)^2 > \frac{1}{6}$? $\left(\frac{\pi\ell}{2\beta}\right)^2 > \frac{1}{6} \Rightarrow \left(\frac{\pi\ell}{\beta}\right)^2 > \frac{2}{3}$, which is sufficient – with no additional inequalities imposed.

If $f(r_+) < 0, 0 < f(r_-)$:

$$-\left\{-\frac{1}{3}\left(\frac{\pi\ell}{\beta}\right)^3 + \frac{2}{9}\left(\frac{\pi\ell}{\beta}\right)\right\}r_- < \frac{1}{18}\left(\frac{\pi\ell}{\beta}\right)^2 - \frac{1}{36} - \frac{1}{3}\left(\frac{Q}{\ell}\right)^2$$

$$\frac{1}{18}\left(\frac{\pi\ell}{\beta}\right)^2 - \frac{1}{36} - \frac{1}{3}\left(\frac{Q}{\ell}\right)^2 < -\left\{-\frac{1}{3}\left(\frac{\pi\ell}{\beta}\right)^3 + \frac{2}{9}\left(\frac{\pi\ell}{\beta}\right)\right\}r_+$$

$$\boxed{0 < r_- < \frac{\left[\frac{1}{18}\left(\frac{\pi\ell}{\beta}\right)^2 - \frac{1}{36} - \frac{1}{3}\left(\frac{Q}{\ell}\right)^2\right]}{\left(\frac{\pi\ell}{\beta}\right)\left[\frac{1}{3}\left(\frac{\pi\ell}{\beta}\right)^2 + \frac{2}{9}\right]} < r_+ < \left(\frac{\pi\ell}{\beta}\right)}$$

For the second minima.

$$\left(\frac{\pi\ell}{\beta}\right)^2_{min} \simeq \frac{2}{3} \Rightarrow \frac{1}{18}\left(\frac{2}{3}\right) - \frac{1}{36} - \frac{1}{3}\left(\frac{Q}{\ell}\right)^2 \simeq 0$$

$$\frac{1}{9} - \frac{1}{12} - \left(\frac{Q}{\ell}\right)^2 \simeq 0$$

$$\frac{1}{36} - \left(\frac{Q}{\ell}\right)^2 \simeq 0 \Rightarrow \underline{\left(\frac{Q}{\ell}\right) \simeq \frac{1}{6}}$$

$$\underline{\left(\frac{\pi\ell}{\beta}\right)} \gg 1 \Rightarrow \left(\frac{Q}{\ell}\right)^2 < \frac{1}{9}\left(\frac{\pi\ell}{\beta}\right)^2$$

$$\underline{\left(\frac{Q}{\ell}\right) < \left(\frac{\pi\ell}{3\beta}\right)}$$

So, given $\left(\frac{\pi\ell}{2\beta}\right)^2 > \frac{1}{6}$, it is generally possible to obtain a second minima if Q falls in a restricted range of values.

C. Lovelock Spacetime Thermodynamics

In this Appendix we describe the Hamiltonian Thermodynamics of a Lovelock Black Hole.

What is a Lovelock Black Hole?
Lovelock described a Lagrangian for a torsion free theory of gravity in D-dim that yielded 2nd order field equations for the metric. In modern language, a lovelock Lagrangian is a sum of a dimensionally extended Euler densities

$$\mathcal{L} = \sum_{m=0}^{k} c_m \mathcal{L}_m$$

$$\mathcal{L}_m = 2^{-m} \delta^{a_1 b_1 ,,, a_m b_m}_{c_1 d_1 ,,, c_m d_m} R^{c_1 d_1}{}_{a_1 b_1} \dots R^{c_m d_m}{}_{a_m b_m},$$

where c_m is an arbitrary constant. If only $c_1 \neq 0$, then we get Einstein's Theory. If $c_0 \neq 0$, then we have a nonzero cosmological constant. If $D \leq 4$, $\mathcal{L}_{m\geq 2} = 0$ (i.e., $\mathcal{L}_k$ vanishes for $2K \geq D$). So, for five or more dimensions we have higher order curvature generalizations (curvature squared for five dimensions). Exact solutions describing black holes have been found for these theories. The thermodynamics of these black holes has been proposed as a model for the possible effects of higher-derivative interactions in the strong gravitational fields present in the final stages of black hole evaporation.

Why Hamiltonian Thermodynamics?
The Path Integral approach developed from the observation that for the Kerr-Neuman family of Blackholes in asymptotically flat space, a saddle point estimate for the path integral yields a partition function which reproduces the black hole entropy [54]. Similar thermodynamical partition functions can be derived from a Lorentzian Hamiltonian quantum theory of black holes in a manner analogous to flat space thermal field theory [68]. A key step in the analysis is provided by the Hamiltonian methods developed by [80]. This approach reveals more of the suspected topological nature of gravitational entropy. The Euclideanization methods central to the Path Integral approach may indicate the existence of a locally stable classical solution but not say whether a well-defined canonical ensemble exists.

By the Lovelock generalization of Birkhoff's theorem, the general solution to the theory (3.1) for $\tilde{\lambda} \neq 0$ can be written in the local curvature coordinates (T;R) as

$$ds^2 = -FdT^2 + F^{-1}dR^2 + R^2 d\Omega_3^2 \ , \tag{A1}$$

where

$$F = 1 + \frac{R^2}{\hat{\lambda}}\left(1 \pm \sqrt{1 + \frac{2\omega\hat{\lambda}}{R^4}}\right) \tag{A2}$$

with $\lambda = \frac{1}{2}\tilde{\lambda}$. The coordinates T and R are respectively the Killing time, whose constant value hypersurfaces are spacelike or timelike depending on the sign of F, and the curvature radius of the three-sphere. For $\tilde{\lambda} = 0$, the solution is the five-dimensional Schwarzschild solution, obtained as the limit $\tilde{\lambda} \to 0$ with the lower sign. In addition to $\tilde{\lambda}$, the only local parameter in the solution is the mass-like quantity ω.

For the reasons discussed in Chapter 7, we take $\tilde{\lambda} \geq 0$. The solution then describes a black hole in asymptotically at space provided we take $\omega > \frac{1}{2}\tilde{\lambda}$ and choose (for $\tilde{\lambda} > 0$) the lower sign [102]. The curvature coordinates are good individually in each region not containing horizons: the horizons are nondegenerate, and the Penrose diagram of the conventional maximal analytic extension is similar to that of Kruskal manifold. The horizon radius is $R_h = \sqrt{\omega - \frac{1}{2}\tilde{\lambda}}$. In our units, the ADM mass is $M = \frac{1}{2}\omega$.

Asymptotically flat infinity

In this appendix we adapt the analysis of the main text to boundary conditions that replace the timelike boundary by an asymptotically at infinity. We shall see that quantization will not lead to a well-defined canonical ensemble.

In the metric theory, we let r take the range $0 \leq r < \infty$. At $r \to \infty$, we introduce the falloff

$$\Lambda(t,r) = 1 + M_+(t)r^{-2} + O^\infty(r^{-2-\epsilon})$$
$$R(t,r) = r + O^\infty(r^{-1-\epsilon})$$
$$P_\Lambda(t,r) = O^\infty(r^{1-\epsilon})$$

$$P_R(t,r) = O^\infty(r^{-\epsilon})$$
$$N(t,r) = N_+(t) + O^\infty(r^{-\epsilon})$$
$$N^r(t,r) = O^\infty(r^{-1-\epsilon})$$

where $N_+(t) > 0$, and ϵ is a parameter that can be chosen arbitrarily in the range $0 < \epsilon \leq 2$. $O^\infty(r^s)$ denotes a term that is bounded at $r \to \infty$ by a constant times r^s, and whose derivatives fall off accordingly. It is straightforward to verify that this falloff makes the coordinates asymptotic to hyperspherical coordinates in Minkowski space, it is consistent with the constraints, and it is preserved by the time evolution. $N_+(t)$ gives the rate at which the asymptotic Minkowski time advances with respect to the coordinate time *t*. When the equations of motion hold, $M_+(t)$ is independent of *t*, and its value is the ADM mass.

The total action takes the form $S = S_\Sigma + S_{\vartheta\Sigma}$, where

$$S_{\partial\Sigma} = \int dt R_0 \left(\frac{1}{3}R_0^2 + \hat{\lambda}\right)(N_1/\Lambda_0) - \int dt N_+ M_+$$

and S_Σ is as in (3.5) except that the upper limit of the r-integral is replaced by infinity. The canonical transformation and Hamiltonian reduction proceed as before. The action of the reduced theory is as in (3.20), but with the Hamiltonian now given by

$$\boldsymbol{h} = N_+ \boldsymbol{m} - N_0 R_0 \left(\frac{1}{3}R_0^2 + \hat{\lambda}\right)$$

The configuration variable **m** has the range $\boldsymbol{m} > \frac{1}{4}\hat{\lambda}$, while the range of **p** is the full real axis. Although in the main text we assumed for presentational simplicity $\tilde{\lambda} > 0$, it is easily verified
that the above derivation extends to the case $\hat{\lambda} = 0$, and the reduced Hamiltonian is valid for all $\hat{\lambda} \geq 0$.

Quantization can now proceed as in the main text. We analytically continue the time evolution operator by $\int N_0 dt = -2\pi i$ and $\int N_+ dt = -i\beta$, interpreting β as the inverse temperature at the infinity. For the renormalized trace of the analytically continued time evolution operator, we obtain formally

$$Z(\beta;\hat{\lambda}) = \mathcal{N} \int_0^\infty \tilde{\mu} dR_0 \exp(-I_*^\infty) \ , \tag{B4}$$

where the effective action I_*^∞ is given by

$$I_*^\infty = \tfrac{1}{2}\beta\left(R_0^2 + \tfrac{1}{2}\hat{\lambda}\right) - 2\pi R_0 \left(\tfrac{1}{3}R_0^2 + \hat{\lambda}\right) \ . \tag{B5}$$

The smooth and slowly varying positive function $\tilde{\mu}(R_0)$ arose from the choice of the inner product, and N is a normalization constant, possibly dependent on $\hat{\lambda}$ but presumably not on _. However, the integral in (B4) is divergent because I_*^{∞} tends to $-\infty$ at large R_0. Thus, the canonical ensemble does not exist under the asymptotically at boundary conditions, neither for $\hat{\lambda} = 0$ nor $\hat{\lambda} > 0$. In this sense, the asymptotically at Lovelock theory is thermodynamically no better behaved than asymptotically at Einstein theory.

The critical points of I_*^{∞} give the (Lorentzian) classical solutions that have the inverse Hawking temperature β at infinity. For $\hat{\lambda} = 0$ there exists exactly one critical point, which is a local maximum: this is similar to what happens with Einstein's theory in four dimensions. For $\hat{\lambda} > 0$, the situation is more versatile. Critical points exist when $\beta \geq 16\pi^2 \hat{\lambda}$, and when the inequality is genuine, there are two critical points. The critical point with the smaller (larger) value of R_0 is a local minimum (maximum, respectively). The local minimum gives the classical solution that was found to be stable against Hawking evaporation in [102]. While the stability of this solution against Hawking evaporation reflects its being a local minimum of I_*^{∞}, the divergence of the integral demonstrates that this local stability is not sufficient to guarantee the existence of the canonical ensemble. The effects of the Lovelock parameter on the asymptotically at thermodynamics are thus qualitatively very similar to those of a fixed charge in asymptotically at four-dimensional Einstein-Maxwell theory [36].

C.1 Metric and Canonical Formulation

Variational formulation

General five-dimensional spherically symmetric ADM metric:

$$dS^2 = -N^2dt^2 + \Lambda^2(dr + N^r dt)^2 + R^2 d\Omega_3^2$$

Bulk contributions to action: Einstein-Hilbert term and 4 dim Euler density:

$$S_{bulk} = \frac{1}{2k}\int d^D x\sqrt{-g}\left[R + \frac{\lambda}{2}(R_{abcd}R^{abcd} - 4R_{ab}R^{ab} + R^2)\right]$$

Consider $D = S$, $\lambda \geq 0$,

Insert metric, integrate over three-sphere, drop a total-derivative, and obtain:

$$S_{\Sigma}^{L} = \int dt \int_0^1 dr\, \mathcal{L}$$

$$\mathcal{L} = -\frac{[\dot{\Lambda} - (N^r\Lambda)'](\dot{R} - N^rR')}{N}\left\{R^2 + \hat{\lambda}\left[1 - \left(\frac{R'}{\Lambda}\right)^2 + \frac{(\dot{R} - N^rR')^2}{3N^2}\right]\right\}$$

$$-\frac{(\dot{R} - N^rR')}{N}\left[\Lambda R - \hat{\lambda}\left(\frac{R'}{\Lambda}\right)'\right] + N\Lambda R\left[1 - \left(\frac{R'}{\Lambda}\right)^2\right]$$

$$-N\left(\frac{R'}{\Lambda}\right)'\left\{R^2 + \hat{\lambda}\left[1 - \left(\frac{R'}{\Lambda}\right)^2\right]\right\}$$

Using the definitions $\mathcal{L} = \mathcal{L}_1 + \mathcal{L}_2$; $x = \frac{\mathcal{L}_2}{2\lambda \sin\theta \sin x}$; $D \equiv \partial_t - N^r\partial_r$, $\nabla\Lambda \equiv \dot{\Lambda} - (N^r\Lambda)$

$x = N^{-3}[(DR)^3(\nabla\Lambda) + R(\nabla\Lambda)^{\cdot}(DR)^2 - (N^r\nabla\Lambda)^{\cdot}(DR)^2 + \Lambda D^2R(DR)^2 + 2R(DR)(\nabla\Lambda)D^2R]$

$+\left(\frac{1}{3}N^{-3}\right)^{\cdot}[R(\nabla\Lambda)(DR)^2 + \Lambda(DR)^3 + 2R(DR)^2(\nabla\Lambda)]$

$+\left(-\frac{1}{3}N^{-3}\right)'[N^rR(\nabla\Lambda)(DR)^2 + \Lambda N^r(DR)^3 + 2R(\nabla\Lambda)(DR)^2N^r]$

$+N^{-3}(N')^2\left[\frac{2R}{\Lambda}(DR)^2\right] - N''N^{-2}\left[\left(\frac{R}{\Lambda}\right)(DR)^2\right]$

$$+N^{-1}\left[\begin{array}{c}(DR)(\nabla\Lambda)\left(1 - \left(\frac{R'}{\Lambda}\right)^2\right) - (DR)^2\left(\frac{R'}{\Lambda}\right)^2 \\ +2\Lambda R\left[D\left(\frac{R'}{\Lambda}\right)\right]^2 + R(\nabla\Lambda)'\left(1 - \left(\frac{R'}{\Lambda}\right)\right) \\ -R(N^r\nabla\Lambda)'\left(1 - \left(\frac{R'}{\Lambda}\right)^2\right) + \Lambda D^2R\left(1 - \left(\frac{R'}{\Lambda}\right)^2\right) - 2RD^2R\left(\frac{R'}{\Lambda}\right)'\end{array}\right]$$

$+(N^{-1})^{\cdot}\left[R(\nabla\Lambda)\left(1 - \left(\frac{R'}{\Lambda}\right)^2\right) + \Lambda(DR)\left(1 - 2\left(\frac{R'}{\Lambda}\right)^2\right) - 2R(DR)\left(\frac{R'}{\Lambda}\right)'\right]$

$$+(N^{-1})'\left[\begin{array}{c}4R(DR)D\left(\frac{R'}{\Lambda}\right) + R\left(\frac{\Lambda'}{\Lambda^2}\right)(DR)^2 + RN^r(\nabla\Lambda)\left(1 - \left(\frac{R'}{\Lambda}\right)^2\right) \\ +\Lambda(DR)N^r\left(1 - \left(\frac{R'}{\Lambda}\right)^2\right) \\ -\left(\frac{R'}{\Lambda}\right)(DR)^2 - 2R\Lambda^{-2}(DR)(\nabla\Lambda)R' - 2R\left(\frac{R'}{\Lambda}\right)'N^r(DR)\end{array}\right]$$

$+N\left[-\left(\frac{R'}{\Lambda}\right)'\left(1 - \left(\frac{R'}{\Lambda}\right)^2\right)\right]$

$+N'\left[R\frac{\Lambda'}{\Lambda^2}\left(1 - \left(\frac{R'}{\Lambda}\right)^2\right) - \left(\frac{R'}{\Lambda}\right)\left(1 - \left(\frac{R'}{\Lambda}\right)^2\right) + \frac{2RR'}{\Lambda^2}\left(\frac{R'}{\Lambda}\right)'\right]$

$+N''\left[-\left(\frac{R}{\Lambda}\right)\left(1 - \left(\frac{R'}{\Lambda}\right)^2\right)\right]$

Hamiltonian form of the action:

$$S_\Sigma = \int dt \int_0^1 dr\,(P_\Lambda\dot{\Lambda} + P_R\dot{R} - NH - N^r H_r)$$

$$H = y\left\{P_R + y\left[\Lambda R - \hat{\lambda}\left(\frac{R'}{\Lambda}\right)'\right]\right\} - \Lambda R\left[1-\left(\frac{R'}{\Lambda}\right)^2\right] + \left(\frac{R'}{\Lambda}\right)'\left\{R^2 + \hat{\lambda}\left(\frac{R'}{\Lambda}\right)^2\right\}$$

$$H_r = R'^{P_R} - \Lambda P'_\Lambda$$

$$0 = \frac{1}{3}\hat{\lambda}y^3 + y\left\{R^2 + \hat{\lambda}\left[1-\left(\frac{R'}{\Lambda}\right)^2\right]\right\} + P_\Lambda$$

Lovelock Generalization of Birkhoff's theorem:

$$\left(\frac{2}{\Lambda}\right)(-R'H + yH_r) = \mathfrak{m}'$$

$$\mathfrak{m} = yP_\Lambda - \frac{1}{6}\hat{\lambda}y^4 + R^2\left(1-\left(\frac{R'}{\Lambda}\right)^2\right) - \hat{\lambda}\left(\frac{R'}{\Lambda}\right)^2\left(1-\frac{1}{2}\left(\frac{R'}{\Lambda}\right)^2\right)$$

Adopt gauge with $y = 0$ and transverse radius $R = r$; $N^r = 0$:

$$m = r^2(1-\Lambda^{-2}) - \hat{\lambda}\Lambda^{-2}\left(1-\frac{1}{2}\Lambda^{-2}\right)$$

From $\dot{P}_\Lambda$ equation $N = \Lambda^{-1}$, and the metric is completely described:

$$ds^2 = -FdT^2 + F^{-1}dR^2 + R^2\partial\Omega_3^2$$

$$F = 1 + \frac{R^2}{\hat{\lambda}}\left(1 \pm \sqrt{1+\frac{2\omega\hat{\lambda}}{R^+}}\right), \qquad \left(\lambda = \frac{1}{2}\hat{\lambda}\right)$$

Want $\hat{\lambda} \geq 0$ for thermodynamic concerns and wont $w > \frac{1}{2}\hat{\lambda}$ and lower sign to have Black Hole solution.

Consider left end of hypersurface that attaches to the bifurcation 3-sphere. We need conditions that guarantee that the classical solutions have a bifurcate horizon, put the left end of the spacelike hypersurfaces at bifurcation sphere, are consistent with constraints, and are preserved by the Hamiltonian evolution.

C.2 Fall-off as $r \to \infty$ and $r \to 0$ bifurcate 2-sphere

For $r \to 0$ we find:

$$\Lambda(t,r) = \Lambda_0(t) + O(r^2)$$
$$R(t,r) = R_0(t) + R_2(t)r^2 + O(r^4)$$
$$P_\Lambda(t,r) = O(r^3)$$
$$P_R(t,r) = O(r)$$
$$N(t,r) = N_1(t)r + O(r^3)$$
$$N^r(t,r) = N_1^r(t)r + O(r^3)$$

Λ_0, R_0 positive, and $N_1 \geq 0$. The above also ensures that the cubic has a unique solution for y near r=0. On the classical solution, the future unit normal vector $n^a(t)$ to the spacelike hypersurface at r=0 evolves according to

$$n^a(t_1)n_a(t_2) = -\cosh\left(\int_{t_1}^{t_2} \Lambda_o^{-1}(t)N_1(t)dt\right)$$

So, we make N_1/Λ_0 a prescribed function of t. At $r = 1$, we make R and $-g_{tt} = N^2 - (\Lambda N^r)^2$ prescribed positive-valued functions of t. For r=0 the boundary term is then

$$\int dt\, R_0\left(\frac{1}{3}R_0^2 + \hat{\lambda}\right)\left(\frac{N}{\Lambda_0}\right)$$

and for r=1 there is:

$$\int dt\left\{N\Lambda^{-1}R^2R' - N^r\Lambda P_\Lambda - \frac{1}{2}\dot{R}(R^2 + \hat{\lambda})\ln\left|\frac{N + \Lambda N^r}{N - \Lambda N^r}\right|\right\}$$

$$+\hat{\lambda}N\left(\frac{R'}{\Lambda}\right)\left[1 - \frac{1}{3}\left(\frac{R'}{\Lambda}\right)^2 - \frac{\dot{R}(\dot{R} - N^rR^r)}{N^2}\right]$$

$$-\frac{\hat{\lambda}N^r\Lambda}{3N}\left[\frac{\dot{R}^3}{N^2 - (\Lambda N^r)^2} + \frac{N^r(R')^3}{\Lambda^2}\right]$$

Let's start with the calculation of surface terms:

$H = \int_0^\infty dr\,\{N\mathcal{H} + N^r\mathcal{H}_r\} + (surface\ terms)$

$$\mathcal{H} = yP_R + y^2\left(\Lambda R - \hat{\lambda}\left(\frac{R'}{\Lambda}\right)'\right) - \Lambda R\left(1 - \left(\frac{R'}{\Lambda}\right)^2\right)$$

$$+\left(\frac{R'}{\Lambda}\right)'\left(R^2 + \hat{\lambda}(1 - \left(\frac{R'}{\Lambda}\right)^2)\right)$$

$\mathcal{H}_r = P_RR' - \Lambda P'_\Lambda$

$y{:}\,\frac{1}{3}\hat{\lambda}y^3 + y\left\{R^2 + \hat{\lambda}\left[1 - \left(\frac{R'}{\Lambda}\right)^r\right]\right\} + P_\Lambda = 0$

$\delta y = ?$

$\hat{\lambda}y^2\delta y + \left\{R^2 + \hat{\lambda}\left[1 - \left(\frac{R'}{\Lambda}\right)^2\right]\right\}\delta y + y\left[2R\delta R \pm 2\left(\frac{R'}{\Lambda}\right)\right]\left[\frac{1}{\Lambda}(\delta R)' - R'\Lambda^{-2}\delta\Lambda\right] + \delta P_\Lambda = 0$

$$\delta y = \frac{\left(-\delta P_\Lambda - 2y\hat{\lambda}\left(\frac{R'}{\Lambda}\right)^2\frac{1}{\Lambda}\delta\Lambda - 2yR\delta R + 2y\hat{\lambda}\left(\frac{R'}{\Lambda}\right)\frac{1}{\Lambda}(\delta R)'\right)}{\left(\hat{\lambda}y^2 + \left[R^2 + \hat{\lambda}\left(1 - \left(\frac{R'}{\Lambda}\right)^2\right)\right]\right)}$$

$$\delta H = \int_0^\infty dr\, W$$

$W = (\delta N)\mathcal{H} + (\delta N^r)\mathcal{H}_r + N[\delta yP_R + y\delta P_R + 2y\delta y\left(\Lambda R - \hat{\lambda}\left(\frac{R'}{\Lambda}\right)'\right)$

$$+y^2\left((\delta\Lambda)R + \Lambda\delta R - \hat{\lambda}\left[\delta\left(\frac{R'}{\Lambda}\right)'\right]\right) - [(\delta\Lambda)R + \Lambda(\Lambda R)]\left(1-\left(\frac{R'}{\Lambda}\right)^2\right)$$
$$+\Lambda R2\left(\frac{R'}{\Lambda}\right)\left[\frac{1}{\Lambda}(\delta R)' - \frac{R'}{\Lambda^2}(\delta\Lambda)\right]$$
$$+\left[\delta\left(\frac{R'}{\Lambda}\right)\right]'\left(R^2+\hat{\lambda}\left(1-\left(\frac{R'}{\Lambda}\right)^2\right)+\left(\frac{R'}{\Lambda}\right)'\left(2R\delta R - 2\hat{\lambda}\left(\frac{R'}{\Lambda}\right)\left[\frac{1}{\Lambda}(\delta R)' -\right.\right.\right.$$
$$\left.\left.\left.R'\Lambda^{-2}(\delta R)\right]\right)\right)$$
$$+N^r[(\delta P_R)R' + P_R(\delta R)' - \delta\Lambda P'_\Lambda - \Lambda(\delta P_\Lambda - \Lambda(\delta P_\Lambda)')]$$

$$\int Ny^2\hat{\lambda}\left[\delta\left(\frac{R'}{\Lambda}\right)\right]' = \int\left[Ny^2\hat{\lambda}\delta\left(\frac{R'}{\Lambda}\right)\right]' - \int(Ny^2\hat{\lambda})\left[\Lambda^{-2}(\delta R)' - R'\Lambda^{-2}\delta\Lambda\right]$$
$$= \int\left[Ny^2\hat{\lambda}\delta\left(\frac{R'}{\Lambda}\right)\right]' - \int\left[(Ny^2\hat{\lambda})'\Lambda^{-1}(\delta R)\right]' + \int\left[\frac{R'}{\Lambda}\right](\delta R) + R'\Lambda^{-2}(\delta\Lambda)$$

Let's now consider δH with the constraints and equations of motion imposed:

$$\delta H|_{constraints} = \int_0^\infty dr \left[\!\left[\frac{N\delta y\hat{\lambda}\left(\frac{R'}{\Lambda}\right)\frac{1}{\Lambda}}{\hat{\lambda}y^2+\left(R^2+\hat{\lambda}\left(1-\left(\frac{R'}{\Lambda}\right)^2\right)\right)}\left(P_R + 2y\left(\Lambda R - \right.\right.\right.\right.$$
$$\left.\left.\hat{\lambda}\left(\frac{R'}{\Lambda}\right)'\right)\right)(\delta R) - \hat{\lambda}Ny^2[\Lambda^{-1}(\delta R)^{-1} - R'\Lambda^{-2}\delta\Lambda] + (Ny^2\hat{\lambda})'\Lambda^{-1}(\delta R) +$$
$$2NR\left(\frac{R'}{\Lambda}\right)(\delta R) + N\left(R^2+\hat{\lambda}\left(1-\left(\frac{R'}{\Lambda}\right)^2\right)\right)\left[\Lambda^{-1}(\delta R)' - R'^{\Lambda^{-2}}\delta\Lambda\right] -$$
$$\left[N\left(R^2+\hat{\lambda}\left(1-\left(\frac{R'}{\Lambda}\right)^2\right)\right)\right]'\Lambda^{-1}(\delta R) - \hat{\lambda}2N\Lambda^{-1}\left(\frac{R'}{\Lambda}\right)\left(\frac{R'}{\Lambda}\right)'(\delta R) +$$
$$\left.\left.NP_R(\delta R) - N^r\Lambda(\delta P_\Lambda)\right]\!\right]'$$

$$= \boldsymbol{S(r)}|_0^\infty \quad \textbf{where}$$

$$S(r)=\underbrace{\left[\frac{N2y\hat{\lambda}\left(\frac{R'}{\Lambda}\right)\frac{1}{\Lambda}\left(P_R+2_y\left(\Lambda R-\hat{\lambda}\left(\frac{R'}{\Lambda}\right)'\right)\right)}{\hat{\lambda}y^2+\left[R^2+\hat{\lambda}\left(1-\left(\frac{R'}{\Lambda}\right)^2\right)\right]}(\delta R)\right]}_{(i)}$$

$$+\underbrace{\left[-\frac{\hat{\lambda}Ny^2}{\Lambda}\left((\delta R)'-\left(\frac{R'}{\Lambda}\right)\delta\Lambda\right)\right]}_{(ii)}$$

$$\underbrace{+\left(Ny^2\hat{\lambda}\right)'\Lambda^{-1}(\delta R)}_{(iii)}+\underbrace{2NR\left(\frac{R'}{\Lambda}\right)(\delta R)}_{(iv)}$$

$$+\underbrace{\left[\frac{N\left(R^2+\hat{\lambda}\left(1-\left(\frac{R'}{\Lambda}\right)^2\right)\right)}{\Lambda}\left((\delta R)'-\left(\frac{R'}{\Lambda}\right)\delta\Lambda\right)\right]}_{(v)}$$

$$-\underbrace{\frac{\left[N\left(R^2+\hat{\lambda}\left(1-\left(\frac{R'}{\Lambda}\right)^2\right)\right)\right]'(\delta R)}{\Lambda}}_{(vi)}+\underbrace{\left[-2N\Lambda^{-1}\hat{\lambda}\left(\frac{R'}{\Lambda}\right)\left(\frac{R'}{\Lambda}\right)'(\delta R)\right]}_{(Vii)}$$

$$\underbrace{+N^rP_R(\delta R)}_{(viii)}+\underbrace{\left[-N^r\Lambda(\delta P_\Lambda)\right]}_{ix}$$

Consider $r\to 0$ contribution with the fall off conditions:
$R=R_0+R_2r^2+O(r^4)$
$R'=2R_2r+O(r^3)$
$\delta R=O(r^0)$
$(\delta R)'=O(r')$
$\Lambda=\Lambda_0+O(r^2)$
$\Lambda'=O(r)$
$\delta\Lambda=O(r^0)$
$N=N_1r+O(r^3)$
$N^r=N_1^rr+O(r^3)$

$P_\Lambda=O(r^3)$
$\delta P_\Lambda=O(r^3)$

$P_R = O(r)$
$y = O(r^3)$

$(i)\ O(r^6)$
$(ii)\ O(r^8)$
$(iii)\ O(r^6)$
$(iv)\ O(r^2)$
$(v)\ O(r^2)$
$(vi)\ O(r^0)$
$(vii)\ O(r^2)$
$(viii)\ O(r^2)$
$(ix)\ O(r^4)$
The $O(r^0)$ part of (vi) is $y = -N'(R^2 + \hat{\lambda})\Lambda^{-1}\delta R$.
Want to re-express $Y = \delta Ƶ$, try $Ƶ = -\left(\frac{1}{3}R^3 + R\hat{\lambda}\right)(N'\lambda^{-1})$
$\delta Ƶ = -\delta\left(\frac{1}{3}R^3 + R\hat{\lambda}\right)\left(N'^{\lambda^{-1}}\right) - \left(\frac{1}{3}R^3 + R\hat{\lambda}\right)\delta(N'\lambda^{-1})$

For $r \to 0$ boundary the action needs the term:

$$S_{\partial\Sigma}|_{r\to 0} = \int dt\left(\left(\frac{1}{3}R_0^3 + R_0\hat{\lambda}\right)(N_1\lambda_0^{-1})\right)$$

Consider $r \to \infty$ contribution:
$\Lambda = 1 + \frac{1}{2}wr^{-2} + O(r^{-2-\epsilon})$
$R = r + O(r^{-1-\epsilon})$
$\left.\begin{matrix} P_\Lambda = (r^{-1-\epsilon}) \\ P_R = (r^{-\epsilon}) \end{matrix}\right\} \to y = O(r^{-1-\epsilon})$
$N = O(r^0)$
$N^r = O(r^{-1-\epsilon})$

$(i)\ O(r^{-4-\epsilon})$
$(ii)\ O(r^{-4-\epsilon})$
$(iii)\ O(r^{-4-\epsilon})$
$(iv)\ O(r^{-\epsilon})$
$(v)\ O(r^0)$
$(vi)\ O(r^{-\epsilon})$
$(vii)\ O(r^{-2-\epsilon})$
$(viii)\ O(r^{-2-\epsilon})$
$(ix)\ O(r^{-2-\epsilon})$

The $O(r^0)$ part of (v) is $Y = -NR^2\Lambda^{-1}\left(\frac{R'}{\Lambda}\right)\delta\Lambda = N\left(\frac{1}{3}R^3\right)'\delta(\Lambda^{-1})$

Express $Y = \delta Z\,, Z = N\left(\frac{1}{3}R^3\right)'\Lambda^{-1} = N_+r^2\left(1-\frac{1}{2}wr^{-2}\right)+O(r^{-\epsilon})$

$\delta Z = -\frac{1}{2}N_+\delta w + O(r^{-\epsilon})$

For $r \to \infty$ boundary

$$S_{\partial\Sigma}|_{r\to\infty} = \int dt\left(-N_+\frac{1}{2}w_+\right)$$

Note that $\delta\left(N, \Lambda_0^{-1}\right) = 0$ and $\delta N_+ = 0$ are the hypersurface restrictions at the boundaries.

Let's now derive the new mass canonical variable and compute the Schwarzschild gauge solution:

$\left[\frac{1}{\Lambda}(-R'H + yH_r)\right] = -\frac{1}{2}P_\Lambda' + \frac{1}{2}y'\left[\hat{\lambda}y^3 + y\left(R^2+\hat{\lambda}\left(1-\left(\frac{R'}{\Lambda}\right)^2\right)\right)\right]$

$+\frac{1}{2}\left[R^2\left(1-\left(\frac{R'}{\Lambda}\right)^2\right)\right]' - \frac{1}{2}\hat{\lambda}\left[\left(\left(\frac{R'}{\Lambda}\right)^2\right)' - \frac{1}{2}\left(\left(\frac{R'}{\Lambda}\right)^4\right)'\right]$

$= -\frac{1}{2}yP_\Lambda' + \frac{1}{2}y'\left[\frac{2}{3}\hat{\lambda}y^3 - P_\Lambda\right] + \frac{1}{2}\left[R^2\left(1-\left(\frac{R'}{\Lambda}\right)^2\right)\right]' - \frac{1}{2}\hat{\lambda}\left[\left(\frac{R'}{\Lambda}\right)^2 - \frac{1}{2}\left(\frac{R'}{\Lambda}\right)^4\right]'$

$= -\frac{1}{2}\left[(-yP_\Lambda) + \frac{1}{6}\hat{\lambda}y^4 + R^2\left(1-\left(\frac{R'}{\Lambda}\right)^2\right) - \hat{\lambda}\left(\frac{R'}{\Lambda}\right)^2\left(1-\frac{1}{2}\left(\frac{R'}{\Lambda}\right)^2\right)\right]'$

$$\left(\frac{-2}{\Lambda}\right)(-R'H + yH_r) = m'$$

$$-m = -yP_\Lambda + \frac{1}{6}\hat{\lambda}y^4 + R^2\left(1-\left(\frac{R'}{\Lambda}\right)^2\right) - \hat{\lambda}\left(\frac{R'}{\Lambda}\right)^2\left(1-\frac{1}{2}\left(\frac{R'}{\Lambda}\right)^2\right)$$

Consider a slice for which y=0, choose for R the transverse radius R=r, what is Λ?

$m = r^2(1-\Lambda^{-2}) - \hat{\lambda}\Lambda^{-2}\left(1-\frac{1}{2}\Lambda^{-2}\right)$

$m - r^2 = (-\hat{\lambda} - r^2)\Lambda^{-2} + \frac{1}{2}\hat{\lambda}\Lambda^{-4}$

$\Lambda^{-2} = \frac{(-\hat{\lambda}-r^2)\pm\sqrt{(-\hat{\lambda}-r^2)^2 - 4\left(\frac{1}{2}-\hat{\lambda}\right)(m-r^2)}}{2\left(\frac{1}{2}-\hat{\lambda}\right)}$

$= 1 + \frac{1}{\hat{\lambda}}r^2 \pm \sqrt{\left(1+\frac{1}{\hat{\lambda}}r^2\right)^2 + 2\left(-\frac{1}{\hat{\lambda}}r^2 - \frac{m}{\hat{\lambda}}\right)}$

From $\dot{P}_\Lambda$ equation $N = \Lambda^{-1}$ and

$$N^2 = 1 + \frac{r^2}{\hat{\lambda}}\left[1 \pm \left(1 + \frac{\hat{\lambda}(\hat{\lambda} + 2m)}{r^4}\right)^{1/2}\right]$$

If negative root taken, the metric is asymptotically flat. For large r, N^2 this becomes

$$N^2 \approx 1 + \frac{r^2}{\hat{\lambda}}\left(1 - \left(1 + \frac{1}{2}\frac{(\hat{\lambda}[\hat{\lambda} - 2m])}{r^4}\right)\right) \approx 1 - r^2\left(\frac{1}{2}[\hat{\lambda} + 2m]\right)$$

$$N^2 = 1 + \frac{r^2}{\hat{\lambda}}\left[1 + \theta\left(1 + \frac{2w\hat{\lambda}}{r^4}\right)^{1/2}\right] \quad ,\theta = \pm 1$$

Take $\theta = -1$ for asymptotically flat solutions.

Fall of conditions at bifurcation 2-sphere in old variables

The metric in curvature coordinates:

$$dS^2 = -f^2(R)dT^2 + f^{-2}(R)dR^2 + R^2 d\Omega_3^2$$

$$f^2(R) = 1 + \frac{R^2}{\hat{\lambda}}\left[1 - \sqrt{1 + \frac{2w\hat{\lambda}}{R^4}}\right]$$

Consider $R = \alpha(1 + p^2)$ such that $(p, \alpha) \to 0$ as $p \to 0$ for appropriate choice of α (interpreted as horizon radius):

$$f^2 = 1 + \frac{\alpha^2(1+p^2)^2}{\hat{\lambda}}\left[1 - \sqrt{1 + \frac{2w\hat{\lambda}}{\alpha^4(4p^2)^4}}\right]$$

$$= 1 + \frac{\alpha^2(1+2p^2)^2}{\hat{\lambda}}\left[1 - \sqrt{1 + \frac{2w\hat{\lambda}}{\alpha^4}(1 - 4p^2)}\right] + O(p^4)$$

$$= 1 + \frac{\alpha^2}{\hat{\lambda}}(1 + 2p^2)^2\left[1 - \left(1 + 1 + \frac{2w\hat{\lambda}}{\alpha^4}\right)^{1/2}\left(1 - \left[1 + \frac{2w\hat{\lambda}}{\alpha^4}\right]^{-1}\left(\frac{2w\hat{\lambda}}{\alpha^4}p^2\right)\right)\right] + O(p^4)$$

$$= 1 + \frac{\alpha^2}{\hat{\lambda}}(1 + 2p^2)\left[\left(1 - \sqrt{1 + \frac{2w\hat{\lambda}}{\alpha^4}}\right) + \frac{1}{\sqrt{1 + \frac{2w\hat{\lambda}}{\alpha^4}}}\left(\frac{4w\lambda}{\alpha^4}\right)p^2\right] + O(p^4)$$

$$= 1 + \frac{\alpha^2}{\hat{\lambda}}\left(1 - \sqrt{1 + \frac{2w\hat{\lambda}}{\alpha^4}}\right) + p^2\left[\frac{2\alpha^2}{\hat{\lambda}}\left(1 - \sqrt{1 + \frac{2w\hat{\lambda}}{\alpha^4}}\right) + \left(\frac{4w}{\alpha^2}\right)\frac{1}{\sqrt{1 + \frac{2w\hat{\lambda}}{\alpha^4}}}\right] + O(p^4)$$

Now, $\alpha = \sqrt{w - \frac{1}{2}\lambda} \to w > \frac{\hat{\lambda}}{2}$ for horizon to exist.

$$f^2 = \left\{2 + \left(\frac{4w}{\alpha^2}\right)\frac{1}{1+\frac{\lambda}{\alpha^2}}\right\}p^2 + O(p^4)$$
$$f^2 = \left[2 + \frac{4w}{\alpha^2+\lambda}\right]p^2 + O(p^4)$$
Consider $p = rp_+(t)$, $T = T(t)$
$$f = \sqrt{2 + \frac{4w}{\alpha^2+\lambda}}\,(rp_t(r)) + O(r^3)$$
$$f^{-1} = \left[p_+(t)\sqrt{1 + \frac{4w}{\alpha^2+\lambda}}\right]^{-1} r^{-1} + O(r)$$
$$dR = 2\alpha(p_+^2 r dr + r^2 p_+ p_+ dt)$$
$$ds^2 = -\left[\sqrt{1 + \frac{4w}{\alpha^2+\lambda}}\,rp_+ + O(r^3)\right]^2 t^2 dt^2 +$$
$$\left[\left(p_+(t)\sqrt{2 + \frac{4w}{\alpha^2+\lambda}}\right)^{-1} r^{-1} + O(r)\right]^2 (2\alpha p_+^2)^2\left(dr + r\frac{\dot{p}}{p_+}dt\right)$$
$$+ [\alpha(1+p^2)]^2 d\Omega_3^2$$

In terms of ADM canonical variables:
$$N = p_+\dot{T}\sqrt{2 + \frac{4w}{\alpha^2+\lambda}}\,r + O(r^3)$$
$$\Lambda = (2\alpha p_+^2 r)\left[\left(p_+\sqrt{2 + \frac{4w}{\alpha^2+\lambda}}\right)^{-1} r^{-1} + O(r)\right]$$
$$N^r = r\left(\frac{\dot{p}_+}{p_+}\right)$$
$$R = \alpha + \alpha p_+^2 r^2$$

Consider the most general gauge obeying the following leading order behaviour:
$$N = N_1(t)r + O_{\geq}(r^2)$$
$$N^r = N_1^r(t)r + O_{\geq}(r^2)$$
$$\Lambda = \Lambda_o(t) + O(r^2)$$
$$R = R_o(t) + R_2(t)r^2 + O_{\geq}(r^3)$$

From $0 = H_r = P_R R' - \Lambda P_\Lambda'$ we get:
$$P_R\left(2R_2 r + O_{\geq}(r^2)\right) = \left(\Lambda_o + O(r^2)\right)P_\Lambda'$$
If

$$\boxed{P_\Lambda = O(r^k) \rightarrow P_R = O(r^{k-2})}$$

From $0 = H$: first need y :

$$\left(1-\left(\frac{R'}{\Lambda}\right)'\right) = 1 - \left[\frac{\left(2R_2 r + O_{\geq}(r^2)\right)}{\Lambda_0 + O(r^2)}\right]^2 = 1 - 4\frac{R_2^2}{\Lambda_0^2}r^2 + O_{\geq}(r^2)$$

$$\left(\frac{R'}{\Lambda}\right)' = \left[\frac{\left(2R_2 r + O_{\geq}(r^2)\right)}{\Lambda_0 + O(r^2)}\right]' = \left(\frac{2R_2}{\Lambda_0}\right) + O_{\geq}(r^2)$$

$$\Lambda R = \Lambda_0 R_0 + O(r^2)$$

$$\Lambda R\left(1-\left(\frac{R'}{\Lambda}\right)^2\right) = \Lambda_0 R_0 + O(r^2)$$

$$y:\quad \frac{1}{3}\hat{\lambda}y^3 + y\left[R^2 + \hat{\lambda}\left(1-\left(\frac{R'}{\Lambda}\right)^2\right)\right] = -P_\Lambda$$

$$\frac{1}{3}\hat{\lambda}y^3 + y\left[R_0^2 + 2R_0R_2r^2 + O_{\geq}(r^3) + \hat{\lambda}\left(1 - 4\frac{R_2^2}{\Lambda_0^2}r^2 + O_{\geq}(r^2)\right)\right] = -P_\Lambda$$

So, $\boxed{O(y) = O(P_\Lambda)}$ or $O(y^3) = O(P_\Lambda)$, the latter case holding only if P_Λ has divergent leading order.

<u>**$0 = H$:**</u>

$$0 = O(r^{2k-2}) + O(r^{2k})\left(O(r^0)\right) - O(r^0) + O(r^2)\left(O(r^0)\right) \rightarrow k \geq 1$$

Suppose $k = 2$, the minimal restriction without structuring the momenta, then the isolated $O(r^0)$ term must be identically zero, only contributing at $O(r^2)$ at next order:

$$\Lambda R\left(1-\left(\frac{R'}{\Lambda}\right)^2\right) = \left(\Lambda_0 R_0 + O(r^2)\right)\left(1 - 4\frac{R_2^2}{\Lambda_0^2}r^2 + O_{\geq}(r^2)\right)$$

$\rightarrow \Lambda_0 R_0 = 0$, but this contradicts the desired leading order behaviour.

The H=0 constraint indicates that there is a problem.

Consider the $\dot{R}$ equation: $\dot{R} = Ny + N^r R'$

$$\dot{R}_0 + \dot{R}_2 r^2 = \left(N_1 r + O_{\geq}(r^2)O(r^k) + \left(N_1^r(t)r + O_{\geq}(r^2)\right)\left(2R_2 r + O_{\geq}(r^2)\right)\right)$$

$\rightarrow \boxed{\dot{R}_0 = 0}$ *ok* $\qquad \dot{R}_2 = 2R_2N_1$

$\rightarrow k \geq 1$ as before

Consider $\dot{\Lambda}$ equation:

$$\dot{\Lambda} = -N\left(P_\Lambda + 2y\left[\Lambda R - \hat{\lambda}\left(\frac{R'}{\Lambda}\right)'\right]\right)\left(R^2 + \hat{\lambda}\left(1-\left(\frac{R'}{\Lambda}\right)^2\right)\right)' + (N^r\Lambda)'$$

$$\dot{\Lambda}_0 + O(r^2) = (-N_1 r + O_{\geq}(r^2))\left(O_{\geq}(r^{k-2}) + O(r^2)\left[\Lambda_0 R_0 - \hat{\lambda}\left(\frac{2R_0}{\Lambda_0}\right) + O(r^1)\right]\right)$$
$$\times\left(R_0 + R_2 r^2 + \hat{\lambda}\left[O(r^k) + 1 - 4\frac{R_0^2}{\Lambda_0}r^2 + O_{\geq}(r^2)\right]\right)^{-1} + \left[(N_1^r r + O_{\geq}(r^2))(\Lambda_0 + O(r^2))\right]'$$
$$= O(r^{k-1}) + N_1^r \Lambda_0 + O_{\geq}(r)$$

The first term indicates $k \geq 3$, the second term gives $\dot{\Lambda}_0 = N_1^r \Lambda_0$, the third term is inconsistent with the order indicate to be $O(r^2)$, setting the order from inequality form to $O(r^2)$ we get:

$$N^r = N_1^r(t)r + O(r^3)$$

So far:

$N = N_1(t)r + O_{\geq}(r^2)$
$N^1 = N_1^r(t)r + O(r^3)$
$\Lambda = \Lambda_0(t) + O(r^2)$
$R = R_o(t) + R_2(t)r^2 + O_{\geq}(r^3)$
$\left.\begin{array}{l} P_R = O(r^1) \\ P_\Lambda = O(r^3) \end{array}\right\} y = O(r^3)$

$$\dot{P}_\Lambda = N\left[\left(\frac{2y\hat{\lambda}\left(\frac{R'}{\Lambda}\right)^2 \frac{1}{\Lambda}}{R^2 + \hat{\lambda}\left[y^2 + \left(1 - \left(\frac{R'}{\Lambda}\right)^2\right)\right]}\right)\left(P_R + 2y\left(\Lambda R - \hat{\lambda}\left(\frac{R'}{\Lambda}\right)'\right)\right) - y^2 R + R\left(1 - \left(\frac{R'}{\Lambda}\right)^2\right)\right]$$
$$+2R\left(\frac{R'}{\Lambda}\right)^2 - 2\hat{\lambda}\Lambda^{-1}\left(\frac{R'}{\Lambda}\right)^2\right] + \hat{\lambda}(Ny^2)'\left(\frac{R'}{\Lambda}\right)\Lambda^{-1} - \left[R^2 + \hat{\lambda}\left(1 - \left(\frac{R'}{\Lambda}\right)^2\right)\right]'\Lambda^{-1} + N^r P'_\Lambda$$

$$O(r^3) = (N_1 r + O_{\geq}(r^2))\left[(O(r^1)) - (O_{\geq}(r^3))^2 + R\left(1 - \left(\frac{R'}{\Lambda}\right)^2\right) + 2R\left(\frac{R'}{\Lambda}\right)^2 - 2\hat{\lambda}\Lambda^{-1}\left(\frac{R'}{\Lambda}\right)^2\left(\frac{R'}{\Lambda}\right)'\right] + O(r^6) - \left[(N_1 r + O_{\geq}(r^2))\left(R^2 + \hat{\lambda}\left(1 - \left(\frac{R'}{\Lambda}\right)^2\right)\right)\right]'\left(\frac{R'}{\Lambda}\right)\Lambda^{-1} + O(r^3)$$
$$= (N_1 r + O_{\geq}(r^2))\left[R_0 + O(r^2) - \frac{2\hat{\lambda}}{\Lambda_0}O(r^2)\right]$$
$$-\frac{(2R_2 r + O_{\geq}(r^2))}{\Lambda_0^2}\left[rN_1(R_0^2 + \hat{\lambda}) + (R_0^2 + \hat{\lambda})O_{\geq}(r^2) + O(r^3)\right]'$$

The $O_{\geq}(r^2)$ can't be eliminated, so it mustn't exist, so we take $O(r^3)$ instead:

$$\boxed{\begin{gathered} N = N_1(t)r + O(r^3) \\ R = R_0(t) + R_2(t)r^2 + O(r^4) \end{gathered}}$$

$$O(r^3) = rN_1\left[R_0 - 2\tfrac{R_2}{\Lambda_0^2}\left(R_0^2 + \hat{\lambda}\right)\right] + O(r^3)$$

$$\left(1 - \frac{2R_0R_2}{\Lambda_0^2}\right) - \frac{2R_0\hat{\lambda}}{\Lambda_0^2 R_0} = 0 \rightarrow R_2 > 0 \ , \qquad \hat{\lambda} > 0$$

$$N = N_1(t)r + O(r^3)$$
$$N^r = N_1^r(t)r + O(r^3)$$
$$\Lambda = \Lambda_0(t) + O(r^2)$$
$$R = R_o(t) + R_2(t)r^2 + O(r^4)$$
$$\left.\begin{aligned} P_R &= O(r^1) \\ P_\Lambda &= O(r^3) \end{aligned}\right\} \ y = O(r^3)$$

$$\dot{P}_R = -N\left[\frac{\left(P_R + 2y\left[\Lambda R - \hat{\lambda}\left(\frac{R'}{\Lambda}\right)'\right]\right)(-2yR)}{\left(R^2 + \hat{\lambda}\left[y^2 + \left(1 - \left(\frac{R'}{\Lambda}\right)^2\right)\right]\right)} + y^2\Lambda - \Lambda\left(1 - \left(\frac{R'}{\Lambda}\right)^2\right)\right]$$
$$+\left[\frac{N\left(P_R + 2y\left[\Lambda R - \hat{\lambda}\left(\frac{R'}{\Lambda}\right)'\right]\right)(2y\hat{\lambda})\left(\frac{R'}{\Lambda}\right)'\Lambda^{-1}}{\left(R^2 + \hat{\lambda}\left[y^2 + \left(1 - \left(\frac{R'}{\Lambda}\right)'^2\right)\right]\right)}\right]' + \left(\frac{(Ny^2\hat{\lambda})'}{\Lambda}\right)' + \left(2NR\left(\frac{R'}{\Lambda}\right)'\right)'$$
$$+\left[\frac{\left(N\left[R^2 + \hat{\lambda}\left(1 - \left(\frac{R'}{\Lambda}\right)^2\right)\right]\right)'}{\Lambda}\right]'$$
$$-\left[2\hat{\lambda}N\left(\frac{R'}{\Lambda}\right)\left(\frac{R'}{\Lambda}\right)'\Lambda^{-1}\right]' + (N^r P_R)'$$

$$O(r^1) = -\left(N_1 r + O(r^3)\right)\left[O(r^6) - \Lambda_0\right] + \left[\left(N_1 r + O(r^3)\right)R_0^{-2}O(r^1)\right]'$$
$$+O(r^5) + O(r^1) + O(r^1) = O(r^1)$$

so consistent.

Consider the mass first integral:

$$m = constant = yP_\Lambda + \tfrac{1}{6}\hat{\lambda}y^4 + R^2\left(1 - \left(\frac{R'}{\Lambda}\right)^2\right) + \hat{\lambda}\left(\frac{R'}{\Lambda}\right)^2\left(1 - \frac{1}{2}\left(\frac{R'}{\Lambda}\right)^2\right)$$
$$= R_0^2 + r^2\left(2R_0R_2 - R_0^2\left(\frac{2R_2}{\Lambda_0}\right)^2 + \hat{\lambda}\left(\frac{2R_2}{\Lambda_0}\right)^2\right) + O(r^4)$$

and we get:

$$\boxed{1 - \frac{2R_0R_2}{\Lambda_0^2} - \frac{2R_2\hat{\lambda}}{{\Lambda_0}^2R_0} = 0}$$

The same relation as required from the $\dot{P}_\Lambda$ equation of motion

Fall off of new variables

Let's succinctly repeat the fall-off analysis given the formulae and leading-order fall-off behaviour obtained thus far:

$F = \left(\frac{R'}{\Lambda}\right)^2 - y^2$

At $r = 0$:

$F(t,r) = 4\left(\frac{R_2}{\Lambda_0}\right)^2 r^2 + O(r^4)$

$M_+ = \frac{\hat{\lambda}}{4} + \frac{1}{2}R_0^2 - \frac{1}{2}R_0^2\left(\frac{2R_2}{\Lambda_0}\right)^2 r^2 - \frac{1}{2}\hat{\lambda}\left(\frac{2R_2}{\Lambda_0}\right)^2 r^2 + O(r^4)$

$= \frac{1}{2}\left(R_0^2 + \frac{1}{2}\hat{\lambda}\right) - \left(2\left(\frac{2R_2}{\Lambda_0}\right)^2\right)(R_0^2 + \hat{\lambda})r^2 + O(r^4)$

$M_+ = M_0(t) + M_2(t)r^2 + O(r^4)$

$M_0 = \frac{1}{2}\left(R_0^2 + \frac{1}{2}\hat{\lambda}\right)$

$M_2 = 2R_2\left(\frac{1}{2}R_0 - R_2\left(\frac{R_0}{\Lambda_0}\right)^2 - \left(\frac{R_2}{{\Lambda_0}^2}\right)\hat{\lambda}\right) = R_0R_2\left(1 - 2\frac{R_0R_2}{\Lambda_0^2} - \frac{2R_2\hat{\lambda}}{\Lambda_0^2R_0}\right)$

$R(t,r) = R_0(t) + R_2(t)r^2 + O(r^4)$

$P_M = O(r^{-1})\ ;\ P_M = -F^{-1}\Lambda y$

$P_{\bar{R}} = \frac{1}{R'}H_r - \frac{1}{R'}P_{M+}M_+' = O(r)$

<u>At $r = \infty$:</u>

$F(t,r) = 1 - w_+r^{-2} + O(r^{-2-\epsilon})$

$$M_+ = \frac{\hat{\lambda}}{4} + r^2\left(\frac{1}{2}w_+r^{-2} + O(r^{-2-\epsilon})\right)$$
$$-\hat{\lambda}\left(1 - \frac{1}{2}wr^{-2} + O(r^{-2-\epsilon})\right)^2\left(1 - \frac{1}{2}\left[1 - \frac{1}{2}wr^{-2} + O(r^{-2-\epsilon})\right]\right)$$
$$+ O(r^{-\epsilon})$$

$$M_+ = \frac{\hat{\lambda}}{4} + \frac{1}{2}w_+ - \hat{\lambda}\left(\frac{1}{2}\right)^2 + O(r^{-\epsilon}) = \frac{1}{2}w_+(t) + O(r^{-\epsilon})$$

$R(t,r) = r + O^\infty(r^{-1-\epsilon})$

$P_M = O(r^{-1-\epsilon})$

$P_R = O(r^{-\epsilon})$

Let's summarize and evaluate the convergence of Action:

$\boxed{r \to \infty}$

$$\Lambda = 1 + \frac{1}{2} w r^{-2} + O(r^{-2-\epsilon})$$
$$R = r + O^{\infty}(r^{-1-\epsilon})$$
$$\left.\begin{array}{l} P_\Lambda = O(r^{1-\epsilon}) \\ P_R = O(r^{-\epsilon}) \end{array}\right\} \; y = O(r^{-1-\varepsilon})$$
$$N = O(r^0)$$
$$N^r = O(r^{-1-\epsilon})$$

$\boxed{r \to 0}$

$$\Lambda = \Lambda_0(t) + O(r^2)$$
$$R = R_0(t) + R_2(t) r^2 + O(r^4)$$
$$\left.\begin{array}{l} P_\Lambda = O(r^3) \\ P_R = O(r^1) \end{array}\right\} \; y = O(r^3)$$
$$N = N_1(t) r + O(r^3)$$
$$N^r = N_1^r(t) r + O(r^3)$$

$$S = \int \left[\Lambda P_\Lambda + \hat{R} P_R - NH - N^r H_r\right] drdt$$

$\underline{\boldsymbol{r \to \infty}}$

$$\lim_{R\to\infty} S_R = \lim_{R\to\infty} \int_0^\infty \begin{bmatrix} O(r^{-2})O(r^{1-\epsilon}) + O(r^{-1-\epsilon})O(r^{-\epsilon}) \\ -O(r^0)O(r^{1-\epsilon}) - O(r^{-1-0})O(r^{-0}) \end{bmatrix} drdt = 0$$

$\underline{\boldsymbol{r \to 0}}$

$$\lim_{\epsilon\to 0} S_\epsilon = \lim_{\epsilon\to 0} \int_0^\epsilon [O(r^0)O(r^3) + O(r^0)O(r^1) - O(r^1)O(r^0) - O(r^1)O(r^2)] drdt = 0$$

Thus, the action integral is convergent. The canonical transformation is, therefore, well defined.

In the following use is made of the relation $\left(P_\Lambda = -\frac{1}{3}\hat{\lambda} y^3 - y\left[R^2 + \hat{\lambda}\left(\frac{R'}{\Lambda}\right)^2\right]\right)$ to get:

$$P_\Lambda \delta\Lambda + P_R \delta R - P_{\bar{R}} \delta R - P_{M_+} \delta M_+$$
$$= \delta\left(\Lambda P_\Lambda + \hat{\lambda} R \left[\left(\frac{R'}{\Lambda}\right) y\right]' + \frac{1}{2}\left(\frac{1}{3}R^3 + \hat{\lambda} R\right)' \ln\left|\frac{R' + y\Lambda}{R' - y\Lambda}\right|\right)$$
$$- \left[\hat{\lambda} R \delta\left[\left(\frac{R'}{\Lambda}\right) y\right] + \frac{1}{2}\delta\left(\frac{1}{3}R^3 + \hat{\lambda} R\right) \ln\left|\frac{R' + y\Lambda}{R' - y\Lambda}\right|\right]'$$

The boundary terms are

$B = \hat{\lambda}R\delta\left[\left(\frac{R'}{\Lambda}\right)y\right] + \frac{1}{2}\delta\left(\frac{1}{3}R^3 + \lambda R\right)\ln\left|\frac{R'+y\Lambda}{R'-y\Lambda}\right|$

For $r \to \infty$

$$B = \hat{\lambda}O(r)O(r^{-1-\epsilon}) + \frac{1}{2}O(r^{1-\epsilon})\ln|1 + O(r^{-1-\epsilon})| = O(r^{-\epsilon})$$

hence vanishes.

For $r \to 0$

$B = \hat{\lambda}O(r^4) + \frac{1}{2}O(r^0)\ln\left|\frac{2R_2r+O(r^3)}{2R_2r-O(r^3)}\right| = \hat{\lambda}O(r^4) + \frac{1}{2}O(r^0)O(r^2) = O(r^2)$,

hence vanishes. So, $\int(P_\Lambda\delta\Lambda + P_\Lambda\delta R - P_R\delta R - P_{M_+})dr = \delta w$, where

$$w = \int_0^\infty dr\left(\Lambda P_\Lambda + \hat{\lambda}R\left[\left(\frac{R'}{\Lambda}\right)y\right]' + \frac{1}{2}\left(\frac{1}{3}R^3 + \lambda R\right)'\ln\left|\frac{R' + y\Lambda}{R' - y\Lambda}\right|\right)$$

If w is to be well defined, the integrand must fall off faster than r^{-1}, i.e., $O(r^{-1-\epsilon})$, as $r \to \infty$. Also, it must fall off at least as fast as r^0 as $r \to 0$:

$\underline{r \to \infty}$

$$\begin{aligned}\text{Integrand} &= \left(P_\Lambda + O(r^{-1-\epsilon})\right) + \hat{\lambda}[O(r^{-2-\epsilon}) + O(r^{-1-\epsilon})] \\ &+\frac{1}{2}(R^2R' + \hat{\lambda}R')\left(\ln\left|1 + \frac{y\Lambda}{R'}\right| - \ln\left|1 - \frac{y\Lambda}{R'}\right|\right) \\ &= -yR^2\Lambda + \frac{1}{2}R^2R'\left(\frac{2y\Lambda}{R'}\right) + O(r^{-1-\epsilon}) \\ &= O(r^{-1-\epsilon})\end{aligned}$$

So, have sufficient fall-off at infinity.

$\underline{r \to 0}$

$$\text{Integrand} = O(r^3) + O(r^1)O(r^2)$$

So, have sufficient fall-off at bifurcate n-sphere. So, the integral expression for w converges, thus w is well defined.

New Lagrange multipliers:

$NH + N^rH_r = N^MM_+' + N^RP_R$

$M_+' = \Lambda^{-1}[yH_r - R'H]$

$P_{\bar{R}} = F^{-1}[-yH + R'\Lambda^{-2}H_r]$

So,

$NH + N^rH_r = N^M\{\Lambda^{-1}[yH_r - R'H]\} + N^R\{F^{-1}[-yH + R'\Lambda^{-2}H_r]\}$

$= \left(-N^M\Lambda^{-1}R'H - N^RF^{-1}yH\right) + \left(N^M\Lambda^{-1}y + N^RF^{-1}R'\Lambda^{-2}\right)H_r$

$$\boxed{\begin{aligned}N &= -N^RF^1y - N^M\Lambda^{-1}R' \\ N^r &= N^M\Lambda^{-1}y + N^RF^{-1}R'\Lambda^{-2}\end{aligned}} \rightarrow N^R = F\Lambda^2(R')^{-1}[N^r - N^M\Lambda^{-1}y]$$

$$N = -F\Lambda^2(R')^{-1}[N^r - N^M\Lambda^{-1}]F^{-1}y - N^M\Lambda^{-1}R'$$
$$N + (y\Lambda)\left(\frac{\Lambda}{R'}\right)N^r = -N^M[-\Lambda(R')^{-1}y^2 + \Lambda^{-1}R']$$
$$\boxed{N^M = -NF^{-1}\left(\frac{R'}{\Lambda}\right) - N^rF^{-1}(y\Lambda)}$$

$$N^R = F\Lambda^2(R')^{-1}\left[N^r + NF^{-1}\left(\frac{R'}{\Lambda}\right)\Lambda^{-1}y + N^rF^{-1}(y\Lambda)(\Lambda^{-1}y)\right]$$
$$= N^r\left(\frac{R'}{\Lambda}\right)^2\Lambda^2(R')^{-1} + Ny$$

$$\boxed{N^R = Ny + N^rR'}$$

Since the linear transformation from (N, N^r) to (N^m, N^R) is non-singular for $r > 0$, we could take the constraint terms as those multiplying N^m, N^R:

<u>At $r \to \infty$</u>

$N^M(t,r) = -\widetilde{N}_+(t) + O(r^{-\epsilon})$
$N^R(t,r) = O(r^{-1-\epsilon})$

<u>At $r \to 0$</u>

$N^M = -\frac{1}{2}N_1\Lambda_0R_2^{-1} + O(r^2)$
$N^R(t,r) = O(r^2)$

From

$$\delta\int drdt\,(-N^MM') \to \int drdt\,(-N^M\delta M)' \to \int dt\,[-N^M\delta M]_{r=0}^{r=\infty}$$
$$= \int dt\left[\widetilde{N}_+(t)\left(\frac{1}{2}\delta w_+(t)\right)\right]_{r=\infty} \text{-}\int dt\left[\left(\frac{1}{2}N,\Lambda_0R_2^{-1}\right)(R_0\delta R_0)\right]_{r=0}$$
$$= \delta\int dt\left[\widetilde{N}_+(t)\left(\frac{1}{2}w_+(t)\right)\right]_{r=\infty} - \int dt\left[(N_1\Lambda_0^{-1})(R_0^2 + \hat{\lambda})\delta R_0\right]_{r=0}$$
$$= \delta\int dt\left[\widetilde{N}_+(t)\left(\frac{1}{2}w_+(t)\right)\right] - \delta\int dt\left[(N_1\Lambda_0^{-1})\left(\frac{1}{3}R_0^2 + \hat{\lambda}\right)\right]$$

with $\widetilde{N}_+(t)$ and $N_1\Lambda_0^{-1}$ fixed. Thus the surface terms:

$$S_{\partial\Sigma} = \int dt\left[-\widetilde{N}_+(t)\left(\frac{1}{2}w_+(t)\right)\right] + \int dt\left[(N_1\Lambda_0^{-1})\left(\frac{1}{3}R_0^3 + \hat{\lambda}\right)\right]$$

Since $N^M = -\frac{1}{2}N_1\Lambda_0R_2^{-1} + O(r^2)$ as $r \to 0$, there is a complication in the expression for the corresponding surface term. Consider a new Lagrange multiplier:

$N^M = -\widetilde{N}^M[(1-g) + gX]$

Where $-\widetilde{N}^M X = -\frac{1}{2} N_1 \Lambda_0 R_2^{-1}$and want X such that $-\widetilde{N}^\mu = N_1 \Lambda_0^{-1}$:

$$X = 2\Lambda_0^2 R_2^{-1}$$

When $M' = 0$: $1 - 2\Lambda_0^{-2}\left(R_o + \frac{\hat{\lambda}}{R_o}\right) = 0$

$\frac{1}{2}\Lambda_0^2 R_0^{-1} = \left(R_0 + \frac{\hat{\lambda}}{R_o}\right)$

Thus,

$X = 4\left(R_0 + \frac{\hat{\lambda}}{R_o}\right)$

and

$$N^m = -\widetilde{N}^\mu\left[(1-g) + 4g\left(R_0 + \frac{\hat{\lambda}}{R_o}\right)\right], g(r=\infty) = 0\,, g(r=0) = 1$$

Where $g \to 1, 0$ consistent with the fall-off of N^m as $r \to 0\,, \infty$.

Total action in the new variables:

$S_\Sigma[M_+, R\,, P_M, P_R, \widetilde{N}^M, N^R]$

$= \int dt \int_0^\infty dr \{P_{M+}\dot{M}_+ + P_R\dot{R} - N^R P_R + \widetilde{N}^\mu[(1-g) + 4gR_0(1 + R_0^{-2}\hat{\lambda})]M'\}$

$+ \int dt \left(\left(\frac{1}{3}R_0^3 + R_0\hat{\lambda}\right)\widetilde{N}_0{}^M - \widetilde{N}_+\left(\frac{1}{2}w_+\right)\right)$

$\dot{M}_+ = 0$

$\dot{R} = N^R$

$\dot{P}_{M_+} = (N^M)'$

$\dot{P}_R = 0$

$M' = 0$

$P_R = 0$

Canonical Transformation

The metric in curvature coordinates:

$$ds^2 = -F(R)dT^2 + F^{-1}(R)dR^2 + R^2 d\Omega_3^2$$

$$F(R) = 1 + \frac{R^2}{\hat{\lambda}}\left(1 - \sqrt{1 + \frac{2w_+\hat{\lambda}}{R^4}}\right)$$

$dT(r,t) = \dot{T}dt + T'dr$

$dR(r,t) = \dot{R}dt + R'dr$

$ds^2 = -(F\dot{T}^2 - F^{-1}\dot{R}^2)dt^2 + 2(-FT'^{\dot{T}} + F^{-1}R'\dot{R})dtdr + (-FT'^2 + F^{-1}R'^2)dr^2 + R^2 d\Omega_3^2$

Comparison with ADM form:

$ds^2 = -(N^2 - \Lambda^2 N^{r2})dt^2 + 2\Lambda^2 N^r dtdr + \Lambda^2 dr^2 + R^2 d\Omega_3^2$

$N^r = \frac{-FT'\dot{T}+F^{-1}R'\dot{R}}{-FT'^2+F'R'^2}$
$N = \frac{R'\dot{T}-T'\dot{R}}{\sqrt{-FT'^2+F'R'^2}}$
$\Lambda^2 = -FT'^2 + F^{-1}R'^2$

Using the $\dot{R}$ equation of motion: $\dot{R} = Ny + N^r R'$

$$y = N^{-1}(\dot{R} - N^r R') = N^{-1}\left[\dot{R}\left(\frac{-FT'^2 + F^{-1}R'^2}{-FT'^2 + F^{-1}R'^2}\right) - R'\left(\frac{-FT'\dot{T}+F^{-1}R'\dot{R}}{-FT'^2+F'R'^2}\right)\right]$$
$$= N^{-1}T'F\frac{\dot{T}R' - \dot{R}T'}{-FT'^2 + F'R'^2} = \Lambda^{-1}FT'$$

$$\boxed{T' = F^{-1}\Lambda y}$$
$$\boxed{F = \left(\frac{R'}{\Lambda}\right)^2 - y^2}$$

$F = 1 + \frac{R^2}{\hat{\lambda}}\left(1 - \sqrt{1 + \frac{2w_+\hat{\lambda}}{R^4}}\right)$
$\left[(F-1)^2 - 2(F-1)\frac{R^2}{\hat{\lambda}} + \left(\frac{R^2}{\hat{\lambda}}\right)^2\right] = \left(\frac{R^2}{\hat{\lambda}}\right)^2 + \frac{2w_+}{\hat{\lambda}}$
$w_+ = \frac{\hat{\lambda}}{2}\left[F^2 - 2F\left(1 + \left(\frac{R^2}{\hat{\lambda}}\right)\right) + \left(1 + 2\left(\frac{R^2}{\hat{\lambda}}\right)\right)\right]$
$= \frac{\hat{\lambda}}{2}\left[\left(\frac{R'}{\Lambda}\right)^4 - 2\left(\frac{R'}{\Lambda}\right)^2 y^2 + y^4 - 2\left[\left(\frac{R'}{\Lambda}\right)^2 - y^2\right]\left(1 + \left(\frac{R^2}{\hat{\lambda}}\right)\right) +\right.$
$\left.\left(1 + 2\left(\frac{R^2}{\hat{\lambda}}\right)\right)\right]$
$\left(w_+ - \frac{\hat{\lambda}}{2}\right) = R^2\left(1 - \left(\frac{R'}{\Lambda}\right)^2\right) - \hat{\lambda}\left(\frac{R'}{\Lambda}\right)^2\left(1 - \frac{1}{2}\left(\frac{R'}{\Lambda}\right)^2\right) + \frac{1}{2}\hat{\lambda}y^4 +$
$y^2\left[R^2 + \hat{\lambda}\left(1 - \left(\frac{R'}{\Lambda}\right)^2\right)\right] = 2m$
$m' = \Lambda^{-1}[yH_r - R'H]$
Consider $\frac{1}{2}w_+$ (a convenient choice with respect to the $r \to \infty$ boundary term in the action):
$\frac{1}{2}w_+ = m + \frac{\hat{\lambda}}{4} = M_+$
$\frac{1}{2}w_+{}' = M_+'$

For new canonical variables try completing $P_R R' - \Lambda P_\Lambda' = P_{\bar{R}}R' + P_{M_+}M_+'$: guess $p_{M_+} = -T'$ for classical solutions and $p_{M_+} = -F^{-1}\Lambda y$

otherwise. Take $\bar{R} = R$. Obtain $P_{\bar{R}}$ from consistency with the supermomentum relation:

$H_r(\Lambda_1 R_1 P_+ P_R) = P_{\bar{R}}\bar{R}' + P_{M+}M'_+$

$P_{\bar{R}} = \frac{1}{R'}H_r - \frac{1}{R'}P_{M_+}M'_+ = \frac{1}{R'}H_r - \frac{1}{R'}(-F^{-1}\Lambda y)\left(-\frac{1}{\Lambda}\right)[R'H - yH_r]$

$$\boxed{P_{\bar{R}} = F^{-1}\left[-yH + R'\Lambda^{-2}H_r\right]}$$

Now, to determine if the difference in Liouville forms is exact:

$$Ƶ = \int_{-\infty}^{\infty} dr\,\{P_\Lambda(r)\delta\Lambda(r) + P_R(r)\delta R(r) - P_{M_+}(r)\delta M_+(r)$$
$$- P_{\bar{R}}(r)\delta\bar{R}(r)\} = \int dr\,X$$
$$? = \delta w[\Lambda, R, P_\Lambda, P_R] + \int_{-\infty}^{\infty} dr\,Y'$$

$P_{M_+} = (R')^{-1}(P_R R' - \Lambda P'_\Lambda) - (R')^{-1}P_{M_+}M'_+$

$\delta M_+ = \frac{1}{2}\delta\left[\frac{1}{2}\hat{\lambda}y^4 + y^2\left(R^2 + \hat{\lambda}\left(1 - \left(\frac{R'}{\Lambda}\right)^2\right)\right) + R^2\left(1 - \left(\frac{R'}{\Lambda}\right)^2\right) - \hat{\lambda}\left(\frac{R'}{\Lambda}\right)^2\left(1 - \frac{1}{2}\left(\frac{R'}{\Lambda}\right)^2\right)\right]$

$$= \hat{\lambda}y^3\delta y + y\delta y\left(R^2 + \hat{\lambda}\left(1 - \left(\frac{R'}{\Lambda}\right)^2\right)\right)$$
$$+ \frac{1}{2}\delta\left[R^2\hat{\lambda}\left(1 - \left(\frac{R'}{\Lambda}\right)^2\right) - \hat{\lambda}\left(\frac{R'}{\Lambda}\right)^2\left(1 - \frac{1}{2}\hat{\lambda}\left(\frac{R'}{\Lambda}\right)^2\right)\right]$$
$$+ \frac{1}{2}y^2\delta\left(R^2 + \hat{\lambda}\left(1 - \left(\frac{R'}{\Lambda}\right)^2\right)\right)$$

y: $\frac{1}{3}\hat{\lambda}y^3 + y\left[R^2 + \hat{\lambda}\left(1 - \left(\frac{R'}{\Lambda}\right)^2\right)\right] + P_\Lambda = 0$

$\left[\hat{\lambda}y^2 + \left(R^2 + \hat{\lambda}\left(1 - \left(\frac{R'}{\Lambda}\right)^2\right)\right)\right]\delta y + y\left[2R\delta R - 2\hat{\lambda}\left(\frac{R'}{\Lambda}\right)\delta\left(\frac{R'}{\Lambda}\right) + \delta\left(\frac{R'}{\Lambda}\right)\right] + \delta P_\Lambda = 0$

$$\delta y = \frac{-y\left[2R\delta R - 2\hat{\lambda} + \delta\left(\frac{R'}{\Lambda}\right)\delta + \delta\left(\frac{R'}{\Lambda}\right)\right] + \delta P_\Lambda}{\hat{\lambda}y^2 + \left(R^2 + \hat{\lambda}\left(1 - \left(\frac{R'}{\Lambda}\right)^2\right)\right)}$$

Let $DV = \frac{V'}{R'}\delta R - \delta V$

$Dy = \frac{y'}{R'}\delta R - \delta y = \frac{-\delta P_\Lambda - 2y\hat{\lambda}\left(\frac{R'}{\Lambda}\right)\delta\left(\frac{R'}{\Lambda}\right)}{\hat{\lambda}y^2 + \left(R^2 + \hat{\lambda}\left(1-\left(\frac{R'}{\Lambda}\right)^2\right)\right)}$

$$DM_+ = \frac{M'_+}{R'}\delta R - \delta M_+ = \left(\hat{\lambda}y^3 + y\left[R^2 + \hat{\lambda}\left(1 - \left(\frac{R'}{\Lambda}\right)^2\right)\right]\right)Dy$$

$$+\left(-R^2\left(\frac{R'}{\Lambda}\right) - \hat{\lambda}\left(\frac{R'}{\Lambda}\right) + \hat{\lambda}\left(\frac{R'}{\Lambda}\right)^3\right)D\left(\frac{R'}{\Lambda}\right) - y^2\hat{\lambda}\left(\frac{R'}{\Lambda}\right)D\left(\frac{R'}{\Lambda}\right)$$

$X = P_R\delta R - P_{\bar{R}}\delta R - P_{M+}\delta M_+ + P_\Lambda\delta\Lambda = (R')^{-1}\Lambda P'_\Lambda\delta R + P_\Lambda\delta\Lambda +$
$P_{M_+}\left(\frac{M'_+}{R'}\delta R - \delta M_+\right)$

Making use of $P_\Lambda = -\frac{1}{3}\hat{\lambda}y^3 - y\left[R^2 + \hat{\lambda}\left(1 - \left(\frac{R'}{\Lambda}\right)^2\right)\right]$:

$$X = \left(\delta(-\Lambda R^2 y) + \Lambda\delta(R^2 y)\right) + \left[-\frac{1}{3}\hat{\lambda}y^3 - y\hat{\lambda}\left(1 - \left(\frac{R'}{\Lambda}\right)^2\right)\right]\delta\Lambda$$
$$+\left(\frac{\Lambda}{R'}\right)\left[(-R^2y)' - \hat{\lambda}y^2y' - \hat{\lambda}y'\left(1 - \left(\frac{R'}{\Lambda}\right)^2\right) + \hat{\lambda}2y\left(\frac{R'}{\Lambda}\right)'\right]\delta R -$$
$$\frac{\Lambda y}{F}\left[\frac{M'_+}{R'}\delta R - \delta M_+{}'\right]$$
$$= \delta(-\Lambda R^2 y) + \Lambda R^2\delta y + 2\Lambda yR\delta R + \left(\frac{\Lambda}{R'}\right)(-R^2y)'\delta R - \frac{\Lambda y^2}{F}R^2 Dy$$
$$+\frac{\Lambda y}{F}R^2\left(\frac{R'}{\Lambda}\right)D\left(\frac{R'}{\Lambda}\right)$$
$$+\hat{\lambda}\left[\begin{array}{c}(\delta\Lambda)\left[-\frac{1}{3}y^3 - y\left(1 - \left(\frac{R'}{\Lambda}\right)^2\right)\right] + \left(\frac{\Lambda}{R'}\right)\left[\begin{array}{c}-y^2y' - y'\left(1 - \left(\frac{R'}{\Lambda}\right)^2\right)\\ +2y\left(\frac{R'}{\Lambda}\right)\left(\frac{R'}{\Lambda}\right)'\end{array}\right]\delta R\\ -\frac{\Lambda y}{F}\left(\left[y^3 + y\left(1 - \left(\frac{R'}{\Lambda}\right)^2\right)\right]Dy + \left(\left(\frac{R'}{\Lambda}\right)^3 - \left(\frac{R'}{\Lambda}\right) - y^2\left(\frac{R'}{\Lambda}\right)\right)D\left(\frac{R'}{\Lambda}\right)\right)\end{array}\right]$$
$$= X_0 + X\hat{\lambda}$$

$$X_0 = \delta(-\Lambda R^2 y) + \Lambda R^2\delta y - \Lambda R^2\left(\frac{y'}{R'}\delta R\right) - \Lambda y^2R^2F^{-1}Dy +$$
$$\frac{yR^2R'}{F\Lambda}D(R') - \left(\frac{R'}{\Lambda}\right)^2\frac{yR^2}{F}D\Lambda$$

$= \delta(-\Lambda R^2 y) - \Lambda R^2 \left[\frac{\left(\frac{R'}{\Lambda}\right)^2 - y^2}{\left(\frac{R'}{\Lambda}\right)^2 - y^2}\right] Dy - \frac{\Lambda R^2 y^2}{\left(\frac{R'}{\Lambda}\right)^2 - y^2} Dy + \frac{y R^2 R'}{F\Lambda} D(R') -$
$\left(\frac{R'}{\Lambda}\right)^2 y \frac{R^2}{F} D\Lambda$
$= \delta(-\Lambda R^2 y) - \left(\frac{R'}{\Lambda}\right)^2 R^2 F^{-1} \Lambda Dy + \left(\frac{R'}{\Lambda}\right) R^2 F^{-1} D(R') -$
$\left(\frac{R'}{\Lambda}\right)^2 R^2 F^{-1} y D\Lambda$
$= \delta(-\Lambda R^2 y) - \left(\frac{R'}{\Lambda}\right) R^2 F^{-1} y D(R') - \left(\frac{R'}{\Lambda}\right)^2 R^2 F^{-1} D(y\Lambda)$
$X_\lambda = (\delta\Lambda)\left[-\frac{1}{3}y^3 - y\left(1 - \left(\frac{R'}{\Lambda}\right)^2\right)\right] + \left(\frac{R\prime}{\Lambda}\right)\left[-\frac{1}{3}y^3 - y\left(1 -\right.\right.$
$\left.\left.\left(\frac{R'}{\Lambda}\right)^2\right)\right]' (\delta R)$
$-\frac{\Lambda y}{F}\left(y\left[y^2 + \left(1 - \left(\frac{R'}{\Lambda}\right)^2\right)\right] D(y) - \left(\frac{R'}{\Lambda}\right)\left[y^2 + \left(1 - \left(\frac{R'}{\Lambda}\right)^2\right)\right] D\left(\frac{R'}{\Lambda}\right)\right)$
$= (\delta\Lambda)\left[-\frac{1}{3}y^3 - y\left(1 - \left(\frac{R'}{\Lambda}\right)^2\right)\right] + \left(\frac{\Lambda}{R\prime}\right)\left[-\frac{1}{3}y^3 - y\left(1 - \left(\frac{R'}{\Lambda}\right)^2\right)\right]' (\delta R)$
$-\frac{\Lambda y}{F}\left[y^2 + \left(1 - \left(\frac{R'}{\Lambda}\right)^2\right)\right]\left(yDy - \left(\frac{R'}{\Lambda}\right) D\left(\frac{R'}{\Lambda}\right)\right)$

<u>In 4-D theory:</u>
$P_R \delta R + P_\Lambda \delta\Lambda - P_{\bar{R}} \delta R - P_M \delta M = \delta\left(\Lambda P_\Lambda + \frac{1}{2} R R' \ln\left|\frac{RR' - \Lambda P_\Lambda}{RR' + \Lambda P_\Lambda}\right|\right) +$
$\left(\frac{1}{2} R \delta R \ln\left|\frac{RR' + \Lambda P_\Lambda}{RR' - \Lambda P_\Lambda}\right|\right)'$ for 5-<u>For 5-D theory guess:</u>
$P_R \delta R + P_\Lambda \delta\Lambda - P_{\bar{R}} \delta R - P_M \delta M = \delta\left(\Lambda P_\Lambda + \frac{1}{2} R^2 R' \ln\left|\frac{R^2 R' - \Lambda P_\Lambda}{R^2 R' + \Lambda P_\Lambda}\right|\right) +$
$\left(\frac{1}{2} R^2 \delta R' \ln\left|\frac{R^2 R' + \Lambda P_\Lambda}{R^2 R' - \Lambda P_\Lambda}\right|\right)'$
In $\hat{\lambda} = 0, 5D$, Lovelock, guess P_Λ is replaced by $-R^2 y$:
$(X_0)_{guess} = \delta\left((1 - \Lambda R^2 y) + \frac{1}{2} R^2 R' \delta \ln\left|\frac{R^2 R' - \Lambda P_\Lambda}{R^2 R' + \Lambda P_\Lambda}\right|\right) +$
$\left(\frac{1}{2} R^2 \delta R \ln\left|\frac{R^2 R' - \Lambda P_\Lambda}{R^2 R' + \Lambda P_\Lambda}\right|\right)$
$= \delta(-\Lambda R^2 y) + \frac{1}{2} R^2 R' \delta\left(\ln\left|\frac{R' + y\Lambda}{R' - y\Lambda}\right|\right) + \frac{1}{2} R^2 \delta R\left(\ln\left|\frac{R' - y\Lambda}{R' + y\Lambda}\right|\right)'$
$= \delta(-\Lambda R^2 y) + \frac{1}{2} R^2 R'\left[\frac{\delta R' + \delta(y\Lambda)}{R' + y\Lambda} - \frac{\delta R' - \delta(y\Lambda)}{R' - y\Lambda}\right] - \frac{1}{2} R^2 \delta R\left[\frac{R'' + (y\Lambda)'}{R' + y\Lambda} -\right.$
$\left.\frac{R'' - (y\Lambda)'}{R' - y\Lambda}\right]$

$$= \delta(-\Lambda R^2 y) + \frac{1}{2} R^2 R' \left(\Lambda^2 F\right)^{-1} [-2(y\Lambda)\delta R' + 2R'\delta(y\Lambda)]$$
$$- \frac{1}{2} R^2 \delta R (\Lambda R)^{-1} [-2(y\Lambda) R" + 2R'(y\Lambda)]$$
$$= \delta(-\Lambda R^2 y) + \left(\frac{R'}{\Lambda}\right)^2 \frac{R^2}{F} \left[\frac{(y\Lambda)}{R'} \left(\frac{R"}{R'} \delta R - \delta R'\right) - \frac{(y\Lambda)'}{R'} \left(\frac{R"}{R'} \delta R - \delta(y\Lambda)\right)\right]$$
$$= \delta(-\Lambda R^2 y) + \left(\frac{R'}{\Lambda}\right)^2 R^2 F^{-1} \left[\frac{(y\Lambda)}{R'} D(R') - D(y\Lambda)\right]$$
$$= X_0$$

So, the guess is correct. Now consider $X_{\hat{\lambda}}$:

$$X_{\hat{\lambda}} = \delta \left[\Lambda \left(\frac{1}{3} y^3 - y \left(1 - \left(\frac{R'}{\Lambda}\right)^2\right)\right)\right] - \Lambda \delta \left(-\frac{1}{3} y^3 - y \left(1 - \left(\frac{R'}{\Lambda}\right)^2\right)\right)$$
$$+ \frac{\Lambda}{R'} \left(\frac{1}{3} y^2 - y \left(1 - \left(\frac{R'}{\Lambda}\right)^2\right)\right) (\delta R) - \frac{\Lambda y}{F} (1 - F) \left(-\frac{1}{2} DF\right)$$

$$X_{\hat{\lambda}} = \delta \left[\left(\frac{1}{3} y^3 - y \left(1 - \left(\frac{R'}{\Lambda}\right)^2\right)\right)\right] + \Lambda D \left[\left(\frac{1}{3} y^3 - y \left(1 - \left(\frac{R'}{\Lambda}\right)^2\right)\right)\right] -$$
$$\frac{\Lambda y}{F} (1 - F) \left(-\frac{1}{2} DF\right)$$
$$= \delta \left[\Lambda \left(\frac{1}{3} y^3 - y \left(1 - \left(\frac{R'}{\Lambda}\right)^2\right)\right)\right] - \Lambda \left[y^2 - \left(\frac{R'}{\Lambda}\right)^2 + 1\right] Dy -$$
$$\Lambda y D \left(\frac{R'}{\Lambda}\right)^2 - \frac{\Lambda y}{F} (1 - F) \left(-\frac{1}{2} DF\right)$$
$$= \delta[\ldots] - \Lambda(1 - F) \left\{Dy + \frac{y}{F} \left(y Dy - \left(\frac{R'}{\Lambda}\right) D \left(\frac{R'}{\Lambda}\right)\right)\right\} + \Lambda y D \left(\frac{R'}{\Lambda}\right)^2$$
$$= \delta[\ldots] - \Lambda(1 - F) \left\{\left(\frac{R'}{\Lambda}\right)^2 - y^2 Dy + \frac{y^2}{F} Dy - y \left(\frac{R'}{\Lambda}\right)^2 D \left(\frac{R'}{\Lambda}\right)\right\} +$$
$$\Lambda y D \left(\frac{R'}{\Lambda}\right)^2$$
$$= \delta[\ldots] - \Lambda \frac{(1-F)}{F} \left\{\frac{\left(\frac{R'}{\Lambda}\right)^2 - y^2}{F} Dy + \frac{y^2}{F} Dy - y \frac{\left(\frac{R'}{\Lambda}\right)^2}{F} D \left(\frac{R'}{\Lambda}\right)\right\} + \Lambda y D \left(\frac{R'}{\Lambda}\right)^2$$
$$= \delta[\ldots] - \frac{\Lambda}{F} \left(\frac{R'}{\Lambda}\right)^2 Dy + \Lambda \left(\frac{R'}{\Lambda}\right)^2 Dy + \frac{\Lambda y}{F} \left(\frac{R'}{\Lambda}\right) D \left(\frac{R'}{\Lambda}\right) -$$
$$\Lambda y \left(\frac{R'}{\Lambda}\right) D \left(\frac{R'}{\Lambda}\right) D \Lambda y \left(\frac{R'}{\Lambda}\right)$$
$$= \delta[\ldots] - \frac{\Lambda}{F} \left(\frac{R'}{\Lambda}\right) \left[\left(\frac{R'}{\Lambda}\right) Dy - y D \left(\frac{R'}{\Lambda}\right)\right] + \Lambda \left(\frac{R'}{\Lambda}\right) \left[\left(\frac{R'}{\Lambda}\right) Dy + y D \left(\frac{R'}{\Lambda}\right)\right]$$
$$= \delta[\ldots] - \left(\frac{R'}{\Lambda}\right)^2 F^{-1} \Lambda Dy - \left(\frac{R'}{\Lambda}\right)^2 F^{-1} y D\Lambda + \left(\frac{R'}{\Lambda}\right) F^{-1} y D(R')$$
$$+ R' D \left(\frac{R'}{\Lambda} y\right)$$

$$= \delta[\ldots] + \left(\frac{R'}{\Lambda}\right) F^{-1} y D(R') - \left(\frac{R'}{\Lambda}\right)^2 F^{-1} D(y\Lambda) + R'D\left(\frac{R'}{\Lambda} y\right)$$

So,

$$X = \delta\left(-\Lambda R^2 y - \hat{\lambda}\Lambda\left(\frac{1}{3}y^3 + y\left(1 - \left(\frac{R'}{\Lambda}\right)^2\right)\right)\right)$$

$$+(R^2 + \hat{\lambda})\left[\left(\frac{R'}{\Lambda}\right) F^{-1} y D(R') - \left(\frac{R'}{\Lambda}\right)^2 F^{-1} D(y\Lambda)\right] + \hat{\lambda}R'D\left(\frac{R'}{\Lambda} y\right)$$

$$= \delta\left(\Lambda\left[-\hat{\lambda}\frac{1}{3}y^3 - y\left(R^2 + \hat{\lambda}\left(\frac{R'}{\Lambda}\right)^2\right)\right]\right)$$

$$+(R^2 + \hat{\lambda})\left(\frac{R'}{\Lambda}\right)^2 F^{-1}\left[\frac{(y\Lambda)}{R'} D(R') - D(y\Lambda)\right] + \hat{\lambda}R'D\left(\frac{R'}{\Lambda} y\right)$$

$$= \delta(\Lambda P_\Lambda) + \left(\frac{R'}{\Lambda}\right)^2 \frac{(R^2+\hat{\lambda})}{F} | \frac{(y\Lambda)}{R'}\left(\frac{R"}{R'}\delta R - \delta R'\right) - \left(\frac{(y\Lambda)}{R'}\delta R - \delta y\Lambda\right) +$$

$$\hat{\lambda}R'D\left(\frac{R'}{\Lambda} y\right)$$

$$= \delta(\Lambda P_\Lambda) + \frac{1}{2}(R^2 + \hat{\lambda})R'\left(\Lambda^2 F\right)^{-1}[-2(y\Lambda)\delta R' + 2R'\delta(y\Lambda)]$$

$$+\frac{1}{2}(R^2 + \hat{\lambda})\delta R\left(\Lambda^2 F\right)^{-1}[-2(y\Lambda)\delta R'' + 2R'\delta(y\Lambda)'] + \hat{\lambda}R'D\left(\frac{R'}{\Lambda} y\right)$$

$$= \delta(\Lambda P_\Lambda) + \frac{1}{2}(R^2 + \hat{\lambda})R'\left[\frac{\delta R' + \delta(y\Lambda)}{R' + y\Lambda} - \frac{\delta R' - \delta(y\Lambda)}{R' - y\Lambda}\right]$$

$$-\frac{1}{2}(R^2 + \hat{\lambda})\delta R\left[\frac{R'' + \delta(y\Lambda)}{R' + y\Lambda} - \frac{R'' - \delta(y\Lambda)'}{R' - y\Lambda}\right] + \hat{\lambda}R'D\left(\frac{R'}{\Lambda} y\right)$$

$$= \delta\left(\left(\Lambda P_\Lambda + \frac{1}{2}\left(\frac{1}{3}R^3 + \hat{\lambda}R\right)' \ln\left|\frac{R' + y\Lambda}{R' - y\Lambda}\right|\right)\right) + \left(\frac{1}{2}\delta\left(\frac{1}{3}R^3 +\right.\right.$$

$$\left.\left.\hat{\lambda}R\right) \ln\left|\frac{R' + y\Lambda}{R' - y\Lambda}\right|\right)' + \hat{\lambda}R'D\left(\frac{R'}{\Lambda} y\right)$$

Consider the $\hat{\lambda}R'D\left(\frac{R'}{\Lambda} y\right)$ term:

$$\hat{\lambda}R'\left[\frac{\left(\frac{R'}{\Lambda}y\right)'}{R'}\delta R - \delta\left(\frac{R'}{\Lambda}y\right)\right] = \hat{\lambda}\left[\left(\frac{R'}{\Lambda}y\right)'\delta R - R'\delta\left(\frac{R'}{\Lambda}y\right)\right]$$

$$= \hat{\lambda}\left[\delta\left(R\left(\frac{R'}{\Lambda}y\right)'\right) - R\delta\left(\frac{R'}{\Lambda}y\right)' - \left(R\delta\left(\frac{R'}{\Lambda}y\right)' + R\delta\left(\frac{R'}{\Lambda}y\right)'\right)\right]$$

$$= \hat{\lambda}\left[\delta\left(R\left(\frac{R'}{\Lambda}y\right)'\right) - \left(R\delta\left(\frac{R'}{\Lambda}y\right)'\right)\right]$$

So,

$$X = \delta\left[\Lambda\left(\frac{1}{3}\hat{\lambda}y^3 - y\left[R^2 + \hat{\lambda}\left(1 - \left(\frac{R'}{\Lambda}y\right)^2\right)\right]\right) + \hat{\lambda}R\left(\frac{R'}{\Lambda}y\right)' +\right.$$

$$\left.\frac{1}{2}\left(\frac{1}{3}R^3 + \hat{\lambda}R\right)' \ln\left|\frac{R' + y\Lambda}{R' - y\Lambda}\right|\right]$$

$$-\left[\hat{\lambda}R\delta\left(\frac{R'}{\Lambda}y\right)+\frac{1}{2}\delta\left(\frac{1}{3}R^3+\hat{\lambda}R\right)\ln\left|\frac{R'+y\Lambda}{R'-y\Lambda}\right|\right]'$$

The transformation from $\{R,\Lambda,P_R,P_\Lambda\}$ to $\{R,M_+,P_R,P_{M+}\}$ is canonical:
$R=R$
$M_+=\frac{1}{2}w_+=R^2\left(1-\left(\frac{R'}{\Lambda}\right)^2\right)-\hat{\lambda}\left(\frac{R'}{\Lambda}\right)^2\left(1-\frac{1}{2}\left(\frac{R'}{\Lambda}\right)^2\right)+\frac{1}{2}\hat{\lambda}y^4+$
$y^2\left[R^2+\hat{\lambda}\left(1-\left(\frac{R'}{\Lambda}\right)^2\right)\right]+\frac{\hat{\lambda}}{4}$
$(M_+'=\Lambda^{-1}[yH_r-R'H])$
$P_R=\frac{1}{R'}H_r(\Lambda,R,P_\Lambda,P_R)-\frac{1}{R'}P_{M+}M_+'=F^{-1}[-yH+R'\Lambda^{-2}H_r]$
$P_{M+}=-F^{-1}\Lambda y$
And $F=\left(\frac{R'}{\Lambda}\right)^2-y^2$ while in on classical solutions $F=1+\frac{R^2}{\hat{\lambda}}\left(1-\sqrt{1+\frac{2w_+\hat{\lambda}}{R^4}}\right)$. So,
$(P_R\delta R-P_{\bar{R}}\delta R-P_{m+}\delta M_++P_\Lambda\delta\Lambda)=$
$\delta\left[(\Lambda P_\Lambda)+\hat{\lambda}R\left(\left(\frac{R'}{\Lambda}\right)y\right)'+\frac{1}{2}\left(\frac{1}{3}R^3+\hat{\lambda}R\right)'\ln\left|\frac{R'+y\Lambda}{R'-y\Lambda}\right|\right]$
$-\left[\hat{\lambda}R\delta\left(\left(\frac{R'}{\Lambda}\right)y\right)+\frac{1}{2}\delta\left(\frac{1}{3}R^3+\hat{\lambda}R\right)\ln\left|\frac{R'+y\Lambda}{R'-y\Lambda}\right|\right]'$

Thus
$\int(P_\Lambda\delta\Lambda+P_R\delta R-P_{M+}\delta M_+)\,dr=\delta w+$ boundary terms. For our fall-off conditions the boundary terms contribute zero, and

$$w=\int_0^\infty dr\left((\Lambda P_\Lambda)+\hat{\lambda}R\left(\left(\frac{R'}{\Lambda}\right)y\right)'+\frac{1}{2}\left(\frac{1}{3}R^3+\hat{\lambda}R\right)'\ln\left|\frac{R'+y\Lambda}{R'-y\Lambda}\right|\right)$$

is convergent when the fall-off conditions are taken into account.

After manipulation of Lagrange multipliers, the new action is:

$$\begin{aligned}S&[M_+,R,P_{M_+},P_R,\tilde{N}^m,N^R]\\&=\int dt\int_0^\infty dr\,\{P_{M_+}\dot{M}_++P_R\dot{R}-N^RP_R\\&\qquad+\tilde{N}^m[(1-g)+4gR_0(1+R_0^{-2}\lambda)]N\}\\&\quad+\int dt\left(\left[\frac{1}{3}R_0^3+R_0\hat{\lambda}\right]\tilde{N}^M-\tilde{N}_+\left(\frac{1}{2}w_+\right)\right)\end{aligned}$$

will fall-off on g defined suitability.

C.3 Constraints, Equations of Motion, Fall-off conditions at $r \to \infty$

The constraints:

$$H = yP_R + y^2\left(\Lambda R - \hat{\lambda}\left(\frac{R'}{\Lambda}\right)'\right) - \Lambda R\left(1 - \left(\frac{R'}{\Lambda}\right)^2\right) + \left(\frac{R'}{\Lambda}\right)'\left(R^2 + \hat{\lambda}\left(1 - \left(\frac{R'}{\Lambda}\right)^2\right)\right)$$

$$H_r = H_r R' - \Lambda P'_\Lambda$$

(and have y: $\frac{1}{3}\hat{\lambda}y^3 + y\left\{R^2 + \hat{\lambda}\left[1 - \left(\frac{R'}{\Lambda}\right)^2\right]\right\} + P_\Lambda = 0$).

The dynamical equations of motion: $H = NH + N^r H_r$:

$$\dot{\Lambda} = \frac{\delta H}{\delta P_\Lambda} = NP_R\frac{\delta y}{\delta P_\Lambda} + (N^r\Lambda)^r$$

$$\hat{\lambda}y^2\frac{\delta y}{\delta P_\Lambda} + \left\{R^2 + \hat{\lambda}\left[1 - \left(\frac{R'}{\Lambda}\right)^2\right]\right\}\frac{\delta y}{\delta P_\Lambda} = -1$$

$$\frac{\delta y}{\delta P_\Lambda} = -\left[R^2 + \hat{\lambda}\left(y^2 + \left(1 - \left(\frac{R'}{\Lambda}\right)^2\right)\right)\right]^{-1}$$

$$\boxed{\dot{\Lambda} = -N\left(P_R + 2y\left[\Lambda R - \hat{\lambda}\left(\frac{R'}{\Lambda}\right)'\right]\right)\left[R^2 + \hat{\lambda}\left(y^2 + 1 - \left(\frac{R'}{\Lambda}\right)^2\right)\right]^{-1} + (N^r\Lambda)'}$$

$$\dot{R} = \frac{\delta H}{\delta P_R} = Ny + N^r R'$$

$$\boxed{\dot{R} = Ny + N^r R'}$$

$$\dot{P}_\Lambda = -\frac{\delta H}{\delta \Lambda} = -N\left\{\frac{\delta y}{\delta \Lambda}P_R + 2y\frac{\delta y}{\delta \Lambda}\left(\Lambda R - \hat{\lambda}\left(\frac{R'}{\Lambda}\right)'\right) + y^2 R - \hat{\lambda}y^2\left(R'\frac{\delta}{\delta\Lambda}\left(\frac{1}{\Lambda}\right)\right)' - R\left(1 - \left(\frac{R'}{\Lambda}\right)^2\right) + \Lambda R_2\left(\frac{R'}{\Lambda}\right)R'\left(-\frac{1}{\Lambda^2}\right) + \left(R'\frac{\delta}{\delta\Lambda}\left(\frac{1}{\Lambda}\right)\right)'\left(R^2 + \hat{\lambda}\left(1 - \left(\frac{R'}{\Lambda}\right)^2\right)\right) + \left(\frac{R'}{\Lambda}\right)'\hat{\lambda}\left(-2\left(\frac{R'}{\Lambda}\right)R'\left(\frac{1}{\Lambda^2}\right)\right)\right\}$$

$$-N^r\{-P'_\Lambda\}$$

$$\frac{\delta y}{\delta \Lambda} = ?$$

$$\hat{\lambda}y^2\frac{\delta y}{\delta \Lambda} + \frac{\delta y}{\delta \Lambda}\left\{R^2 + \hat{\lambda}\left[1 - \left(\frac{R'}{\Lambda}\right)^2\right]\right\} - 2y\hat{\lambda}\left(\frac{R'}{\Lambda}\right)^{R'}\left(\frac{1}{\Lambda^2}\right) = 0$$

$$\frac{\delta y}{\delta \Lambda} = \frac{-2y\hat{\lambda}\left(\frac{R'}{\Lambda}\right)^2\left(\frac{1}{\Lambda}\right)}{R^2 + \hat{\lambda}\left\{y^2 + \left(1 - \left(\frac{R'}{\Lambda}\right)^2\right)\right\}}$$

$$\dot{P}_{\Lambda} = N\left[\begin{array}{c}\left(\frac{2y\hat{\lambda}\left(\frac{R'}{\Lambda}\right)^{2}\frac{1}{\Lambda}}{R^2+\hat{\lambda}\left\{y^2+\left(1-\left(\frac{R'}{\Lambda}\right)^{2}\right)\right\}}\right)\left(P_R + 2y\left(\Lambda R - \hat{\lambda}\left(\frac{R'}{\Lambda}\right)'\right)\right) \\ -y^2 R + R\left(1-\left(\frac{R'}{\Lambda}\right)^2\right) + 2R\left(\frac{R'}{\Lambda}\right)^2 \\ -2\hat{\lambda}\Lambda^{-1}\left(\frac{R'}{\Lambda}\right)^2\left(\frac{R'}{\Lambda}\right)' + \hat{\lambda}(Ny^2)'\left(\frac{R'}{\Lambda}\right)\Lambda^{-1} \\ -\left[N\left(R^2+\hat{\lambda}\left(1-\left(\frac{R'}{\Lambda}\right)^2\right)\right)\right]'\left(\frac{R'}{\Lambda}\right)\Lambda^{-1} + N^r P'_{\Lambda}\end{array}\right]$$

$$\dot{P}_R = -\frac{\delta H}{\delta R} = -N\left\{\frac{\delta y}{\delta R}P_R + 2y\frac{\delta y}{\delta R}\left(\Lambda R - \hat{\lambda}\left(\frac{R'}{\Lambda}\right)'\right) + y^2\Lambda - \right.$$

$$y^2\Lambda\hat{\lambda}\left(\frac{\left(\frac{\delta}{\delta R}R\right)'}{\Lambda}\right)' - \Lambda\left(1-\left(\frac{R'}{\Lambda}\right)^2\right) + \Lambda 2R\left(\frac{R'}{\Lambda}\right)\left(\frac{\left(\frac{\delta}{\delta R}R\right)'}{\Lambda}\right) +$$

$$\left(\frac{\left(\frac{\delta}{\delta R}R\right)'}{\Lambda}\right)'\left(R^2 + \hat{\lambda}\left(1-\left(\frac{R'}{\Lambda}\right)^2\right) + 2R\left(\frac{R'}{\Lambda}\right) + \right.$$

$$\left.\left.\left(\frac{R'}{\Lambda}\right)'\left(-2\hat{\lambda}\left(\frac{R'}{\Lambda}\right)\left(\frac{\left(\frac{\delta}{\delta R}R\right)'}{\Lambda}\right)\right)\right)\right\} - N^r\left(P_R\left(\left(\frac{\delta}{\delta R}R\right)'\right)\right)$$

$$\frac{\delta y}{\delta R} = ?$$

$$\hat{\lambda}y^2\frac{\delta y}{\delta R} + \frac{\delta y}{\delta R}\left(R^2 + \hat{\lambda}\left[1-\left(\frac{R'}{\Lambda}\right)^2\right]\right) + y\left(2R - 2\hat{\lambda}\left(\frac{R'}{\Lambda}\right)\left(\frac{\left(\frac{\delta}{\delta R}R\right)'}{\Lambda}\right)\right) = 0$$

$$\frac{\delta y}{\delta R} = \frac{2y\hat{\lambda}\left(\frac{R'}{\Lambda}\right)\Lambda^{-1}\left(\frac{\delta}{\delta R}R\right)' - 2yR}{\left(R^2+\hat{\lambda}\left\{y^2+\left(1-\left(\frac{R'}{\Lambda}\right)^2\right)\right\}\right)}$$

$$\dot{P}_R = -N\left[\begin{array}{c}\left(P_R + 2y\left[\Lambda R - \hat{\lambda}\left(\frac{R'}{\Lambda}\right)\right]\right)\frac{(-2yR)}{\left(R^2+\hat{\lambda}\left[y^2+\left(1-\left(\frac{R'}{\Lambda}\right)^2\right)\right]\right)} \\ +y^2\Lambda - \Lambda\left(1-\left(\frac{R'}{\Lambda}\right)^2\right)\end{array}\right]$$

$$-\left(-N\left(P_R+2y\left(\Lambda R-\hat{\lambda}\left(\frac{R'}{\Lambda}\right)'\right)\right)\frac{\left(2y\hat{\lambda}\left(\frac{R'}{\Lambda}\right)\Lambda^{-1}\right)}{\left(R^2+\hat{\lambda}\left[y^2+\left(1-\left(\frac{R'}{\Lambda}\right)^2\right)\right]\right)}\right)'+2R\left(\frac{R'}{\Lambda}\right)'$$

$$+\left[\frac{(Ny^2\hat{\lambda})'}{\Lambda}\right]'$$

$$-\left(-N2R\left(\frac{R'}{\Lambda}\right)\right)'-\left[\left[N\left(R^2+\hat{\lambda}\left(1-\left(\frac{R'}{\Lambda}\right)^2\right)\right)\right]'/\Lambda\right]'-$$

$$\left[2\hat{\lambda}N\left(\frac{R'}{\Lambda}\right)\left(\frac{R'}{\Lambda}\right)'\Lambda^{-1}\right]'+(N^rP_R)'$$

Determination of the most general fall-off conditions at asymptotic spatial infinity for which the generalized Schwarzschild metric is s member:

$$ds^2=-f^2dt^2+f^{-2}dr^2+r^2d\Omega_3^2$$

$f^2=1+\frac{r^2}{\hat{\lambda}}\left[1-\sqrt{1+\frac{2w\hat{\lambda}}{r^4}}\right]$ (let $\hat{\lambda}-2m=2w$ from previous notes)

$\underline{\boldsymbol{r\to\infty}}$

$$f^2=1+\frac{r^2}{\hat{\lambda}}\left(r-\left[r+\frac{1}{2}\left(\frac{2w\hat{\lambda}}{r^4}\right)-\frac{1}{8}\left(\frac{2w\hat{\lambda}}{r^4}\right)+\cdots\right]\right)$$

$$=1-\frac{w}{r^2}+\frac{w^2\hat{\lambda}}{2r^6}-\cdots=1-\frac{1}{r^2}+O(r^{-6})$$

$$f^{-1}=1+\frac{\left(\frac{1}{2}w\right)}{r^2}+O(r^{-6})$$

$\rightarrow\Lambda=1+\frac{1}{2}wr^{-2}+O^\infty(r^{-2-\epsilon})\qquad 0<\epsilon\leq 1$

$R=r+O^\infty(r^{-2-\epsilon})$?

Take $ds^2=-f^2(R)dt^2+f^{-2}(R)dr^2+R^2d\Omega_3^2$

$$f^{-1}(R)=1+\frac{\left(\frac{1}{2}w\right)}{R^2}+O(R^{-6})$$

$$=1+\frac{1}{2}w\frac{1}{(r+O^\infty(r^{-2-\epsilon}))}+O(r^{-6})$$

$$=1+\frac{1}{2}wr^2(1-O^\infty(r^{-2-\epsilon}))$$

$$=1+\frac{1}{2}wr^2+O^\infty(r^{-2-\epsilon})$$

Schwarzschild type solution : $(r\rightarrow\infty)$

$\Lambda=1+\frac{1}{2}wr^{-2}+O^\infty(r^{-4})$

$R=r$

Most generalized gauge compatible with this:

$$\Lambda = 1 + \frac{1}{2} w r^{-2} + O(r^{-2-\epsilon}) \qquad 0 < \epsilon \leq 1$$
$$R = r + O^{\infty}(r^{1-\epsilon})$$

Consider $ds^2 = -f^2(R)dT^2 + f^{-2}(R)dR^2 + R^2 d\Omega_3^2$

$f^2(R) = 1 + \frac{1}{\hat{\lambda}} R^2 \left[1 - \sqrt{1 + \frac{2w\hat{\lambda}}{R^4}}\right]$

Take $r \to \infty$

$f^2(R) = 1 - \frac{w}{R^2} + O(R^{-6})$

$\Lambda = f^{-1} = 1 + \frac{1}{2} w R^{-2} + O(R^{-4})$

Try $R = r + O^{\infty}(r^{1-\epsilon})$

$dR = dr\left(1 + O^{\infty}(r^{-\epsilon})\right)$

$\Lambda = \left\{1 + \frac{1}{2} w\left(r + O^{\infty}(r^{1-\epsilon})\right)1 + O^{\infty}(r^{-\epsilon})\right\}$

$= 1 + O^{\infty}(r^{-\epsilon}) + \frac{1}{2} w r^{-2}\left(1 + O^{\infty}(r^{-\epsilon})\right) + O(r^{-4})$

$= 1 + O^{\infty}(r^{-\epsilon}) + \frac{1}{2} w r^{-2} + O(r^{-2-\epsilon})$

Due to the $O^{\infty}(r^{-\epsilon})$ contribution in the above, the minimal requirement on fall-off for R, compatible with that specified for Λ, is that

$$R = r + O^{\infty}(r^{-1-\epsilon})$$

For the Schwarzschild type curvature coordinate gauge to be in the class of fall-off conditions considered the minimal condition imposed on Λ is:

$\Lambda = 1 + \frac{1}{2} w r^{-2} + O(r^{-2-\epsilon})$

For this choice of Λ to be consistent with its relation $\Lambda(R)$ in the curvature coordinate gauge, the adjoining minimal condition on R is:

$R = r + O^{\infty}(r^{-1-\epsilon})$

Consider $H_r = P_R R' - \Lambda P_\Lambda' = 0$

$P_R\left(1 + O(r^{-2-\epsilon})\right) = P_\Lambda'\left(1 + \frac{1}{2} w r^{-2} + O(r^{-2-\epsilon})\right)$

So, if $P_\Lambda = O(r^{-k+1-\epsilon})$ *then* $P_R = O(r^{-k-\epsilon})$

Now consider $H = yP_R + y^2\left(\Lambda R - \hat{\lambda}\left(\frac{R'}{\Lambda}\right)'\right) - \Lambda R\left(1 - \left(\frac{R'}{\Lambda}\right)^2\right) + \left(\frac{R'}{\Lambda}\right)'\left(R^2 + \hat{\lambda}\left(1 - \left(\frac{R'}{\Lambda}\right)^2\right)\right)$

$$\left(1-\left(\frac{R'}{\Lambda}\right)^2\right)=1-\frac{\left(1+O\left(r^{-2-\epsilon}\right)\right)}{\left(1+\frac{1}{2}wr^{-2}+O\left(r^{-2-\epsilon}\right)\right)}=1-\left[1-\frac{1}{2}wr^{-2}+\right.$$
$$\left.O\left(r^{-2-\epsilon}\right)\right]=\frac{1}{2}wr^{-2}+O\left(r^{-2-\epsilon}\right)$$
$$\left(\frac{R'}{\Lambda}\right)'=\left[\frac{\left(1+O\left(r^{-2-\epsilon}\right)\right)}{\left(1+\frac{1}{2}wr^{-2}+O\left(r^{-2-\epsilon}\right)\right)}\right]'=\left(1-\frac{1}{2}wr^{-2}+O\left(r^{-2-\epsilon}\right)\right)'=wr^3+$$
$$O\left(r^{-3-\epsilon}\right)$$
$$\Lambda R=r+\frac{1}{2}wr^{-1}+O^{\infty}\left(r^{-1-\epsilon}\right)$$
$$\Lambda R\left(1-\left(\frac{R'}{\Lambda}\right)^2\right)=\frac{1}{2}wr^{-1}+O\left(r^{-1-\epsilon}\right)$$

$$H=yP_R+y^2\left(r+\frac{1}{2}wr^{-1}+O\left(r^{-1-\epsilon}\right)\right)-\frac{1}{2}wr^{-1}+O\left(r^{-1-\epsilon}\right)+$$
$$wr^{-1}+O\left(r^{-1-\epsilon}\right)$$
$$H=y\left[P_R+y\left(r+\frac{1}{2}wr^{-1}+O\left(r^{-1-\epsilon}\right)\right)\right]+O\left(r^{-1-\epsilon}\right)$$
Consistency is now composed with $P_\Lambda=O(r^{-k+1-\epsilon})$ and $P_R=O(r^{-k-\epsilon})$ and the least restricted k determined.
$$\frac{1}{3}\hat{\lambda}y^3+y\left\{R^2+\hat{\lambda}\left[1-\left(\frac{R'}{\Lambda}\right)^2\right]\right\}=-P_\Lambda$$
$$\frac{1}{3}\hat{\lambda}y^3+y\left(r^2+O^{\infty}\left(r^{-\epsilon}\right)\right)=O\left(r^{-k+1-\epsilon}\right)$$
$$y=O\left(r^{-m-\epsilon}\right)$$
So, $3m=k-1$ or $-m+2=-k+1$ or both $(3m=m-2\rightarrow m=1)$
:
$H=yP_R+y^2r+O(r^{-1-\epsilon})=yO(r^{-k-\epsilon})+y^2r+O(r^{-1-\epsilon})$:
Case 1: $3m=k-1\rightarrow O(r^{-2m-\epsilon})r=O(r^{-1-\epsilon})$, $m=1\rightarrow k=4$, consistent with $y=O(r^{-k-\epsilon})$.
Case 2: $m=k+1\rightarrow O(r^{-2m-\epsilon})r=O(r^{-1-\epsilon})$, $m=1\rightarrow k=0$ (again this is consistent).

For $k=0$: $P_\Lambda=O(r^{1-\epsilon})$, $P_R=O(r^{-\epsilon})$ are consistent with the Λ, R fall-off conditions and the constraints. Whether $P_\Lambda=O(r^{1-\epsilon})$ is consistent with the other fall-off conditions when the equations of motion are considered is yet to be seen.

$\dot{\Lambda}$; $\dot{R}$ will be used to find consistent fall-off behaviour for N ; N^r.
Since $P_\Lambda=O(r^{1-\epsilon})$ seems unlikely, use
$$\begin{matrix}P_\Lambda=O(r^{-k+1-\epsilon})\\ P_R=O(r^{-k-\epsilon})\end{matrix}\rightarrow y=O(r^{-k-1-\epsilon})$$

Then

$$\dot{\Lambda} = -N\left[R^2 + \hat{\lambda}\left(y^2 + \left(1 - \left(\frac{R'}{\Lambda}\right)^2\right)\right)\right]^{-1}\left(P_R + 2y\left[\Lambda R - \hat{\lambda}\left(\frac{R'}{\Lambda}\right)'\right]\right) +$$
$$(N^r\Lambda)'$$
$$O(r^{-2-\epsilon}) = -NO(r^{-k-\epsilon})r^{-2} + (N^r)r^{-3} + (N^r)'$$

$\dot{w} = 0$ is indicated

$\dot{R} = Ny + N^r R'$ indicates

$$O(r^{-2-\epsilon}) = -NO(r^{-k-\epsilon}) + N^r \to O(N^r) \le O(r^{-1-\epsilon})$$

$$\boxed{N^r \le O(r^{-1-\epsilon})}$$

Similarly, $N \le O(r^{k-\epsilon})$. Choose:

$$\boxed{\begin{array}{c} N^r = O(r^{-1-\epsilon}) \\ N = O(r^{k-\epsilon}) \end{array}}$$

It now remains to be seen what values of k are consistent with the remaining two equations of motion.

Consider

$$\dot{P}_\Lambda = N\left[\begin{array}{c}\left(\dfrac{2y\hat{\lambda}\left(\frac{R'}{\Lambda}\right)^2\frac{1}{\Lambda}}{R^2 + \hat{\lambda}\left[y^2 + \left(1 - \left(\frac{R'}{\Lambda}\right)^2\right)\right]}\right)\left(P_R + 2y\left(\Lambda R - \hat{\lambda}\left(\frac{R'}{\Lambda}\right)'\right)\right) \\ -y^2R + R\left(1 - \left(\frac{R'}{\Lambda}\right)^2\right) + 2R\left(\frac{R'}{\Lambda}\right)^2 - 2\hat{\lambda}\Lambda^{-1}\left(\frac{R'}{\Lambda}\right)^2\left(\frac{R'}{\Lambda}\right)'\end{array}\right]$$
$$+\hat{\lambda}(Ny^2)'\left(\frac{R'}{\Lambda}\right)\Lambda^{-1} - \left[N\left(R^2 + \hat{\lambda}\left(1 - \left(\frac{R'}{\Lambda}\right)^2\right)\right)\right]'\left(\frac{R'}{\Lambda}\right)\Lambda^{-1} + N^r P'_\Lambda$$

$$\Lambda = 1 + \tfrac{1}{2}wr^{-2} + O(r^{-2-\epsilon})$$
$$R = r + O^\infty(r^{-1-\epsilon})$$
$$P_\Lambda = O(r^{-k+1-\epsilon}) \to y = O(r^{-k-1-\epsilon})$$
$$P_R = O(r^{-1-\epsilon})$$
$$N = O(r^{k-\epsilon})$$
$$N^r = O(r^{-1-\epsilon})$$
$$O(r^{-k+1-\epsilon}) =$$
$$O(r^{k-\epsilon})\left[\begin{array}{c}\left(\frac{O(r^{-k-1-\epsilon})}{r^2}\right)(O(r^{-k-\epsilon}) + (r^{-k-1-\epsilon})r) \\ -r[O(r^{-k-1-\epsilon})]^2 + \tfrac{1}{2}wr^{-1} + O(r^{-1-\epsilon}) + 2r + O(r^{-3})\end{array}\right]$$

$+\hat{\lambda}[O(r^{k-\epsilon})O(r^{-k-1-\epsilon})]' - [O(r^{k-\epsilon})r^2]' + O(r^{-2-\epsilon})[O(r^{-k-1-\epsilon})]$
$= [O(r^{-k-1-\epsilon}) + 2rO(r^{k-\epsilon})] + O(r^{-2-\epsilon}) + O(r^{k+1-\epsilon}) + O(r^{k+1-\epsilon})$
If $k \geq 0$ then $k+1 = k+1 \rightarrow k = 0$ is the only solution.
If $k < 0$ then $k+1 = -k+1 \rightarrow k = 0$, no solution.
So, $\boxed{k=0}$ is indicated. Hopefully further restrictions don't conflict with this in considering $\dot{P}_R$:

$$\dot{P}_R = -N\left[\frac{\left(P_R+2y\left[\Lambda R-\hat{\lambda}\left(\frac{R'}{\Lambda}\right)'\right]\right)}{\left(R^2+\hat{\lambda}\left[y^2+\left(1-\left(\frac{R'}{\Lambda}\right)^2\right)\right]\right)} + y^2\Lambda - \Lambda\left(1-\left(\frac{R'}{\Lambda}\right)^2\right)\right]$$

$$+\left[N\frac{\left(P_R+2y\left[\Lambda R-\hat{\lambda}\left(\frac{R'}{\Lambda}\right)'\right]\right)\left(2y\hat{\lambda}\left(\frac{R'}{\Lambda}\right)\Lambda^{-1}\right)}{\left(R^2+\hat{\lambda}\left[y^2+\left(1-\left(\frac{R'}{\Lambda}\right)^2\right)\right]\right)}\right]' + \left[\frac{(Ny^2\hat{\lambda})'}{\Lambda}\right]'$$

$$+\left(2NR\left(\frac{R'}{\Lambda}\right)\right)' + \left[\frac{\left[NR^2+\hat{\lambda}\left(1-\left(\frac{R'}{\Lambda}\right)^2\right)\right]'}{\Lambda}\right]'$$

$$-\left[2\hat{\lambda}N\left(\frac{R'}{\Lambda}\right)\left(\frac{R'}{\Lambda}\right)'\Lambda^{-1}\right]' + (N^R P_R)'$$

$$O(r^{-k-\epsilon}) = O(r^{k-\epsilon})\left[\frac{O(r^{-k-\epsilon})}{r^2} + \left(O(r^{-k-1-\epsilon})\right)^2 - \frac{1}{2}wr^{-2}\right]$$

$$+\left[\frac{O(r^{k-\epsilon})O(r^{-k-\epsilon})O(r^{-k-1-\epsilon})}{r^2}\right]' + [O(r^{k-\epsilon})[O(r^{-k-1-\epsilon})]^2]$$

$$+[O(r^{k-\epsilon})r]' + [[O(r^{k-\epsilon})r^2]']$$

$$+[O(r^{k-\epsilon})r^3]' + \left(O(r^{-1-\epsilon})O(r^{-k-\epsilon})\right)'$$

$$= O(r^{-k-2-\epsilon}) + O(r^{k-2-\epsilon}) + O(r^{-k-4-\epsilon}) + O(r^{-k-4-\epsilon}) + O(r^{k-\epsilon}) + O(r^{k-\epsilon}) + O(r^{-k-4-\epsilon}) + O(r^{-k-2-\epsilon})$$

$$= O(r^{-k-2-\epsilon}) + O(r^{k-\epsilon})$$

Only solution, again, is $k=0$, so good.

At $r \rightarrow \infty$ leading order fall-off behaviour is:
$\Lambda = 1 + \frac{1}{2}wr^{-2-\epsilon} + O(r^{-2-\epsilon})$
$R = r + O^{\infty}(r^{-1-\epsilon})$
$P_\Lambda = O(r^{-1-\epsilon})$
$P_R = O(r^{-\epsilon})$ $\rightarrow (y = O(r^{-1-\epsilon}))$
$N = O(r^{-\epsilon})$
$N^r = O(r^{-1-\epsilon})$
However, with this form it is required that $\dot{w} = 0$. In order to not impose this restriction some of the variables must have $O(r^{-k})$ terms at leading order rather than $(r^{-k-\epsilon})$. In particular, consider $\dot{\Lambda}$.

$\dot{\Lambda} = -N\left(P_R + 2y\left[\Lambda R - \hat{\lambda}\left(\frac{R'}{\Lambda}\right)'\right]\right)\left[R^2 + \hat{\lambda}\left(y^2 + \left(1 - \left(\frac{R'}{\Lambda}\right)^2\right)\right)\right]^{-1} +$
$(N^r \Lambda)'$
$\dot{\Lambda} = \frac{1}{2} r^2 \dot{w}(t) + O(r^{-2-\epsilon})$
If $\dot{w}(t) \neq 0$ then $\dot{\Lambda} = O(r^{-2})$, which requires the following:

(i) $N = O(r^0)$; $P_R = O(r^0) \rightarrow P_\Lambda = O(r^1)$
(ii) $N = O(r^0)$; $y = O(r^{-1})$ (thus, equivalent to (i))
(iii) $N^r = O(r^{-1})$

Cases (i) + (ii) indicate (iii) and that $R = r + O(r^{-1})$, from $\dot{R} = Ny + N^r R'$, and then N^r follow from the $\dot{R}$ equation as well. So (i) or (ii) unravel the ϵ formalism as needed.
The change $N^r = O(r^{-1})$ leads to $R = r + O(r^{-1})$. Through the H=0 constraint this requires either y ; P_R similarity.
Once y and P_R have $O(r^{-k})$ then so does P_Λ (from $H_r = 0$) and then so does N, from the leading order in $\dot{P}_\Lambda$. This matches the conditions begun with earlier that "unravelled" the ϵ formulation. So, $\dot{w}(t) \neq 0$ is not compatible with any of the leading order terms having form $O(r^{-k-\epsilon})$, nor with the second order term in the case of R.

Do the fall-off conditions for 4D Schwarzschild, given by Kuchar, satisfy the constraints and equations of motion?
$\Lambda = 1 + M(t)r^{-1} + O^\infty(r^{-1-\epsilon})$
$R = r + O(r^{-\epsilon})$
$P_\Lambda = O(r^{-\epsilon})$
$P_R = O(r^{-1-\epsilon})$
$N = N(t) + O(r^{-\epsilon})$
$N^r = O(r^{-\epsilon})$
$\dot{\Lambda} = N[R^{-2}\Lambda P_\Lambda - R^{-1}P_R] + (\Lambda N^r)'$
$\dot{M}r^{-1} = N(t)[r^{-2}(1 + Mr^{-1})O(r^{-\epsilon}) - r^{-1}O(r^{-1-\epsilon})] + \big((1 + Mr^{-1})O(r^{-\epsilon})\big) = O(r^{-1-\epsilon})$
here again an alteration of $N^r = O(r^0)$ is indicated. The analysis of Kuchar suffers from the same set-back, and yet it carries through correctly as our own derivation indicates.

Starting with

$$\boxed{\Lambda = 1 + \frac{1}{2} w_+(t) r^{-2} + O(r^{-3})}$$

Consider $ds^2 = -f^2(R)dT^2 + f^2(R)dR^2 + R^2 d\Omega_3^2$

$f^2(R) = 1 + \frac{1}{\hat{\lambda}} R^2 \left[1 - \sqrt{1 + \frac{2w\hat{\lambda}}{R^4}}\right]$

As $R \to \infty$, $f^2 R = 1 - \frac{w}{R^2} + O(R^{-6})$

$f^{-1} = 1 + \frac{1}{2} w R^{-2} + O(R^{-4})$

Try $\boxed{R = r + O(R^{-1})}$

$dR = dr\left(1 + O(r^{-2})\right)$

$f^{-1} = \left\{1 + \frac{1}{2} w r^{-2}\left(1 + O(r^{-2})\right)^{-2} + O(r^{-4})\right\}$

$\Lambda dr = f^{-1} dR$

$\Lambda = \left[1 + \frac{1}{2} w r^{-2} + O(r^{-4})\right]\left(1 + O(r^{-2})\right)$

$\Lambda = 1 + O(r^{-2})$

So, the most general R fall-off, to accompany that given for Λ, is $R = r + O(r^{-1})$.

Together with $0 = H_r = P_R R' - \Lambda P'_\Lambda$

$P_R\left(1 + O(r^{-2})\right) = \left(1 + \frac{1}{2} w r^{-2} + O(r^{-3})\right) P'_\Lambda$

Suppose

$$\boxed{\begin{array}{l} P_R = O(r^{-k}) \\ P_\Lambda = O(r^{-k+1}) \end{array}}$$

Try

$\Lambda = 1 + \frac{1}{2} w_+ r^{-2} + O(r^{-3})$

$R = r + \frac{1}{2} p r^{-1} + O(r^{-2})$

Consider $dS^2 = -f^2(R) dT^2 + f^2(R) dR^2 + R^2 d\Omega_3^2$

$f^2(R) = 1 + \frac{1}{\hat{\lambda}} R^2 \left[1 - \sqrt{1 + \frac{2w\hat{\lambda}}{R^4}}\right]$

As $r \to \infty$, $f^2(R) = 1 - \frac{w}{R^2} + O(R^{-6})$

$f^{-1} = 1 + \frac{1}{2} w R^{-2} + O(R^{-4})$

$= 1 + \frac{1}{2} w r^{-2} + O(r^{-4})$

$dR = dr\left(1 + \frac{1}{2} p r^2 + O(r^{-3})\right)$

$\Lambda = \left(1 + \frac{1}{2} w r^{-2} + O(r^{-4})\right)\left(1 + \frac{1}{2} p r^2 + O(r^{-3})\right)$

$= 1 + \frac{1}{2}(w + p) r^{-2} + O(r^{-3})$

So, $w_+ = w + p \rightarrow w = w_+ - p$ when the equations of motion hold w is independent of t and $\dot{w}_+(t) = \dot{p}(t)$ is expected.

So far we've arrived at likely fall-off in the form:

$\Lambda = 1 + \frac{1}{2} w_+ r^{-2} + O(r^{-3})$

$R = r + \frac{1}{2} p r^{-1} + O(r^{-2})$

Consider the H=0 constraint:

$$H = yP_R + y^2\left(\Lambda R - \hat{\lambda}\left(\frac{R\prime}{\Lambda}\right)'\right) - \Lambda R\left(1 - \left(\frac{R'}{\Lambda}\right)^2\right) + \left(\frac{R'}{\Lambda}\right)^2\left(R^2 + \hat{\lambda}\left(1 - \left(\frac{R'}{\Lambda}\right)^2\right)\right)$$

$$y: \frac{1}{3}\hat{\lambda} y^3 + y\left[R^2 + \hat{\lambda}\left(1 - \left(\frac{R'}{\Lambda}\right)^2\right)\right] = -P_\Lambda$$

$$\left(1 - \left(\frac{R'}{\Lambda}\right)^2\right) = 1 - \frac{\left(1 - \frac{1}{2} p r^{-2}\right)^2}{\left(1 + \frac{1}{2} w_+ r^{-2}\right)^2} = 1 - (1 - pr^2 - w_+ r^{-2}) + O(r^{-3})$$

$$= (p + w_+) r^{-2} + O(r^{-3})$$

$$\left(\frac{R'}{\Lambda}\right)' = \left[\frac{\left(1 - \frac{1}{2} p r^{-2} + O(r^{-3})\right)}{1 + \frac{1}{3} w_+ r^{-2} + O(r^{-3})}\right]' = (p + w_+) r^{-3} + O(r^{-3})$$

$$\Lambda R = r + \frac{1}{2}(p + w_+) r^{-1} + O(r^{-2})$$

$$\Lambda R\left(1 - \left(\frac{R'}{\Lambda}\right)^2\right) = (p + w_+) r^{-1} + O(r^{-2})$$

$$y: \frac{1}{3}\hat{\lambda} y^3 + y r^2 = -P_\Lambda = O(r^{-k+1})$$

$$H = yP_R + y^2 r + O(r^{-1}) = yO(r^{-k}) + y^2 r + O(r^{-1})$$

$0 = H$ indicates (i) $y = O(r^{-1})$ or (ii) $y = O\left(r^{-|k|-1}\right)$, $(k < 0)$ and either way a positive order on y is not indicated. Take most general, $\boxed{y = O(r^{-k-1})}$, for now.

The $\dot{\Lambda}$ and $\dot{R}$ equations of motion will be used to obtain suitable fall-off conditions for N and N^r, where, as the most general form we obtain:

$$N = \mathcal{O}(r^k), \quad N^r = \mathcal{O}(r^{-1}).$$

Similarly, let's consider $\dot{P}_\Lambda$.

$\Lambda - 1 + \frac{1}{2} w_+ r^{-2} + O(r^{-3})$

$R = r + \frac{1}{2} p r^{-1} + O(r^{-2})$

$$\left.\begin{aligned} P_\Lambda &= O(r^{-k-1}) \\ P_R &= O(r^{-k}) \end{aligned}\right\} \rightarrow y = O(r^{-k-1})$$

$N = O(r^k)$

$N^r = O(r^{-1})$

$\dot{P}_\Lambda = O(r^{-k-1}) = O(r^k)\left[\left(\frac{O(r^{-k-1})}{r^2}\right)O(r^{-k}) - O(r^{-2k-1}) + (p+w_+)r^{-1} + 2r + O(r^{-3})\right]$

$+\hat{\lambda}O(r^{-3}) + O(r^{k+1}) + O(r^{-k-1})$ the solution for k is k=0.

Lastly, consider $\dot{P}_R$:

$$\dot{P}_R = -N\left[\frac{\left(P_R + 2y\left[\Lambda R - \hat{\lambda}\left(\frac{R'}{\Lambda}\right)'\right](-2yR)\right)}{\left(R^2 + \hat{\lambda}\left[y^2 + \left(1 - \left(\frac{R'}{\Lambda}\right)^2\right)\right]\right)} + y^2\Lambda - \Lambda\left(1 - \left(\frac{R'}{\Lambda}\right)^2\right)\right] - 2NR\left(\frac{R'}{\Lambda}\right)'$$

$$+\left[\frac{N\left(P_R + 2y\left[\Lambda R - \hat{\lambda}\left(\frac{R'}{\Lambda}\right)'\right]\right)\left[2y\hat{\lambda}\left(\frac{R'}{\Lambda}\right)\Lambda^{-1}\right]}{\left(R^2 + \hat{\lambda}\left[y^2 + \left(1 - \left(\frac{R'}{\Lambda}\right)^2\right)\right]\right)}\right]' \left[\frac{(Ny^2\hat{\lambda})'}{\Lambda}\right]'$$

$$+\left(2NR\left(\frac{R'}{\Lambda}\right)\right)' + \left[\frac{\left(N\left[R^2 + \hat{\lambda}\left(1 - \left(\frac{R'}{\Lambda}\right)^2\right)\right]\right)'}{\Lambda}\right]'$$

$$-\left[2\hat{\lambda}N\left(\frac{R'}{\Lambda}\right)\left(\frac{R'}{\Lambda}\right)'\Lambda^{-1}\right]' + (N^r P_R)'$$

where similar analysis again shows k=0.

The bulk Hamiltonian via Legendre transformation

$$L = N\left\{\left\{\Lambda R\left(1 - \left(\frac{R'}{\Lambda}\right)^2\right)\right\} - \left(\frac{R'}{\Lambda}\right)'\left(R^2 + \hat{\lambda}\left(1 - \left(\frac{R'}{\Lambda}\right)^2\right)\right)\right\}$$

$$-N^{-1}\left\{(D\Lambda)(DR)\left(R^2 + \hat{\lambda}\left(1 - \left(\frac{R'}{\Lambda}\right)^2 + \frac{(DR)^2}{3N^2}\right)\right) + (DR)^2\left(\Lambda R - \hat{\lambda}\left(\frac{R'}{\Lambda}\right)'\right)\right\}$$

Adopt convention $\nabla\Lambda \equiv D\Lambda$

$$P_\Lambda = \frac{\delta L}{\partial R} = -N^{-1}(D\Lambda)\left(R^2 + \hat{\lambda}\left(1 - \left(\frac{R'}{\Lambda}\right)^2\right)\right) - N^{-1}(D\Lambda)\hat{\lambda}N^{-2}(DR)^2$$

$$+N^{-1}2(DR)\left(\Lambda R - \hat{\lambda}\left(\frac{R'}{\Lambda}\right)'\right)$$

$$P_\Lambda = \frac{\delta L}{\partial \Lambda} = -N^{-1}(DR)\left(R^2 + \hat{\lambda}\left(1 - \left(\frac{R'}{\Lambda}\right)^2 + \frac{(DR)^2}{3N^2}\right)\right)$$

$$P_\Lambda + \left[\frac{1}{3}\hat{\lambda}\left(\frac{(DR)}{N}\right)^3 + \left(\frac{(DR)}{N}\right)\left(R^2 + \hat{\lambda}\left[1 - \left(\frac{R'}{\Lambda}\right)^2\right]\right)\right] = 0$$

Denote $y = \left(\frac{Dr}{N}\right)$ following Jorma's notation

$0 = \frac{1}{3}\hat{\lambda}y^3 + y\left[R^2 + \hat{\lambda}\left(1 - \left(\frac{R'}{\Lambda}\right)^2\right)\right] + P_\Lambda$ N is chosen such that the relation between y and P_Λ is invertable?

$$P_R = -\frac{(D\Lambda)}{N}\left(R^2 + \hat{\lambda}\left(1 - \left(\frac{R'}{\Lambda}\right)^2\right)\right) - \frac{(D\Lambda)}{N}\hat{\lambda}y^2 + 2y\left(\Lambda R - \hat{\lambda}\left(\frac{R'}{\Lambda}\right)'\right)$$

$$\frac{(D\Lambda)}{N} = \frac{-P_R + 2y\left(\Lambda R - \hat{\lambda}\left(\frac{R'}{\Lambda}\right)'\right)}{\left[R^2 + \hat{\lambda}\left(1 - \left(\frac{R'}{\Lambda}\right)^2\right) - \hat{\lambda}y^2\right]}$$

$$L = P_\Lambda\dot{\Lambda} + P_R\dot{R} - NH - sH_r$$
$$H = H_{k.m} + H_{pot}$$

$$H_{pot} = -N^{-1}L_{pot} = -\Lambda R\left[1 - \left(\frac{R'}{\Lambda}\right)^2\right] + \left(\frac{R'}{\Lambda}\right)'\left\{R^2 + \hat{\lambda}\left(1 - \left(\frac{R'}{\Lambda}\right)^2\right)\right\}$$

$$L_{kin} = P_\Lambda\dot{\Lambda} + P_R\dot{R} - NH_{kin} - sH_r$$

$$L_{kin} = -(D\Lambda)\frac{(D\Lambda)}{N}\left\{R^2 + \hat{\lambda}\left(1 - \left(\frac{R'}{\Lambda}\right)^2 + \frac{(DR)^2}{3N^2}\right)\right\} - (DR)\frac{(D\Lambda)}{N}\left\{\Lambda R - \hat{\lambda}\left(\frac{R'}{\Lambda}\right)'\right\}$$

$$= -N\left[\left(\frac{-P_R + 2y\left(\Lambda R - \hat{\lambda}\left(\frac{R'}{\Lambda}\right)'\right)}{\left[R^2 + \hat{\lambda}\left(1 - \left(\frac{R'}{\Lambda}\right)^2\right) - \hat{\lambda}y^2\right]}\right)y\left\{R^2 + \hat{\lambda}\left(1 - \left(\frac{R'}{\Lambda}\right)^2\right) - \hat{\lambda}\frac{1}{3}y^2\right\}\right]$$

$$+y^2\left\{\Lambda R - \hat{\lambda}\left(\frac{R'}{\Lambda}\right)'\right\}$$

$$+P_\Lambda\dot{\Lambda} + P_R\dot{R}$$

$$-P_\Lambda\left[N\left(\frac{-P_R + 2y\left(\Lambda R - \hat{\lambda}\left(\frac{R'}{\Lambda}\right)'\right)}{R^2 + \hat{\lambda}\left(1 - \left(\frac{R'}{\Lambda}\right)^2\right) - \hat{\lambda}y^2}\right) + (N^r\Lambda)'\right]$$

$$-P_R[Ny + N^rR']$$

$= P_\Lambda \dot{\Lambda} + P_R \dot{R} - N\left\{y\left[P_R + y\left(\Lambda R - \hat{\lambda}\left(\frac{R'}{\Lambda}\right)'\right)\right]\right\} - NH_{kin} -$
$N^r(P_R R' - \Lambda P'_\Lambda)$
$L = P_\Lambda \dot{\Lambda} + P_R \dot{R} - N(H_{kin} + H_{pot}) - sH_r$
$H_{kin} = y = \left[P_R + y\left(\Lambda R - \hat{\lambda}\left(\frac{R'}{\Lambda}\right)'\right)\right]$
$H_r = P_R R' - \Lambda P'_\Lambda$

$\delta H|_{\delta R} = N\Bigg((\delta y)P_R + 2y(2y)\left(\Lambda R - \hat{\lambda}\left(\frac{R'}{\Lambda}\right)'\right) + y^2\Big(\Lambda \delta R -$

$\hat{\lambda}\left[\frac{(\delta R)'}{\Lambda}\right]'\Big) - \Lambda(\delta R)\left(1 - \left(\frac{R'}{\Lambda}\right)^2\right) + 2\Lambda R\left(\frac{R'}{\Lambda}\right)\frac{(\delta R)'}{\Lambda} + \frac{(\delta R)'}{\Lambda}\Bigg(R^2 +$

$\hat{\lambda}\left(1 - \left(\frac{R'}{\Lambda}\right)^2\right)\Bigg) + \left(\frac{R'}{\Lambda}\right)'\left(2R\delta R - 2\hat{\lambda}\left(\frac{R'}{\Lambda}\right)\left(\frac{(\delta R)'}{\Lambda}\right)\right)\Bigg)dr$

$$\hat{\lambda}y^2(\delta y) + \left[R^2 + \hat{\lambda}\left(1 - \left(\frac{R'}{\Lambda}\right)^2\right)\right](\delta y)$$
$$+y\left[2R(\delta R) - 2\hat{\lambda}\left(\frac{R'}{\Lambda}\right)\left(\frac{R'}{\Lambda}\right)'\right] = 0$$
$$\delta y = \frac{-Ry(\delta R) + 2\hat{\lambda}y\left(\frac{R'}{\Lambda}\right)\Lambda^{-1}(\delta R)'}{\left[\hat{\lambda}y^2 + \left(R^2 + \hat{\lambda}\left(1 - \left(\frac{R'}{\Lambda}\right)^2\right)\right)\right]}$$

$-\frac{\delta H}{\delta R} = -N\left[\left(P_R + 2y\left[\Lambda R - \hat{\lambda}\left(\frac{R'}{\Lambda}\right)'\right]\right)\left(\frac{\partial y}{\partial R}\right)\right] + \left[N\left(\frac{\partial H}{\partial R}\right)\left(\frac{\partial y}{\partial R}\right)\right]'$
$-Ny^2\Lambda + \left[\frac{(Ny^2\hat{\lambda})^2}{\Lambda}\right]' + N\Lambda\left(1 - \left(\frac{R'}{\Lambda}\right)^2\right) + \left[2NR\left(\frac{R'}{\Lambda}\right)\right]'$
$-\left[\frac{\left(N\left[R^2 + \hat{\lambda}\left(1 - \left(\frac{R'}{\Lambda}\right)^2\right)\right]\right)'}{\Lambda}\right]' - 2R\left(\frac{R'}{\Lambda}\right)' N$
$-2\hat{\lambda}\left[N\Lambda^{-1}\left(\frac{R'}{\Lambda}\right)\left(\frac{R'}{\Lambda}\right)'\right]'(N_r P_R)'$

$\dot{P}_R = -N\left(\frac{\partial H}{\partial y}\right)\left(\frac{\partial y}{\partial R}\right) + \left[N\left(\frac{\partial H}{\partial y}\right)\left(\frac{\partial y}{\partial R'}\right)\right]' - Ny^2\Lambda + \hat{\lambda}\left[\frac{1}{\Lambda}(Ny^2)'\right]'$

$$+N\Lambda\left[1-\left(\frac{R'}{\Lambda}\right)'\right]+(N_r P_R)'-2NR\left(\frac{R'}{\Lambda}\right)'-\left[\frac{N\prime}{\Lambda}\left(R^2+\hat{\lambda}\left(1-\left(\frac{R'}{\Lambda}\right)^2\right)\right)\right]$$
$$+\left[2NR\left(\frac{R'}{\Lambda}\right)'\right]-2\hat{\lambda}\left[N\Lambda^{-1}\left(\frac{R'}{\Lambda}\right)\left(\frac{R'}{\Lambda}\right)'\right]'-\left[\frac{N}{\Lambda}\left(2RR'-2\hat{\lambda}\left[\left(\frac{R'}{\Lambda}\right)\left(\frac{R'}{\Lambda}\right)'\right]\right)\right]$$

which is in agreement with previous results.

C.4 Obtaining the Lovelock Lagrangian

The method of exterior forms permits straightforward determination of Riemann. The Ricci scalar may be verified through a separate, geodetic path, calculation. Define

$$D \equiv \partial_+ - N^r\partial_r$$
$$\nabla\Lambda \equiv \dot{\Lambda} - (N^r\Lambda)$$

For the exterior forms an orthonormal basis is chosen for which the following is obtained:

$$R_{0101} = -(N\Lambda)^{-1}\left[(N^{-1}\nabla\Lambda)^{\cdot} - \left(\Lambda^{-1}\mathrm{N}' + N^{-1}N^r(\nabla\Lambda)\right)'\right]$$
$$R_{0202} = R_{0303} = R_{0404} = -(NR)^{-1}[N^{-1}D^2R - N^2(DR)(DN) - \Lambda^{-2}R'N']$$
$$R_{0202} = R_{0303} = R_{0404} = (RN)^{-1}\left[(\Lambda N)^{-1}N'^{(DR)} - \left(\frac{R\prime}{\Lambda}\right)\right]$$
$$R_{1212} = R_{1313} = R^1{}_{414} = R^{-2}\left[1 + N^2(DR)(\nabla\Lambda) - \left(\frac{R\prime}{\Lambda}\right)'\right]$$
$$R_{2323} = R_{2424}R^3{}_{434} = R^{-2}\left[1 + N^{-2}(DR)^2 - \Lambda^{-2}(R')^2\right]$$
$$R_{\mu\nu} = R^\alpha{}_{\mu\alpha\nu}$$
$$R_{00} = R_{0101} + 3R_{0202}$$
$$R_{01} = 3R_{0212}$$
$$R_{11} = -R_{0101} + 3R_{1212}$$
$$R_{22} = R_{33} = R_{44} = -R_{0202} + R_{1212} + 2R_{2323}$$
$${}^{(5)}R = g^{\mu\nu}R_{\mu\nu} = -2(R_{0101} + 3R_{0202}) + 6(R_{1212} + R_{2323})$$
$$\left(\frac{\sqrt{-g^{(5)}}R}{\sin\theta\sin^2 x}\right) = 2R^3\left[(N^{-1}\nabla\Lambda)' - \left(\Lambda^{-1}N' + N^{-1}N^r\nabla\Lambda\right)'\right]$$
$$+6R^2\Lambda\left[N^{-1}D^2R - N^{-2}(DR)(DN) - \Lambda^{-2}R'N'\right]$$
$$+6R^2N\left[N^{-2}(DR)(\nabla\Lambda) - \left(\frac{R'}{\Lambda}\right)'\right] + 6N\Lambda R\left[1 + N^{-2}(DR)^2 - \Lambda^{-2}(R')^2\right]$$

$$= -(2R^3)^{\cdot}(N^{-1}\nabla\Lambda) + (2R^3)'\left(\Lambda^{-1}N' + N^{-1}N^r\nabla\Lambda\right)$$

$$+6R^2N^{-1}\Lambda((DR)' - N^r(DR)') - 6R^2\Lambda N^{-2}(DR)(DN) - 6R^2\Lambda^{-1}R'N'$$
$$+ 6R^2N^{-1}(DR)(\nabla\Lambda) - 6R^2N\left(\frac{R'}{\Lambda}\right)'$$
$$+ 6N\Lambda R\left[1 + N^{-2}(DR)^2 - \Lambda^{-2}(R^r)^2\right]$$
$$+[2R^3(N^{-1}\nabla\Lambda)]^{\cdot} - \left[2R^3\left(\Lambda^{-1}N' + N^{-1}N'\nabla\Lambda\right)\right]'$$

Dropping the surface terms:

$$+[2R^3(N^{-1}\nabla\Lambda)]^{\cdot} - \left[2R^3\left(\Lambda^{-1}N' + N^{-1}N^r\nabla\Lambda\right)\right]'$$
$$+[6R^2N^{-1}\Lambda(DR)]^{\cdot} - [6R^2N^{-1}\Lambda N^r(DR)]'$$

$\mathcal{L}_1^{(1)} = \left(\frac{\sqrt{-g}^{(s)}R}{\sin\theta\sin^2 x}\right) = -6R^2N^{-1}(DR)(\nabla\Lambda) + 6R^2\Lambda^{-1}N'R' -$
$N^{-1}(6R^2\Lambda)'(DR) + \frac{\dot{N}}{N^2}(R\hat{\lambda})(DR) + (6R^2N^{-1}\Lambda N^r)'(DR) -$
$6R^2\Lambda N^{-2}(DR)\left(\dot{N} - N^rN'\right) - 6R^2\Lambda^{-1}R' + 6R^2N^{-1}(DR)(\nabla\Lambda) -$
$6R^2N\left(\frac{R'}{\Lambda}\right)' + 6N\Lambda R[\ldots]$

$\mathcal{L}_1^{(1)} = -N^{-1}(DR)[(6R^2\Lambda)^{\cdot} - N^r(6R^2\Lambda)'] + (6R^2\Lambda)\left(\frac{N^r}{N}\right)'(DR) +$
$(6R^2\Lambda)N^r\left(\frac{N^r}{N^2}\right)'$
$-6R^2N\left(\frac{R'}{\Lambda}\right)' + 6N\Lambda R\left[1 + N^{-2}(DR)^2 - \Lambda^{-2}(R')^2\right]$
$= -N^{-1}(DR)[6R^3(D\Lambda) + 12R\Lambda(DR)] + (6R^2\Lambda)N^{-1}(N^{r\prime})(DR)$
$-6R^2N\left(\frac{R'}{\Lambda}\right)' + 6N\Lambda R\left[1 + N^2(DR)^2 - \Lambda^{-2}(R')^2\right]$
$= N^{-1}(DR)(6R^2)\left[-\left(\dot{\Lambda} - N^r\Lambda'\right) + \Lambda(N^r)'\right] - 12N^{-1}R\Lambda(DR)^2 -$
$6R^2N\left(\frac{R'}{\Lambda}\right)'$
$+6N\Lambda R\left[1 - \Lambda^{-2}(R')^2\right] + 6N^{-1}R\Lambda(DR)^2$
$= -N^{-1}[(6R^2)(DR)(-\nabla\Lambda) - (6R\Lambda)(DR)^2] + N\Big[(6\Lambda R) -$
$6\left(\frac{R}{\Lambda}\right)(R')^2 - 6R^2\left(\frac{R'}{\Lambda}\right)'\Big]$

$$= 6\left\{\begin{matrix} N^{-1}[-(DR)(\nabla\Lambda)R^2 - (\Lambda R)(DR)^2] \\ +N\left[\Lambda R\left(1 - \left(\frac{R'}{\Lambda}\right)^2\right) - R^2\left(\frac{R'}{\Lambda}\right)'\right] \end{matrix}\right\}$$

In order to obtain the Lagrangian in quadratic form for the first derivatives, integration by parts was used. Similar technique will be needed in the analysis of $\mathcal{L}_2^{\square}$:

$$S_\Sigma = \frac{1}{6}\int \sqrt{-g^{(5)}}\,[\mathcal{L}_1^{\square} + \mathcal{L}_2^{\square}]d^5x = \frac{(2\pi^2)}{6}\int dt\,dr\left[\mathcal{L}_1^{(1)} + \mathcal{L}_2^{(1)}\right]d^5x$$

In order to arrive at $\mathcal{L}_2^{(1)}$ from $\mathcal{L}_2^{\square}$, by dropping surface terms, it is necessary that $\mathcal{L}_2^{\square}$ not be quadratic in those Riemann tensor contributions that have second order time derivatives. In other words, the R_{0101} and R_{0202} terms must group in such a way that the higher order time derivative terms cancel.

$$\mathcal{L}_2^{\square} = \frac{\lambda}{2}(R^2 - 4R_{ab}R^{ab} + R_{abcd}R^{abcd})\sqrt{-g^5}$$

$${}^{(5)}_{\square}R^2 = 4[-(R_{0101} + 3R_{0202}) + 3(R_{1212} + R_{2323})]^2$$
$$= 4[R_{0101}^2 + 6R_{0101}R_{0202} + 9R_{0202} - 6(R_{0101}R_{1212}R_{2323} +$$
$$3R_{0202}R_{1212} + 3R_{0202}R_{2323} + 9R_{1212}^2 + 18R_{1212}R_{2323} + 9R_{2323}^2)]$$
$${}^{(5)}_{\square}R^2 = 4R_{0101}^2 + 24R_{0101}R_{0202} + 36R_{0202} - 24R_{0101}(R_{1212} +$$
$$R_{2323}) - 72R_{0202}(R_{1212} + R_{2323}) + 36R_{1212} + 72R_{1212}R_{2323} +$$
$$36R_{2323}^2$$

$$4R_{ab}R^{ab} = 4[R_{00}^2 - 2R_{01}^2 + R_{11}^2 + 3R_{22}^2]$$
$$= 4[(R_{0101}^2 + 6R_{0101}R_{0202} + 9R_{0202}^2) - 18R_{0212}^2 + (R_{0101}^2 -$$
$$6R_{0101}R_{1212} + 9R_{1212}^2)]$$
$$+3R_{0202}^2 - 2R_{0202}R_{1212} - 4R_{0202}R_{2323} + R_{1212}^2 + 4R_{1212}R_{2323} +$$
$$4R_{2323}^2]$$
$$= 8R_{0101}^2 + 24R_{0101}R_{0202} + 48R_{0202}^2 - 72R_{0212}^2 - 24R_{0101}R_{1212} -$$
$$24R_{0202}R_{1212}$$
$$-48R_{0202}R_{2323} + 48R_{1212}^2 + 48R_{1212}R_{2323} + 48R_{2323}^2$$

$$R_{abcd}R^{abcd} = 4R_{0101}^2 + 4\cdot 3R_{0202}^2 - 8\cdot 3R_{1202}^2 + 4\cdot 3R_{1212}^2 + 4\cdot$$
$$3R_{2323}^2$$

$$\frac{\mathcal{L}_2^{\square}}{\sqrt{-g^{(5)}}} =$$

$$\frac{\lambda}{2}\begin{bmatrix} R_{0101}^2(4-8-4) + R_{0202}^2(36-48+12) + R_{0101}R_{0202}(24-24) \\ +R_{0101}R_{1212}(-24+24) + R_{0101}R_{2323}(-24) + R_{0202}R_{1212}(-72+24) \\ +R_{0202}R_{2323}(-72+48) + R_{1212}R_{2323}(72-48) + R_{2323}^2(36+48+12) \\ +R_{1212}^2(36-48+12)R_{0212}^2(72-24) \end{bmatrix}$$

$= \frac{\lambda}{2}[-24R_{0101}R_{2323} - 48R_{0202}R_{1212} - 24R_{0202}R_{2323} + 24R_{1212}R_{2323} + 48R_{0212}^2]$
$= 12\lambda[(R_{1212}R_{2323} + 2R_{0212}^2) - (R_{0101}R_{2323} + R_{0202}R_{2323} + 2R_{0202}R_{1212})]$

$R_{1212}R_{2323} = (\Lambda R^3)^{-1}\left[N^{-2}(DR)(\nabla\Lambda) - \left(\frac{R\prime}{\Lambda}\right)'\right][1 + N^{-2}(DR)^2 - \Lambda^{-2}(R')^2]$
$2R_{0212}^2 = 2(RN)^{-2}\left[(\Lambda N)^{-1}N'^{(DR)} - D\left(\frac{R\prime}{\Lambda}\right)\right]^2$
$R_{0101}R_{2323} = -R^{-2}(N\Lambda)^{-1}\left[(N^{-1}\nabla\Lambda)^{\cdot} - \left(\Lambda^{-1}N' + N^{-1}N^r(\nabla\Lambda)\right)'\right]$
$\times\left[1 + N^{-2}(DR)^2 - \Lambda^2(R')^2\right]$
$R_{0202}R_{2323} = -(NR^3)^{-1}[N^{-1}D^2R - N^{-2}(DR)(DN) - \Lambda^{-2}R'N'][1 + N^{-2}(DR)^2 - \Lambda^{-2}(R')^2]$
$2R_{0202}R_{1212} = -2(N\Lambda R^2)^{-1}\left[N^{-2}(DR)(\nabla\Lambda) - \left(\frac{R\prime}{\Lambda}\right)'\right][N^{-1}D^2R - N^{-2}(DR)(DN) - \Lambda^{-2}R'N']$

$$X = \frac{\mathcal{L}_2^{\square}}{2\lambda\sin\theta\sin^2 x} =$$
$$\left\{\begin{array}{c} N\left(\left(\frac{(DR)(\nabla\Lambda)}{N^2}\right) - \left(\frac{R'}{\Lambda}\right)'\right)\left(1 + \frac{(DR)^2}{N^2} - \left(\frac{R'}{\Lambda}\right)^2\right) \\ + \frac{2\Lambda R}{N}\left(\frac{N'^{(DR)}}{N\Lambda} - D\left(\frac{R'}{\Lambda}\right)\right)^2 \\ +R\left(\left(\frac{\nabla\Lambda}{N}\right)^{\cdot} - \left(\frac{N'}{\Lambda} + \frac{N^r}{N}(\nabla\Lambda)\right)'\right)\left(1 + \frac{(DR)^2}{N^2} - \left(\frac{R'}{\Lambda}\right)^2\right) \\ +\Lambda\left(\frac{D^2R}{N} - \frac{(DR)(DN)}{N^2} - \frac{R\prime N\prime}{\Lambda^2}\right)\left(1 + \frac{(DR)^2}{N^2} - \left(\frac{R'}{\Lambda}\right)^2\right) \\ +2R\left(\frac{(DR)(\nabla\Lambda)}{N^2} - \left(\frac{R'}{\Lambda}\right)'\right)\left(\frac{D^2R}{N} - \frac{(DR)(\nabla\Lambda)}{N^2} - \frac{R\prime N\prime}{\Lambda^2}\right) \end{array}\right\}$$

As in the case of $\mathcal{L}_1^{\square}$, integration by parts is done to express the lapse without any derivative operations. Terms for which this doesn't appear possible are expected to be identically zero.
$X = N^{-3}[(DR)^3(\nabla\Lambda) + R(\nabla\Lambda)^{\cdot}(DR)^2 - (N^r\nabla\Lambda)'(DR)^2 + \Lambda D^2R(DR)^2 + 2R(DR)(\nabla\Lambda)D^2R]$
$+\left(-\frac{1}{3}N^{-3}\right)^{\cdot}[-R(\nabla\Lambda)(DR)^2 - \Lambda(DR)^3 - 2R(DR)^3(\nabla\Lambda)]$

$$+\left(-\frac{1}{3}N^{-3}\right)'\left[N^rR(\nabla\Lambda)(DR)^2+\Lambda N^r(DR)^3+2R(\nabla\Lambda)(DR)^2N^r\right]$$
$$+N^{-3}(N')^2\left[\frac{2R}{\Lambda}(DR)^2\right]-N^HN^{-2}\left[\left(\frac{R}{\Lambda}\right)(DR)^2\right]$$
$$+N^{-1}\begin{bmatrix}(DR)(\nabla\Lambda)\left(1-\left(\frac{R'}{\Lambda}\right)^2\right)-(DR)^2\left(\frac{R'}{\Lambda}\right)'+2\Lambda R\left[D\left(\frac{R'}{\Lambda}\right)\right]^2\\+R(\nabla\Lambda)^{\cdot}\left(1-\left(\frac{R'}{\Lambda}\right)^2\right)\\-R(N^r\nabla\Lambda)'\left(1-\left(\frac{R'}{\Lambda}\right)^2\right)+\Lambda D^2R\left(1-\left(\frac{R'}{\Lambda}\right)^2\right)-2RD^2R\left(\frac{R'}{\Lambda}\right)'\end{bmatrix}$$
$$+(-N^{-1})'\left[-R(\nabla\Lambda)\left(1-\left(\frac{R'}{\Lambda}\right)^2\right)-\Lambda(DR)\left(1-\left(\frac{R'}{\Lambda}\right)^2\right)+\right.$$
$$\left.2R(DR)\left(\frac{R'}{\Lambda}\right)^2\right]$$
$$+(-N^{-1})\begin{bmatrix}-4R(DR)D\left(\frac{R'}{\Lambda}\right)+R\left(\frac{\Lambda'}{\Lambda^2}\right)(DR)^2+RN^r(\nabla\Lambda)\left(1-\left(\frac{R'}{\Lambda}\right)^2\right)\\+\Lambda(DR)N^r\left(1-\left(\frac{R'}{\Lambda}\right)^2\right)-\left(\frac{R'}{\Lambda}\right)(DR)^2\\-2P\Lambda^{-2}(DR)(\nabla\Lambda)R'-2R\left(\frac{R'}{\Lambda}\right)'N^r\end{bmatrix}$$
$$+N\left[-\left(\frac{R'}{\Lambda}\right)'\left(1-\left(\frac{R'}{\Lambda}\right)^2\right)\right]$$
$$+N'\left[R\frac{\Lambda\prime}{\Lambda^2}\left(1-\left(\frac{R'}{\Lambda}\right)^2\right)-\left(\frac{R'}{\Lambda}\right)\left(1-\left(\frac{R'}{\Lambda}\right)^2\right)+\frac{1RR\prime}{\Lambda^2}\left(\frac{R'}{\Lambda}\right)'\right]$$
$$+N''\left[-\left(\frac{R}{\Lambda}\right)\left(1-\left(\frac{R'}{\Lambda}\right)^2\right)\right]$$

Consider $N^{-3}(N')^2\left[\frac{2R}{\Lambda}(DR)^2\right]-N''N^{-2}\left[\frac{2R}{\Lambda}(DR)^2\right]$
$=N^{-3}(N')^2\left[\frac{2R}{\Lambda}(DR)^2\right]+N'\left(N^{-2}\left[\frac{2R}{\Lambda}(DR)^2\right]\right)'$ (using integration by parts)
$=N'N^{-2}\left[\frac{2R}{\Lambda}(DR)^2\right]'=(-N^{-1})'\left[\frac{2R}{\Lambda}(DR)^2\right]'$
The last term to be grouped with other $(-N^{-1})'$ terms.

$X=N^{-3}\{(DR)^3(\nabla\Lambda)+R(\nabla\Lambda)'(DR)^2-(N^r\nabla\Lambda)'(DR)^2+$
$\Lambda D^2R(DR)+2R(DR)(\nabla\Lambda)\}$
$-\frac{1}{3}\left[\dot{R}(\nabla\Lambda)(DR)^2+R(\nabla\Lambda)^{\cdot}(DR)^2+R(\nabla\Lambda)2(DR)^{\cdot}(DR)+\dot{\Lambda}(DR)^3+\right.$
$3\Lambda(DR)^2(DR)^{\cdot}+2\dot{R}(DR)^2(\nabla\Lambda)+4R(DR)(DR)'(\nabla\Lambda)+$
$\left.2R(DR)^2(\nabla\Lambda)'\right]$

$$+\frac{1}{3}\left[\begin{array}{c}(N^r\nabla\Lambda)'R(DR)^2+R'(N^r\nabla\Lambda)(DR)^2+R(N^r\nabla\Lambda)2(DR)' \\ +\Lambda'N^r(DR)^3-\Lambda(N^r)'(DR)^3(DR)' \\ +2R'^{(\nabla\Lambda)(DR)^2N^r}+2R(\nabla\Lambda)'(DR)^2N^r+2R(\nabla\Lambda)2(DR)(DR)'N^r\end{array}\right]\Bigg\}$$

$$+N^{-1}\left\{\begin{array}{c}(DR)(\nabla\Lambda)\left(1-\left(\frac{R'}{\Lambda}\right)^2\right)-(DR)^2\left(\frac{R'}{\Lambda}\right)'+2\Lambda R\left(D\left(\frac{R'}{\Lambda}\right)\right)^2 \\ +R(\nabla\Lambda)'\left(1-\left(\frac{R'}{\Lambda}\right)^2\right) \\ -R(N^r\nabla\Lambda)'\left(1-\left(\frac{R'}{\Lambda}\right)^2\right)+AD^2R\left(1-\left(\frac{R'}{\Lambda}\right)^2\right)+RD^2R\left(\frac{R'}{\Lambda}\right)'\end{array}\right\}$$

$$+N^{-1}\left[\begin{array}{c}-\dot{R}(\nabla\Lambda)\left(1-\left(\frac{R'}{\Lambda}\right)^2\right)-R(\nabla\Lambda)^{\cdot}\left(1-\left(\frac{R'}{\Lambda}\right)^2\right)+R(\nabla\Lambda)2\left(\frac{R'}{\Lambda}\right)\left(\frac{R'}{\Lambda}\right)^{\cdot} \\ -\dot{\Lambda}(DR)\left(1-\left(\frac{R'}{\Lambda}\right)^2\right)-\Lambda(DR)^{\cdot}\left(1-\left(\frac{R'}{\Lambda}\right)^2\right)+\Lambda(DR)\,2\left(\frac{R'}{\Lambda}\right)\left(\frac{R'}{\Lambda}\right)^{\cdot} \\ +2\dot{R}(DR)\left(\frac{R'}{\Lambda}\right)'+2R(DR)^{\cdot}\left(\frac{R'}{\Lambda}\right)+2R(DR)\left(\frac{R'}{\Lambda}\right)^{\cdot}\end{array}\right]$$

$$+N^{-1}\left[\begin{array}{c}-4R'^{(DR)}D\left(\frac{R'}{\Lambda}\right)-4R(DR)'D\left(\frac{R'}{\Lambda}\right)-4R(DR)\left(D\left(\frac{R'}{\Lambda}\right)\right)' \\ +R'\left(\frac{\Lambda'}{\Lambda^2}\right)(DR)^2+R\left(\frac{\Lambda'}{\Lambda^2}\right)(DR)^2+R\left(\frac{\Lambda'}{\Lambda^2}\right)2(DR)(DR)' \\ +(RN)'(\nabla\Lambda)\left(1-\left(\frac{R'}{\Lambda}\right)^2\right)+RN^r(\nabla\Lambda)'\left(1-\left(\frac{R'}{\Lambda}\right)^2\right) \\ -2NR(\nabla\Lambda)\left(\frac{R'}{\Lambda}\right)\left(\frac{R'}{\Lambda}\right)' \\ +\Lambda'(DR)N^r\left(1-\left(\frac{R'}{\Lambda}\right)^2\right)+\Lambda(DR)'N^r\left(1-\left(\frac{R'}{\Lambda}\right)^2\right) \\ +\Lambda(DR)N^r\left(1-\left(\frac{R'}{\Lambda}\right)^2\right) \\ -2\Lambda(DR)N^r\left(\frac{R'}{\Lambda}\right)\left(\frac{R'}{\Lambda}\right)'-\left[\left(\frac{R'}{\Lambda}\right)'(DR)^2+2\left(\frac{R'}{\Lambda}\right)(DR)(DR)'\right] \\ -2\left(\frac{R'}{\Lambda}\right)'\left(\frac{R'}{\Lambda}\right)(DR)(\nabla\Lambda)-2\left(\frac{R'}{\Lambda}\right)\left(\frac{R'}{\Lambda}\right)'(DR)(\nabla\Lambda) \\ -2\left(\frac{R'}{\Lambda}\right)\left(\frac{R'}{\Lambda}\right)[(DR)'\nabla\Lambda+DR]+\left[\left(\frac{R}{\Lambda}\right)(DR)^2\right]'' \\ -2R'^{\left(\frac{R'}{\Lambda}\right)'N^r(DR)}-2R\left(\frac{R'}{\Lambda}\right)'N^r(DR)-2R\left(\frac{R'}{\Lambda}\right)'(N^r)'(DR) \\ -2R\left(\frac{R'}{\Lambda}\right)'N^r(DR)'\end{array}\right]$$

$$+N\left\{\begin{array}{c} -\left(\frac{R'}{\Lambda}\right)'\left(1-\left(\frac{R'}{\Lambda}\right)^2\right) - R'^{\frac{\Lambda'}{\Lambda^2}\left(1-\left(\frac{R'}{\Lambda}\right)^2\right)} - R\left(\frac{\Lambda'}{\Lambda^2}\right)'\left(1-\left(\frac{R'}{\Lambda}\right)^2\right) \\ +R\left(\frac{\Lambda'}{\Lambda^2}\right)2\left(\frac{R'}{\Lambda}\right)\left(\frac{R'}{\Lambda}\right)' + \left(\frac{R'}{\Lambda}\right)'\left(1-\left(\frac{R'}{\Lambda}\right)^2\right) - 2\left(\frac{R'}{\Lambda}\right)\left(\frac{R'}{\Lambda}\right)\left(\frac{R'}{\Lambda}\right)' \\ -2\frac{(R')^2}{\Lambda^2}\left(\frac{R'}{\Lambda}\right)' + 2RR'\left(\frac{2\Lambda'}{\Lambda^3}\right)\left(\frac{R'}{\Lambda}\right)' - \frac{2RR'}{\Lambda^2}\left(\frac{R'}{\Lambda}\right)' \\ -\frac{2RR'}{\Lambda^2}\left(\frac{R'}{\Lambda}\right) - \left[\left(\frac{R}{\Lambda}\right)\left(1-\left(\frac{R'}{\Lambda}\right)^2\right)\right]'' \end{array}\right\}$$

$$X =$$
$$N^{-3}\left\{\begin{array}{c} (DR)^3(\nabla\Lambda) - \frac{1}{3}(DR)^3\left(\dot{\Lambda} - (\Lambda N^r)\right)\frac{1}{3}(\nabla\Lambda)(DR)^2\left(\dot{R} - N^r R'\right) \\ -\frac{1}{3}2(DR)^2(\nabla\Lambda)\left(\dot{R} - N^r R'\right) \end{array}\right\}$$
$$+N^{-1}\{\ldots\}$$
$$+N\{\ldots\}$$

$$X = N^{-3}\left\{-\frac{1}{3}(DR)^3(\nabla\Lambda)\right\}$$
$$+N^{-1}\left\{\begin{array}{c} \left(1-\left(\frac{R'}{\Lambda}\right)^2\right)\left[(DR)(\nabla\Lambda) + R(\nabla\Lambda)^{\cdot} - R(N^r\nabla\Lambda)' + \Lambda D^2 R - \hat{R}(\nabla R)\right] \\ -R(\nabla\Lambda)' - \dot{\Lambda}(DR) - \Lambda(DR)' + (N^r)'(\nabla\Lambda)R + N^r(\nabla\Lambda)R \\ +N^r R'^{(\nabla\Lambda)} + \Lambda'^{N^r(DR)''} + \Lambda(DR)'N^r + \Lambda(DR)(N^r)' \\ +\cdots \end{array}\right.$$
$$+N\{\ldots\}$$

$$X = N^{-3}\left\{-\frac{1}{3}(DR)^3(\nabla\Lambda)\right\}$$
$$+N^{-1}\left\{-\left(1-\left(\frac{R'}{\Lambda}\right)^2\right)(DR)(\nabla\Lambda) + other\right\}$$
$$+N\left\{\left(1-\left(\frac{R'}{\Lambda}\right)^2\right)\left[-\left(\frac{R'}{\Lambda}\right)' - R'\left(-\frac{1}{\Lambda}\right)' - R\left(-\frac{1}{\Lambda}\right)'' + \left(\frac{R'}{\Lambda}\right)'\right]\right\}$$
$$+\left(\frac{R'}{\Lambda}\right)'\left[2R\left(-\frac{1}{\Lambda}\right)'\left(\frac{R'}{\Lambda}\right) - 4\left(\frac{R'}{\Lambda}\right)^2 + 2RR'\left(-\frac{1}{\Lambda}\right)'\right] - 2\frac{2RR'}{\Lambda^2}\left(\frac{R'}{\Lambda}\right)' -$$
$$\frac{2RR'}{\Lambda^2}\left(\frac{R'}{\Lambda}\right)'$$
$$-\left[\left(\frac{R'}{\Lambda}\right)\left(1-\left(\frac{R'}{\Lambda}\right)^2\right)\right]''$$

$$= N^{-3}\{\ldots\}$$
$$+N^{-1}\{\ldots\}$$

$$+N\left\{\begin{array}{c}\left(1-\left(\frac{R'}{\Lambda}\right)^2\right)\left[\left(R\left(\frac{1}{\Lambda}\right)'\right)'-\left(\frac{R}{\Lambda}\right)''\right]+4\left(\frac{R'}{\Lambda}\right)\left(\frac{R'}{\Lambda}\right)\left(\frac{R'}{\Lambda}\right)' \\ +\left(\frac{R'}{\Lambda}\right)'\left[2R\left(-\frac{1}{\Lambda}\right)'\left(\frac{R'}{\Lambda}\right)-4\left(\frac{R'}{\Lambda}\right)^2+2RR'^{\left(-\frac{1}{\Lambda^2}\right)'}\right]-2\frac{RR'}{\Lambda^2}\left(\frac{R'}{\Lambda}\right)'' \\ +\left(\frac{R}{\Lambda}\right)2\left[\left(\frac{R'}{\Lambda}\right)\left(\frac{R'}{\Lambda}\right)'\right]'\end{array}\right\}$$

$$X=N^{-3}\{\ldots\}$$
$$+N^{-1}\{\ldots\}$$
$$+N\left\{\left(1-\left(\frac{R'}{\Lambda}\right)^2\right)\left[-\left(\frac{R'}{\Lambda}\right)'\right]+\left(\frac{R'}{\Lambda}\right)'\left[4\left(\frac{R'}{\Lambda}\right)^2+4R\left(\frac{1}{\Lambda}\right)'\left(\frac{R'}{\Lambda}\right)\right]\right\}$$
$$+2R\left(-\frac{1}{\Lambda}\right)'\left(\frac{R'}{\Lambda}\right)-4\left(\frac{R'}{\Lambda}\right)^2+2RR'\left(-\frac{1}{\Lambda}\right)'\Bigg]-2\frac{RR''}{\Lambda^2}\left(\frac{R'}{\Lambda}\right)''$$
$$+2\left(\frac{R}{\Lambda}\right)\left[\left(\frac{R'}{\Lambda}\right)'\right]^2+2\left(\frac{R}{\Lambda}\right)\left(\frac{R}{\Lambda}\right)\left(\frac{R'}{\Lambda}\right)''-2\frac{RR''}{\Lambda^2}\left(\frac{R'}{\Lambda}\right)'$$

$$X=N^{-3}\{\ldots\}$$
$$+N^{-1}\{\ldots\}$$
$$+N\left\{\left(1-\left(\frac{R'}{\Lambda}\right)^2\right)\left(-\left(\frac{R'}{\Lambda}\right)'\right)\right\}$$

$$X=N^{-3}\{\ldots\}+N\{\ldots\}$$
$$+N^{-1}\left\{1-\left(\frac{R'}{\Lambda}\right)^2(DR)(\nabla\Lambda)-(DR)^2 1-\left(\frac{R'}{\Lambda}\right)'+2\Lambda R\left(D\left(1-\right.\right.\right.$$
$$\left.\left.\left.\left(\frac{R'}{\Lambda}\right)\right)\right)^2\right\}$$
$$-2RD^2R\left(\frac{R'}{\Lambda}\right)'-4R'^{(DR)}D\left(\frac{R'}{\Lambda}\right)-4R(DR)'D\left(\frac{R'}{\Lambda}\right)-4R(DR)\left(D\left(\frac{R'}{\Lambda}\right)\right)$$
$$+R'\left(\frac{\Lambda'}{\Lambda^2}\right)(DR)^2+R\left(\frac{\Lambda'}{\Lambda^2}\right)(DR)^2+2R\left(\frac{\Lambda'}{\Lambda^2}\right)(DR)(DR)'$$
$$-\left(\frac{R'}{\Lambda}\right)'(DR)^2-2\left(\frac{R'}{\Lambda}\right)(DR)(DR)'-2\left(\frac{R}{\Lambda}\right)\left(\frac{R'}{\Lambda}\right)(DR)(\Lambda R)'-$$
$$2\left(\frac{R}{\Lambda}\right)\left(\frac{R'}{\Lambda}\right)'(DR)(\Lambda R)'-2\left(\frac{R}{\Lambda}\right)\left(\frac{R'}{\Lambda}\right)[(DR)(\Lambda R)']$$
$$+2R\left(\frac{R'}{\Lambda}\right)(\nabla\Lambda)D\left(\frac{R'}{\Lambda}\right)+2R'^{(DR)}D\left(\frac{R'}{\Lambda}\right)+2(DR)^2\left(\frac{R'}{\Lambda}\right)'$$
$$+2R\left(\frac{R'}{\Lambda}\right)'D^2R+2R(DR)D\left(\left(\frac{R'}{\Lambda}\right)'\right)-2R(DR)(N^r)'\left(\frac{R'}{\Lambda}\right)'$$
$$\left.+\left[\left(\frac{R}{\Lambda}\right)(DR)^2\right]''\right\}$$

$$X = N^{-3}\{\dots\} + N\{\dots\} + N^{-1}\{-\left(1-\left(\frac{R'}{\Lambda}\right)^2\right)(DR(\nabla\Lambda)) +$$

$$+\left[R\left(\frac{-1}{\Lambda}\right)\right]'(DR)^2 + 2\left(\frac{R}{\Lambda}\right)' - 2\left[\left(\frac{R}{\Lambda}\right)\left(\frac{R'}{\Lambda}\right)(DR)(\nabla\Lambda)\right]'$$

+ terms involving $D\left(\frac{R'}{\Lambda}\right), D\left(\left(\frac{R'}{\Lambda}\right)\right), and\ (N^r)'$

$$D\left(\frac{R'}{\Lambda}\right) = \left[\left(\frac{R'}{\Lambda}\right)' - N^r\left(\frac{R'}{\Lambda}\right)'\right]$$

Using: $(DR)' = \left(\dot{R} - N^r R'\right)' = \dot{R}' - (N^r)R' - N = (DR') - (N^r)'R'$

$$D\left(\frac{R'}{\Lambda}\right) = \frac{R''}{\Lambda} + R'\left(\frac{1}{\Lambda}\right)^{\cdot} - N^r\left(\frac{R''}{\Lambda} + R'\left(\frac{1}{\Lambda}\right)'\right)$$

$$= \frac{1}{\Lambda}(R'' - N^r R'') + \frac{1}{\Lambda}(-(N^r)'R' + (N^r)'R') - \frac{R'}{\Lambda^2}(\dot{\Lambda} - N^r\Lambda')$$

$$= \frac{1}{\Lambda}(DR)' - \frac{R'}{\Lambda^2}(\dot{\Lambda} - [N^r\Lambda' - (N^r)'\Lambda])$$

$$= \frac{1}{\Lambda}(DR)' - \frac{R'}{\Lambda^2}(\nabla\Lambda)$$

$$D\left(\left(\frac{R'}{\Lambda}\right)'\right) = \left[\left(\frac{R'}{\Lambda}\right)' - N^r\left(\frac{R'}{\Lambda}\right)''\right] = \left[\left(\frac{R'}{\Lambda}\right)' - N^r\left(\frac{R'}{\Lambda}\right)'\right] + (N^r)'\left(\frac{R'}{\Lambda}\right)'$$

$$= \left[\frac{1}{\Lambda}(DR)' - \frac{R'}{\Lambda^2}(\nabla\Lambda)\right]' + (N^r)'\left(\frac{R'}{\Lambda}\right)'$$

$$X = N^{-3}\{\dots\} + N\{\dots\} + N^{-1}\left\{\left[\left(\frac{R}{\Lambda}\right)(DR)^2\right]''\right\}$$

$$+N^{-1}\left\{-\left(1-\left(\frac{R'}{\Lambda}\right)^2\right)(DR)(\nabla\Lambda) - \left(R\left(\frac{1}{\Lambda}\right)'\right)(DR)^2 -\right.$$

$$\left.2\left(\frac{R'}{\Lambda}\right)(DR)(DR)'\right\}$$

$$-2\left[\left(\frac{R}{\Lambda}\right)\left(\frac{R'}{\Lambda}\right)(DR)(\nabla\Lambda)\right]' + 2\Lambda R\left[D\left(\frac{R'}{\Lambda}\right)\right]^2 - 2R'(DR)D\left(\frac{R'}{\Lambda}\right)$$

$$-4R\left[(DR)D\left(\frac{R'}{\Lambda}\right)\right]' + 2RD\left(\frac{R'}{\Lambda}\right)\nabla\Lambda DD\left(\frac{R'}{\Lambda}\right) + 2R(DR)\left[D\left[\frac{R'}{\Lambda}\right]'\right] -$$

$$(N^r)'\left(\frac{R'}{\Lambda}\right)'$$

$$X = N^{-3}\{\dots\} + N\{\dots\}$$

$$+N^{-1}\left\{(DR)(\nabla\Lambda)\left[-\left(1-\left(\frac{R'}{\Lambda}\right)^2\right) - 2\left(\left(\frac{R}{\Lambda}\right)\left(\frac{R'}{\Lambda}\right)\right)'\right] - \left(R\left(\frac{1}{\Lambda}\right)\right)'(DR)^2\right\}$$

$$-2\left(\frac{R'}{\Lambda}\right)(DR)(DR)' - 2\left(\frac{R}{\Lambda}\right)\left(\frac{R'}{\Lambda}\right)[(DR)(\nabla\Lambda)]$$

$$+2\left(\frac{R}{\Lambda}\right)\left[(DR)' - \left(\frac{R'}{\Lambda}\right)(\nabla\Lambda)\right]^2 - 2(\mathrm{DR})\left(\frac{R'}{\Lambda}\right)\left[(DR)' - \left(\frac{R'}{\Lambda}\right)(\nabla\Lambda)\right]$$

$$-4\left(\frac{R'}{\Lambda}\right)(DR)'\left[(DR)'-\left(\frac{R'}{\Lambda}\right)(\nabla\Lambda)\right]-4R(DR)\left(\frac{1}{\Lambda}\right)'\left[(DR)'-\left(\frac{R'}{\Lambda}\right)(\nabla\Lambda)\right]$$
$$4\left(\frac{R'}{\Lambda}\right)(DR)\left[(DR)'-\left(\frac{R'}{\Lambda}\right)(\nabla\Lambda)\right]'+2\left(\frac{R}{\Lambda}\right)\left(\frac{R'}{\Lambda}\right)(\nabla\Lambda)\left[(DR)'-\right.$$
$$\left.\left(\frac{R'}{\Lambda}\right)(\nabla\Lambda)\right]$$
$$+2\left(\frac{R}{\Lambda}\right)(DR)'\left[(DR)'-\left(\frac{R'}{\Lambda}\right)(\nabla\Lambda)\right]'+\left[\left(\frac{R}{\Lambda}\right)''(DR)^2+\right.$$
$$\left.4\left(\frac{R'}{\Lambda}\right)(DR)(DR)'+2\left(\frac{R'}{\Lambda}\right)(DR)(DR)'\right]+2R(DR)\left(\frac{1}{\Lambda}\right)'\left[(DR)'-\right.$$
$$\left.\left(\frac{1}{\Lambda}\right)(\nabla\Lambda)\right]$$

$$X=N^{-3}\{\dots\}+N\{\dots\}$$
$$+N^{-1}\left\{(DR)(\nabla\Lambda)\left[-\left(1-\left(\frac{R'}{\Lambda}\right)^2\right)-2\left(\left(\frac{R'}{\Lambda}\right)\left(\frac{R'}{\Lambda}\right)\right)'+2\left(\frac{R'}{\Lambda}\right)^2+\right.\right.$$
$$\left.\left.4R\left(\frac{1}{\Lambda}\right)'\left(\frac{R'}{\Lambda}\right)-2R\left(\frac{1}{\Lambda}\right)'\right]\right\}$$
$$\left.+2\left(\frac{R'}{\Lambda}\right)\left(\frac{R'}{\Lambda}\right)'\right]$$
$$+(DR)^2\left[-R\left(\frac{1}{\Lambda}\right)'+\left(\frac{R}{\Lambda}\right)''\right]+(DR)(DR)'\left[-2\left(\frac{R'}{\Lambda}\right)'-2\left(\frac{R'}{\Lambda}\right)-\right.$$
$$\left.4R\left(\frac{1}{\Lambda}\right)'+2R\left(\frac{1}{\Lambda}\right)+4\left(\frac{R'}{\Lambda}\right)\right]$$
$$+(DR)'(\nabla\Lambda)\left[-2\left(\frac{R}{\Lambda}\right)\left(\frac{R'}{\Lambda}\right)-4\left(\frac{R}{\Lambda}\right)\left(\frac{R'}{\Lambda}\right)+4\left(\frac{R}{\Lambda}\right)\left(\frac{R'}{\Lambda}\right)+2\left(\frac{R}{\Lambda}\right)\left(\frac{R'}{\Lambda}\right)\right]$$
$$+(DR)(\nabla\Lambda)'\left[-2\left(\frac{R}{\Lambda}\right)\left(\frac{R'}{\Lambda}\right)+4\left(\frac{R}{\Lambda}\right)\left(\frac{R'}{\Lambda}\right)-2\left(\frac{R}{\Lambda}\right)\left(\frac{R'}{\Lambda}\right)\right]$$
$$+[(DR)]^2\left[2\left(\frac{R}{\Lambda}\right)-4\left(\frac{R}{\Lambda}\right)+2\left(\frac{R}{\Lambda}\right)\right]+(DR)(DR)''\left[-4\left(\frac{R}{\Lambda}\right)+2\left(\frac{R}{\Lambda}\right)+\right.$$
$$\left.2\left(\frac{R}{\Lambda}\right)\right]$$
$$+(\nabla\Lambda)^2\left[\left(\frac{R}{\Lambda}\right)\left(\frac{R'}{\Lambda}\right)^2-2\left(\frac{R}{\Lambda}\right)\left(\frac{R'}{\Lambda}\right)^2\right]$$

$$X=N^{-3}\{\dots\}+N\{\dots\}$$
$$+N^{-1}\left\{(DR)(\nabla\Lambda)\left[\begin{matrix}-1\left(\frac{R'}{\Lambda}\right)^2-2\left(\frac{R'}{\Lambda}\right)^2-2R\left(\frac{1}{\Lambda}\right)'\left(\frac{R'}{\Lambda}\right)+2\left(\frac{R'}{\Lambda}\right)^2-2\left(\frac{R'}{\Lambda}\right)\\+2R\left(\frac{1}{\Lambda}\right)'\left(\frac{R'}{\Lambda}\right)+2\left(\frac{R'}{\Lambda}\right)\left(\frac{R'}{\Lambda}\right)'\end{matrix}\right]\right\}$$
$$+(DR)^2\left[-\left(R\left(\frac{1}{\Lambda}\right)'\right)'+\left(R\left(\frac{1}{\Lambda}\right)'\right)'+\left(\frac{1}{\Lambda}\right)'\right]$$
$$+(DR)(DR)'\left[\left(\frac{R'}{\Lambda}\right)(-2-2+4)+R\left(\frac{1}{\Lambda}\right)'(-2-2+4)\right]$$

$$X = N^{-3}\{\dots\} + N\{\dots\} + N^{-1}\left\{(DR)(\nabla\Lambda)\left[-1+\left(\frac{R'}{\Lambda}\right)^2\right] + (DR)^2\left[\left(\frac{R'}{\Lambda}\right)'\right]\right\}$$

So,

$$X = N^{-3}\left\{-\frac{1}{3}(DR)^3(\nabla\Lambda)\right\} + N^{-1}\left\{-(DR)(\nabla\Lambda)\left[1-\left(\frac{R'}{\Lambda}\right)^2\right] + (DR)^2\left[\left(\frac{R'}{\Lambda}\right)'\right]\right\} + N\left\{\left(1-\left(\frac{R'}{\Lambda}\right)^2\right)\left(1-\left(\frac{R'}{\Lambda}\right)'\right)\right\}$$

$$\begin{aligned} S_\Sigma = (2\pi^2)\int dtdr\Bigg\{N\Bigg[\Lambda R\left(1-\left(\frac{R'}{\Lambda}\right)^2\right) - R^2\left(\frac{R'}{\Lambda}\right)' \\ + 2\lambda\left(1-\left(\frac{R'}{\Lambda}\right)^2\right)\left(-\left(\frac{R'}{\Lambda}\right)'\right)\Bigg] \\ + N^{-1}\Bigg[-(DR)(\nabla\Lambda)R^2 - (\Lambda R)(DR)^2 \\ + 2\lambda(DR)(\nabla\Lambda)\left(1-\left(\frac{R'}{\Lambda}\right)^2\right) + 2\lambda(DR)^2\left(\frac{R'}{\Lambda}\right)'\Bigg] \\ + N^{-3}\left[(2\lambda)\left(-\frac{1}{3}(DR)^3(\nabla\Lambda)\right)\right]\Bigg\} \end{aligned}$$

Let's write

$$S_\Sigma = (2\pi^2)\int dtdr\{L_{kin} + L_{pot.}\}; \quad \hat{\lambda} = \lambda(D-3)(D-4) = 2\lambda$$

$$N^{-1}L_{pot} = \Lambda R\left[1-\left(\frac{R'}{\Lambda}\right)^2\right] - \left(\frac{R'}{\Lambda}\right)'\left\{R^2 + \hat{\lambda}\left[1-\left(\frac{R'}{\Lambda}\right)^2\right]\right\}$$

$$NL_{kin} = -(D\Lambda)(DR)\left\{R^2 + \hat{\lambda}\left[1-\left(\frac{R'}{\Lambda}\right)^2 + \frac{(DR)^2}{3N^2}\right]\right\} - (DR)^2\left\{\Lambda R - \hat{\lambda}\left(\frac{R'}{\Lambda}\right)'\right\}$$

$DR = \dot{R} - N^r R'$
$\nabla\Lambda = \dot{\Lambda} - (N^r\Lambda)'$
$N^r \equiv S$

Exterior form calculation

Need to calculate Riemann Tensor for
$ds^2 = -N^2dt^2 + L^2(dr + N^r dt)^2 + R^2 d\Omega_3^2$

$d\Omega_3^2 = dx^2 + \sin^2 x\,(d\theta^2 + \sin^2\theta\, d\varphi^2)$
$0 \le x \le \pi$
$0 \le \theta \le \pi$
$0 \le \varphi \le 2\pi$
$\int (dx)(\sin x\, d\theta)(\sin x \sin\theta d\varphi) = 2\pi^2$
$(N^r = g_{rk}N^k = g_{rr}N^r = L^2 N^r)$

The method of exterior differential forms is employed. Choose basis 1-forms:
$w^0 = Ndt\,, w^1 = L(dr + N^r dt)\,, w^2 = Rd\,, w^3 = R\sin x d\theta\,, w^4 = R \sin x \sin\theta\, d\varphi$

Define $D \equiv \partial_t - N^r \partial_r$:
$w^0 = Ndt$
$w^1 = L(dr + N^r dt)$
$w^2 = Rdx$
$w^3 = R\sin x d\theta$
$w^4 = R\sin x \sin\theta d\varphi$

$dw^0 = N' dr \wedge dt = -N'(LN)^{-1} w^0 \wedge w^1$

$dw^1 = L' dr \wedge (dr + N^r dt) + L' dt \wedge (dr + N^r dt) + L(N^r)' dr \wedge dt$
$= (L' - N^r L') dt \wedge dr - (N^r) L dt \wedge dr$
$= (DL - (N^r)' L)(LN^{-1}) w^0 \wedge w^1$

$dw^2 = R' dr \wedge dx + \dot{R} dt \wedge dx = R'(dr + N^r dt) \wedge dx + \big(\dot{R} - N^r R'\big) dt \wedge dx$
$= (DR/(NR) w^0 \wedge w^0 + R'/(RL) w' \wedge w^2)$

$dw^3 = \big(R' dr \wedge d\theta + \dot{R} dt \wedge d\theta\big) \sin X + R\cos x dx \wedge d\theta$
$= (DR/(NR) w^0 \wedge w^3 + R'/(RL) w' \wedge w^3 + R^{-1} \cot x\, w^2 \wedge w^3)$

$dw^4 = \big(R' dr \wedge d\theta + \dot{R} dt \wedge d\theta\big) \sin x \sin\theta + R\cos x \sin\theta\, dx \wedge d\varphi + R \sin x \cos\theta d\theta \wedge d\varphi$
$= (DR/(NR) w^0 \wedge w^4 + R'/(RL) w' \wedge w^4 + R^{-1} \cot x\, w^2 \wedge w^4 + (R\sin x)^{-1} \cot\theta\, w^3 \wedge)$

$dw^\alpha = -w^\alpha{}_\beta \wedge w^\beta$
$w_1^0 = w_0^1 = N'(LN)^{-1} w^0 + (DL - (N^r)' L)(LN)^{-1} w^1$
$w_0^2 = w_2^0 = (DR)/(NR)\, w^2$

$w_0^3 = w_3^0 = (DR)/(NR)\, w^3$
$w_0^4 = w_4^0 = (DR)/(NR)\, w^4$
$w_1^2 = -w_2^1 = R'(RL)^{-1} w^2$
$w_1^3 = -w_3^1 = R'(RL)^{-1} w^3$
$w_1^4 = -w_4^1 = R'(RL)^{-1} w^4$
$w_2^3 = -w_3^2 = R^{-1} \cot x\, w^3$
$w_2^4 = -w_4^2 = R^{-1} \cot x\, w^4$
$w_3^4 = -w_4^3 = (R \sin x)^{-1} \cot \theta\, w^4$

$R_v^\mu = dw_v^\mu + w_\alpha^\mu \wedge w_v^\alpha$
$R_1^0 = dw^0{}_1 + w_\alpha^0 \wedge w_v^\alpha$
$w_1^0 = N'^{L^{-1}} dt + (L' - (LN^r)')N^{-1}(dr + N^r dt)$
$dw_1^0 = \left[\frac{N^r}{L} + \frac{N^r}{N}(L' - (LN^r)')\right]' dr \wedge dr + [N^{-1}(L' - (LN^r)')]^{\cdot} dt \wedge dr$

$\boxed{R_1^0} = (LN)^{-1} \left[-\left(\frac{N^r}{L} + \frac{N^r}{N}(L' - (LN^r)')\right)' + \left(N^{-1}(L' - (LN^r)')\right)^{\cdot}\right] w^0 \wedge w^1$

$R_2^0 = dw_2^0 + w_\alpha^0 \wedge w_2^\alpha = dw_2^\alpha + w_1^0 \wedge w_2^1$
$w_2^0 = (N^{-1}(DR)d\theta)$
$dw_2^0 = (DR)d(N^{-1}) \wedge d\theta + N^{-1} d\left(\dot{R} - N^r R'\right) \wedge d\theta$
$= -N^{-1}(DR)\left(\frac{N'}{N}\right) dr \wedge d\theta + N^{-1}(DR)' dr \wedge d\theta$
$= -N^{-1}(DR)\left(\frac{\dot{N}}{N}\right) dr \wedge d\theta + N^{-1}(DR)^{\cdot} dr \wedge d\theta$

$dr = L^{-1} w' - N^r dt\, , d\theta = R^{-1} w^2$
$dw_2^0 = (N^2 R)^{-1}\left[(DR)^{\cdot} - N^r(DR') - N^{-1}(DR)\left(\dot{N} - N^r N'\right)\right] w^0 \wedge w^2$
$+(LNR)^{-1}\left[-(DR)\left(\frac{N'}{N}\right) + (DR)'\right] w' \wedge w^2$
$= (N^2 R)^{-1}[D^2 R - N^{-1}(DR)(DN)] w^0 \wedge w^2 + (LNR)^{-1}\left[(DR)' - (DR)\left(\frac{N'}{N}\right)\right] w^1 \wedge w^2$

$w_1^0 \wedge w_2^1 = -(RNL^2)^{-1}\left[R' N'^{w^0} \wedge w^2 + R'(DL - (N^r)'L) W^1 \wedge w^2\right]$
$\boxed{R_2^0}_* = (NR)^{-1}[N^{-1} D^2 R - N^{-2}(DR)(DN) - L^{-2} R' N'] w^0 \wedge w^2$
$+(LNR)^{-1}\left[(DR)' - \left(\frac{N'}{N}\right)(DR) - L^{-1} R'(DL - (N^r)L)\right] w^1 \wedge w^2$

$R^0_3 = dw^0_3 + w^0_1 \wedge w^1_3 + w^0_2 \wedge w^2_5$
$w^0_3 = \left(\frac{DR}{N}\right) \sin x d\theta$
$dw^0_3 = \left(\frac{DR}{N}\right)^{\cdot} \sin x dt \wedge d\theta + \left(\frac{DR}{N}\right)' \sin x dr \wedge d\theta + \left(\frac{DR}{N}\right) \cos x dx \wedge d\theta$
$= (NR)^{-1} D \left(\frac{DR}{N}\right) w^0 \wedge w^3 + (LR)^{-1} \left(\frac{DR}{N}\right)' w' \wedge w^3 +$
$R^{-2} \left(\frac{DR}{N}\right) \cot x w^2 \wedge w^3$
$w^0_1 \wedge w^1_3 = \{w^0_1 \wedge w^1_3 | w^2 \to w^3\}$
$= -(RNL^2)^{-1}\left[R' N'^{w^0} \wedge w^3 + R'^{(DL-(N^r)^1 L)} w' \wedge w^3\right]$
$w^0_2 \wedge w^2_3 = -(NR^2)^{-1}(DR) \cot \theta \, w^2 \wedge w^3$
$\boxed{R^0_3}_* = \{R^0_2 | w^2 \to w^3\}$ with the $w^2 \wedge w^3$ term cancelling.

$R^0_4 = dw^0_4 + w^0{}_1 \wedge w^1_4 + w^0_2 \wedge w^2_4 + w^0_3 \wedge w^3_4$
$\boxed{R^0_4}_* = \{R^0_2 | w^2 \to w^4\}$ with the $w^2 \wedge w^3$ and $w^3 \wedge w^4$ terms cancelling.

$R^1_2 = dw^1_2 + w^1_0 \wedge w^0_2$
$w^1_2 = -L^{-1} R' dx$
$dw^1_2 = \left(\frac{R'}{L}\right)' dr \wedge dx + \left(-\frac{R'}{L}\right)^{\cdot} dt \wedge dx$
$= (RN)^{-1} D \left(-\frac{R'}{L}\right) w^0 \wedge w^2 + (RL)^{-1} \left(-\frac{R'}{L}\right)' w' \wedge w^2$
$w^1_0 \wedge w^0_2 = (LN^2R)^{-1}\left[N'^{(DR) w^0} \wedge w^2 + (DL - (N^r) L (DR) w^1 \wedge w^2)\right]$
$\boxed{R^1_2}_* = (RN)^{-1} \left[(LN)^{-1} N'^{(DR)} - D\left(-\frac{R'}{L}\right)\right] w^2 \wedge w^2$
$+ (RL)^{-1} \left[N^{-2}(DL - (N^r)')(DR) - \left(-\frac{R'}{L}\right)'\right] w' \wedge w^2$

$R^1_3 = dw^1_3 + w^1_0 \wedge w^0_3 + w^1_2 \wedge w^2_3$
$w^1_3 = -L^{-1} R' \sin x d\theta$
$dw^1_3 = (RN)^{-1} D \left(-\frac{R'}{L}\right) w^0 \wedge w^3 + (RL)^{-1} \left(-\frac{R'}{L}\right)^{-1} w' \wedge w^3 -$
$L^{-1} R' \cot x \, R^{-2} w^2 \wedge w^3$
$w^1_0 w^0_3 = (LN^2R)^{-1}\left[N'^{(DR) w^0} \wedge w^3 + (DL - (N^r) L)(DR) w^1 \wedge w^2\right]$
$w^1_2 \wedge w^2_3 = (R^2 L)^{-1} R' \cot x \, w^2 \wedge w^3$
$\boxed{R^1_3}_* = \{R^1_2 | w^2 \to w^3\}$ with the $w^2 \wedge w^3$ term cancelling.

$\boxed{R^1_4}_* = \{R^1_2 | w^2 \to w^4\}$ with the $w^2 \wedge w^3$ and $w^3 \wedge w^4$ terms cancelling.

$R^2_3 = dw^2_3 + w^2_0 \wedge w^0_3 + w^2 \wedge w^1_3$

$w_3^2 = -\cos x d\theta$
$dw_3^2 = \sin x dx \wedge d\theta = R^{-2} w^2 \wedge w^3$
$w^2{}_0 \wedge w_3^0 + w^2{}_1 \wedge w_3^1 = (NR)^{-2}(DR)^2 w^2 \wedge w^3 - (RL)^{-2}(R')^2 w^2 \wedge w^3$
$\boxed{R_3^2}_* = R^{-2}\{1 + N^{-2}(DR)^2 - L^{-2}(R')^2 w^2 \wedge w^3\}$

$R_4^2 = dw_4^2 + w_0^2 \wedge w_4^0 + w_1^2 \wedge w_4^1 + w_3^2 \wedge w_4^3$
$w_4^2 = -\cos x \sin\theta d\varphi$
$dw_4^3 = R^{-2} w^2 \wedge w^4 - (R^2 \sin x)^{-1} \cot x \cot\theta\, w^3 \wedge w^4$
$\boxed{R_4^2} = \{R_3^2 | w^3 \to w^4\}$ with the $w^3 \wedge w^4$ term cancelling.

$R_4^3 = dw_4^3 + w_0^3 \wedge w_4^0 + w_1^3 \wedge w_4^1 + w_2^3 \wedge w_4^2$
$w_4^3 = -\cos\theta d\varphi$
$dw_4^3 = \sin\theta d\theta \wedge d\varphi = (R \sin x)^{-2} w^3 \wedge w^4$
$+w_1^3 + w_2^3 \wedge w_4^2 = (NR)^{-2}(DR)^2 w^3 \wedge w^4 - (RZ)^{-2}(R')^2 w^3 \wedge w^4 -$
$R^{-2} \cot^2 x\, w^3 \wedge w^{\square}$
$\boxed{R_4^3} = \{R_3^2 | w^2 \wedge w^3 \to w^3 \wedge w^4\}$

$\boldsymbol{R_v^\mu = R_{v|\alpha\beta|}^\mu w^\alpha \wedge w^\beta}$ **:**
$R_{101}^0 = (LN)^{-1}\left[\left(N^{-1}\left(\dot{L} - (LN^r)\right)\right)^{\cdot} - \left(\frac{N\prime}{L} + \frac{\prime\prime^r}{L}\left(\dot{L} - (LN^r)'\right)\right)\right]$
$R_{202}^0 = R_{303}^0 = R_{404}^0 = (NR)^{-1}\left[\frac{D^2R}{N} - \frac{(DR)(DN)}{N^2} - \frac{R\prime N\prime}{L^2}\right]$
$R_{212}^0 = R_{313}^0 = R_{414}^0 = (LNR)^{-1}\left[(DR)' - \left(\frac{N'}{N}\right)(DR) - L^{-1}R'(DL -\right.$
$\left.(N^r)'L)\right]$
$R_{202}^1 = R_{303}^1 = R_{404}^1 = (RN)^{-1}\left[(LN)^{-1}N'^{(DR)} - D\left(\frac{R\prime}{L}\right)\right]$
$R_{202}^1 = (LNR)^{-1}\left[\left(\frac{N'}{N}\right)(DR) - DR' + (N)'R' - (N)'R' + R'\frac{DL}{L}\right]$
$= -(LNR)^{-1}\left[(DR)' - \left(\frac{N'}{N}\right)(DR) - L^{-1}R'(DL) - (N^r)'L\right]$
$R_{212}^1 = R_{313}^1 = R_{414}^1 = (RL)^{-1}\left[N^{-2}(DR)(DL - (N^r)'L) - \left(\frac{R'}{L}\right)'\right]$
$R_{323}^2 = R_{424}^2 = R_{434}^3 = R^{-2}[1 + N^{-2}(DR)^2 - L^{-2}(R')^2]$

$\boldsymbol{R_{\mu v} = R_{\mu\alpha v}^\alpha;\ \left(R_{\alpha\beta\gamma\delta} = R_{[\alpha\beta][\gamma\delta]} = R_{[\gamma\delta][\alpha\beta]}\right)}$ **:**
$R_{00} = R_{010}^1 + 3R_{020}^2 = -(R_{101}^0 + 3RR_{202}^0)$
$R_{01} = 3R_{021}^2 = -3R_{212}^0$
$R_{11} = R_{101}^0 + 3R_{121}^2 = R_{101}^0 + 3R_{212}^1$
$R_{22} = R_{202}^0 + R_{212}^1 + 2R_{232}^3 = R_{202}^0 + R_{212}^1 + 2R_{323}^2$

$R_{33} = R_{22}$
$R_{44} = R_{22}$
$R = g^{\mu\nu}R_{\mu\nu} = -(R^0_{101} + 3R^0_{202}) + R^0_{101} + 3R^1_{212} + 3(R^0_{202} + R^1_{212} + 2R^2_{323})$

$$R = 6(R^1_{212} + R^2_{323})$$
$$= 6\left[(RL)^{-1}\left(N^{-2}(DR)(DL - (N^r)'L) - \left(\frac{R'}{L}\right)'\right) + R^{-2}(1 + N^{-2}(DR)^2 - L^{-2}(R')^2)\right]$$
$$= 6\left[\frac{(\dot{R}-N^rR')(\dot{L}-N^rL'-(N^r)'L)}{(LRN^2)} - \frac{\left(\frac{R'}{L}\right)'}{(RL)} + \frac{1}{R^2} + \frac{(\dot{R}-N^rR')^2}{(NR)^2} - \frac{(R')^2}{(LR)^2}\right]$$
$$= \left\{6R^{-1} - 6\left(\frac{R'}{L}\right)^2 L^{-2} + 6L^{-2}\left(\frac{R'}{L}\right)^2\left(\frac{1'}{L}\right) - 6L^{-2}\left(\frac{R''}{R}\right)\right\}$$
$$+6\left[\frac{(\dot{R}-N^rR')(\dot{L}-N^rL'-(N^r)'L)}{(LRN^2)} + \frac{(\dot{R}-N^rR')^2}{(NR)^2}\right]$$

This result checks with the same derivation using a 4+1 split and the geodetic path method.

For no "t" dependence, $N^r = 0\,, i = 2,3,4$, and $N = f,\ L = h^{-1}\,, R = r$:
$R^0_{101} = (LN)^{-1}\left(-\frac{N'}{L}\right)' = (h^{-1}f)^{-1}(= f'h)' = -f'h(f'h)'$
$R^0_{i0j} = (NR)^{-1}\left[-\frac{R'N'}{L^2}\right]\delta_{ij} = -r^{-1}(f^{-1}h^2f')\delta_{ij}$
$R^1_{i1j} = -(RL)^{-1}\left(\frac{R'}{L}\right)'\delta_{ij} = -\frac{h}{r}h'\delta_{ij}$
$R^1_{jkm} = r^{-2}(1 - h^2)(\delta^i_k\delta_{ijm} - \delta^i_m\delta_{jk})$
This checks with Wheeler's results [95].

$\mathcal{L}_1 = \frac{1}{2}R$
$\mathcal{L}_2 = \frac{1}{4}(R_{abcd}R^{abcd} - 4R_{ab}R^{ab} + R^2)$
No derivative appears at higher than second order.

Now to calculate the contributions to $\mathcal{L}_2$:
$R = 6(R_{1212} + R_{2323})$
$R^2 = 36((R_{1212})^2 + 2(R_{1212})(R_{2323}) + (R_{2323})^2)$
$R_{ab}R^{ab} = (R_{00})^2 - 2(R_{01})^2 + (R_{11})^2 + 3(R_{22})^2$
$= (R^0_{101} + 3R^0_{202})^2 - 2(-3R^0_{212})^2 + (R^0_{101} + 3R^1_{212})^2 + 3(R^0_{202} + R^1_{212} + 2R^2_{323})^2$

$= 2(R^0_{101})^2 + 6R^0_{101}R^0_{202} + 12(R^0_{202})^2 - 18(R^0_{212})^2 + 6R^0_{101}R^1_{212}$
$+12(R^1_{212})^2 + 12(R^2_{323})^2 + 6R^0_{202}R^0_{212} + 12R^0_{202}R^0_{323} + 12R^1_{212}R^2_{323}$

$R_{ab}R^{ab} = R_{0101}R^{0101} + R_{0110}R^{0110} + R_{1010}R^{1010} + R_{1001}R^{1001} \rightarrow$
$-4(R_{0101})^2$
$+3R_{0202}R^{0202} \ldots. \qquad \rightarrow -12(R_{0202})^2$
$+3(R_{0212}R^{0212} + R_{0221}R^{0221} + R_{2021}R^{2021} + R_{2012}R^{2012} + \cdots) \rightarrow$
$-24(R_{0212})^2$
$+3(R_{1212}R^{1212} + \cdots \qquad) \rightarrow \; 12(R_{1212})^2$
$+3(R_{2323}R^{2323} + \cdots \qquad) \rightarrow \; 12(R_{2323})^2$
$= -4(R_{0101})^2 - 12(R_{0202})^2 - 24(R_{0212})^2 + 12(R_{1212})^2 + 12(R_{2323})^2$
$R_{ab}R^{ab} - 4R_{ab}R^{ab} + R^2 =$
$(R_{0101})^2[-4 - 8]$
$+(R_{0101}R_{0202})[-24]$
$+(R_{0101}R_{1212})[24]$
$+(R_{0202})^2[-12 - 48]$
$+(R_{0202}R_{1212})[24]$
$+(R_{0202}R_{2323})[48]$
$+(R_{0212})^2[-24 + 72]$
$+(R_{1212})^2[12 - 48 + 36]$
$+(R_{1212}R_{2323})[-48 + 72]$
$+(R_{2323})^2[12 - 48 + 36]$
$= -12(R_{0101})^2 - 24(R_{0101}R_{0202}) + 24(R_{0101}R_{1212}) - 60(R_{0202})^2 + 24(R_{0202}R_{1212}) + 48(R_{0202}R_{2323}) + 48(R_{0212})^2 + 24(R_{1212}R_{2323})$
The $(R_{0101})^2$ term alone presents serious difficulties. How to deal with the $\{[\dot{L} - (LN^r)']^{\cdot}\}^2$ term? Eventually a Hamiltonian formulation would be obtained, but in practice the computation becomes virtually impossible. It is necessary to work with the Hamiltonian formulation directly. The Hamiltonian form given by [98,99] is essential.

In 4-D spherical symmetric theory it was possible to express $L = L + \partial_\mu \alpha^\mu$. The Legendre transformation of the Lagrangian to Hamiltonian form is then straightforward. This is not a practical approach in the $L_1 + L_2$ theory.

Geodesic path calculation
Evaluation of ${}^{(5)}R$ using 4+1 split and geodesic path method:

$${}^{(5)}R = {}^{(4)}R - K_{ab}K^{ab} + (K^a_a)^2 + D$$

where D are (tensorial) divergences.

$$K_{ab} = -\frac{1}{2}N^{-1}\left[\dot{h}_{ab} - D_a N_b - D_b N_a\right]$$
$$ds^2 = -N^2 dt^2 + L^2(dr + N^r dt)^2 + R^2 d\Omega_3^2 + R^2 d\Omega_3^2$$

In the $\{t, r, \theta, \varphi, \varphi\}$ coordinate frame:

$$\begin{pmatrix} {}^{(5)}g_{00} & {}^{(5)}g_{0k} \\ {}^{(5)}g_{i0} & {}^{(5)}g_{ik} \end{pmatrix} = \begin{pmatrix} (N_r N^r - N^2) & N^r & 0 \\ N^r & & \\ 0 & & \left({}^{(4)}g_{ik} \right) \end{pmatrix}$$

$$\begin{pmatrix} {}^{(5)}g^{00} & {}^{(5)}g^{0m} \\ {}^{(5)}g^{ko} & {}^{(5)}g^{km} \end{pmatrix} = \begin{pmatrix} -{}^{1}/_{N^2} & {}^{N_r}/_{N^2} & 0 & 0 \\ {}^{N_r}/_{N^2} & \left({}^{(5)}g^{rr} - \left(\frac{N_r}{N}\right)^2 \right) & 0 & 0 \\ 0 & 0 & \left({}^{(4)}g^{km} \right) & \end{pmatrix}$$

${}^{(4)}R$ is based on ${}^{(4)}g_{ik} = g_{ik} : dl^2 = L^2 dr^2 + R^2 d\Omega_3^2$ and ${}^{(4)}g \to {}^{(4)}\Gamma \to {}^{(4)}R$ is the calculation, with the Γ'^s evaluated by extremal path method.

Extremal path method

$\left(dl^2 = L^2 dr^2 + R^2 d\Omega_3^2 \ , \ddot{x}^\mu + \Gamma^\mu{}_{\alpha\beta}\dot{x}^\alpha \dot{x}^\beta = 0\right)$

$I =$

$$\int \frac{1}{2}\left(\frac{dl}{d\lambda}\right)^2 d\lambda = \int \frac{1}{2} \underbrace{\left\{ L^2 \left(\frac{dr}{d\lambda}\right)^2 + R^2 \left(\left(\frac{dx}{d\lambda}\right) + \sin^2 x \left(\frac{d\theta}{dN}\right) + \sin^2 \theta \left(\frac{d\theta}{dN}\right) \right) \right\}}_{\left(\frac{d\Omega_2}{d\lambda}\right)^2} d\lambda$$

$$(\delta I)_r = \int \frac{1}{2}\left\{ 2LL'\left(\frac{dr}{d\lambda}\right)^2 dr + 2L\left(\frac{dr}{d\lambda}\right)\frac{d}{d\lambda}(\delta r) + 2RR'\left(\frac{d\Omega}{d\lambda}\right)^2 \delta r \right\} d\lambda$$

$$= \int \left\{ LL'\left(\frac{dr}{d\lambda}\right)^2 - \frac{d}{d\lambda}\left(L^2\left(\frac{dr}{d\lambda}\right)\right) + RR'\left(\frac{d\Omega_3}{d\lambda}\right) \right\} \delta r d\lambda$$

$$0 = -L^2\left(\frac{d^2 r}{d\lambda^2}\right) - 2L\left(\frac{dL}{d\lambda}\right)\left(\frac{dr}{d\lambda}\right) + LL'\left(\frac{dr}{d\lambda}\right)^2 + RR'\left(\left(\frac{dx}{d\lambda}\right)^2 + \right.$$
$$\left. \sin^2 x \left(\frac{dx}{d\lambda}\right)^2 + \sin^2\theta\left(\frac{d\varphi}{d\lambda}\right)\right)$$

$$\Gamma^r{}_{rr} = \left(\frac{L\prime}{L}\right)$$
$$\Gamma^r{}_{tr} = \left(\frac{\dot{L}}{L}\right)$$
$$\Gamma^r{}_{xx} = -\left(\frac{RR'}{L^2}\right)\sin^2 x$$
$$\Gamma^r{}_{\theta\theta} = -\left(\frac{RR'}{L^2}\right)\sin^2 x$$
$$\Gamma^r{}_{\varphi\varphi} = -\left(\frac{RR'}{L^2}\right)\sin^2 x \sin^2\theta$$

$$(\delta I)_x = \int \left\{ R^2 \left(\frac{dx}{d\lambda}\right) \frac{d}{d\lambda}(\delta x) + R^2 \sin x \cos x \left(\left(\frac{d\theta}{d\lambda}\right) + \sin^2\theta \left(\frac{d\varphi}{d\lambda}\right)^2 \right)^2 \delta x \right\} d\lambda$$

$$-\frac{d}{d\lambda}\left(R^2 \left(\frac{dx}{d\lambda}\right)\right) + R^2 \sin x \cos x \left(\left(\frac{d\theta}{d\lambda}\right)^2 + \sin^2\theta \left(\frac{d\varphi}{d\lambda}\right)^2 \right) = 0$$

$$-\left(\frac{d^2x}{d\lambda^2}\right) - 2R^{-1}\left(R'^{\left(\frac{dr}{d\lambda}\right)} + \dot{R}\left(\frac{dt}{d\lambda}\right)\right)\left(\frac{dx}{d\lambda}\right) + \sin x \cos x \left(\left(\frac{d\theta}{d\lambda}\right)^2 + \sin^2\theta \left(\frac{d\theta}{d\lambda}\right)^2 \right) = 0$$

$$\Gamma^x_{\theta\theta} = -\cos x \sin x$$

$$\Gamma^x_{\varphi\varphi} = -\cos x \sin x \sin^2\theta$$

$$\Gamma^x_{rx} = \left(\frac{R'}{R}\right)$$

$$(\delta I)_\theta = \int R^2 \sin^2 x \left\{ \left(\frac{d\theta}{d\lambda}\right) \frac{d}{d\lambda}(\delta\theta) + \sin\theta\cos\theta \left(\frac{d\varphi}{d\lambda}\right)^2 \delta\theta \right\} d\lambda$$

$$\left(\frac{d^2\theta}{d\lambda^2}\right) + 2R^{-1}\left(R'\left(\frac{dr}{d\lambda}\right) + \dot{R}\left(\frac{dt}{d\lambda}\right)\right)\left(\frac{d\theta}{d\lambda}\right) + 2\cot x \left(\frac{dx}{d\lambda}\right)\left(\frac{d\theta}{d\lambda}\right) - \sin\theta\cos\theta\left(\frac{d\varphi}{d\lambda}\right)^2 = 0$$

$$\Gamma^\theta_{x\theta} = \cot x$$

$$\Gamma^\theta_{r\theta} = \left(\frac{R'}{R}\right)$$

$$\Gamma^\theta_{\varphi\varphi} = -\sin\theta\cos\theta$$

$$(\delta I)_\theta = \int R^2 \sin^2 x \sin^2\theta \left(\frac{d\varphi}{d\lambda}\right) \frac{d}{d\lambda}(\delta\varphi) d\lambda$$

$$\frac{d}{d\lambda}\left\{R^2 \sin^2 x \sin^2\theta \left(\frac{d\varphi}{d\lambda}\right)\right\} = 0$$

$$\frac{d^2\varphi}{d\lambda} + 2R^{-1}\left(R'^{\left(\frac{\partial r}{\partial\lambda}\right)} + \dot{R}\left(\frac{\partial t}{\partial\lambda}\right)\right)\left(\frac{d\varphi}{d\lambda}\right) + 2\cot x\left(\frac{dx}{d\lambda}\right)\left(\frac{d\varphi}{d\lambda}\right) + 2\cot\theta\left(\frac{d\theta}{d\lambda}\right)\left(\frac{d\varphi}{d\lambda}\right) = 0$$

$$\Gamma^\varphi_{r\varphi} = \left(\frac{R'}{R}\right)$$

$$\Gamma^\varphi_{\varphi\varphi} = \cot\theta$$

$$\Gamma^\varphi_{x\varphi} = \cot x$$

So,

$$^{(4)}R = g^{ik}\left[\Gamma^{j}_{ik,j} - \Gamma^{j}_{ijk} + \Gamma^{j}_{ik}\Gamma^{l}_{jl} - \Gamma^{j}_{il}\Gamma^{l}_{kj}\right]$$

$$\Gamma^{r}_{rr} = \left(\frac{L\prime}{L}\right)$$

$$\Gamma^{r}_{xx} = -\left(\frac{RR'}{L^2}\right) = \sin^2 x\, \Gamma^{r}_{\theta\theta} = \sin^{-2} x \sin^{-2}\theta\, \Gamma^{r}_{\varphi\varphi}$$

$$\Gamma^{x}_{rx} = \Gamma^{\theta}_{r\theta} = \Gamma^{\varphi}_{r\varphi} = \left(\frac{R\prime}{R}\right)$$

$$\Gamma^{x}_{\varphi\varphi} = -\cos x \sin x \sin^2\theta$$

$$\Gamma^{x}_{\theta\theta} = -\cos x \sin x$$

$$\Gamma^{\theta}_{\varphi\varphi} = -\cos\theta \sin\theta$$

$$\Gamma^{\varphi}_{\theta\varphi} = \cot\theta$$

$$\Gamma^{\varphi}_{x\varphi} = \Gamma^{\theta}_{x\theta} = \cot x$$

$$^{(4)}R = \underbrace{g^{ik}\left[\Gamma^{ij}_{il}\Gamma^{l}_{kj} - \Gamma^{j}_{ik}\Gamma^{l}_{jl}\right]}_{4G_L} + \underbrace{\frac{1}{\sqrt{4g}}\left\{\sqrt{4g}\left(g^{ik}\Gamma^{l}_{ik} - g^{jl}\Gamma^{k}_{jk}\right)\right\}_{,l}}_{\frac{1}{\sqrt{4g}}\left(\sqrt{4g}w^l\right)_{,l}}$$

$$4G_L = L^{-2}\left[3\left(\frac{R\prime}{R}\right)^2 - 3\left(\frac{R\prime}{R}\right)\left(\frac{L\prime}{L}\right)\right] + R^{-2}\left[2\Gamma^{x}_{xx}\Gamma^{x}_{rx} + \left(\Gamma^{\varphi}_{x\theta}\right)^2 -\right.$$

$$\left.\Gamma^{r}_{xx}\left(\Gamma^{r}_{rr} + 3\Gamma^{x}_{rx}\right)\right]$$

$$+R^{-2}\sin^{-2} x\left[2\Gamma^{\theta}_{\theta\theta}\Gamma^{\theta}_{r\theta} + 2\Gamma^{\theta}_{\theta\theta}\Gamma^{\theta}_{x\theta} + \left(\Gamma^{\varphi}_{\theta\theta}\right)^2 - \Gamma^{r}_{\theta\theta}\left(\Gamma^{r}_{rr} + 3\Gamma^{x}_{rx}\right)\right]$$

$$-\Gamma^{r}_{\theta\theta}\left(\Gamma^{\theta}_{x\theta} + \Gamma^{\varphi}_{x\varphi}\right)$$

$$+R^2\sin^{-2} x \sin^{-2}\theta\left[2\Gamma^{r}_{\varphi\varphi}\Gamma^{\varphi}_{r\varphi} + 2\Gamma^{\theta}_{\varphi\varphi}\Gamma^{\varphi}_{\theta\varphi} - \Gamma^{r}_{\varphi\varphi}\left(\Gamma^{r}_{rr} + 3\Gamma^{x}_{rx}\right) -\right.$$

$$\left.\Gamma^{\theta}_{\varphi\varphi}\left(\Gamma^{\varphi}_{\theta\varphi}\right)\right]$$

$$-\Gamma^{x}_{\varphi\varphi}\left(\Gamma^{\theta}_{x\theta} + \Gamma^{\varphi}_{x\varphi}\right)$$

$$= 3\left(\frac{R\prime}{R}\right)L^{-2}\left(\frac{R\prime}{R} - \frac{L\prime}{L}\right) + R^{-2}\left[\cot^2 x + \cot^2 x + \left(\frac{RR\prime}{L^2}\right)\left(\frac{L\prime}{L}\right) + \left(\frac{R\prime}{R}\right)\right]$$

$$= 6L^{-2}\left(\frac{R\prime}{R}\right)^2 + 2R^2\cot^2 x$$

$$\sqrt{^4g}w^r = \left(LR^3\sin^2 x \sin\theta\right)\left(g^{ik}\Gamma^{r}_{ik} - g^{jr}\Gamma^{k}_{jk}\right)$$

$$= \left(LR^3\sin^2 x \sin^2\theta\right)\left(g^{rr}\Gamma^{r}_{rr} + 3g^{xx}\Gamma^{r}_{xx} - g^{rr}\left(\Gamma^{r}_{rr} + 3\Gamma^{x}_{rx}\right)\right)$$

$$= \left(LR^3\sin^2 x \sin^2\theta\right)\left(L^{-2}\left(\frac{L'}{L}\right) + 3R^{-2}\left(-\frac{RR'}{L^2}\right) - L^{-2}\left(\frac{L'}{L}\right) + 3\left(\frac{R'}{R}\right)\right)$$

$$= 6R^2R'^{L^{-1}}\sin^2 x\, sin\,\theta$$

$\left(\sqrt{{}^4g}w^r\right)_{,r} = -6L^{-1}\sin^2 x\sin\theta\,[2R(R')^2 + R^2R''] +$
$6R^2R'^{L^{-2}}L'\sin^2 x\sin^2\theta$

$\sqrt{{}^4g}w^x = (LR^3\sin^2 x\sin^2\theta)\left(2g^{\theta\theta}\Gamma^x_{\theta\theta} - g^{xx}(2\Gamma^\theta_{x\theta})\right)$
$= (LR^3\sin^2 x\sin^2\theta)\left(2R^{-2}\sin^2 x\,(\cos x\sin x) - R^{-2}(\cot x)\right)$
$= -4LR(\cos x\sin x\sin\theta)$

$\left(\sqrt{{}^4g}w^x\right)_{,x} = 4LR(\sin^2 x\sin\theta - \cos^2 x\sin\theta)$

$\sqrt{{}^4g}w^\theta = (LR^3\sin^2 x\sin^2\theta)\left(g^{\varphi\varphi}\Gamma^\theta_{\varphi\varphi} - g^{\varphi\varphi}\Gamma^\varphi_{\varphi\varphi}\right)$
$= (LR^3\sin^2 x\sin^2\theta)(-2R^{-2}\sin^{-2} x\cot\theta)$
$= -2LR\cos\theta$

$\left(\sqrt{{}^4g}w^\theta\right)_{,\theta} = 2LR\sin\theta$

$\left(\sqrt{{}^4g}w^\varphi\right)_{,\varphi} = 0$

$\frac{1}{\sqrt{{}^4g}}\left(\sqrt{{}^4g}w^l\right)_{,l} = R^{-2}(4 - 4\cot^2 x + 2\sin^{-2} x) - 6L^{-2}\left(\frac{R''}{R} + 2\left(\frac{R'}{R}\right)^2 - \left(\frac{R'}{R}\right)\left(\frac{L'}{L}\right)\right)$

$= R^{-2}(6 - 2\cot^2 x) - 6L^{-2}\left(\frac{R''}{R} + 2\left(\frac{R'}{R}\right)^2 - \left(\frac{R'}{R}\right)\left(\frac{L'}{L}\right)\right)$

${}^{(4)}R = 4G_L + \frac{1}{\sqrt{{}^4g}}\left(\sqrt{4g}w^l\right)_{,l} = -6L^{-2}\left(\frac{R'}{R}\right)^2 + 6R^{-2} - 6L^2\left(\frac{R''}{R}\right) +$
$6L^{-2}\left(\frac{R'}{R}\right)\left(\frac{L'}{L}\right)$

$$\boxed{{}^{(4)}R = 6R^{-2} - 6\left(\frac{R'}{R}\right)^2 L^{-2} + 6L^{-2}\left(\frac{R'}{R}\right)\left(\frac{L'}{L}\right) - 6L^{-2}\left(\frac{R''}{R}\right)}$$

This approach will be incomplete when attention turns to $R_{ab}R^{ab}, R_{abcd}R^{abcd}$, since need to know all the metrical components (for this the method of exterior differential forms will be employed). However, the result of this approach provides a good check for ${}^5R$:

$K_{ab} = -\frac{1}{2}N^{-1}\left[\dot{L}_{ab} - D_aN_b - D_bN_a\right]$
$N_{i|k} = \frac{\partial N_i}{\partial x^k} - {}^4\Gamma^m{}_{ik}N_m\ , N_m = 4gmlN^l$

$$K_N = \frac{1}{2N}(-(2LL') + 2[(N_r)_r - {}^4\Gamma^r_{rr}N_r])$$
$$= \frac{1}{2N}\left(-(2LL') + 2\left[(L^2N^r)_r - \left(\frac{L\prime}{L}\right)L^2N^r\right]\right)$$
$$= -N^{-1}L\big(\dot{L} - (LN^r)'\big)$$
$$K_{xx} = (2N)^{-1}\big(-(2R\dot{R}) - 2\Gamma^r_{xx}N_r\big) = (2N)^{-1}\Big(-2R\dot{R} -$$
$$2\left(-\frac{RR\prime}{L^2}\right)(L^2N^r)\Big)$$
$$= -N^{-1}R\big(\dot{R} - R'N^r\big)$$
$$K_{\theta\theta} = \sin^2 x\, K_{xx}$$
$$K_{\varphi\varphi} = \sin^2 x \sin^2\theta\, K_{xx}$$

$$K^{ab}K_{ab} - K^2 = -2\big[K^r_r\big(K^x_x + K^\theta_\theta + K^\varphi_\varphi\big) + K^x_x\big(K^\theta_\theta + K^\varphi_\varphi\big)K^\theta_\theta + K^\varphi_\varphi\big]$$
$$= -2R^{-2}[3K_{rr}K_{xx} + 3R^{-2}(K_{xx})^2]$$
$$= -6R^{-2}\left[\left(\frac{DR}{N^2}\right)\big(L - (LN^r)\big)\big(\dot{R} - R'N^r\big) + N^{-2}\big(\dot{R} - R'N^r\big)\right]$$

$${}^5R = {}^4R - K^{ab}K_{ab} + (K^a_a)^2$$
$$= \left[6R^{-2} - 6\left(\frac{R'}{R}\right)^2 L^{-2} + 6L^{-2}\left(\frac{R'}{R}\right)\left(\frac{L'}{L}\right) - 6L^{-2}\left(\frac{R''}{R}\right)\right]$$
$$-6R^{-2}N^{-2}\left[(LR)\big(\dot{L} - (LN^r)'\big)\big(\dot{R} - R'N^r\big) + \big(\dot{R} - R'N^r\big)^2\right]$$
$$= \left[6R^{-2} - 6\left(\frac{R'}{R}\right)^2 L^{-2} + 6L^{-2}\left(\frac{R'}{R}\right)\left(\frac{L'}{L}\right) - 6L^{-2}\left(\frac{R''}{R}\right)\right]$$
$$+6\left[\frac{(\dot{R} - N^rR\prime)(\dot{L} - L\prime N^r)'}{(LRN^2)} + \frac{(\dot{R} - R\prime N^r)^2}{(NR)^2}\right]$$

Appendix D. Emanator Theory Synopsis

D.1 Introduction

In quantum physics unitary propagation is a standard part of the description. Efforts to move to algebras to describe such propagation leads to formulations based on the normed division algebras (real, complex, quaternion, and octonion). In an effort to achieve maximal information propagation we relax the unitarity condition and show that multiplication (right) on a unit norm trigintaduonion base by a unit norm chiral trigintaduonions emanator results in a new unit norm product [126]. A path is comprised of repeated (right) multiplications. Each step of the 'emanation' arrived at is a multiplication by a chiral trigintaduonion. Use of methods from noise budget analysis, a constructive perturbation analysis, as well as analysis relating to maximal perturbation according to the Kato Rellich theorem, show that the chiral trigintaduonion with maximal perturbation has magnitude α, precisely the fine structure constant. A relation between α and π results. Suppose repeated achiral emanation steps can be described as an iterative mapping, with unit-norm constraint resulting in a quadratic relation on components, we then expect the Feigenbaum universal bifurcation parameter, C_∞, to appear according to the number of independent dimensions in a chiral trigintaduonion emanation step and the precise form of the "emanator" construction. The number of effective dimensions is shown to be 29 plus a little more, and a relation between α, π and C_∞results that is in agreement with the choice of emanator examined in computational studies shown here. The computational studies with the emanator have also been explored via "random walks" in the trigintaduonion space during emanation and to explore noise additivity effects. Component-level evolution is seen to behave like a random walk, with random walk asymptotics (established computationally). This helps to establish that the Emanation process is Martingale, since random walk processes are Martingale.

Just from the propagation structure on one path we already see core emergent structure that results in a universal emanation with structural parameters 10,22,78,137 and perturbation maximum α=~1/137. The central notion in the universal emanation hypothesis is that there should be *maximal information flow*, where this is accomplished by finding the highest theoretical dimensionality of unit-norm 'propagation', here called an emanation, which turns out to be 10, then add the maximal perturbation that still allows unit-norm propagation, where that perturbation is into the space the 10D motion is embedded in, here a 32 dimensional (trigintaduonion algebra) space.

The existing Standard Model, and reasonable extensions for the massive neutrino, cannot explain the parameters of the model. Why they are 19, or so, in number, and why the local gauge structure exists with the odd-looking product form: $U(1) \otimes SU(2)_L \otimes SU(3)$.

At the heart of the quantum formulation underlying the standard model is the fundamental theoretical element known as the quantum propagator (corresponds to a complex unitary matrix). Most notably, if the wavefunction has unit norm at the outset, after unitary propagation it remains unit norm.

Efforts to generalize the quantum formulation by considering hypercomplex but non-unitary matrices have been stymied for a variety of reasons [127]. Given the existing close agreement of experimental observation and quantum predictions, it is hard to work within the unitary propagator-type theoretical framework and arrive at anything other than an equivalent quantum formulation. Instead of working within the existing theory, in Emanator Theory the objective is to obtain a generalized theory that can project the quantized Lagrangian and Standard Model (or closely related extended Standard Model). To seek a larger theory that projects the existing theory is an excellent way to break free of the Godel-Incompleteness Trap that occurs when working within the existing standard physics and standard model. The drawback, of course, is it's likely to be a far-fetched idea, whatever it is, so it will face a higher bar to be seen as even interesting, not to mention valid. In what follows a synopsis of emanator theory will be given, as well as a few of the latest results, that will hopefully be convincing on both accounts.

If trying to generalize to a theory that can project to a renormalizable quantum field theory (with the Standard Model parameters), there is the issue of why it should project the particular formulation described. Of all the possible projections, how do you argue why it should be the unitary propagator-type physics seen? Here the aforementioned difficulty in generalizing the existing propagator theory (from the inside out) is a strength as it explains why the projection should be the standard physics and standard model as seen. That being the case, there is still the matter of finding the missing structure elsewhere (the Path Integral Action-Lagrangian formulation itself, for example, and the Standard Model parameters) and then projecting that as the renormalizable quantum field theory with Standard Model parameters as seen experimentally.

D.1.1 Dirac used Lorentz Invariance

If trying to guess the mathematical basis of a generalized quantum theory, that would project the existing Lorentz Invariant quantum field theory, then Dirac provides a powerful lesson. Recall that Dirac's guess and derivation for the relativistic wave equation was purely based on seeking a representation that was Lorentz invariant. In doing this, Dirac discovered the Dirac Equation that describes spin-1/2 fermionic matter (all the fundamental particles).

In Dirac's approach 4-vectors, V^a, were used to describe the spinors (spin ½ fermions), where the length of the 4-vector is constant under Lorentz Transformation L:

$$L\left\{\frac{1}{2}\eta_{ab}V^aV^b\right\} = \frac{1}{2}\eta_{ab}V^aV^b.$$

In Penrose's book on spinors [128] we see how to write a 4-vector as a 2x2 Hermitian matrix, $\psi(V^a)$, where the length of the four vector is equal to the determinant of the equivalent 2x2 matrix:

$$det[\psi(V^a)] = \frac{1}{2}\eta_{ab}V^aV^b.$$

Suppose the length of the 4-Vector is 1 (it's a spinor probability amplitude), then the associated 2x2 Hermitian matrix will have determinant equal to 1, thus $SL(2,\mathbb{C})$, and we have a direct representation of the Lorentz Group.

Let's now consider a similar process involving transformational invariance but instead of encoding the Lorentz transform in the form of matrix transformation invariance let's use elements of the Cayley algebras instead. Specifically, consider the following transformation:

$$q' = aqa_c^*, \quad where \quad aa_c = 1,$$

where a is a (unitary) complex bi-quaternion: $H(\mathbb{C}) \times H(\mathbb{C})$, and $q = (ct, ix, iy, iz)$ (note this notation is for $q = (ReH_1, iImH_1, iReH_2, iImH_2)$). The $q' = (ct', ix', iy', iz')$ that results will correspond to a proper orthochronous Lorentz transform [129-131].

This is a remarkable result, but is there better (higher dimensionality/complexity)? Consider that a complex bi-quaternion is isomorphic to a complex octonion which is isomorphic to an 'achiral' sedenion, thus a theory for achiral sedenion emanation is indicated from this result as far back as 1917. The 'halting condition' on the generalization in 1917 seems to be that octonions are the highest-order of the division algebras that have inverses defined (which is necessary to have $aa_c = 1$ be defined). But they have already extended past octonions since these are complex octonions ≅ sedenions, so how are they guaranteed to have $aa_c = 1$ be defined? This is possible for the *chiral* sedenions, as shown in [126], if restricted to be unit norm. So now we have our answer based on the results from [126] – we can go one complexation order higher:

$$Q' = AQA_c^*, \; where \;\; AA_c = 1,$$

where A is a (unitary) bi-complex bi-quaternion: $H(\mathbb{C} \times \mathbb{C}) \times H(\mathbb{C} \times \mathbb{C})$, which is isomorphic to a unit norm quaternionic bi-quaternion, $H(\mathbb{H}) \times H(\mathbb{H})$, and $Q = (ct, ix, iy, iz)$.
where $Q = (ReH_1, iImH_1, iReH_2, iImH_2)$ as before, except $H_1 = H \times \mathbb{H}$ not $H_1 = H \times \mathbb{C}$
The $Q' = (ct', ix', iy', iz')$ that results will again correspond to a proper orthochronous Lorentz transform. Again, how do we know the critical operation $AA_c = 1$ can always be satisfied? Previously we saw that a complex bi-quaternion was equivalent to a 'chiral' sedenion (in the sense described in [126]), thus a bi-complex bi-quaternion is isomorphic to a complex chiral sedenion, which is isomorphic to a 'doubly chiral' trigintaduonion. This is precisely the construct examined in [126], so a generalization of the 1917 result to unitary quaternionic bi-quaternions appears possible. As shown in [126], however, there is no higher order construct. This latter form (on doubly-chiral trigintaduonions) establishes the Emanator as Lorentz Invariant.

Thus, in terms of Cayley algebras, we can generalize to encoding the Lorentz transformation into split-Cayley algebras as long as the Cayley transform has an inverse, and doing this for the highest order Cayley algebra possible. The Cayley algebras with inverses, the division algebras, consist of the first four Cayley Algebras: reals (1 parameter), complex numbers (2 parameter), quaternions (4 parameters), and octonions (8 parameters). Another aspect of the division algebras is that they have norm, and this then means that their unitary evolution can be described as unit norm propagation. Suppose we focus on this latter feature, unit norm propagation alone. Can we extend to even higher order Cayley Algebras in the sense of starting with unit norm and for a subset of that higher order algebra still effect a unit norm, invertible, propagation? In other words for the next higher algebras, sedenions (16 parameter) and trigintaduonion (32 parameter), etc., does there exist a subspace allowing unit norm propagation? This is equivalent to considering the maximal Cayley subalgebra order for which unit norms exist. This is the actual starting point of the Maximal Information Emanation Hypothesis

D.1.2 The Maximum Information Emanation (MIE) Hypothesis and Prior Results

Emanator theory stems from the Maximum Information Emanation (MIE) Hypothesis. The definition of information is context dependent, so how the MIE hypothesis will manifest depends on circumstance. We start with the fundamental notion of the quantum propagator, for which mathematical 'propagators' satisfy unitarity. We seek to extend this foundational element so start by asking what is the highest Cayley algebra that can remain unitary – where the answer comes down to the highest order division algebra, which are the octonions. What if we change the desired property from unitary to unit-norm preserving? Then we can extend 2 more dimensions beyond the 8D octonion algebra to a chiral subspace of the trigintaduonions that is 10D [126].

The 10D chiral trigintaduonions are then identified as the maximal information 'carriers' or emanators, operationally like the quantum propagators in a larger theory, where evolution of the system will be shown to result from sums on paths of emanators (similar to quantum evolution in terms of a path integral on propagators 'steps'). Having posited this maximal construct we see the classic signs of "asking the right question" since we get a variety of clear results:

(1) the chiral emanator is manifestly Lorentz Invariant as indicated above, and the emanation step or propagation is simply multiplication by a unit-norm 'emanator' according to the Cayley algebra multiplication rules.

(2) chiral emanation involves a 10D element in a 32D space (trigintaduonions) for which maximum perturbation is determined computationally (still permitting unit-norm transmission) to be ~1/137, e.g., 'alpha' – we therefore have the mysterious alpha by a computational definition.

(3) The emanation process involves multiplication on the current unit norm trigintaduonion base element by a unit norm chiral trigintaduonion

emanator to arrive at a new, unit norm, trigintaduonion base (and then it repeats infinitely). When the trigintaduonion multiplication steps are expanded to octonionic level, we find that there are 137 independent tri-octonionic terms that occur in the emanation process. Due to complex noise contributions, the effective number is 137*, which is slightly greater than 137. Thus, theoretically, the maximum perturbation that is allowed is 1/137* which happens to be exactly 'alpha'. We, thus, have a theoretical derivation for alpha, referred to as the $\{\alpha, \pi\}$ relation:

$$\alpha^{-1} = \frac{137}{\cos\beta}\cos\theta\frac{\theta}{\sin\theta},$$

where

$$\beta = \frac{\pi}{137} \; and \; \theta = \frac{\pi}{137x29}$$

and

$$\alpha^{-1} = 137.03599978669910\,,$$

which agrees with 2002 experimental observation:

$$\alpha^{-1} = 137.03599976(50).$$

(4) Chiral trigintaduonion (32D) emanation with perturbation does not have effective dimension 32D due to chiral and other constraints -- noise budget analysis or (equivalently) Kato Rellich operator analysis, both indicate effective dimension slightly greater than 29 referred to as "29*". It is hypothesized that the maximal level of information flow dimensionally should relate to the Universal fractal constant C_∞ according to:

$$\alpha^{-1} = (C_\infty)^\gamma, \;\; \gamma \equiv 29^*,$$

where $\gamma$is estimated by [132]:

$$\gamma \cong \frac{1}{2}\left(29 + \left(\frac{4\pi}{72}\right)\left[1 + \left(\frac{\pi}{137 \times 29}\right)\left\{\frac{\pi}{72} + \frac{3}{72}\right\}\right]\right)$$

thus

$$\alpha^{-1} \cong 137.035999206\,...$$

in agreement with the exact alpha with nine digit accuracy.

In the process of estimating 29* we must explore the definition of the emanation process in the sense of is it one step then normalization, or a chain of steps (analogous to hands greater than one in size being dealt) then normalization, with infinite repeat. The answer appears to be two steps (since this suffices to 'flood' or max-out the noise channels, so no further steps needed) and working within this construct we obtain the $\gamma \cong 29^*$ estimate mentioned above. It's not an exact match to 16 decimal places and known measurement, but at a 9-decimal place match its pretty good, certainly indicating that there may well be a relation $\alpha^{-1} = (C_\infty)^{29^*}$ as hypothesized (sometimes called the $\{\alpha, \pi, C_\infty\}$ relation). And, if there is such a relation, it would indication that not only is evolution at maximum perturbation in the quantum sense, but it is also evolution at the edge-of-chaos in the thermal/statistical mechanical sense.

(5) the chiral emanator indicates 'motion' in a 10D subspace of 32D, suggesting 22 constants of the motion. This is a naïve analysis, but it turns out to be true upon deeper analysis within the emanator formalism as there are 22 types of emanation that result in no change to the base trigintaduonion describing the system. This, in turn, shows that the emanator theory will have 22 fixed parameters.

(6) By using the split form of the trigintaduonions, we not only have manifest Lorentz invariance, we also have an exact algebraic split to a space that is simply the direct product of 29* real dimensions (not a local approximation to such). This is important because the fundamental existence of a complex structure means that we trivially have the extension $\mathbb{R}^{29^*} \rightarrow \mathbb{C}^{29^*}$. Suppose we have point-like singular elements in $\mathbb{C}^{29^*}$, such will occur due to zero-divisors (zd's) in the 32D trigintaduonion space. To achieve a maximal domain of analyticity (an application of MIE), we must remove the zd-singularities. In doing so we obtain point-like matter in the theory and a small-h constant that enters the sum on emanator paths just as Planck's constant in the sum on propagator paths in the quantum formulation – suggesting that these small-h numbers are related.

(7) Three derivations of alpha are thus obtained: (i) Computational: $\{\alpha\}$ based on the maximal perturbation for which chiral emanation retains the unit-norm property; (ii) Theoretical: $\{\alpha, \pi\}$ based on the maximal noise transmission on a chiral emanation path; and (iii) Approximate: $\{\alpha, \pi, C_\infty\}$ based on the maximal noise transmission on an achiral emanation path (where maximal emanation is at "the edge of chaos" which is defined according to Feigenbaum Universality [133]).

(8) At component level in the emanation product, using 100's of millions of computational steps, we see an excellent asymptotic fit to random walk behavior. Since a random walk is a Martingale process, this strongly suggests that the achiral emanation process is Martingale (between normalization steps involved with zero crossings in their values). In turn,

the projected quantum process (standard theory) would retain the imprint of that Martingale process.

(9) The achiral emanator, a sum of achiral emanation paths, can be shown to have the mathematical form $\sum \exp(i\mathbb{H} \times \mathbb{O}) \to \mathbb{C} \times \mathbb{H} \times \mathbb{O}$, which can be shown to give the gauge theory of the standard model: $U(1) \times SU(2)_L \times SU(3)$. Thus, there is no grand unified theory in terms of gauges that is fundamental (although a GUT may approximately occur at early times cosmologically, at high temperature, where there is conformal flatness). The 'ugly' product gauge that is observed is precisely what is predicted by emanator theory.

(10) Universal thermality is indicated by application of the MIE hypothesis to the choice of whether 'effective' achiral emanation is associative or not, at least at the sedenion-level of propagation. In effect, the mechanism to extend unit norm (chiral) propagation on a 10D subspace of the 32D trigintaduonions can be repeated for other mathematical properties from the lower-level Cayley algebras. Thus, the mechanism identified for extending 'nice' properties of lower-order Cayley algebras to higher order (no more than two orders higher to be precise), can be used for other than existence of a norm. By the chiral extension mechanism, associativity can be extended to the chiral octonions and chiral sedenions. This provides the basis for the universal thermality relation that appears to be ubiquitous (e.g., analytic time). It also provides the basis for an associative matching of terms for cancellation, e.g., renormalization. Thus provides the mechanism to correct the projected quantum field theory such that it has the renormalization counter-terms needed.

(11) In the emanation process there is a clear separation between spinorial elements and manifold elements. Manifold elements include geometry and thermality, have no alpha-perturbation effects, and appear to be part of the 'apparatus' from the perspective of the quantum theory.

D.2 Synopsis of Methods and Prior Results

D.2.1 The Cayley Algebras

The list representation for hypercomplex numbers will make things clearer in what follows so will be introduced here for the first seven Cayley algebras:

Reals: $X_0 \rightarrow (X_0)$.
Complex: $(X_0 + X_1 \boldsymbol{i}) \rightarrow (X_0, X_1)$ with one imaginary number.
Quaternions: $(X_0 + X_1 \boldsymbol{i} + X_2 \boldsymbol{j} + X_3 \boldsymbol{k}) \rightarrow (X_0, X_1, X_2, X_3)$ with three imaginary numbers.
Octonions: $(X_0, \ldots, X_7)$ with seven imaginary numbers.
Sedenions: $(X_0, \ldots, X_{15})$ with fifteen imaginary numbers.
Trigintaduonions (Bi-Sedenions): $(X_0, \ldots, X_{31})$ with 31 imaginary numbers.
Bi-Trigintaduonions: $(X_0, \ldots, X_{63})$ with 63 types of imaginary number.

Consider how the familiar complex numbers can be generated from two real numbers with the introduction of a single imaginary number '$\boldsymbol{i}$', $\{X_0, X_1\} \rightarrow (X_0 + X_1 \boldsymbol{i})$. This construction process can be iterated, using two complex numbers, $\{Z_0, Z_1\}$, and a new imaginary number '$\boldsymbol{j}$':

$$(Z_0 + Z_1 \boldsymbol{j}) = (A+B\boldsymbol{i}) + (C+D\boldsymbol{i})\boldsymbol{j} = A+B\boldsymbol{i} + C\boldsymbol{j} + D\boldsymbol{ij} = A+B\boldsymbol{i} + C\boldsymbol{j} + D\boldsymbol{k},$$

where we have introduced a third imaginary number '$\boldsymbol{k}$' where '$\boldsymbol{ij=k}$'. In list notation this appears as the simple rule ((A,B),(C,D)) = (A,B,C,D). This iterative construction process can be repeated, generating algebras doubling in dimensionality at each iteration, to generate the 1,2,4,8,16, 32, and 64 dimensional algebras listed above. The process continues indefinitely to higher orders beyond that, doubling in dimension at each iteration, but we will see that the main algebras of interest for physics are those with dimension 1,2,4,and 8, and sub-spaces of those with dimension 16 and 32 dimensional algebras.

Addition of hypercomplex numbers is done component-wise, so is straightforward. For hypercomplex multiplication, list notation makes the freedom for group splittings more apparent, where any hypercomplex product ZxQ to be expressed as (U,V)x(R,S) by splitting Z=(U,V) and Q=(R,S). This is important because the product rule, generalized by Cayley, uses the splitting capability. The Cayley algebra multiplication rule is:

$$\mathbf{(A,B)(C,D) = ([AC-D^*B],[BC^*+DA])},$$

where conjugation of a hypercomplex number flips the signs of all of its imaginary components:

$$(A,B)^* = \mathrm{Conj}(A,B) = (A^*,-B)$$

The specification of new algebras, with addition and multiplication rules as indicated by the constructive process above, is known as the Cayley-

Dickson construction, and this gives rise to what is referred to as the Cayley algebras in what follows.

If a Split Cayley algebra is used, then the multiplication rule has a single sign difference:

$$\mathbf{(A,B)(C,D) = ([AC+D^*B],[BC^*+DA])}.$$

D.2.2 Unit-norm propagation

For a physical system, a unit norm object can be used to represent a system, and by repeated transformation to other unit norm objects, it thereby evolves. Mathematical objects that can effect this 'transformation' simply by the rule of multiplication would be objects like division algebras, ideals, and what I'll simply call projections or emanations. In the universal propagator we have a unit norm trigintaduonion (32D) and perform a right multiplication with a chiral (10D) unit norm 'alpha-step' (defined by a max perturbation α into the 29 free dimensions, given by 32 minus one for each chiral choice, and one for the unit normalization overall). Consider multiplication of a given (starting) trigintaduonion from the right with a chiral bi-sedenion as a 'projection' through the (chiral) step indicated. The repeated application and repeated 'chiral steps' thereby arriving at a path describing a chiral propagation. The resulting universal propagation consists of a 32D unit norm trigintaduonion with propagation via right multiplication using a unit-norm, chiral bi-sedenion, with max-α perturbation.

We thereby arrive at a 'Universe Propagator' that takes on the physics parameters desired (notably the fine-structure constant) and imprints them onto the evolution as seen from the 'internal reference frame' where we reference an object in the 4D spacetime with Standard Model gauge field, and where the standard Lagrangian emerges as the necessary 'propagate-able' structure (where Hilbert space must be complex, not real, quaternionic or octonionic, etc. [127]). From maximum information flow with the constructs, and the required emergent complex Hilbert space (thus complex path integral, thus standard quantum operator formalism) we arrive back at the familiar results with justification of their core mathematical representations (e.g., complex Hilbert space), and now with justification of all parameters, all from the emanation hypothesis.

Emanator Theory uses unit-norm propagation

Unit-norm right product propagation is trivial for the division algebras since norm(XY) = norm(X) $\times$ norm(Y). From this it is apparent that we have an automorphism group given by the norm itself (since an automorphism if A(XY)=A(X)A(Y)), and in the case of the octonions this automorphism group is G2 [134]. It can be shown that SU(3) is in G2 [98]. Let's now consider the situation with a higher-order Cayley algebra, the Sedenions, 'S'. We obviously don't have norm(S_1S_2) = norm(S_1) × norm(S_2) in general, as this would then allow S to join the ranks of the division algebras, and it is proven that such don't exist above the Octonions [135]. Can we still have a propagation structure? Is it possible to have a 'base' sedenion for which norm(S_{base})=1, and to have a right propagator (product) sedenion also norm(S_{right})=1, such that norm(S_{base} x S_{right}) =1? The answer is yes (see [126]), when the sedenion has the (chiral) form of an octonion crossed with a real octonion: S_{chiral} = (O,O_{real}) or S_{chiral} = (O_{real},O). Can we continue this to arrive at a propagation structure on the Trigintaduonions? Again the answer is yes, with the chiral form generalizing off the chiral Sedenion as might be expected: T_{chiral} = (S_{chiral}, S_{real}) or (S_{real}, S_{chiral}) [126]. It is proven that this extension process will go no further [126]. What happens is that due to the chiral form we are still able to re-express all T products (or S) as collections of terms involving tri-octonionic products (which have nice properties as described in [126]), and this can no longer occur above the (chiral) trigintaduonion level.

Thus, we have achiral emanation, and to get achiral there must be a way to sum over all chiral to get an achiral result to arrive at the full emanator process. These details are described next.

D.2.3 Chiral T-emanation has 78 generators of change and 137 independent octonion terms

We begin with constructing the theoretical expression for a general element of the trigintaduonion algebra after two chiral trigintaduonion multiplicative propagation steps. A simple analysis of the number of terms in this expression, when reduced to three-element algebraic 'braid-level', results in a count on algebraic braids of 137, plus a little extra (e.g. some lagniappe for the best 'cooking') of a contribution towards a 138th braid when the "noise analysis" is done [126,7].

Consider a general Norm=1 (32D) Trigintaduonion (Bi-Sedenion): (A,B), where A and B are sedenions (16D). Then have (A,B) = ((a,b), (c,d)), where {a,b,c,d} are octonions. Slightly different than a propagator, we have an 'emanator' with the following notation and properties, where the emanator describes a 10D multiplicative step. The emanator is a chiral bi-sedenion: a trigintaduonion whose first sedenion half is itself a chiral bi-octonion, and the second sedenion half is a pure real (as is the second octonion half): $(\tilde{A},\beta)$, $\tilde{A} = (\tilde{a},\alpha)$, where the norm is 1, α is a real octonion, and β is a real sedenion. Thus:

Emanator: $(\tilde{A},\beta) = ((\tilde{a},\alpha), \beta)$.
Note: $\tilde{A}^* = (\tilde{a}^*,-\alpha)$.

Let's set up a description of the Universal 'Emanation' along a 'chiral path' resulting from a few emanation steps. To begin, suppose we have already arrived at, or received, a unit norm trigintaduonion (32D) state 'T', and suppose our emanations are the result of right multiplication with a chiral trigintaduonion (bi-sedenion) 'step', and suppose we consider one such path after just a few steps. Here's the notation to begin:

T = (A,B), a unit norm trigintaduonion.
$\boldsymbol{\tau} = (\tilde{A},\beta) = ((\tilde{a},\alpha), \beta)$, the 'emanator' above (so named to distinguish from a 'propagator').

Universal Emanation from T on single path with three steps: $((\mathbf{T} \bullet \boldsymbol{\tau}_1) \bullet \boldsymbol{\tau}_2) \bullet \boldsymbol{\tau}_3)$ …

Consider the first emanation step:
$\mathbf{T} \bullet \boldsymbol{\tau}_1 = (A,B) \bullet (\tilde{A},\beta) = ([A\bullet\tilde{A}-\beta^*\bullet B] , [B\bullet\tilde{A}^*+\beta\bullet A])$. (Standard Cayley algebra multiplication rules.)
$A\bullet\tilde{A} = (a,b) \bullet (\tilde{a},\alpha) = ([a\bullet\tilde{a}-\alpha^*\bullet b] , [b\bullet\tilde{a}^*+\alpha\bullet a])$
$B\bullet\tilde{A}^* = (c,d) \bullet (\tilde{a}^*,-\alpha) = ([c\bullet\tilde{a}^*+\alpha^*\bullet d] , [d\bullet\tilde{a}-\alpha\bullet c])$
Thus,
$\mathbf{T} \bullet \boldsymbol{\tau}_1 = (A,B) \bullet (\tilde{A},\beta) = ([(a\bullet\tilde{a}-\alpha^*\bullet b-\beta c) , (b\bullet\tilde{a}^*+\alpha\bullet a-\beta d)] , [(c\bullet\tilde{a}^*+\alpha^*\bullet d+\beta a) , (d\bullet\tilde{a}-\alpha\bullet c+\beta b)])$.

At the lowest octonion level, that covers the pure real trigintaduonion, we have:

$(a\bullet\tilde{a}-\alpha^*\bullet b-\beta c)$ → 8x8 + 8 + 8 – 2 = 64+14 = 78 independent octonion terms (78 independent generators of motion). The –2 comes from the unit norm constraints on T and τ.

Now consider the second propagation step:

$(\mathbf{T} \bullet \boldsymbol{\tau}_1) \bullet \boldsymbol{\tau}_2 = ([(a\bullet\tilde{a}-\alpha^*\bullet b-\beta c) , (b\bullet\tilde{a}^*+\alpha\bullet a-\beta d)] , [(c\bullet\tilde{a}^*+\alpha^*\bullet d+\beta a)$ $, (d\bullet\tilde{a}-\alpha\bullet c+\beta b)]) \bullet (\tilde{A},\beta),$
where $\boldsymbol{\tau}_2 = (\tilde{A}',\beta') = ((\tilde{a}',\alpha'), \beta')$.

Let $(\mathbf{T} \bullet \boldsymbol{\tau}_1) \bullet \boldsymbol{\tau}_2 = ([Z_{11},Z_{12}] , [Z_{21}, Z_{22}])$.
$Z_{11} = (a\bullet\tilde{a}-b\alpha-c\beta)\bullet\tilde{a}' - (b\bullet\tilde{a}^*+\alpha a-\beta d)\, \alpha' - (c\bullet\tilde{a}^*+d\alpha+a\beta)\beta'$.

In Z_{11} we can replace the octonions with their unit component forms:

$$a = a_1e_1 + a_2e_2 + \ldots + a_8e_8 ,$$

where $\{e_1, e_2, \ldots, e_8\}$ are the unit octonions (one real, seven imaginary), while 'α'$=\alpha e_9$ and 'β'$=\beta e_{17}$, originally, but in expressions, are reduced to just their real part. All expressions, thus, involve 10 components: $\{e_1, e_2, \ldots, e_8, e_9, e_{17}\}$, and as the equations for Z_{11} shows, grouped in factors of three (three-element octonionic 'braids'). We don't have associativity but we do have alternativity and the braid rules on three-element octonionic products that allows their regrouping. Applying these rules to have only ordered $e_i\bullet e_j\bullet e_k$ products in a simplified expression, we will then have 10x9x8/3! = 120 independent terms when the products involve different components. We have 8 independent terms when the first product are on the same component (equals 1), have 8 independent terms when the second product involves the same component, and have 1 independent term when the three-way product equals 1 (further details on this and the properties of the exponentiation map on hypercomplex numbers is given in the next section. There are, thus, 137 independent terms in Z_{11}, where each term has norm less than unity (since each octonionic component has norm less than one and the norm of a product of octonions is the product of their norms). The terms involving products with the same component, or with the components three-way product equal unity, correspond to the 'telescoping terms' in what follows.

When $\mathbf{T=((a,b),(c,d))} \rightarrow ((\mathbf{T} \bullet \boldsymbol{\tau}_1) \bullet \boldsymbol{\tau}_2)=((Z_{11},Z_{12}),(Z_{21}, Z_{22}))$. we have $a\rightarrow Z_{11}$ and the terms involving 'a' in Z_{11} are referred to as 'telescoping' due to their simple math properties with further emanation steps. In particular, the terms involving 'a' are:

$Z_{11}[\text{a terms}]= a\bullet\tilde{a}\bullet\tilde{a}' - a\alpha\alpha' - a\beta\beta'$.

We can see that the original 'a' information is passed along three (telescoping) channels, one involving repeated full octonionic factors $\tilde{a}$, one involving repeated real-octonion α factors, and one involving repeated real-octonion β factors:

(1) a → $(a\bullet\tilde{a})\bullet\tilde{a}'$, if this product is continued indefinitely, then we have ***the random product of a collection of octonions***, all of which have norm less than one (although their norms can be quite close to one). If their norms were perfectly equal to one, then the addition of their random 'phases' would tend to cancel to zero, giving only a real octonionic component (same argument for phase cancelation on S1 as on S7 or S15). What results is a 'mostly' real octonion, having some imaginary part. A more precise, and lengthy, derivation is given in the next section.

(2) a → $a\alpha\alpha'$, if this product is continued indefinitely, 'telescoped' with repeated α products, we see that the original 8 independent terms arising from 'a' are passed forward with an overall real octonion product, giving rise to 8 independent terms.

(3) a → $a\beta\beta'$, as with (2), we have 8 independent terms.

From the above, we see an alternative accounting of the extra 17 independent terms to go with the 120 for a total of 137 independent terms in the propagation of the octonionic sectors of the universal emanation. A benefit of the telescoping analysis is it clarifies how in (1) an imaginary component may arise, and in perturbation expansions it will then be natural to refer to an overall imaginary component.

There are 137 terms in the dually chiral 'emanation', each with norm bounded by unity, with total bi-sedenion norm equal to unity. In the analysis that led to the computational discovery of α [126], an imaginary (non 10D) component was added of growing magnitude until unit-norm propagation failed. In essence, a maximum perturbation, from propagation strictly in the 10D subspace of the 32D trigintaduonions, was sought.

We identify maximal perturbation by doing an independent term analysis, and by adding a maximum perturbation term that implicitly identifies a definition of maximum antiphase. From this definition of maximum antiphase, there results the parameter π.

D.2.3.1 Exponential Map Properties when using hypercomplex numbers

For what follows, it helps to recall some important properties of the exponential, particularly its well-defined properties with hypercomplex numbers [7]. Important map relations:

(1) exponential map on Im(T) gives unit norm object: $\exp(\mathrm{Im}(T)\theta) = \cos\theta + \mathrm{Im}(T) \sin\theta$.

(2) exponential map on iT gives $C \times T$:
$\exp(iT) = \exp(i\mathrm{Re}(T)) \times \exp(i\mathrm{Im}(T)) = (\cos\theta + i\mathrm{Re}(T)\sin\theta) \times (\cos\varphi + i\mathrm{Im}(T)\sin\varphi) = C \times T$

Use (1) to focus on fluctuations in imaginary parameters free of normalization concerns.

Use (2) to get complex structure $C \times$ (object). Note that exponentiation into phase terms is precisely what occurs in the path integral propagator formalism, and will occur here as well for the emanator formalism, thus the "$C \times$" complex factor. When drawn upon in the emanator formalism, this method of achieving additional "$C \times$" complex structure will be forced by the zero-divisor handling (that will give rise to point-like matter with very small phase coupling, thus a highly oscillatory integral, and ties over to foundational aspects of the path integral formalism).

D.2.3.2 Alternate 137th count using Exponential Map

The derivation below follows [7], but with a more succinct accounting of the independent terms.
Consider a general norm=1 bisedenion in list notation: (A,B), where A and B are sedenions. Consider a propagator bisedenion (C,β), $C = (c,\alpha)$, where c is an octonion and α is shorthand for the real octonion $(\alpha,0,0,0,0,0,0,0)$, where α is a real number, and β is shorthand for the real sedenion $(\beta,0,0,0,0,0,0,0,0,0,0,0,0,0,0,0)$, where β is a real number. Using A=(a,b), B=(u,v), and the multiplication rule from Section 2, we have:
$(A,B)(C,\beta) = ([AC-\beta^*B], [BC^*+\beta A])$, where
$AC = (a,b)(c,\alpha) = ([ac-\alpha^*b],[bc^*+\alpha a])$; $BC^*=(u,v)(c^*,-\alpha) = ([uc^*+\alpha^*v],[vc^*-\alpha u])$.
Thus, we have:
$(A,B)(C,\beta) = ([(ac-\alpha^*b , bc^*+\alpha a)-\beta^*(u,v)] , [(uc^*+\alpha^*v , vc-\alpha u)+\beta(a,b)])$, so,

$(A,B)(C,\beta) = ([ac-\alpha^*b-\beta^*u\ ,\ bc^*+\alpha a-\beta^*v]\ ,\ [uc^*+\alpha^*v+\beta a\ ,\ vc-\alpha u+\beta b])$. Now consider another propagator bisedenion (C',β'), $C' = (c',\alpha')$, and form the product corresponding to the next multiplicative step: $(\,(A,B)(C,\beta)\,)\,(C',\beta') = (\,[(ac)c' - \alpha^*bc' - \beta^*uc' - \alpha'^*(bc^*+\alpha a-\beta^*v)\ ,\ \ldots]\ ,\ [\ldots\ ,\ \ldots]\,)$, where only the first expression at octonionic level ($T=(O_1,O_2,O_3,O_4)$) is shown:

$$O_1 = (ac)c' - \alpha^*bc' - \beta^*uc' - \alpha'^*(bc^*+\alpha a-\beta^*v).$$

At octonionic-level there are 10x9x8/3x2=120 independent terms for 8 octonionic components (labeled a, b, c) plus a separate octonion component (α) and one sedenion component (β), e.g., have 10 choose 3. Also have telescoping terms with repeated real octonion factors, such as with the $a\alpha\alpha'^*$ term (think $a\alpha(\alpha'^*)^n$), which gives an additional 8 independent terms. Also have telescoping terms with alternating real octonion factors and real sedenion factors, such as with the $v\beta^*\alpha'^*$ term (think $v(\beta^*\alpha'^*)^n$), which gives another 8 independent terms. There is one other 'telescoping' term due to repeated octonion right products seen in $(ac)c'$ (now think $((ac)c')c'.....c')$). The change in this term corresponds to an element of the automorphism group on octonions, G2, and as such provides one last independent term, for a total of 137 independent terms at octonion level.

All of the octonion products involve octonions with norms at most unity, and by the normed division algebra rules on octonions, their norm is simply the norm of the individual octonions multiplied together, all of which are bounded by unity, thus their product is bounded by unity. The overall bound for the expression, each individual term being bounded by unity, is therefore simply the counting on the independent terms.

The maximum magnitude of each component of the octonion in the product term is given with a 'channel multiplier' of 137. Also, in seeking the maximum information propagation we require that the real chiral component never cross zero (e.g., stay in its connected $\{\alpha,\beta\}$ quadrant), thus the strictest condition on evaluating evolution might be intuited to be when the imaginary components combine to have real component contribution that is antiphase, e.g., the total imaginary angle is π. The choice of antiphase will used in what follows and will be justified when "C ×" allows the antiphase to be understood in the context of the Universal Mandelbrot set [7] position on the negative real axis that gives the maximal magnitude of displacement from the origin: C_∞. We limit the maximum perturbation allowable by the antiphase worst case. At octonionic-level there is thus the channel multiplier: $137 + i\pi$.

D.2.3.3 The {α, π} relation using Exponential Map

The maximum perturbation, referred to as maximum noise in what follows, is first evaluated for a chiral emanation where we take a norm=1 $T_{base}=(A,B)$ and take the right product with T_{chiral} in the form $T_{chiral}=(C,\beta)$, with product $(A,B)(C,\beta)$ proven to be unit norm above [91]. In the prior section we saw that there are 137 independent octonion terms at the octonion sub-level of the new unit norm trigintaduonion that results, which leads to 137 independent terms at component level. In order to use the map rules mentioned in the previous section, it is necessary to move from the trigintaduonion, T, space, to the $C \times T$ space. This is done in later sections anyway where we consider sums on exp(iT). The exponential function (map) provides a well-defined 'lift' of a hypercomplex (Cayley) algebra from T to $C \times T$. The exponential map also provides a very useful maneuver when working with unit-norm hypercomplex numbers via the generalized deMoivre theorem $\exp(\mathrm{Im}(T))=\cos(\theta)+\mathrm{Im}(T)\sin(\theta)$, with the real part recoverable from $\cos(\theta)$. More details on this follow later but for now, in evaluating the maximum noise allowed we have three structures to adopt: (1) the noise is generalized to be complex (as will be the case for the components themselves once the $T \rightarrow C \times T$ structure is adopted). (2) At component-level, the noise (for maximum noise) is equipartitioned in both real and imaginary parts. (3) Total imaginary noise magnitude is π for maximal antiphase (to be justified later).

(I) Chiral emanation noise: have 137 terms with max unit norm each, for the real part, and for the imaginary part have a "phase angle" β such that $137\beta=\pi$ (here referred to as a phase angle in the sense that the exp(Im(T)) map is being used). The noise magnitude at octonionic-component level is then given by the right triangle with real part = 137 and angle $\beta=\pi/137$, thus maximum chiral emanation noise magnitude is:

$$H = 137/\cos(\pi/137)$$

(II) Achiral emanation noise: now have 29 "free" components, each with 137 independent terms. For maximum achiral emanation we thus have 137 x 29 independent terms that are built from the aforementioned chiral emanation terms (to make achiral). If we equipartition as before, with noise magnitude Hc, we have a "noise triangle" with magnitude (hypotenuse) Hc and with angle $\theta = \pi/137x29$. The imaginary part is then (Hc)sin(θ). As regards the H magnitude separated form (separating out the 'H' factor for now), we have for the imaginary part sin(θ)c. As before, we take maximal noise transmission when all the imaginary parts add to maximal antiphase. Given the equipartitioning assumption, we then simply have the factor 137x29:

$$\sin(\theta)\, c\, (137x29) = \pi \rightarrow c = \theta/\sin\theta.$$

The maximum real noise perturbation that the system can have is then α, where:

$$\alpha^{-1} = \frac{137}{\cos\beta}\cos\theta\,\frac{\theta}{\sin\theta}, \qquad where\ \beta = \frac{\pi}{137}\ and\ \theta = \frac{\pi}{137x29}$$

$$\alpha^{-1} = 137.03599978669910\,,$$

where the evaluation was done at WolframAlpha to high precision [136] (e.g., higher precision than that reported in earlier work). This matches the experimentally observed value to all 11 decimal places currently known. As of 2002, the measured value of α is:

$$\alpha^{-1} = 137.03599976(50).$$

Note that in quantum field theory the parameters are renormalized at a particular energy scale. Thus choice of energy scale impacts the value of α (as a coupling constant in the classical theory or a perturbation expansion factor in the quantum theory). At 0K we have the extreme low-energy end of the renormalization group (with the largest α value). We are at the 2.7K CMBR, so we have the max α to very high precision. (In studies at high energy scale at LEP, at the energy scale of the Z-boson (91GeV), we get the renormalized value to be: $\alpha^{-1}[M_Z] \cong 127.5$. Note that 91GeV is way above the energy scale of the familiar Hagedorn temperature at ~pion mass=150MeV or 1.7x10^12 K) [137], where hadronic matter 'evaporates' into quark matter.)

D.2.4 Trigintaduonion Emanation: achirality from chirality

There are four chiralities, and for a given chirality (with unit norm) there are 29 dimensions of freedom (10D + 19D of chiraly consistent perturbation). When analytic extension is taken to give maximal information flow, the effective dimension for each of the four chiralities is 29* (detailed in [132]). This clear decomposition into 29* independent effective dimensions is then revealed in the {α,π,C_∞} relation in [132].

The Mandelbrot Set is one of many that encounter the universal constant C_∞. The Mandelbrot set also describes a boundary with 2D fractal dimension at its "edge of chaos" [132]. If driven to similar optimality in approaching a zero-value (a zero-divider issue), we see a two-value zero-crossing specification effectively like a double zero. The parameterization of the zeros of the Emanator at chiral zero-divisor points will thus be as double-zeros.

For what follows we use the simple description of the emanator:

$$T^{(k)}_{chiral} = \begin{cases} ((0,\alpha),\beta) \\ ((\alpha,0),\beta) \\ (\beta,(0,\alpha)) \\ (\beta,(\alpha,0)) \end{cases}.$$

$$\text{Emanation}(\mathbf{T}) = \frac{1}{N} \sum_{k\in\{4x72\}^n} \mathbf{T} \bullet \boldsymbol{T}^{(k)}_{\boldsymbol{chiral}} = \frac{1}{N} \sum_{K\in 4\ chiralities} \mathbf{T} \bullet \overline{\boldsymbol{T}}^{(\boldsymbol{K})}_{\boldsymbol{chiral}}$$

If working with non-split T's, then we restrict to emanations that are perturbations of unity:

$$T^{(k)}_{chiral} = \begin{cases} ((0,\alpha),\beta) \\ ((\alpha,0),\beta) \\ (\beta,(0,\alpha)) \\ (\beta,(\alpha,0)) \end{cases}, where\ T^{(k)}_{chiral} = \mathbf{1} + i\boldsymbol{\delta}.$$

From unit norm we have $\alpha^2 = 1 - 0^2 - \beta^2$, with ± sign choice on α, similarly for β.

If working with split T's (bi-sedenions, etc.), then we have manifest Lorentz Invariance (shown in Chapter 7). So, it is often convenient to work with split-T' since this is manifest from the outset.

Issues with zero-divisors

Suppose we add the rule that emanation may not proceed when a particular chirality is zeroed-out, in other words:

$$\mathbf{T} \bullet \overline{\boldsymbol{T}}^{(\boldsymbol{K})}_{\boldsymbol{chiral}} \neq \mathbf{0}.$$

For 'normal' numbers this goes without saying, since for real numbers if we have $r_1 \times r_2 = r_3$ then $r_3 \neq 0\ if\ neither\ r_1 = 0\ or\ r_2 = 0$. This holds true for the Real, Complex, Quaternion, and Octonion numbers. This does not hold true for Sedenions or higher. For sedenions the dimensionality of the zero-divisor event is mostly constrained, while for trigintaduonions it is significant. If such zeros were eliminated from the emanator description by using analytic extension component-wise (on 29* effective components) we see how a description devoid of matter (pure static field with no source or sink) might acquire matter by way of extending to a maximal domain of analyticity be removing zero-divisor events (a Wick transformation from real dimensionless action to pure imaginary action that is dimensionless but consisting of a dimensionful ratio). For further discussion along these lines, see Book 7 [7].

D.2.4.1 Achiral T-emanation has 29* effective dimensions

Let's estimate of the effective dimension of information transmission in an achiral T-emanation process. There are 4 chiralities, so to get an achiral emanator candidate, minimally need a "4-card deck" to emanate in the four chiralities, with emanator equal to normalized sum. The actual deck appears to require a normalized sum over sub-chiralities, as will be explicitly enumerated in what follows. Here are the four chiralities with real fluctuation noise shown:

$$
\begin{gathered}
(\,(\,(\mathrm{O}[0] \pm \delta, \dots), \alpha \pm \delta), \beta \pm \delta) \\
(\,(\,\alpha \pm \delta, (\mathrm{O}[0] \pm \delta, \dots)), \beta \pm \delta) \\
(\,\beta \pm \delta, (\,(\mathrm{O}[0] \pm \delta, \dots), \alpha \pm \delta)\,) \\
(\,\beta \pm \delta, (\,\alpha \pm \delta, (\mathrm{O}[0] \pm \delta, \dots)\,)\,)
\end{gathered}
$$

where α is a real octonion and β is a real sedenion, and Tem is an equal weight sum of the action of each of the sub-chiral propagations on the base T, with the fluctuations indicated each done separately. We have the constraints $\alpha \neq 0$, $\beta \neq 0$, and common octonion O not pure real.

Each of the δ's is an independent fluctuation corresponding to its own sub-chiral emanation, but no subscripting on δ's is used or shown. There are thus 9x2x4=72 independent *imaginary* noise fluctuations to consider in the exp(Im(Tem)) evaluation (that automatically provides unit-norm). The real noise fluctuations in the real (first) component are, thus, not counted. If our definition for Tem entails only one card being dealt, then the sum over those possibilities is the sum

$$
\mathbf{T} \bullet \mathrm{T}_{\mathrm{em}} \equiv \mathrm{Emanation}(\mathbf{T}) = \frac{1}{72} \sum_{k \in \{72\}} \mathbf{T} \bullet \boldsymbol{T}_{\boldsymbol{chiral}}^{(\boldsymbol{k})}
$$

For one-card, or a one-step, emanation, with real components and real noise, this makes sense from the counting shown, and it's what we use going forward. Using this will allow an entirely separate method for evaluating α (here at the one-card hand approximation). This will be done by determining the effective dimension 29*>29 of maximal information propagation (or maximal noise fluctuation). Before moving on, however, let's examine what happens when we allow complex noise fluctuations as this will trivially be allowed when we consider C × T via exp(iT) in later discussion anyway.

D.2.4.2 Effective Deck size is 72, which is consistent with the {α,π,C_∞} relation

Maximum information transmission involves a complex extension to the T components and their noise fluctuations, but in doing this it must retain emanation structures such as the octonionic triple that occurs in previous expressions (starting with the proof of the T_{chiral} solution itself), which leads to the counting that gives 137 independent terms, etc. Thus, the maximal complex extension on the noise is that it remain real in the octonion components:

$$(((\mathrm{O}[0] \pm \delta, \ldots), \alpha \pm i\delta), \beta \pm i\delta)$$
$$((\alpha \pm i\delta, (\mathrm{O}[0] \pm \delta, \ldots)), \beta \pm i\delta)$$
$$(\beta \pm i\delta, ((\mathrm{O}[0] \pm \delta, \ldots), \alpha \pm i\delta))$$
$$(\beta \pm i\delta, (\alpha \pm i\delta, (\mathrm{O}[0] \pm \delta, \ldots)))$$

The first chiral T component is where new imaginary terms might arise (the others are already counted since in imaginary components). We see there are six more, so the deck is now78. In application, as we will see, those added six are precluded due to constraints such that the effective deck size (impacting the sums and numbers of independent terms in the emanator definition) remains 72.

All noise terms will be treated additively, including terms in different imaginary components as well as imaginary noise terms in the real component. The criterion for max noise (in-phase constructive interference) gives the extreme of linear additivity. (Not like Gaussian statistical noise that adds in quadrature.) Also note that the discussion in terms of "noise transmission" and "information transmission" will be used almost interchangeably, whenever one description or the other best suits the analysis it will be used. Note that with this kind of noise analysis we can effectively shift around T noise terms associatively. Also note that application of the Kato-Rellich theorem [138,139] is related to the noise budget analysis done here focusing on first order terms.

There are 137 independent tri-octonionic terms in each of 29 free components indicated by a particular chirality (within the 32 components of a general trigintaduonion). This is a nontrivial result since (T_{chiral} • T_{chiral}) is no longer T_{chiral} type (but still T_{norm1} type), so direct expansions are needed to identify the number of independent terms and this is briefly described below, with more detail in [7].

Obtaining an achiral emanation from a collection of chiral emanations requires that all chiralities be summed over (there are four) as well as sub-chiralities (there are 72). Noise analysis requires collecting of first-order terms. Analysis of noise transmission indicates 29* dimensions, where:

$$29^* \cong 29 + \left(\frac{4\pi}{72}\right)\left[1 + \left(\frac{\pi}{137 \cdot 29}\right)\left(\left(\frac{\pi}{72}\right) + \left(\frac{3}{72}\right)\right)\right]$$

The above result was obtained in [132] to describe the 72-card chiral 'deck' of chiral emanation products for a "single-step" emanation. In the new Results to follow this is reviewed and elaborated further.

D.2.4.3 'Edge of chaos' maximal perturbation hypothesis [132]
Consider the 'edge of chaos' maximal perturbation in each of the 29* dimensions to be at position C_∞, which is on the negative real axis, i.e., at π rotation to have −1 factor, ***thus at maximal antiphase***. This results in the relation for maximal perturbation at maximal antiphase (maximum reference angle with sign chosen positive by convention) has a lower bound on α given by:

$$\alpha_{\square}^{-1} = \left(\sqrt{C_\infty}\right)^{29^*}.$$

where

$$C_\infty = 1.4011551890920506004\ldots$$

This ties $1/\alpha$ to the second Feigenbaum constant C_∞ in the context of the Mandelbrot set. It is well known that the Feigenbaum constants are universal, and part of a description of a universal transition to chaos regime. The Mandelbrot set is also universal [140], and maximal in that its fractal boundary has maximal fractal dimension of 2 [140], a detail that will be important in the meromorphic matter description given later.

For C_∞, most references only provide $C_\infty = 1.401155189$ ..., and a higher precision tabulation is not readily found, so use is made of the relation

$$C_n = a_n(a_n - 2)/4,$$

together with the tabulation on a_∞ [141]:

$$a_\infty = 3.56994567187094490 18 \ldots$$

The resulting C_∞ is:

$$C_\infty = 1.4011551890920506004 \ldots$$

The resulting $\alpha_{\square}^{-1}$ is:

$$\alpha_{\square}^{-1} = 137.03599933370198263 \ldots$$

D.3 Implication of MIE

D.3.1 MIE requires a complex Hilbert Space

As mentioned previously, according to [127], a complex Hilbert space is selected by the quantum deFinetti theorem, since it is required for information propagation (and thereby consistent with the maximum information propagation concept in its selection). Because it's a complex Hilbert space, this explains why the path integral operates in a complex space, even though the underlying universal algebraic construct from which it is emergent is hypercomplex to the level of the trigintaduonions.

From [127], a simple derivation shows why the quantum deFinetti Theorem requires amplitudes to be complex. Suppose f(n) is the number of real parameters to specify an n-dimensional mixed state. For real amplitudes f(n)=n(n+1)/2, for complex amplitudes f(n)=n^2, and for quaternionic f(n) = n(2n-1). For propagation, etc., need f(n1n2)=f(n1)f(n2), which only works for complex amplitudes.

D.3.2 Time is analytic and Matter is meromorphic

A variety of efforts have been made to find a definition of time that is somehow implicit to the main QFT and GR formalisms, whether it be a choice of vacuum for QFT in curved spacetime (and even if the spacetime is not curved [142]) which is indirectly a choice of time. Or seeking an internal time-reference in a full-GR quantum minisuperspace analysis of dust-shell collapse [143]. Or in seeking a notion of time in full general relativistic (GR) models, in the equilibrium sense, with an assumption of euclideanizability [36,91]. For the latter, the self-consistent stable solutions that were indicated showed the general utility of the euclideanizability hypothesis on emanation/propagation solutions in general (that is especially relevant, or interpretable, when the system is in equilibrium). In none of these efforts, however, was there success in identifying some internal notion of time, time, it seems, is an added construction, and this is consistent with the results shown here, where we find that time is likely analytic and an emergent construct.

In Book 7 [7], we see matter as meromorphic residue precipitation, in amounts of one quantum given by a precursor to Planck's constant h*. The meromorphic residue winding number is also notable in that it gives an integer that stays constant in the meromorphic region. This raises the possibility that elementary particle attributes might encode by way of different winding numbers, with reference to their different winding numbers at residues, but this will not be discussed further here.

D.3.3 Emergent Evolution and Emergent Universal Learning

We see that the definition of the emanator process is not known, but that consistency arguments (such as achirality constructed from a collection of chiral emanations) lead to a certain set of forms. And that consistency with the $\{\alpha, \pi\ C_\infty\}$ relation imposes further constraints on the form of the emanator. What is hypothesized is that the emanator is selected for maximal information transmission, thus emergent itself under that criterion. Let's now consider the maximal information transmission idea from the receiving end, e.g., maximal information receiving, or learning, in this context. If we turn to the information geometry analysis of learning in neural nets [10-12,144,145] (which uses differential geometry) we obtain a fundamental origin for statistical entropy (Shannon Entropy), and we identify optimal learning processes, based on expectation/maximization, that involves two-steps (as the name suggests) that may be done according to two fundamentally different conventions, e.g. the optimal learning involves four types of step, or is doubly chiral, consistent with the emanator 4-chiral processes described [7]. The potential applications of these results to Trigintaduonion encoded neuromanifolds is beyond the scope of this paper.

D.3.4 Objective Reduction, Zero-Divisors, and possible origins of Planck 's constant

A new mechanism for objective reduction [146,147] is also indicated by the way π enters the theory as a maximum anti-phase amount comprising part of the maximal perturbation propagation. Consider in the context where there is a 'classical' trigintaduonion path in a congruence of paths (a flow-line description). On the classical path in the congruences, we have α calculated using a $+\pi$ maximal anti-phase, but this could also occur with $-\pi$ maximal anti-phase as well, thus we could have a $\pm\pi$ phase toggle when a zero divisor is encountered in the 32D propagation (given the perturbations extending outside the 10D somewhat into the entire 32D). The zero-divisor discontinuity requires the field to reformulate a new 'consistency' with the 32D algebraic propagation (and 64D and higher, as well), which could have the result that since the prior π phase had the discontinuity, then it must toggle to the other, negative, phase, e.g., objective reduction may occur as a zero-divisor phase-toggle event.

D.3.5 Where's the geometry?

The geometry side of emanation theory does not result from the action of the repeated emanator product directly, but from the accumulated product in the T base that results. Geometry is, in effect, emergent (projected) on the T_{base} 'space' of the T_{em} product action. Geometry appears as a manifold construct in both space-time curvature (where it is locally given with the standard model action) and as an intrinsic entropic property, via neuromanifold 'geodesic' motion being equivalent to, and possibly the origin of, the minimization of the relative entropy (and maximization of entropy, the 2nd Law). Setting aside thermodynamic issues in this discussion, this puts the Lagrangian formulation with standard model terms, and Hilbert action for GR, into better perspective. The representation of the geometry via the Hilbert action for GR suffices with maximal extension in whatever causally connected domain of interest. So, we've got the existing QFT in CST space-time formulations in the black hole exterior, for example. We may have the resolution at the black hole horizon causal boundary via String Theory on the surface (using Ads/CFT relation and related holographic hypothesis [17,18]).

We describe repeated chiral product action on the trigintaduonion spinor space. The emanation process, consisting of a chain of chiral trigintaduonion products, leads to a Lagrangian variational formalism *with the standard model*. The origins of the parameters of the model are beginning to be understood as well. Apparently state information memory/inertia is carried via the manifold curvature response to the matter density, where 'G' is the linkage for the balance on this 'learning' process. Presumably the G learning rate is set for optimal learning, e.g., maximal information flow, and as such its value may eventually be clarified theoretically.

D.4 Results

D.4.1 Two-card Hand consistent with 137 and Neuromanifold Two-step Learning

MIE may guide our choice of 'hand' size dealt. Consider that one chiral emanation step (multiplication) off a base trigintaduonion will only have 78 independent tri-octonionic noise terms introduced. With a second step we 'flood the channels' and get the full assortment of 137 tri-octonionic noise terms (as shown in the derivation that obtained the 137 terms in [132]). Thus, a minimum of a two card hand must be considered. Having achieved a maximal perturbation scenario, what need for further cards in the hand, i.e., or further multiplicative factors in the pre-achiral normalization step?

Previously, when speaking of "Single-step achiral" we presumed initial conditions for the noise that effectively performed an average two-card 'hand' emanation analysis. If not in such an approximate form the ***{α, π}*** relation is recovered. The approximate form, however, expresses a new hypothesized relationship: $\alpha_{\square}^{-1} = \left(\sqrt{C_\infty}\right)^{29^*}$, that is itself verified to many decimal places.

Two-card-hand and two-step emanation jibes with two-step neuromanifold 'learning via 'em' and indicating relative entropy as optimal in this process (similar to choice of Euclidean distance in Riemann spatial change, here we have relative entropy as local entropic measure (with fixed reference, this then gives Boltzmann's entropy.

D.4.2 The Chiral-Extension Cayley-Family relation

Recall that the unit-norm propagation was extended from octonion-level to trigintaduonion-level by introduction of chiral trigintaduonion emanators. The mathematical construct at the trigintaduonion-level that results can be represented in terms of tri-octonion products, thus all of the octonion properties are thereby inherited with such a construct since it is represented in terms of octonions, most notably a norm (here unit norm is of interest physically, for repetitive 'propagation'). The same double-chiral extension can be done at any Cayley level, and we thereby have a Cayley-Family extension property, via use of chiral Cayley emanation only, that is general.

Associativity Extension most likely Candidate for Renormalization and Thermality reification

Application of the Chiral-Extension Cayley-Family relation is first considered for associativity. In the Cayley Algebras, the highest algebra that retains the property of associativity is the quaternions. If we consider extending this associativity to higher -order algebras in the context of chiral propagation we can do this up to algebras at the level of the octonions and the sedenions. Consider that ***unit-norm*** trigintaduonion emanation will reduce to sedenion propagation in a general sense, and it is in this space that we can now add another sedenion element that is associative.

D.4.2.1 Sedenion Associative Element and the Axiom of (Renormalization) Choice

The description of the renormalization process involves counter-terms. The emanation/selection of the standard gauge field group, and other elements (described in Section X), seems consistent with the Standard Model (and certain extended versions), but makes no mention of counterterms. Rather than add counter-terms as an external regularization element to the theory, here we can consider them as a more complete manifestation of the existing Emanator theory (again the MIE hypothesis justifies this). So, let's suppose there really is chiral trigintaduonion emanation, we can now see where a side-version of the theory presents that is sufficient to provide a (reified) version of the renormalization terms needed to renormalize the theory. Note that the application of the renormalization 'cancellation' is a fundamental application of the Axiom of Choice. More accurately, the universe that is hypothesized to be MIE, implicitly selects for the existence of the Axiom of Choice (and well-ordering foundation of mathematics) in its choice of renormalization cancellation.

D.4.2.2 Octonion Associative Element and Complex Periodic Time

If extension of associativity on a sub-class of chiral-trigintaduonion emanations was possible, via the aforementioned effective sedenion associativity on chiral propagation, what of the indicated octonion-associativity extension? At the level of the octonion we can see a more direct manifestation of Lorentz invariance, including terms that are spatial or temporal (see generalized Lorentz invariance representations in Book 4 [4]), so suppose it is here that the theory creates ties between the real time quantum dynamics on a manifold and the complex-time statistical dynamics on a neuromanifold. These ties are shown to exist during equilibrium analyses in a number of applications (ss Book 6 [6]), and in thermal quantum field theory applications (see Book 5 [5]), but here we speak to this being a more fundamental attribute, revealed in the indicated circumstances, yes, but also present in general (albeit perhaps not in a useful form).In essence, we use our second wish/demand (we only get two) of the associative-extension genie that it explain the recurrence of the appearance of a fundamental thermality by way of time having a complex periodicity proportional to inverse system temperature.

'Contact' Associativity from Higher-order Cayley Algebras (64 and 128 orders for example) was suggested previously, involving use of limit processes and infinitesimal processes on the non-propagatable orders (even chiraly) above 32-order trigintaduonion, to be at zero displacement from the 32-algebra (thus 'contact'), but still imparting possible nontrivial contributions, such that the above renormalizability and complex-time relations might be obtained. Here we find a better route to this result via the chiral-extension property.

D.4.2.3 Octonion Commutative Element and Analytic Time
If we are considering operations reduced to octonion-level, there are also chiral extensions on the commutative complex numbers to arrive as a subset of commutative octonion-level operations in the unit-norm chiral emanation process. This allows further modification to the octonions as regards the internal time parameter analyticity, not just its imaginary-time periodicity. Together these chiral extensions may provide for the analytic time that is observed in the many successful applications of such a notion (required in path integral descriptions of bound states, for example).

The multiplication properties of the Cayley algebras are: Commutativity, Associativity, Alternativity, and Power Associativity. As you go to higher order algebras you lose properties. The Complex numbers are the last to have commutativity, the quaternions the last to have associativity, and the octonions the last to have alternativity. All the algebras have power associativity. Thus, the chiral extensions described are now exhausted since the interesting cases to extend, from complex, quaternionic, and octonionic, have now been considered.

D.4.3 The Extended Standard Model predicted by Emanator Theory
According to the Standard model of Particle Physics, the Universe consists of point-like spin-1/2 fermions that reside in two families: the Leptons and the Quarks. These particles exist in a geometry with a gauge field, where they interact through that gauge field according to local gauge invariance and indirectly through the geometry according to general relativity. Even more indirectly, the gravitational interaction via mass has mass itself determined through exchange of spin-0 Higgs particles.

The local gauge invariance of the Standard model is:

$$U(1)_Y \otimes SU(2)_L \otimes SU(3)_C.$$

Let's consider each part separately:

$U(1)_Y$→ gives rise to the electromagnetic force between charged particles.

$SU(2)_L$→ gives rise to the force between left-handed particles.

$SU(3)_C$→ gives rise to the strong force that operates on particles (hadrons) that carry color charge (three types).

In Book 5 [5] we see that each gauge group listed above derives from the general local gauge invariant group form:

$$SU(N)_L \otimes SU(N)_R.$$

For $N = 1$, we have the trivial case (no left or right): $U(1)$.
For $N = 2$, we have the asymmetric case of only left: $SU(2)_L$.
For $N = 3$, we have the symmetric case of left-right, or simply a fixed ratio: $SU(3)$.

In Book 5 we see that there is mixing between the $U(1)$ and $SU(2)$ parts to give the actual electroweak theory observed:

$$U(1) \otimes SU(2) \rightarrow \; U(1)_Y \otimes SU(2)_L.$$

The mixing introduces universal (globally gauge invariant) 'mixing angles' that number among the fundamental parameters of the theory (along with particle masses and a few other constants).

In Book 7 [7] we see that Emanator Theory predicts the form of the local gauge invariance as a general form, to be exactly that observed: $U(1)_Y \otimes SU(2)_L \otimes SU(3)_C$. Emanator theory also predicts a theory governed by 22 parameters, not 19 as currently listed for the Standard Model.

As strange as the form $U(1)_Y \otimes SU(2)_L \otimes SU(3)_C$ might seem, it is so far only applied to the interaction element of te theory. Let's now consider an interaction with what? The answer resides in the representations of the indicated groups (the left-handed particles interacting according to $SU(2)_L$ will be seen as interacting visa the weak force, etc.). So, saying there is a local gauge invariance is to say that there will be particles according to the irreducible, independent, representations of these groups. Thus, for any chosen representation there will be a collection of particles predicted. This is precisely what is observed experimentally. Thus, we not only know the groups $SU(2)_L$ and $SU(3)_C$, say, we also know their specific representations to obtain particle numbers and groupings observed experimentally.

So, the Standard Model is actually indicting the Model pair of Local Gauge Field and Representation $\mathcal{R}$:

$$\{U(1)_Y \otimes SU(2)_L \otimes SU(3)_C;\ \mathcal{R}\},$$

and it (the Standard Model) then says "times three", where the representation has there generations or copies:

$$\{U(1)_Y \otimes SU(2)_L \otimes SU(3)_C;\ 3 \times \mathcal{R}\}.$$

For the fundamental Leptons and Quarks this is shown in tabular form with the symbols for the particles as:

	electroweak left-handed Leptons	electro-only right-handed Leptons	Electro-only Quarks
1st Generation	$[e^-{}_L, \nu_{e,L}(m=0)]$	$[e^-{}_R, ---]$	d, u
2nd Generation	$[\mu^-{}_L, \nu_{\mu,L}(m=0)]$	$[\mu^-{}_R, ---]$	s, c
3rd Generation	$[\tau^-{}_L, \nu_{\tau,L}(m=0)]$	$[\tau^-{}_R, ---]$	b, t

Note that there are no right-handed neutrinos in the Standard Model and the left-handed neutrinos are massless. We already have ample evidence that the left-handed neutrinos are massive, so we know the Standard model will extend in this regard, which leaves the supposedly non-existent right-handed neutrinos. There probably are right-handed neutrino but if "electro-only" and having no charge, what results is a particle completely decoupled from the other matter, except gravitationally, precisely what describes dark matter. More on this will follow once we go from the 19 parameter model to the hypothesized 22 parameter model, as this suggests limits on extensions to the theory, as does cosmological data, which suggests a possible 4th neutrino and no more scenario consistent with the cosmological evolution, unless the other neutrinos are very massive (all probably the case).

With the Standard Model quantum field theory structure, with scatter results according to the indicated local gauge field and particle representation, we find additional conservation rules. One of these rules describes the conservation of Lepton number at each generation, in other words:

$$L_e = N(e^-) + N(\nu_e) - N(e^+) + N(\bar{\nu}_e)$$

is a constant, and the same for constants for the muon and tau generations. There is also conservation of baryon number and the familiar conservation of charge.

The Standard Model is a renormalizable/renormalized quantum field theory, and while this poses no difficulty with the $SU(2)$ elements alone or $SU(3)$ alone, here we have a product gauge and, what's more, it has handedness with use of $SU(2)_L$ not $SU(2)$. Not surprisingly, anomalous terms arise in he renormalization effort and for the renormalization to work, additional constraints are imposed on the structure of the theory. Most notable is that at a generational level across both leptons and baryons there should be a charge sum of zero. This is observed and was used to predict missing quarks (charm) in the early discovery process.

The structure identified by the standard model and the high decimal number agreement on key results from quantum field theory together provide a theory that can't be displaced directly by anything better, and only minimally augmented (with addition of right-handed neutrino), so trying to find a unified theory is impossible from 'within' the theory, and Godel's incompleteness theorem also suggests the hopelessness of such a task. This, then leaves no choice but to look for an encompassing theory that projects the quantum theory with the Standard Model structure in its entirety. And this is what is attempted with emanator theory (Book 7 [7]).

The Standard Model can't explain the three generations of matter, the origin of the local gauge group with its odd product form, the dimensionless constant alpha, and certainly can't explain the extension to massive(light) neutrinos, or the possible extension to massive (dark) right-handed neutrinos. In Emanator Theory (Book 7 [7]) we have:
(1) The local gauge $U(1) \otimes SU(2) \otimes SU(3)$ is projected by the theory. Why it should have three generations and why it should by asymmetric with use of $SU(2)_L$ is due to those choices providing the maximal packing of the Light matter sector (for maximal complexity information flow, an aspect of the MIE hypothesis), within he constraint of a 22-parameter emanation process.
(2) The 22 parameter, constants of the motion (emanation), result helps to constrain the particle representations such as to stay at 22-parameters, yet have maximally complex interaction.
(3) The MIE Hypothesis indicates the given generational structure given the constraint of working with the indicated product algebra. It also suggests any 'fine-tuning' on gravitational constant G might be

dominated by the dark matter sector and its contribution to the evolution of the universe. Such fine-tuning would be most powerful if it indicated a fractal scale invariance property (see fractal G discussion in [148]).

Before considering the 22-parameter Emanator theory prediction of an extended Standard Model, let's first recount the 19-parameter Standard Model, which I'll separate into four groups:

> (I) 9 Yukawa coupling constants (masses) for the charged fermions
> (II) 5 constants for Weinberg Angle and the CKM matrix (with three mixing angles and CP-violating phase)
> (III) 3 Constants for electromagnetic coupling (alpha), for strong interaction (g3), and strong CP-violating phase ($\theta_3 \approx 0$).
> (IV) 2 Higgs parameters: Mass and Vacuum Expectation

If we allow for the left-handed neutrinos to have mass, then we get 3 more masses and another 4 constants for the PMNS matrix (three mixing angles and a CP-violating phase):

> (V) Extended model: 7 more constants → We, thus, have 26 parameters.

Let's update out table with this extended version of the theory:

	electroweak left-handed Leptons	electro-only right-handed Leptons	Electro-only Quarks
1st Generation	$[e^-{}_L, \nu_{e,L}(m \neq 0)]$	$[e^-{}_R, ---]$	d, u
2nd Generation	$[\mu^-{}_L, \nu_{\mu,L}(m \neq 0)]$	$[\mu^-{}_R, ---]$	s, c
3rd Generation	$[\tau^-{}_L, \nu_{\tau,L}(m \neq 0)]$	$[\tau^-{}_R, ---]$	b, t

There is now a problem. In order to maintain renormalizability, if there are left-handed neutrinos with mass there must be right handed neutrinos with mass [149]. So, just how sure are we about neutrinos having mass? There is strong evidence not only for neutrino mass, but for neutrino family dynamics (here seen as oscillation between neutrino mass states). We'll see more evidence of generational family kinematics/dynamics in a later section. Neutrino family dynamics was first seen in measurements indicating that neutrinos spontaneously changed flavor (the Solar neutrino experiments [150]), which not only indicates they have mass, but allows us to determine the mass differences.

Clearly the Extended Standard Model needs to be extended further, to allow for the massive right-handed neutrino that is hypothesized to have no weak interaction like its right-handed electron cousin (and no electric interaction since no electric charge and no strong interaction since no color charge, thus 'dark' matter). Such a right-handed neutrinos (with no charge) can act as their own antiparticle (which is consistent with formation of "Bright" supermassive Black Holes in the early universe as seen with Webb). Furthermore, in addition to the Dirac mass relation to the left-handed neutrino it also has a Majorana mass term not tied to the Higgs mechanism [151] What results is that instead of mass $e^{-}{}_{L}$ equal to mass $e^{-}{}_{R}$ here we have

$$m_{\nu_{e,L}} \propto \frac{1}{m_{\nu_{e,R}}},$$

which is known as the see-saw mechanism [152] Thus the very low-mass left-handed neutrinos indicate very large mass right-handed neutrinos. Large mass neutrinos are an excellent candidate for cold dark matter, precisely what is needed to complete the Cosmological Standard Model.

Given the renormalization constraints and the observation of neutrino mass we arrive at the following updated Model:

	electroweak left-handed Leptons	electro-only right-handed Leptons	Electro-only Quarks
1st Generation	$[e^{-}{}_{L}, \nu_{e,L}(m \ll m_e)]$	$[e^{-}{}_{R}, \nu_{e,R}(m \gg m_e)]$	d, u
2nd Generation	$[\mu^{-}{}_{L}, \nu_{\mu,L}(m \ll m_e)]$	$[\mu^{-}{}_{R}, \nu_{\mu,R}(m \gg m_e)]$	s, c
3rd Generation	$[\tau^{-}{}_{L}, \nu_{\tau,L}(m \ll m_e)]$	$[\tau^{-}{}_{R}, \nu_{\tau,R}(m \gg m_e)]$	b, t

which is described by the 26 parameter theory indicated for massive left-handed neutrinos (if we assume that the right handed neutrino masses can be determined from the left-handed neutrino masses).

The standard counting to arrive at the 19 parameters (or, now, 26) includes $\theta_{QCD} \cong 0$ as a parameter of theory and its nearness to exactly zero is often referred to as the Strong CP problem. Going forward this is removed as a concern. The 'fine-tuning' that selects $\theta_{QCD} = 0$ is considered to be the same as that which selects the generational number, to arrive at 22 parameters while respecting the local gauge field constraint. Thus t is part of the MIE optimal selection.

Also included in the 19 parameter count is the U(1) gauge coupling g_1 and the SU(2) gauge coupling g_1, and these coupling are related to alpha by:

$$\alpha = \frac{1}{4\pi}\frac{{g_1}^2{g_2}^2}{{g_1}^2 + {g_2}^2}$$

Now, the emanation's projection of the quantizable Lagrangian theory involves various constants (22) in the resulting quantization. In that process, however, alpha already exists, in fact, from the derivation of alpha, we see that it establishes maximal perturbation starting at the level of the emanation process itself. Thus, alpha doesn't number among the 22, and the relation between coupling constants involving alpha means only one of those coupling constants is independently specifiable, so only one should be counted. At tis juncture we've reduced the count on 26 parameters to 24.

Notice in the prior counting, to arrive at the count of 26, we extended the count to be the same with dark neutrinos included since the dark neutrino mass was assumed to have some fixed (inverse) relation to the light neutrino mass. Let's not make this assumption, but retain the same mixing angle matrix, to now have 3 more masses in the theory (adding to the 24 count we now have 27).

In order to reduce the 27 parameter count to 22 we need to remove an 'overcount' of 5. Such an overcount would occur if there was a family-relationship generationally, for all masses. Consider such a mass relationship for the electron-muon-tau family, it is already known to exist and is known as the Koide relation. The Koide relation [153] was first observed for the three massive leptons currently known:

$$\frac{m_e + m_\mu + m_\tau}{\left(\sqrt{m_e} + \sqrt{m_\mu} + \sqrt{m_\tau}\right)^2} = \frac{2}{3}$$

To a lesser extent this relation is satisfied for the quarks as well, particularly for the three most massive, where the value is 0.6695. The problem with a simple application to the quark masses is that they are dependent on energy scale. A theoretical explanation for the Koide relation describes how this relation might exist for the masses of a given generation (or family group) [154]. As mentioned with neutrino oscillations, their oscillations involve the existence of such a family relationship in the quantum field theory implementation, so the hypothesis that it exists for each column in the table is not such a stretch. There are five columns of (independent) masses (the right handed electrons not independent of the left handed electrons removes them from the count). Thus, the five groups of three represent five groups with only two independent mass parameters, and the overall parameter count is thereby reduced by 5 from 27 to 22, as needed to be in agreement with Emanator theory. Thus Emanator theory, with the MIE hypothesis, would select for family group dynamics as observed with the neutrinos since it is what allows the maximal particle set within the 22-parameter constraint.

D.5 Conclusion

Maximal information propagation as an emergent construct appears to require two forms of propagation, an early hypercomplex 'emanation' that involves a chiral 10D propagation in a 32D trigintaduonion space, and standard propagation with complex propagators (consistent with the quantum deFinetti relation) operating inside the geometry and gauge field that is projected. From the 'emanation' stage we see the maximum dimensionality and fractal limits provide the fundamental constants that then imprint upon the emergent geometry and gauge field, including giving rise to the constants α. The origin of α has been a long-standing mystery. So much so that the central role of α in modern physics is literally engraved in stone, the tombstones of Sommerfeld (which displays $e^2/\hbar c$, which is α) and Schwinger (which displays $\alpha/2\pi$) for example. Its origin has eluded physics for over a century, and appears to reside in the algebra of trigintaduonions.

Emanator Theory results from a hypothesized maximal information propagation and this means maximal analyticity, maximal domain, etc. As a process, Emanator theory is also hypothesized to operate up to "the edge of chaos" to permit maximal perturbation (noise) domain. When taken with the results showing that Emanator theory is Martingale, thus has well-defined limits, we then must wonder if there are well-defined multi-scale (fractal) limits. In other words, is there a relation that would tie the micro scale constant to fundamental constants (as they are counted in the 22) with the cosmological scale 'constants' that have settled out, at macro scale, in the current evolution of the Universe? In this context, the Gravitational constant G is hypothesized to be a multiscale fractal coupling parameter.

In seeking a deeper theory we build on the sum-on-paths with propagator formulation to arrive at a sum-on-emanations with emanator formulation. Propagation in a complex Hilbert space, however, in a standard QM/QFT formulation, requires the propagator function to be a complex number (not real or quaternionic, etc., [127]). This prohibits what would otherwise be an obvious generalization to hypercomplex algebras. In order to achieve this generalization, we have to introduce a new layer to the theory, one with universal emanation involving hypercomplex algebras (trigintaduonions) that is hypothesized to project to the familiar complex Hilbert space propagation with associated fixed elements (e.g., the emanator formalism projects out the observed constants and group structure of the standard model). The 'projection' is an induced mathematical construct, like having SU(3) on products of octonions, but here it we be the standard model U(1)xSU(2)xSU(3) on products of emanator trigintaduonions. Thus, a unified variational formulation is posed, one that arrives at alpha as a natural structural element, among other things, uniquely specified by the condition of maximal information emanation.

A 'deeper' phase of universal evolution is described by a theory of emanations, where mathematically invariant emergent structures appear:

emanation → propagation → trajectory

At the emanation to propagation emergence, one of the emergent constructs is the familiar path integral based on standard (unitary) propagators in a complex Hilbert space.

We have α, 10,22,78,137 as parameters resulting from analysis on a single path maximal information flow construct, where the number 22 corresponds to the number of emergent parameters in the description of the propagating construct (exact derivation of the 22-parameter in [7]). In addition, the time choice is emergent via a multi-path construct, along with the propagator construct, and is coupled in both time step and imaginary time increment. The formulation is inherently embedded in a higher dimensional complex space, thus all of the QFT complex analysis analyticity methods are valid as the assumptions made are now part of the maximal information flow emergent construct.

References

[1] Winters-Hilt, S. Classical Mechanics and Chaos. 2023. (Physics Series: "Physics from Maximal Information Emanation" Book 1.)

[2] Winters-Hilt, S. The Dynamics of Fields, Fluids, and Gauges. 2023. (Physics Series: "Physics from Maximal Information Emanation" Book 2.)

[3] Winters-Hilt, S. The Dynamics of Manifolds. 2023. (Physics Series: "Physics from Maximal Information Emanation" Book 3.)

[4] Winters-Hilt, S. Quantum Mechanics, Path Integrals, and Algebraic Reality. 2024. (Physics Series: "Physics from Maximal Information Emanation" Book 4.)

[5] Winters-Hilt, S. Quantum Field Theory and the Standard Model. 2024. (Physics Series: "Physics from Maximal Information Emanation" Book 5.)

[6] Winters-Hilt, S. Thermal & Statistical Mechanics, and Black Hole Thermodynamics. 2024. (Physics Series: "Physics from Maximal Information Emanation" Book 6.)

[7] Winters-Hilt, S. Emanation, Emergence, and Eucatastrophe. 2023. (Physics Series: "Physics from Maximal Information Emanation" Book 7.)

[8] Winters-Hilt, S. Informatics and Machine Learning. Wiley Publishing. 9781119716747, Sept. 2021.

[9] Winters-Hilt S. Topics in Quantum Gravity and Quantum field Theory in Curved Spacetime. UWM PhD Dissertation, 1997.

[10] Amari S; Dualistic Geometry of the Manifold of Higher-Order Neurons. Neural Networks, Vol. 4(4), 1991:443-451.

[11] Amari S: Information Geometry of the EM and em Algorithms for Neural Networks. Neural Networks, Vol. 8(9), 1995:1379-1408.

[12] Amari S and Nagaoka H: Methods of Information Geometry. 2000. Translations of Mathematical Monographs Vol. 191.

[13] Marsland R., Brown H.R., Valente G., Time and irreversibility in axiomatic thermodynamics (American Journal of Physics, 2015, 83:7, 628-634).

[14] Brown H.R. and Jos Uffink, “The origins of time asymmetry in thermodynamics: The minus first law,” Stud. His. Philos. Mod. Phys. 32(4), 525–538 (2001).

[15] Uffink, Jos. “Bluff your way in the second law of thermodynamics,” Stud. Hist. Philos. Mod. Phys. 32(3), 305–394 (2001).

[16] Carathedory, Constantin. “Untersuchungen €uber die Grundlagen der Thermodynamik,” Math. Ann. 67, 355–386 (1909). English translation by Joseph Kestin: “Investigation into the Foundations of Thermodynamics,” in the Second Law of Thermodynamics: Benchmark Papers on Energy, edited by Joseph Kestin (Dowden, Hutchinson and Ross, Stroudsberg, Pennsylvania, 1976), Vol. 5, pp. 229–256.

[17] Maldacena, J.. The Large N limit of superconformal field theories and supergravity.
Advances in Theoretical and Mathematical Physics. 2 (4): 231–252.

[18] Witten, Edward (1998). "Anti-de Sitter space and holography". Advances in Theoretical and Mathematical Physics. 2 (2): 253–291.

[19] Feynman, R.P. Statistical Mechanics: A set of lectures. Addison Wesley 1972.

[20] Morse, Philip M. Thermal Physics. New York,: W. A. Benjamin. 1961.

[21] Goodstein, David L. States of Matter. Prentice-Hall. 1975.

[22] Landau, L.D. and E.M Lifshitz. Statistical Physics, 3rd Ed. Part 1. Pergamon Press. 1980.

[23] Plischke, Michael and Birger Bergersen. Equilibrium Statistical Physics. World Scientific. 1989.

[24] Chandler, David. Introduction to Modern Statistical Mechanics. Oxford University Press. 1987.

[25] Stauffer, Dietrich. Introduction to Percolation Theory. Taylor & Francis. 1985.

[26] Kubo, Ryogo. Statistical Mechanics. North-Holland Publishing Company - Amsterdam; Reprint. edition (January 1, 1967)

[27] Donald A. McQuarrie. Statistical Mechanics. University Science Books; First Edition (June 16, 2000)

[28] Chentsov, N.N. Statistical Decision Rules and Optimal Inference. 1972.

[29] Onsager, Lars. "Crystal statistics. I. A two-dimensional model with an order-disorder transition", *Physical Review*, Series II, **65** (3–4): 117–149, 1944.

[30] Peierls, R.E. (1936). Proc. Cambridge Philos. Soc. 32:477.

[31] Wannier, G.H. (1966). Statistical Physics. New York: Wiley.

[32] Mayer, J.E. and Mayer, M.G. Statistical Mechanics. New York: Wiley. 1940.

[33] Schultz, T. et al. Rev. Mod. Phys. 36:856. 1964.

[34] Kadanoff, Leo P. (1966). "Scaling laws for Ising models near " (https://doi.org/10.1103%2F PhysicsPhysiqueFizika.2.263). Physics Physique Fizika. 2 (6): 263.

[35] Wilson, K.G. (1975). "The renormalization group: Critical phenomena and the Kondo problem". Rev. Mod. Phys. 47 (4): 773.

[36] Louko, J. and S.N. Winters-Hilt, Phys. Rev. D 54, 2647 (1996).

[37] S. W. Hawking, Commun. Math. Phys. 43, 199 (1975).

[38] J. B. Hartle and S. W. Hawking, Phys. Rev. D 13, 2188 (1976).

[39] W. G. Unruh, Phys. Rev. D 14, 870 (1976).

[40] W. Israel, Phys. Lett. 57A, 107 (1976).

[41] J. D. Bekenstein, Nuovo Cimento Lett. 4, 737 (1972).

[42] J. D. Bekenstein, Phys. Rev. D 9, 3292 (1974).

[43] P. C. W. Davies, Proc. R. Soc. London A 353, 499 (1977).

[44] P. C. W. Davies, Rep. Prog. Phys. 41, 1313 (1978).

[45] G. W. Gibbons and M. J. Perry, Proc. R. Soc. London A 358, 467 (1978).

[46] L. Smolin, Gen. Relativ. Gravit. 17, 417 (1985).

[47] D. Garfinkle and R. M. Wald, Gen. Relativ. Gravit. 17, 461 (1985).

[48] G. W. Gibbons and S. W. Hawking, Phys. Rev. D 15, 2752 (1977).

[49] S. W. Hawking, in General Relativity: An Einstein Centenary Survey, edited by S. W. Hawking and W. Israel (Cambridge University Press, Cambridge, 1979).

[50] J. W. York, Phys. Rev. D 33, 2092 (1986).

[51] J. D. Brown and J. W. York, Phys. Rev. D 47, 1407 (1993).

[52] J. D. Brown and J. W. York, in Quantum Mechanics of Fundamental Systems 4, The Black Hole Twenty Five Years After, edited by C. Teitelboim and J. Zanelli (Plenum, New York, in press). (gr-qc/9405024)

[53] S. Carlip and C. Teitelboim, Phys. Rev. D 51, 622 (1995). (gr-qc/9405070)

[54] D. N. Page, in Black Hole Physics, edited by V. D. Sabbata and Z. Zhang (Kluwer Academic Publishers, Dordrecht, 1992).

[55] R. M. Wald, Quantum Field Theory in Curved Spacetime and Black Hole Thermodynamics (The University of Chicago Press, Chicago, 1994).

[56] S. Carlip, Class. Quantum Grav. 12, 283 (1995). (gr-qc/9506079)

[57] L. E. Reichl, A Modern Course in Statistical Physics (University of Texas Press, Austin, Texas, 1980).

[58] R. D. Sorkin, Int. J. Theor. Phys. 18, 309 (1979).

[59] H. W. Braden, B. F. Whiting, and J. W. York, Phys. Rev. D 36, 3614 (1987).

[60] B. F. Whiting and J. W. York, Phys. Rev. Lett. 61, 1336 (1988).

[61] B. F. Whiting, Class. Quantum Grav. 7, 15 (1990).

[62] H. W. Braden, J. D. Brown, B. F. Whiting, and J. W. York, Phys. Rev. D 42, 3376 (1990).

[63] J. D. Brown, G. L. Comer, E. A. Martinez, J. Melmed, B. F. Whiting, and J. W. York, Class. Quantum Grav. 7, 1433 (1990).

[64] J. D. Brown, E. A. Martinez and J. W. York, Phys. Rev. Lett. 66, 2281 (1991).

[65] J. Louko and B. F. Whiting, Class. Quantum Grav. 9, 457 (1992).

[66] G. L. Comer, Class. Quantum Grav. 9, 947 (1992).

[67] J. Melmed and B. F. Whiting, Phys. Rev. D 49, 907 (1994).

[68] J. Louko and B. F. Whiting, Phys. Rev. D 51, 5583 (1995). (gr-qc/9411017)

[69] S. R. Lau, Class. Quantum Grav. 13, 1541 (1996). (gr-qc/9508028)

[70] S. Bose, J. Louko, L. Parker, and Y. Peleg, Phys. Rev. D 53, 5708 (1996). (gr-qc/9510048)

[71] S. W. Hawking and D. N. Page, Commun. Math. Phys. 87, 577 (1983).

[72] D. N. Page and K. C. Phillips, Gen. Relativ. Gravit. 17, 1029 (1985).

[73] M. Banados, C. Teitelboim, and J. Zanelli, Phys. Rev. D 49, 975 (1994). (gr-qc/9307033)

[74] J. D. Brown, J. Creighton, and R. B. Mann, Phys. Rev. D 50, 6394 (1994).

[75] O. B. Zaslavskii, Class. Quantum Grav. 11, L33 (1994).

[76] B. Hoffmann, Quart. J. Math. (Oxford) 4, 179 (1933).

[77] B. Carter, Commun. Math. Phys. 10, 280 (1968).

[78] B. Carter, in Black Holes, Proceedings of the 1972 session of Les Houches Ecole d'ete de Physique Theorique, edited by C. DeWitt and B. S. DeWitt (Gordon and Breach, New York, 1973).

[79] D. Kramer, H. Stephani, E. Herlt, and M. MacCallum, Exact Solutions of Einstein's Field Equations, edited by E. Schmutzer (Cambridge University Press, Cambridge, England, 1980), Sec. 13.4.

[80] K. V. Kuchar, Phys. Rev. D 50, 3961 (1994). (gr-qc/9403003)

[81] T. Thiemann, Int. J. Mod. Phys. D 3, 293 (1994); Nucl. Phys. B436, 681 (1995).

[82] A. Ashtekar and A. Magnon, Class. Quantum Grav. 1, L39 (1984).

[83] M. Henneaux and C. Teitelboim, Commun. Math. Phys. 98, 391 (1985).

[84] C. W. Misner, K. S. Thorne, and J. A. Wheeler, Gravitation (Freeman, San Francisco, 1973), Exercise 32.1.

[85] P. Kraus and F. Wilczek, Nucl. Phys. B437, 231 (1995). (hep-th/9411219)

[86] M. Henneaux and C. Teitelboim, Quantization of Gauge Systems (Princeton University Press, Princeton, New Jersey, 1992).

[87] N. M. J. Woodhouse, Geometric Quantization (Clarendon Press, Oxford, 1980).

[88] C. J. Isham, in Relativity, Groups and Topology II: Les Houches 1983, edited by B. S. DeWitt and R. Stora (North-Holland, Amsterdam, 1984).

[89] A. Ashtekar, Lectures on Non-Perturbative Canonical Gravity (World Scientific, Singapore, 1991).

[90] M. Reed and B. Simon, Methods of Modern Mathematical Physics (Academic, New York, 1975), Vol. II.

[91] Louko, J., J.Z. Simon, and S.N. Winters-Hilt, Phys. Rev. D (1997)

[92] D. Lovelock, J. Math. Phys. 12, 498 (1971).

[93] D. Lovelock, J. Math. Phys. 13, 874 (1972).

[94] D. G. Boulware and S. Deser, Phys. Rev. Lett. 55, 2656 (1985); Phys. Lett. 175B, 409 (1986).

[95] J. T. Wheeler, Nucl. Phys. B268, 737 (1986); B273, 732 (1986).

[96] D. L. Wiltshire, Phys. Lett. 169B, 36 (1986).

[97] R. C. Myers, Phys. Rev. D 36, 392 (1987). [Equation (7) is in error; see Ref. [12].]

[98] C. Teitelboim and J. Zanelli, Class. Quantum Grav. 4, L125 (1987).

[99] C. Teitelboim and J. Zanelli, in Constraints Theory and Relativistic Dynamics, edited by G. Longhi and L. Lusanna (World Scienti_c, Singapore, 1987).

[100] D. L. Wiltshire, Phys. Rev. D 38, 2445 (1988).

[101] B. Whitt, Phys. Rev. D 38, 3000 (1988).

[102] R. C. Myers and J. Z. Simon, Phys. Rev. D 38, 2434 (1988).

[103] J. Melmed, Ph.D. Thesis, University of North Carolina, Chapel Hill (1989,unpublished).

[104] E. Poisson, Class. Quantum Grav. 8, 639 (1991).

[105] E. Poisson, Phys. Rev. D 43, 3923 (1991).

[106] K. Maeda, Phys. Lett. 166B, 59 (1986); J. Madore, Phys. Lett. 110A, 289 (1985); Phys. Lett. 111A, 283 (1985); Class. Quantum Grav. 3, 361 (1986); N. Deruelle and J. Madore, Phys. Lett. 114A, 185 (1986); Phys. Lett. 186B, 25 (1987); H. Ishihara, Phys. Lett. 179B, 217 (1986); W. Puszkarz, Phys. Lett. 226B, 39 (1989); N. Deruelle and L. Farina-Busto, Phys. Rev. D 41, 3696 (1990).

[107] T. Jacobson and R. C. Myers, Phys. Rev. Lett. 70, 3684 (1993). (hepth/ 9305016)

[108] T. Jacobson, G. Kang, and R. C. Myers, Phys. Rev. D 49, 6587 (1994). (grqc/ 9312023)

[109] T. Jacobson, G. Kang, and R. C. Myers, Phys. Rev. D 52, 3518 (1995). (grqc/ 9503020).

[110] G. Oliveira-Neto, Phys. Rev. D 53, 1977 (1996).

[111] M. Varadarajan, Phys. Rev. D 52, 7080 (1995). (gr-qc/9508039)

[112] S. Bose, J. Louko, L. Parker, and Y. Peleg, Phys. Rev. D 53, 5708 (1996). (gr-qc/9510048)

[113] J. D. Romano and C. G. Torre, Phys. Rev. D 53, 5634 (1996). (gr-qc/9509055)

[114] K. V. Kuchar, J. D. Romano, and M. Varadarajan, \Dirac Constraint Quantization of a Dilatonic Model of Gravitational Collapse", e-print gr-qc/9608011.

[115] T. Thiemann and H. A. Kastrup, Nucl. Phys. B399, 211 (1993). (grqc/ 9310012)

[116] H. A. Kastrup and T. Thiemann, Nucl. Phys. B425, 665 (1994). (grqc/ 9401032)

[117] T. Thiemann, Int. J. Mod. Phys. D 3, 293 (1994).

[118] T. Thiemann, Nucl. Phys. B436, 681 (1995).

[119] J. Gegenberg and G. Kunstatter, Phys. Rev. D 47, R4192 (1993). (grqc/ 9302006)

[120] J. Gegenberg, G. Kunstatter, and D. Louis-Martinez, Phys. Rev. D 51, 1781 (1995). (gr-qc/9408015)

[121] D. Louis-Martinez and G. Kunstatter, Phys. Rev. D 52, 3494 (1995). (grqc/ 9503016)

[122] Planck, Max (1901), "Ueber das Gesetz der Energieverteilung im Normalspectrum", *Ann. Phys.*, **309** (3): 553–63,

[123] Aharonov, Y; Bohm, D (1959). "Significance of electromagnetic potentials in quantum theory". *Physical Review*. **115** (3): 485–491.

[124] A. S. Wightman, "Fields as Operator-valued Distributions in Relativistic Quantum Theory," *Arkiv f. Fysik, Kungl. Svenska Vetenskapsak.* **28**, 129–189 (1964).

[125] Osterwalder, Konrad; Schrader, Robert (1975). "Axioms for Euclidean Green's functions II". *Communications in Mathematical Physics*. **42** (3). Springer Science and Business Media LLC: 281–305.

[126] Winters-Hilt, S. Feynman-Cayley Path Integrals select Chiral Bi-Sedenions with 10-dimensional space-time propagation. Advanced Studies in Theoretical Physics, Vol. 9, 2015, no. 14, 667 – 683.

[127] Caves, C.M., C.A., Fuchs, R. Schack. Unknown quantum states: The Quantum de Finetti Representation. J. Math. Phys. 43, 4537 (2002).

[128] Penrose, R., W. Rindler (1984) Volume 1: Two-Spinor Calculus and Relativistic Fields, Cambridge University Press, United Kingdom.

[129] Cailler, C. 1917. Archs. Sci. Phys. Nat. ser. 4, 44 p. 237.

[130] Girard, P.R.. The Quaternion group and modern physics. Eur. J. Phys. 5 (1984): 25-32.

[131] Synge, J.L. Quaternions, Lorentz Transformations and the Conway-Dirac-Eddington Matrices.

[132] Winters-Hilt, S. Fiat Numero: Trigintaduonion Emanation Theory and its Relation to the Fine-Structure Constant α, the Feigenbaum Constant C∞, and π. Advanced Studies in Theoretical Physics Vol. 15, 2021, no. 2, 71 - 98.

[133] Feigenbaum, M. J. (1976) "Universality in complex discrete dynamics", Los Alamos Theoretical Division Annual Report 1975-1976 .

[134] Gunaydin, M. and F. Gursey. Quark structure and the octonions. J. Math. Phys., 14, 1973.

[135] Hurwitz, A. (1923), "Über die Komposition der quadratischen Formen", Math. Ann., 88 (1–2): 125.

[136] https://www.wolframalpha.com

[137] Gaździcki, Marek; Gorenstein, Mark I. (2016), Rafelski, Johann (ed.), "Hagedorn's Hadron Mass Spectrum and the Onset of Deconfinement", Melting Hadrons, Boiling Quarks – From Hagedorn Temperature to Ultra-Relativistic Heavy-Ion Collisions at CERN, Springer International Publishing, pp. 87–92.

[138] Kato T. Fundamental Properties of Hamiltonian Operators of Schrodinger Type, Transactions of the American Mathematical Society, 1951, Pg. 195-211.

[139] Kato, T. Perturbation theory for linear operators. Springer 1980.

[140] McMullen, Curtis T. 2000. The Mandelbrot set is universal. In The Mandelbrot Set, Theme and Variations, ed. T. Lei, 1–18. Cambridge U.K.: Cambridge Univ. Press. Revised 2007.

[141] Briggs, K. A precise calculation of the Feigenbaum constants. Mathematics of Computation, Vol. 57, Num.195, July 1991, pages 435-439.

[142] Winters-Hilt S, I. H. Redmount, and L. Parker, "Physical distinction among alternative vacuum states in flat spacetime geometries," Phys. Rev. D 60, 124017 (1999).

[143] Friedman, J.L., J. Louko, and S. Winters-Hilt. Reduced phase space formalism for spherically symmetric geometry with a massive dust shell. Phys Rev. D Vol. 56, Num 12 (1997).

[144] Nielsen, Frank (2022). "The Many Faces of Information Geometry". Notices of the AMS. 69 (1). American Mathematical Society: 36-45.

[145] Nielsen, Frank (2018). "An Elementary Introduction to Information Geometry". Entropy. 22 (10).

[146] Penrose, Roger (May 1996). "On Gravity's role in Quantum State Reduction". *General Relativity and Gravitation*. **28** (5): 581–600.

[147] Diósi, L. (1989). "Models for universal reduction of macroscopic quantum fluctuations". Physical Review A. 40 (3): 1165–1174.

[148] Winters-Hilt S. Emanator Theory is shown to be an optimal Martingale process at the fractal edge of chaos, where the Gravitational constant G is hypothesized to be a multiscale fractal coupling parameter. Advanced Studies in Theoretical Physics, 2023.

[149] Peskin, M.E., Schroeder, D.V. (1995). An Introduction to Quantum Field Theory.

[150] Haxton, W.C.; Hamish Robertson, R.G.; Serenelli, Aldo M. (18 August 2013). "Solar Neutrinos: Status and Prospects". *Annual Review of Astronomy and Astrophysics*. **51** (1): 21–61.

[151] Sonjanovic, G. 2011. Probing the origin of neutrino mass: from GUT to LHC.

[152] Grossman, Y. 2003. TAST 2002 lectures on neutrinos.

[153] Koide, Y., Nuovo Cim. A 70 (1982) 411 [Erratum-ibid. A 73 (1983) 327].

[154] Sumino, Y. (2009). "Family Gauge Symmetry as an Origin of Koide's Mass Formula and Charged Lepton Spectrum". Journal of High Energy Physics. 2009 (5): 75. arXiv:0812.2103.

[155] Jackson, J.D. Classical Electrodynamics, 2nd Edition. Wiley 1975.

[156] Lorentz, Hendrik Antoon (1899), "Simplified Theory of Electrical and Optical Phenomena in Moving Systems" , Proceedings of the Royal Netherlands Academy of Arts and Sciences, 1: 427–442.

[157] D'Alembert, Jean Le Rond (1743). Traité de dynamique.

[158] Laplace, P S (1774), "Mémoires de Mathématique et de Physique, Tome Sixième" [Memoir on the probability of causes of events.], Statistical Science, 1 (3): 366–367.

[159] Winters-Hilt, S. Theory of Trigintaduonion Emanation and Origins of α and π. Researchgate 05/24/20.

[160] Winters-Hilt, S. Chiral Trigintaduonion Emanation Leads to the Standard Model of Particle Physics and to Quantum Matter. Advanced Studies in Theoretical Physics, Vol. 16, 2022, no. 3, 83-113.

[161] Winters-Hilt, S. Meromorphic precipitation of quantum matter with dimensionful action. May 2021. DOI:10.13140/RG.2.2.32294.24640.

[162] Winters-Hilt, S. Emanator Theory using split octonions is Manifestly Lorentz Invariant and reveals why the fundamental constant $\hbar$ should be so small. Advanced Studies in Theoretical Physics, 2023.

[163] Landau, Lev D.; Lifshitz, Evgeny M. (1969). Mechanics. Vol. 1 (2nd ed.). Pergamon Press.

[164] Goldstein, Herbert (1980). Classical Mechanics (2nd ed.). Addison-Wesley.

[165] Fetter, A.L and J.D Walecka, Theoretical Mechanics of Particles and Continua, Dover (2003).

[166] Percival, I.C. and D. Richards. Introduction to Dynamics. (1983) Cambridge University Press.

[167] Arnold, V.I. Ordinary Differential Equations. MIT Press. (1978).

[168] Arnold, Vladimir I. (1989). Mathematical Methods of Classical Mechanics (2nd ed.). New York: Springer.

[169] Woodhouse, N.M.J. Introduction to Analytical Dynamics. Springer, 2nd Edition. 2009.

[170] Bender, C.M. and S.A. Orszag. Advanced Mathematical Methods for Scientists and Engineers: Asymptotic Methods and Perturbation Theory. Springer. 1999.

[171] Robert L. Devaney. An Introduction to Chaotic Dynamical Systems. Addison -Wesley.

[172] Landau, Lev D.; Lifshitz, Evgeny M. (1971). The Classical Theory of Fields. Vol. 2 (3rd ed.). Pergamon Press.

[173] Penrose, Roger (1965), "Gravitational collapse and space-time singularities", Phys. Rev. Lett., 14 (3): 57.

[174] Hawking, Stephen & Ellis, G. F. R. (1973). The Large Scale Structure of Space-Time. Cambridge: Cambridge University Press.

[175] Peebles, P. J. E. (1980). Large-Scale Structure of the Universe. Princeton University Press.

[176] B. Abi et al. Measurement of the Positive Muon Anomalous Magnetic Moment to 0.46 ppm. Phys. Rev. Lett. 126, 141801 (2021).

[177] Einstein, A. “On a heuristic point of view concerning the production and transformation of light” . Ann. Phys., Lpz 17 132-148.

[178] Balmer, J. J. "Notiz über die Spectrallinien des Wasserstoffs" [Note on the spectral lines of hydrogen]. Annalen der Physik und Chemie. 3rd series (in German). 25: 80–87. (1885).

[179] Heisenberg, Werner. "Über quantentheoretische Umdeutung kinematischer und mechanischer Beziehungen". Zeitschrift für Physik (in German). 33 (1): 879–893. ("Quantum theoretical re-interpretation of kinematic and mechanical relations"). (1925).

[180] Schrödinger, E. (1926). "An Undulatory Theory of the Mechanics of Atoms and Molecules". Physical Review. 28 (6): 1049–1070.

[181] Born, Max; J. Robert Oppenheimer. "Zur Quantentheorie der Molekeln" [On the Quantum Theory of Molecules]. Annalen der Physik (in German). 389 (20): 457–484. (1927).

[182] Dirac, P. A. M. "The Quantum Theory of the Electron". Proceedings of the Royal Society A: Mathematical, Physical and Engineering Sciences. 117 (778): 610–624. (1928).

[183] Dirac, P. A. M. . The Principles of Quantum Mechanics. Oxford: Clarendon Press. (1930).

[184] Dirac, Paul A. M. (1933). "The Lagrangian in Quantum Mechanics". Physikalische Zeitschrift der Sowjetunion. 3: 64–72.

[185] Feynman, Richard P. (1942). The Principle of Least Action in Quantum Mechanics (PhD). Princeton University.

[186] Feynman, R.P. Space-Time Approach to Non-Relativistic Quantum Mechanics. Rev. Mod. Phys. 20, 367 – Published 1 April 1948.

[187] Erdeyli, A. Asymptotic Expansions. 1956 Dover.

[188] Erdeyli, A. Asymptotic Expansions of differential equations with turning points. Review of the Literature. Technical Report 1, Contract Nonr-220(11). Reference no. NR 043-121. Department of Mathematics, California Institute of Technology, 1953.

[189] Hawking, S. W. (1974-03-01). "Black hole explosions?". Nature. 248 (5443): 30–31.

[190] Sommerfeld, A., Atombau und Spektrallinien. Friedrich Vieweg und Sohn, Braunschweig, 1919.

[191] Tolkien, J.R.R. (1990). The Monsters and the Critics and Other Essays. London: Harper Collins Publishers.

Index

www.ingramcontent.com/pod-product-compliance
Lightning Source LLC
Chambersburg PA
CBHW050453190326
41458CB00005B/1262
* 9 7 9 8 8 9 4 8 7 0 0 9 0 *